Descriptions of Medical Fungi

Fourth Edition

Descriptions of Medical Fungi

Fourth Edition

Sarah Kidd[1,3], Catriona Halliday[2] and David Ellis[3]

[1]National Mycology Reference Centre, SA Pathology, Adelaide, South Australia.

[2]Clinical Mycology Reference Laboratory, Centre for Infectious Diseases and Laboratory Services, Institute of Clinical Pathology and Medical Research, NSW Health Pathology, Sydney, Australia.

[3]School of Biological Sciences, University of Adelaide, Adelaide, Australia.

CABI

CABI is a trading name of CAB International

CABI
Nosworthy Way
Wallingford
Oxfordshire OX10 8DE
UK

Tel: +44 (0)1491 832111
E-mail: info@cabi.org
Website: www.cabi.org

CABI
200 Portland Street
Boston
MA 02114
USA

Tel: +1 (617)682-9015
E-mail: cabi-nao@cabi.org

A catalogue record for this book is available from the British Library, London, UK.

ISBN-13: 9781800622326 (hardback)
9781836995265 (paperback)
9781800622333 (ePDF)
9781800622340 (ePub)

DOI: 10.1079/9781800622340.0000

Commissioning Editor: Rebecca Stubbs
Editorial Assistant: Emma McCann
Production Editor: James Bishop

Typeset by Straive, Pondicherry, India
Printed in the USA

Contents

Preface

The first edition of this book entitled *Descriptions of Medical QAP Fungi* was published in 1992 to provide all laboratories participating in the Mycology module of the Royal College of Pathologists of Australasia Quality Assurance Program (QAP) with a set of description sheets covering medically important fungi. Subsequent editions were published by the authors in 2007 and 2016. We now provide an updated fourth edition which includes new and revised descriptions. We have endeavoured to reconcile current morphological descriptions with more recent genetic data, however nomenclature changes in mycology will be on-going. To search for current accepted fungal names, go to Index Fungorum (www.indexfungorum.org) and/or Mycobank (www.mycobank.org).

Morphological descriptions: These descriptions have by necessity been kept brief and many have been based on previous descriptions by other authors. For further information regarding any of the mycoses or pathogenic fungi mentioned, the reader is referred to the references cited. A glossary of the mycological terminology used is provided, otherwise the reader is referred to Ainsworth and Bisby's *Dictionary of the Fungi* (Kirk *et al.*, 2008).

Fungal phylogeny continues to be refined and nine phylum-level clades are now recognised in the Kingdom Eumycota: Opisthosporidia, Chytridiomycota, Neocallimastigomycota, Blastocladiomycota, Zoopagomycota, Mucoromycota, Glomeromycota, Basidiomycota and Ascomycota (Ruggiero *et al.*, 2015; Naranjo-Ortiz and Gabaldon, 2019; de Hoog *et al.*, 2020). Key fungal phyla, orders and genera of medical importance are listed in Table p1.

Table p1. Key fungal phyla, orders and genera of medical importance.

Kingdom: Chromista	
Phylum: Oomycota	
Order: Peronosporales	*Pythium*
Kingdom: Eumycota	
Phylum: Chytridiomycota	
Order: Chytridiales	*Batrachochytrium*
Phylum: Zoopagomycota	
Order: Entomophthorales	*Basidiobolus, Conidiobolus*
Phylum: Mucoromycota	
Order: Mucorales	*Apophysomyces, Cunninghamella, Lichtheimia, Mucor, Rhizomucor, Rhizopus, Saksenaea, Syncephalastrum*
Order: Mortierellales	*Mortierella*
Phylum: Ascomycota	
Order: Pneumocystales	*Pneumocystis*
Order: Saccharomycetales	*Candida, Clavispora, Cyberlindnera, Debaryomyces, Diutina, Kluyveromyces, Magnusiomyces, Meyerozyma, Pichia, Saccharomyces, Torulaspora, Wickerhamomyces, Yarrowia*
Order: Onygenales	Dermatophytes - *Arthroderma, Epidermophyton, Lophophyton, Microsporum, Nannizzia, Paraphyton, Trichophyton* Dimorphic fungi - *Blastomyces, Coccidioides, Emergomyces, Histoplasma, Paracoccidioides*; and *Malbranchea, Chrysosporium*
Order: Arachnomycetales	*Onychocola*
Order: Eurotiales	*Arthrographis, Aspergillus, Penicillium, Talaromyces*

Continued

Table p1. Coninued.

Order: Chaetothyriales	*Cladophialophora, Exophiala, Fonsecaea, Phialophora, Rhinocladiella, Veronaea*
Order: Dothideales	*Aureobasidium, Hortaea, Lasiodiplodia, Neoscytalidium*
Order: Capnodiales	*Cladosporium*
Order: Pleosporales	*Alternaria, Bipolaris, Curvularia, Drechslera, Exserohilum, Phoma, Trematosphaeria*
Order: Clavicipitales	*Beauvaria, Paecilomyces, Purpureocillium*
Order: Microascales	*Scopulariopsis, Lomentospora, Scedosporium*
Order: Ophiostomatales	*Sporothrix*
Order: Glomerellales	*Colletotrichum*
Order: Diaporthales	*Phaeoacremonium, Pleurostoma*
Order: Sordariales	*Acrophialophora, Chaetomium, Lecythophora, Madurella, Myrmecridium, Phialemonium*
Order: Hypocreales	*Acremonium, Cylindrocarpon, Fusarium, Sarocladium, Trichoderma*
Order: Venturiales	*Ochroconis, Verruconis*
Phylum: Basidiomycota	
Order: Microstromatales	*Quambalaria*
Order: Malasseziales	*Malassezia*
Order: Sporidiobolales	*Rhodotorula*
Order: Tremellales	*Cryptococcus, Naganishia, Papiliotrema*
Order: Trichosporonales	*Apiotrichum, Cutaneotrichosporon, Trichosporon*
Order: Polyporales	*Schizophyllum*

Key morphological characteristics for the identification of moulds

Mucoromycota hyphal characteristics: Mucormycetous fungi have characteristic thin walled, infrequently septate, multi-nucleate hyphae. Many refer to mucormycetous hyphae as non-septate because they don't have true septa with pores, they are one continuous cell. Older cells are separated by solid cross walls and there is no flow of cytoplasmic material between cells. Thus, coenocytic hyphae are very susceptible to death if damaged (Fig. p1).

For example, in a biopsy sample there may be only a few hyphal cells ramifying through the tissue. During the surgical removal and subsequent laboratory processing all may be damaged and become non-viable (Fig. p2).

Frequently the Histopathology laboratory reports numerous non-septate hyphae present consistent with mucormycosis, and the Microbiology laboratory can't grow it. More than likely the specimen has been chopped up, damaging all the hyphae making them non-viable. Mucormycetous fungi rely on rapid growth and sporulation for survival, most will fill a petri dish within 24 hours, providing there are viable cells.

Ascomycota hyphal morphology: Includes most common conidial moulds, for example *Aspergillus, Fusarium, Scedosporium* etc. Ascomycetous hyphae are multi-nucleate and compartmentalised by septa with simple pores where cytoplasmic and nuclear migration is not inhibited, i.e. all cytoplasmic contents including nuclei may flow up and down the hyphal strand. Simple septal pores have woronin bodies that can block the pore if a hyphal compartment becomes damaged. This is an important survival mechanism for these fungi (Figs. p3-4). Unlike mucormycetous fungi, you can chop up a mould like *Aspergillus*, and enough hyphal compartments will remain viable for culture.

Basidiomycota hyphal morphology: Basidiomycetes have septate hyphae with "complex" septal pores called dolipores, which allow cytoplasmic but not nuclear migration (Fig. p5). Hyphae are dikaryotic (two nuclei per hyphal compartment - one from each parent cell). Basidiomycetous fungi can often be recognised using a light microscope by the presence of characteristic hyphal clamp connections over the septa. Clamp connections are specialised hyphal bridges that allow the simultaneous mitosis of two nuclei to occur in such a position that the dikaryons of compatible nuclei are duplicated in the proper relationship to each other (Fig. p5).

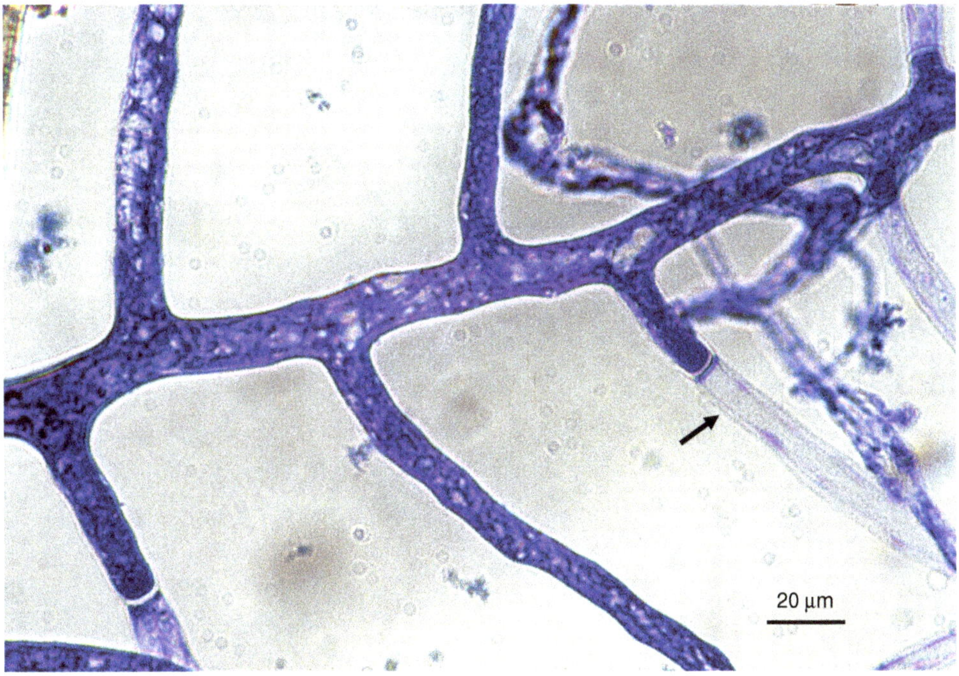

Fig. p1. Coenocytic hyphae are large single cells, that are prone to death if damaged (arrow).

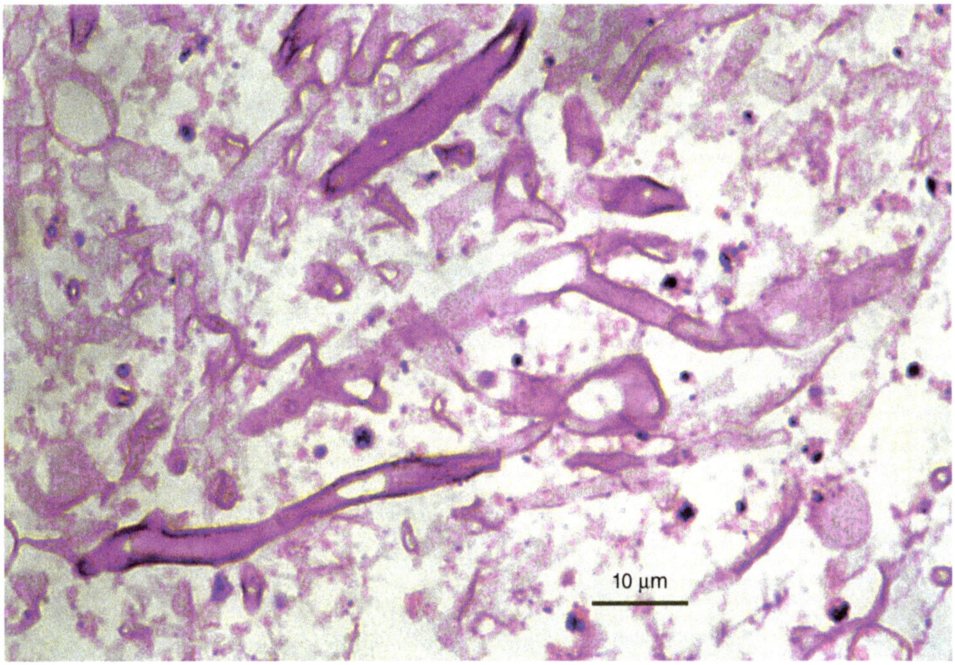

Fig. p2. Haematoxylin and eosin-stained section of lung tissue showing coenocytic hyphae ramifying through the tissue. **Note:** These cells could be easily damaged by laboratory processing and rendered non-viable.

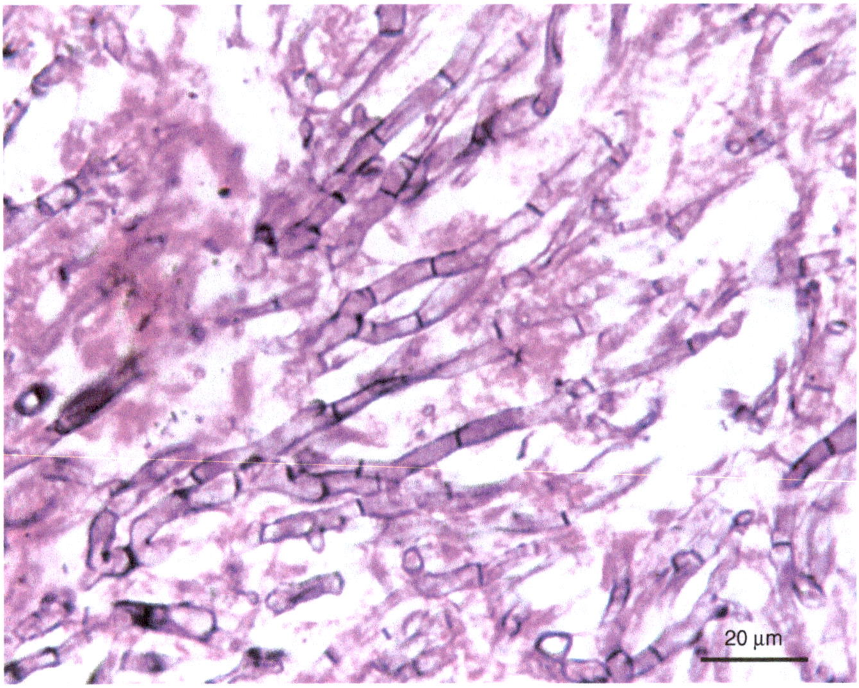

Fig. p3. Septate hyphae of *Aspergillus* in lung tissue stained by Grocott-Gomori's methenamine silver. Like in a submarine the hyphae are compartmentalised by cross walls (septa) with simple pores that can be closed if one section becomes damaged.

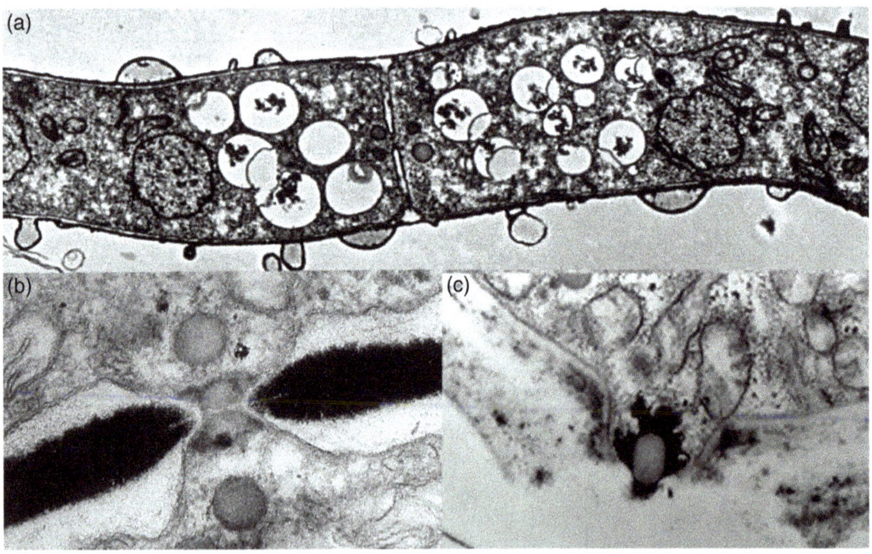

Fig. p4. Transmission electron micrographs of an ascomycetous fungus showing (a) hyphae with typical cytoplasmic organelles, nuclei, vacuoles, mitochondria, and a septum with a central septal pore and associated woronin bodies; (b) a "simple" septal pore showing characteristic woronin bodies; and (c) a woronin body plugging a pore due to damage of the adjacent hyphal segment.

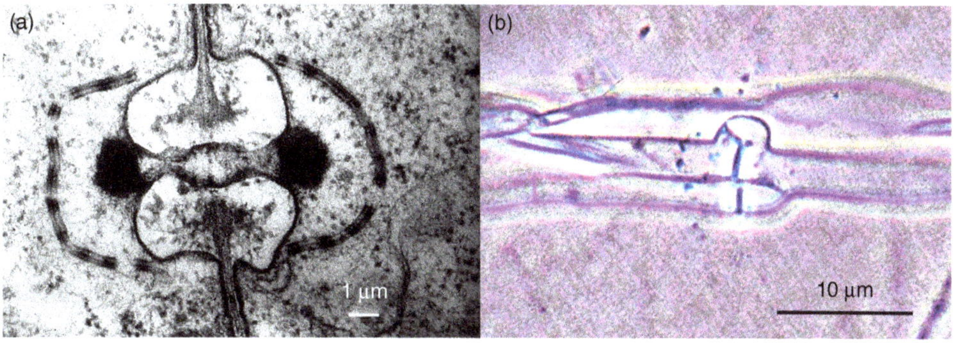

Fig. p5. Basidiomycetous fungal hyphae showing, (a) transmission electron micrograph of a "complex" septal pore (dolipore) with a "parenthesome", a barrier to prevent nuclear migration (photograph courtesy of Stanley L. Flegler, Michigan State University, USA); and (b) a light microscopic image of a hyphal clamp connection over a septum.

Yeast pseudohyphal morphology: Pseudohyphae are a string of elongated blastoconidia formed by some yeasts that resemble a hypha-like filament, each segment or compartment is just a single yeast cell (Fig. p6).

Culture characteristics: These are the least reliable as the media and growth conditions play an important part (Fig. p7).

- Surface texture [glabrous, suede-like, powdery, granular, fluffy, downy, cottony]
- Surface topography [flat, raised, heaped, folded, domed, radial grooved]
- Surface pigmentation [white, cream, yellow, brown, pink, grey, black, etc]
- Reverse pigmentation [none, yellow, brown, red, black, etc]
- Growth rate [colony diameter <5 cm in 14 days or >5 cm in 14 days]
- Growth at 37°C, 40°C, 45°C

Mucoromycota sporangia characteristics:

- Arrangement of sporangiospores [multi-spored, sporangiola, merosporangium] (Fig. p8)
- Arrangement of sporangiophores [unbranched often in groups or frequently branched]
- Sporangium shape [pyriform, spherical, flask-shaped, etc]
- Sporangium size [<100 µm diam. or >100 µm diam.]
- Columella [present or absent] (Fig. p8)
- Apophyses [present or absent [(Fig. p8)
- Sporangiophore height [<0.8 mm or >1 mm]
- Rhizoids [present or absent] (look in the agar)
- Sporangiospore size [<6 µm or >6 µm]

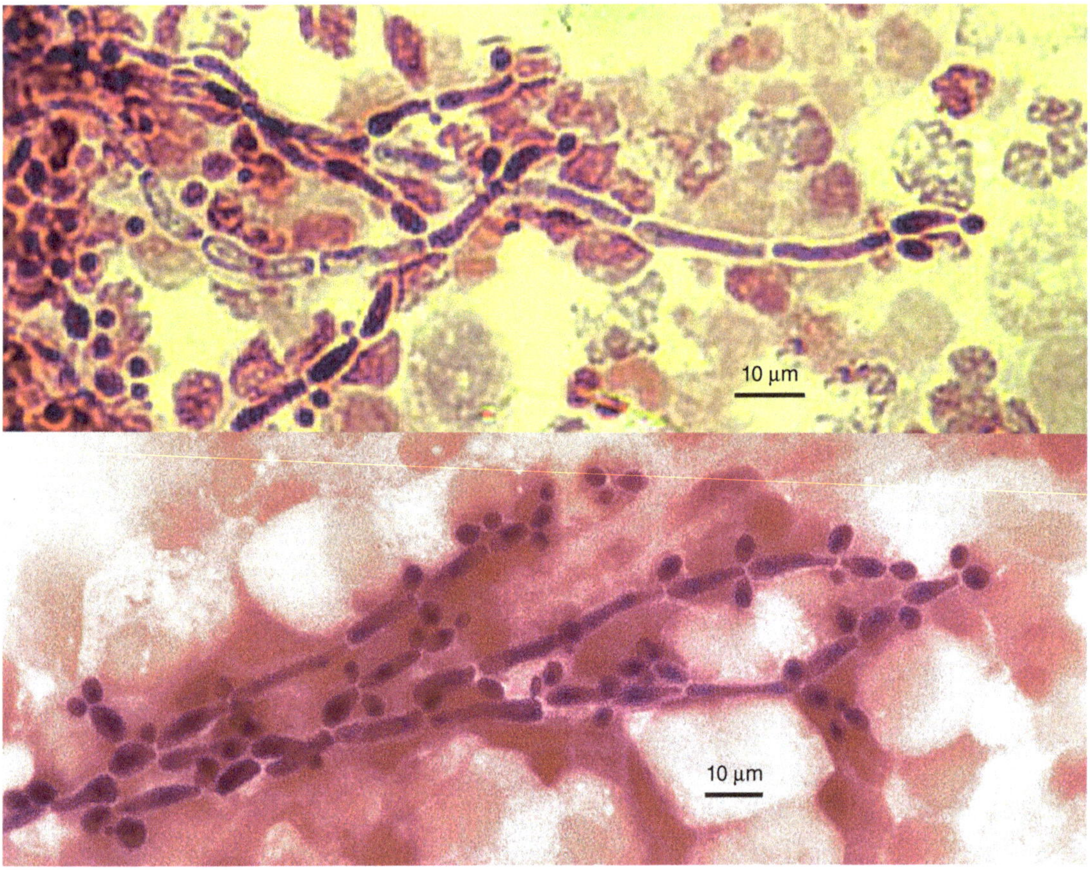

Fig. p6. Budding yeast cells of *Candida albicans*, elongating to form a filament. Each segment is just a single yeast cell.

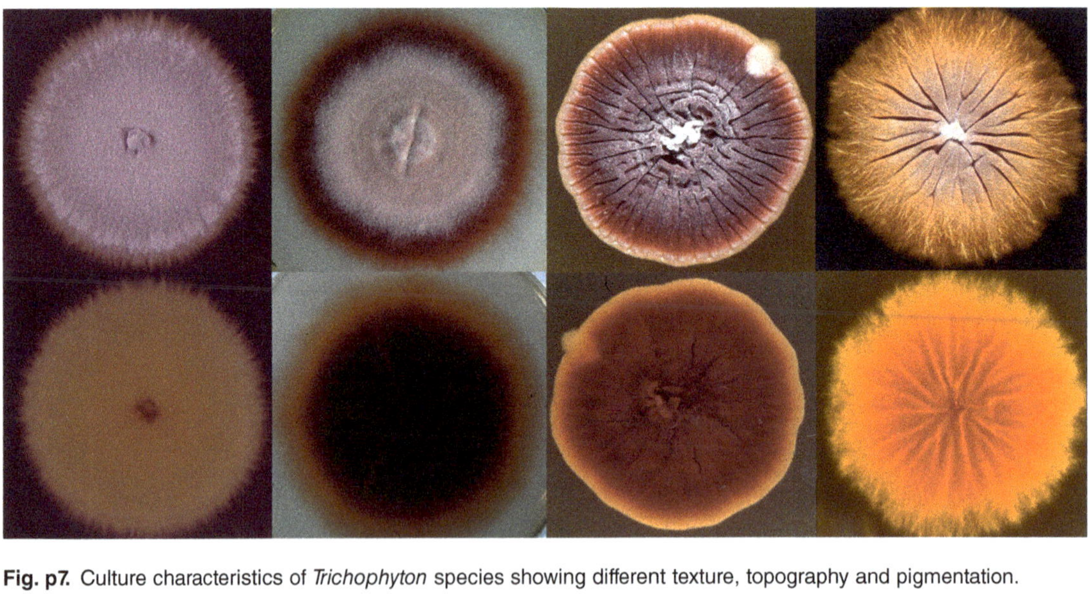

Fig. p7. Culture characteristics of *Trichophyton* species showing different texture, topography and pigmentation.

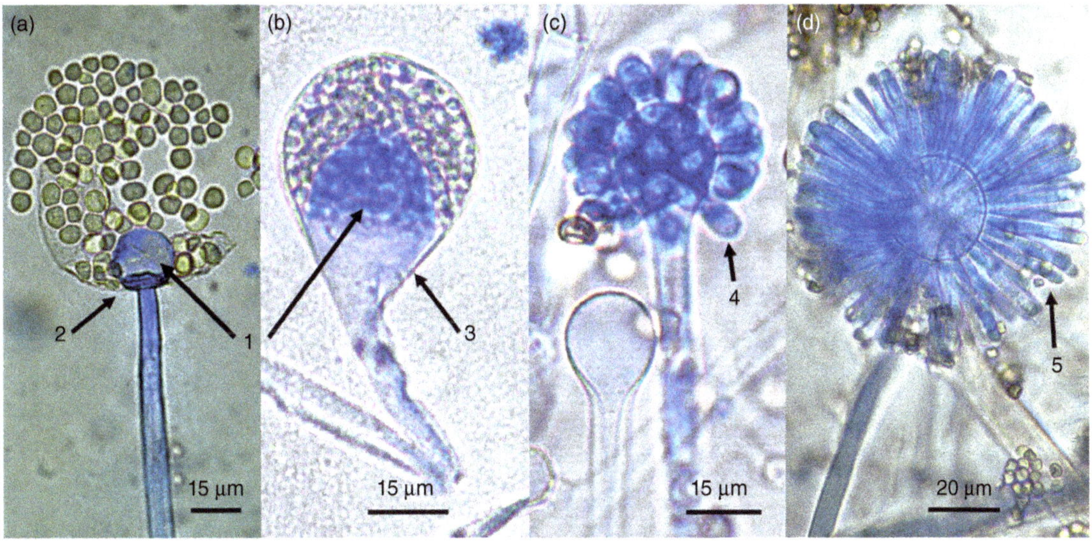

Fig. p8. The presence or absence of columellae (1) and apophyses (3) are key sporangium characteristics: (a) *Mucor* has a multi-spored sporangia with a columella (1) with inconspicuous collarettes (2), but no apophysis, whereas (b) *Lichtheimia* has a multi-spored sporangium with both a columella and apophysis (funnel-shaped swelling of the sporangiophore, immediately below the columella); (c) *Cunninghamella* forms single-celled sporangiola (4); (d) *Syncephalastrum* produces merosporangia with the sporangiospores aligned in a row (5).

Ascomycota - Conidial moulds

1. Conidial characteristics:

- Septation [one-celled, two-celled, multi-celled with transverse septa only, or multi-celled with both transverse and longitudinal septa] (Fig. p9)
- Shape [spherical, sub-spherical, pyriform, clavate, ellipsoidal, etc.]
- Size, need a graduated eyepiece [length <10 µm or >10 µm]
- Colour [hyaline or darkly pigmented]
- Wall texture [smooth, rough, verrucose, echinulate]
- How many conidial types present? [i.e. micro and macroconidia]

2. Arrangement of conidia as they are borne on the conidiogenous cells:

- Solitary [single or in balls]
- Catenulate (in chains) [acropetal, youngest conidium at the tip or basipetal, youngest conidium at the base] (Fig. p10)

3. Growth of the conidiogenous cell:

- Determinant [no growth of the conidiophore after the formation of conidia]
- Sympodial [a mode of conidiogenous cell growth which results in the development of conidia on a geniculate or zig-zag rachis] (Fig. p11)

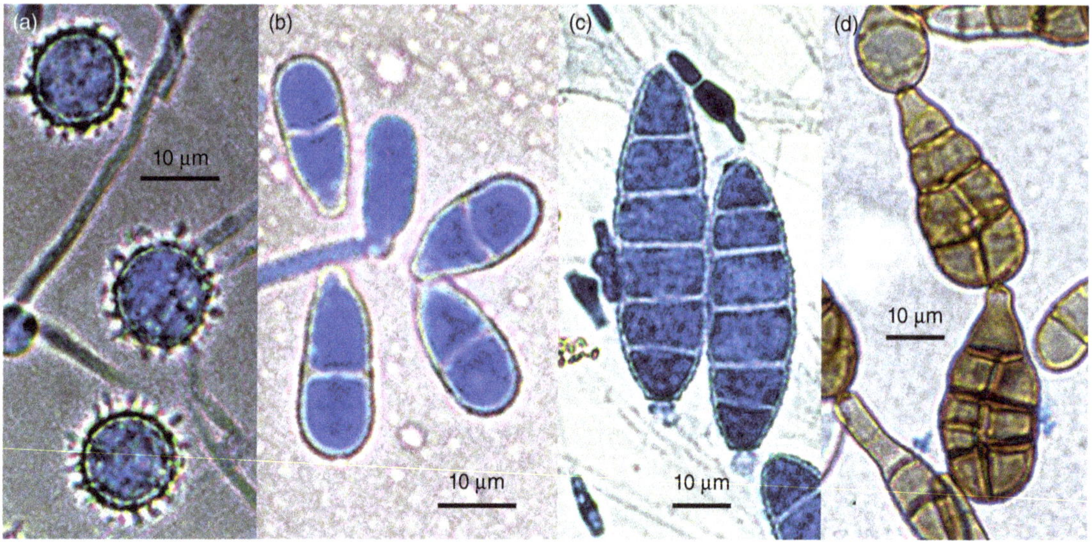

Fig. p9. (a) one-celled conidia (ameroconidia), (b) two-celled conidia (didymoconidia), (c) multi-celled conidia with transverse septa only (phragmoconidia), (d) multi-celled conidia with transverse and longitudinal septa (dictyoconidia). Also note shape, size, pigmentation, and wall texture.

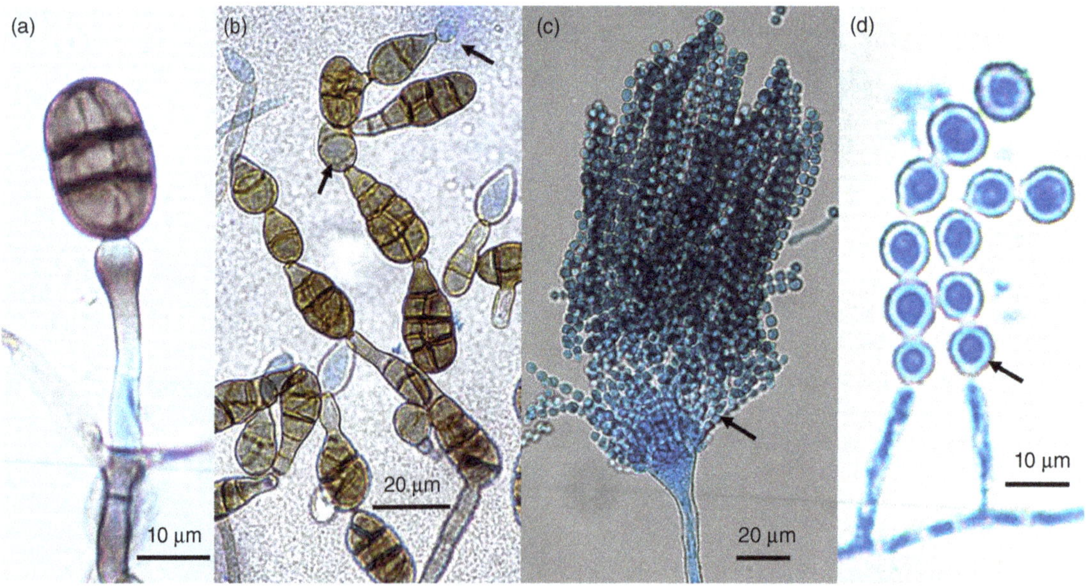

Fig. p10. Arrangement of conidia (a) solitary; (b) acropetal chain with youngest conidia at the tip (arrows); (c) basipetal chain from a phialide with the youngest at the base (arrow); (d) basipetal chain from an annellide with the youngest at the base (arrow).

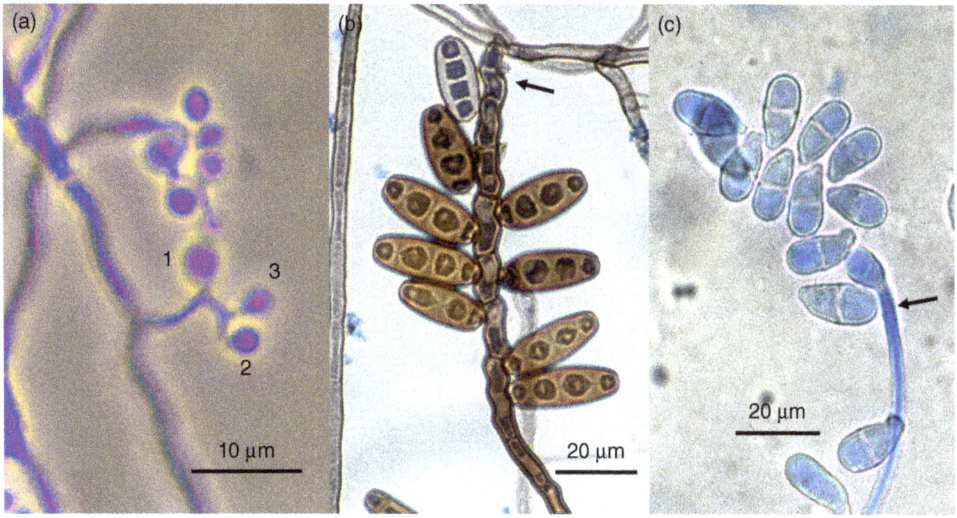

Fig. p11. Sympodial (1 - first conidium, 2 - second conidium, 3 - third conidium etc), (a) *Beauvaria* showing narrow base sympodial (rachiform) development, (b) *Curvularia* showing broad base sympodial (raduliform) development (note the growing conidiophore tip (arrow) with the youngest conidium at the tip), and (c) *Trichothecium* showing retrogressive growth, where the conidiophore (arrow) becomes shorter with the successive formation of conidia (youngest conidium at the base).

4. Type of conidiogenous cell present:

- Non-specialised
- Phialide [specialised conidiogenous cells that produces conidia in basipetal succession without increasing in length] (Fig. p12)
- Annellide [specialised conidiogenous cell producing conidia in basipetal succession by a series of short percurrent proliferations (annellations). The tip of an annellide increases in length and becomes narrower as each subsequent conidium is formed] (Fig. p12)

5. Thallic conidiogenesis:

A mode of conidial ontogeny where a conidium is formed from a pre-existing hyphal segment or cell, for example arthroconidia and chlamydospores (Fig. p13).

6. Additional features:

- Hyphal structures [clamps, spirals, nodular bodies, etc.]
- Synnemata (Fig. p14), sporodochia, chlamydospores (Fig. p15), pycnidia and sclerotia (Fig. p16)
- Confirmatory tests for dermatophytes

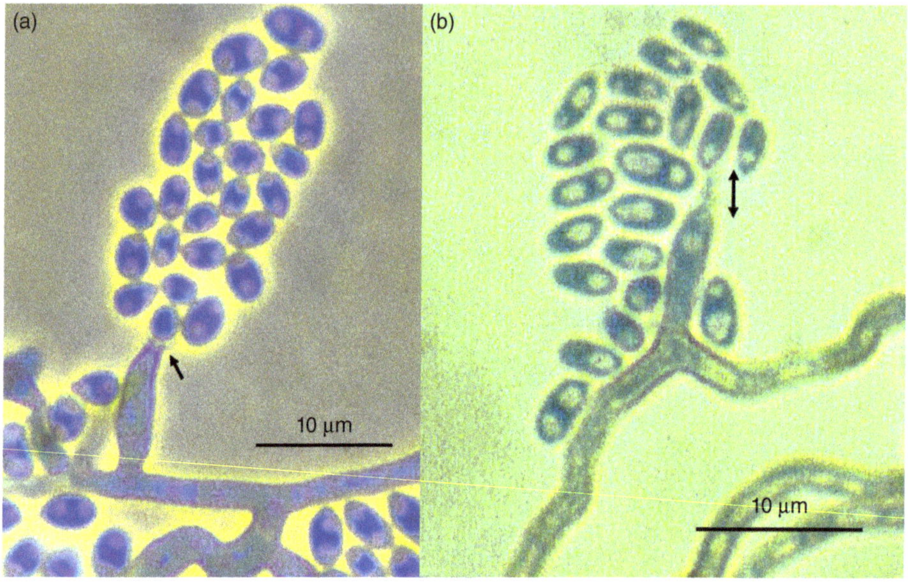

Fig. p12. (a) *Phialophora* showing a typical phialide with a collarette (arrow); (b) *Exophiala* showing a typical annellide. **Note:** It is often very difficult to see the collarettes or annellations using a light microscope; however, phialides do not grow in length whereas annellides grow in length and taper with the formation of successive conidia, which is a useful character (arrows).

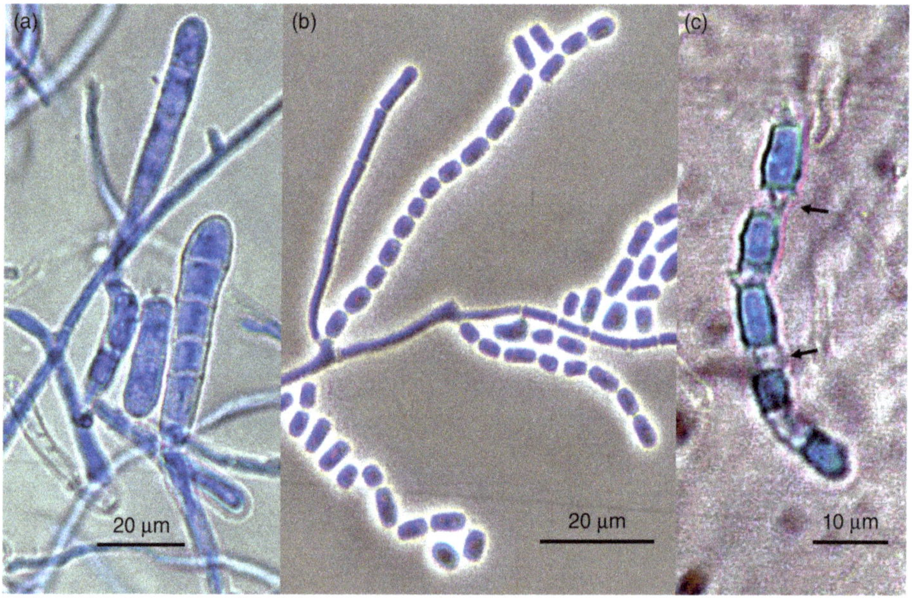

Fig. p13. (a) Macroconidia of *Epidermophyton* where a hyphae segment becomes compartmentalised and then expands to form a conidium, (b) *Geotrichum candidum* arthroconidium formation where hyphal elements are progressively compartmentalised by fragmentation of septa, (c) *Coccidioides immitis* showing arthroconidium formation where hyphal elements are progressively compartmentalised, and arthroconidia are separated from each other by a disjunctor cell (arrows).

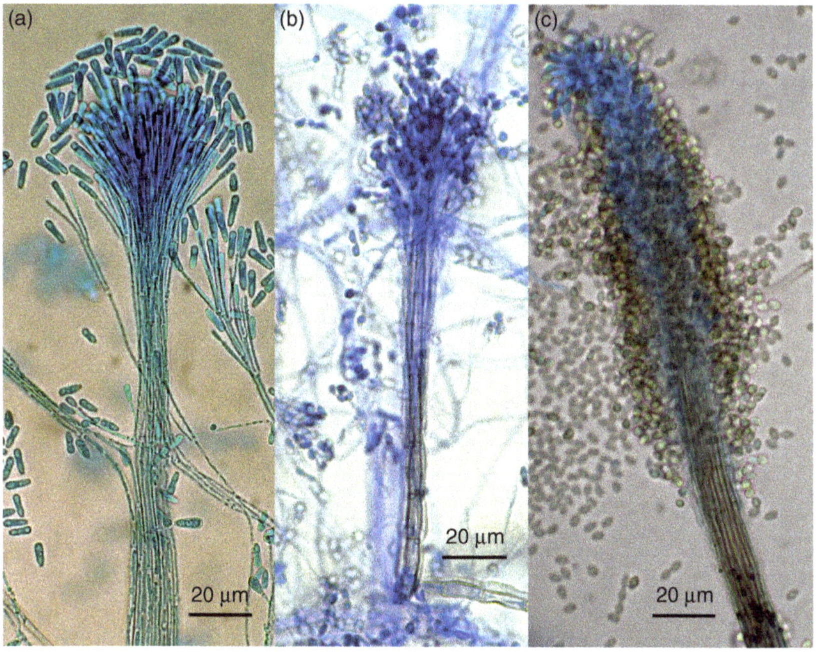

Fig. p14. Synnemata are a group of erect conidiophores that are cemented together producing conidia at the apex and/or along the sides. (a) *Scedosporium* (Photograph courtesy Mathew O'Sullivan, Centre for Infectious Diseases and Microbiology, NSW Health Pathology, Westmead, Australia), (b) *Graphium*, (c) *Doratomyces* (Photograph courtesy Karen Rogers, Mycology, LabPlus, Auckland, New Zealand).

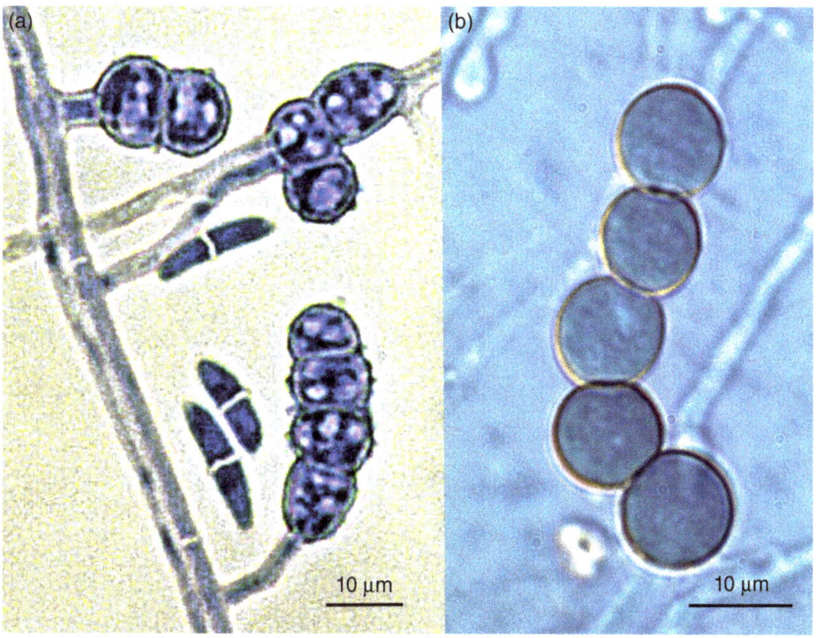

Fig. p15. Chlamydospores (a) *Fusarium* and (b) *Humicola*. Chlamydospores are thick-walled, thallic conidia formed within the vegetative hyphae. Chlamydospores function as organs of perennation (long term survival structures) rather than dissemination.

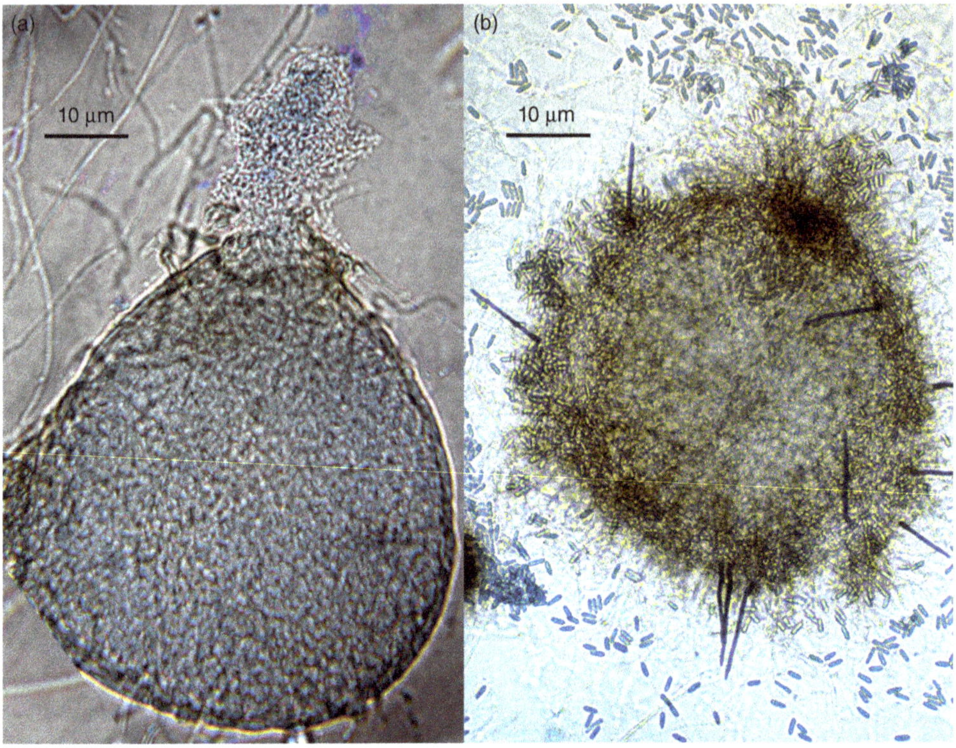

Fig. p16. Pycnidia are an asexual fruiting bodies containing conidia that are formed by coelomycetous fungi. These are conidial fungi where the conidia form in either pycnidia with an apical ostiole or in disc-shaped acervuli; (a) pycnidium of *Phoma* spp. showing an apical ostiole discharging conidia (Photograph courtesy Ann McPhee, Mycology, Sullivan Nicolaides Pathology, Brisbane, Australia). Sclerotia are a mass of thick-walled cells formed by the vegetative hyphae that function as organs of perennation; (b) sclerotia with hair-like setae of *Colletotrichum coccodes*.

Molecular and/or MALDI-ToF MS identification: The use of PCR-based assays, DNA sequencing, and other molecular methods, including those incorporating proteomic approaches such as matrix assisted laser desorption ionization time of flight mass spectroscopy (MALDI-ToF MS) have a clear role in accurate species identification of fungal cultures, although methods and databases are still being developed and improved in some groups of fungi. These are used mainly to complement conventional methods since they require standardisation before widespread implementation can be recommended (Halliday *et al.*, 2015). Molecular-based fungal identification is particularly helpful for fungi that lack distinguishing morphological features, e.g. *Apophysomyces elegans*, or to distinguish between species of the *Aspergillus fumigatus* complex. Comparative sequence analysis is now the "gold standard" for identification of fungi. Methods are referenced where available and in many instances are recommended for more definitive identifications (Fig. p17, Table p2).

Regardless of the genetic locus selected, accurate sequence-based identification is dependent upon database accuracy and adequate species representation. Genbank is well known to contain numerous errors in sequences and the species names attributed to the sequences, which are rarely corrected. Therefore, caution must be used when interpreting sequencing comparisons against this database, and the use of multiple sequence databases is encouraged. Well-curated databases that are helpful for fungal species identification include: (1) Westerdijk Fungal Biodiversity Institute (formerly CBS-KNAW) (https://wi.knaw.nl/page/Pairwise_alignment); and (2) MycoBank Database (https://www.mycobank.org/page/Pairwise_alignment).

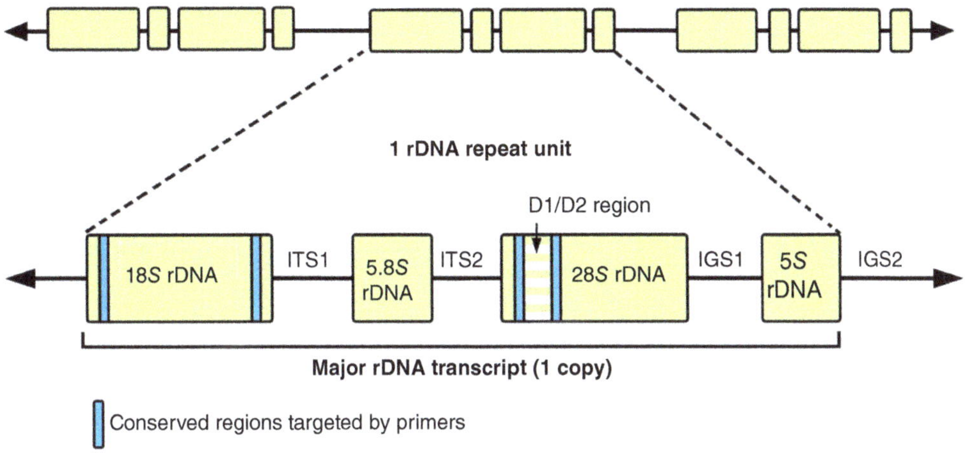

Fig. p17. Schematic diagram of the fungal rDNA gene cluster (adapted from CLSI MM18 and Halliday *et al.*, 2015). The 18*S*, 5.8*S* and 28*S* rRNA genes are separated by the two internal transcribed spacer regions, ITS1 and ITS2. The 28*S* and 5*S* rDNA genes are separated by the intergenic spacer 1 (IGS1). The intergenic spacer 2 (IGS2) separates the rDNA repeat units from each other.

Table p2. Frequently used molecular targets for species identification.

Molecular Target		Application
ITS	Internal transcribed spacer regions (ITS1-5.8*S*-ITS2)	Species level identification of wide range of fungi
D1/D2	D1/D2 variable domains of the 28*S* rRNA gene	Species identification of many of the Mucorales
β-tubulin	Beta tubulin II	Accurate species resolution of *Aspergillus*
Cal	Calmodulin	Species discrimination of *Alternaria, Aspergillus*
TEF-1	Translation elongation factor alpha subunit (also referred to as *EF-1α*)	Species complex identification within genera of *Aureobasidium, Cladosporium* and *Fusarium*
RPB1RPB2	RNA polymerase I subunitRNA polymerase II subunit	Species complex identification within genera of *Fusarium*, *Penicillium* and *Talaromyces*
ACT	Actin	Species discrimination of *Aspergillus, Cladosporium, Coniochaeta, Verticillium* and *Verruconis*
GAPDH (GDP)	Glyceraldehyde-3-phosphate dehydrogenase	Species discrimination of *Bipolaris, Colletotrichum, Curvularia* and *Verticillium*
CHS	Chitin synthase	Species discrimination of *Sporothrix*
Chi18-5	Chitinase 18-5	Species discrimination of *Trichoderma*

Antifungal susceptibility: For many species, antifungal susceptibility data has also been provided. This has been derived from both the literature and in-house data from Australian clinical isolates generated by using the CLSI M27 and M27M44S (formerly M60) protocols for yeasts and the CLSI M38 and M38M51S (formerly M61) protocols for filamentous fungi. This composite data is provided as a guide only. Minimum Inhibitory Concentrations (MICs) or Minimum Effective Concentrations (MECs) for *Aspergillus, Candida, Cryptococcus* and *Scedosporium/Lomentospora* species are provided from large Australian studies based predominantly on primary isolates. In many cases the clinical relevance of *in vitro* antifungal susceptibility results remains difficult to interpret, and expert advice from a consulting microbiologist or infectious disease specialist is recommended.

Due to longstanding concerns about the reliability and reproducibility of caspofungin MICs, both the CLSI and EUCAST recommend this antifungal not be tested or its MICs reported (Espinel-Ingroff *et al.,* 2013). Anidulafungin or micafungin MICs may be used as a proxy for caspofungin. Caspofungin MICs have not been included in this edition.

Abbreviations: Amphotericin B (AmB), Fluconazole (FLU), Isavuconazole (ISAV), Voriconazole (VORI), Posaconazole (POSA), Itraconazole (ITRA), Ketoconazole (KETO), Anidulafungin (ANID), Micafungin (MICA), 5-fluorocytosine (5FC), Terbinafine (TERB).

Risk group (RG): Recommendations are based on published data and on current definitions in accordance with the Australian/New Zealand Standard AS/NZS 2243.3:2010. Safety in laboratories Part 3: Microbiological safety and containment. **Note:** International biosafety guidelines vary in their risk ratings of fungal species.

1 *Acremonium* Link ex Fries

The genus *Acremonium* contains many species; most are saprophytic being isolated from dead plant material and soil. Several species have been reported as opportunistic pathogens of man and animals, causing mycetoma, mycotic keratitis and onychomycosis, however molecular studies have shown that only *Sarocladium kiliense* and *Acremonium egyptiacum* have affected a significant number of patients (Perez-Cantero and Guarro, 2020).

Recently, several *Acremonium*-like species previously reported as opportunistic pathogens have been transferred to other genera; *Xenoacremonium recifei* (formerly *A. recifei*), *Neocosmospora falciforme* (formerly *A. falciforme*), *Gilomastix roseogrisea* (formerly *A. roseogriseum*), *Sarocladium kiliense* (formerly *A. kiliense*) and *Sarocladium strictum* (formerly *A. strictum*) (Glenn *et al.*, 1996; Summerbell *et al.*, 2011; Lombard *et al.*, 2015).

RG-2 organism for species isolated from humans

Morphological description: Colonies are usually slow growing, often compact, and moist at first, becoming powdery, suede-like, or floccose with age, and may be white, grey, pink, rose or orange in colour. Hyphae are fine and hyaline and produce mostly simple awl-shaped erect phialides with inconspicuous collarettes (Fig. 1.1). Conidia are usually one-celled, hyaline or rarely pigmented, globose to cylindrical, and mostly aggregated in slimy heads at the apex of each phialide. Chlamydospores may be present.

Key features: Solitary, erect, hyaline, awl-shaped phialides producing one-celled, globose to cylindrical conidia, mostly in slimy heads.

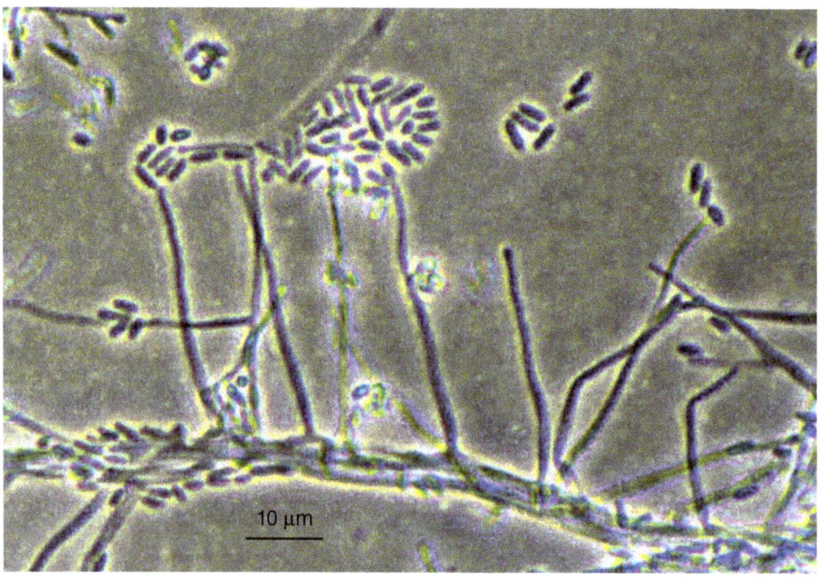

Fig. 1.1. *Acremonium* spp. showing long awl-shaped phialides producing cylindrical, one-celled conidia mostly aggregated in slimy heads at the apex of each phialide.

Note: The microconidia of *Fusarium* species may be confused with *Acremonium*, but they usually grow faster and have colonies with a characteristic fluffy appearance. *Phialemonium* species differ by having short, tapering phialides, mostly lacking a basal septum. *Coniochaeta* is characterised by having sessile phialidic collarettes that are formed directly on the hyphae.

Molecular identification: Summerbell *et al.* (2011) revised the genus based on 18*S* and D1/D2 sequence phylogeny. Sequence based identification may be performed using the D1/D2 or the ITS region. Caution must be exercised though in the interpretation of database sequence comparisons due to the scarcity of database sequences from well-characterised strains, and some sequences may have been attributed to species that have been reclassified (Perdomo *et al.*, 2011a).

Antifungal susceptibility: (Table 1.1). For additional species-specific data see Guarro *et al.* (1997) and Perdomo *et al.* (2011a).

Table 1.1. *Acremonium* spp. (Australian national data); MIC µg/mL.

Antifungal	No.	≤0.016	0.03	0.06	0.125	0.25	0.5	1	2	4	8	16	32	≥64
AmB	7					1	2	1	1	2				
ISAV	1						1							
VORI	7			1	3	3								
POSA	6				1	1	1	3						
ITRA	7				1		4				1			

References: Gams (1971); Samson *et al.* (1995); Domsch *et al.* (2007); Summerbell *et al.* (2018); de Hoog *et al.* (2020); Perez-Cantero and Guarro (2020).

DOI: 10.1079/9781800622340.0001

2 *Acrophialophora fusispora* (S.B. Saksena) Samson

The genus *Acrophialophora* contains 16 species, which are most commonly associated with soil, especially from India (Zhang *et al.*, 2015a). Two species have been reported as human pathogens, *A. fusispora* and *A. levis* (Sandoval-Denis *et al.*, 2015a). *A. levis* conidia lack spiral bands and the phialides have verruculose walls; ITS sequencing is recommended for species identification.

RG-1 organism

Morphological description: Colonies fast growing, greyish-brown with a black reverse. Conidiophores arising singly, terminally, and laterally from the hyphae, erect, straight, or slightly flexuose, tapering towards the apex, pale brown, rough-walled, up to 15 µm long, 2-5 µm wide, with whorls of phialides on the upper part. Phialides flask-shaped with a swollen base and a long, narrow neck, hyaline, smooth-walled or echinulate, 9-15 x 3-4.5 µm in the broadest part. Conidia in long chains, limoniform, one-celled, pale brown 5-12 x 3-6 µm, smooth to finely echinulate with indistinct spiral bands (Fig. 2.1). Temperature: optimum 40°C; maximum 50°C.

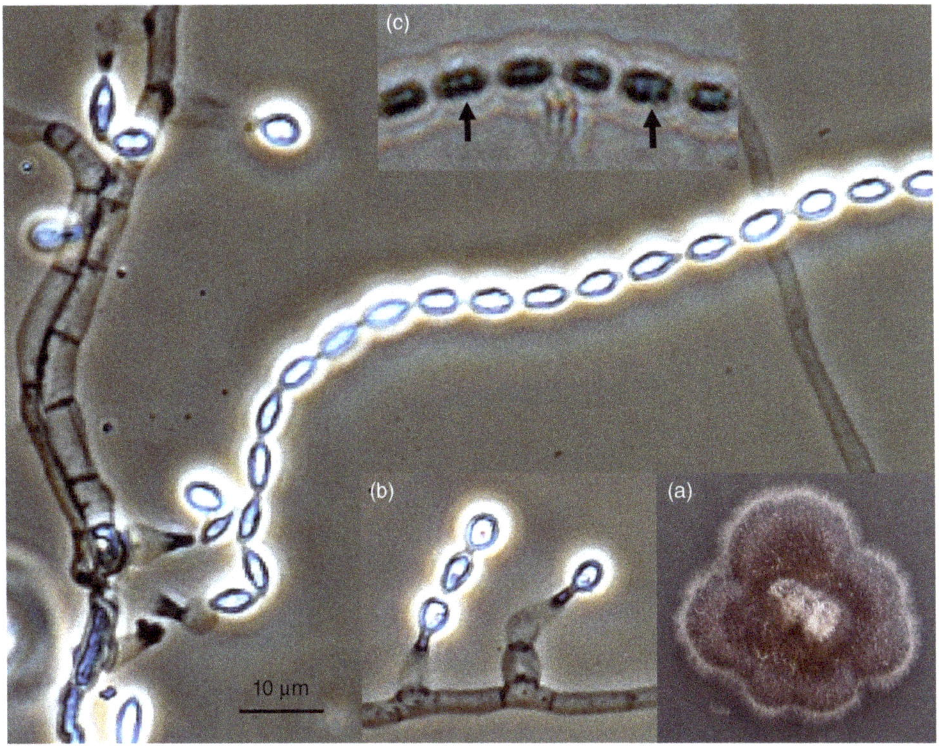

Fig. 2.1. *Acrophialophora fusispora* showing (a) culture, (b) flask-shaped phialides producing long chains of conidia and (c) conidia with spiral striations (arrows).

Key features: Flask-shaped phialides producing long chains of single-celled, limoniform, pale brown conidia, with indistinct spiral bands. **Note:** *A. levis* conidia lack spiral bands and the phialides have verruculose walls.

Molecular identification: D1/D2, ITS, 18*S* rDNA and β-tubulin sequences have been reported by Sandoval-Denis *et al.* (2015a) and Zhang *et al.* (2015a).

Antifungal susceptibility: (Table 2.1).

Table 2.1. *Acrophialophora* spp. (Sandoval-Denis *et al.*, 2015a; and Australian national data); MIC, *MEC μg/mL.

Antifungal	Range	MIC_{90}	Antifungal	Range	MIC_{90}
AmB	1-32	16	ITRA	0.125-4	1
VORI	0.06-0.5	0.25	ANID*	0.25-8	4
POSA	0.25-1	1	MICA*	0.125-32	32

References: Domsch *et al.* (2007); Guarro *et al.* (2007); de Hoog *et al.* (2015); Al-Mohsen *et al.* (2000).

DOI: 10.1079/9781800622340.0002

3 *Alternaria* Nees ex Fries

A ubiquitous genus containing common saprophytes in soil and air, and plant pathogens. *Alternaria alternata* is a well-known cause of leaf spot, rot and blight on numerous plant hosts. *Alternaria infectoria* is the most common clinical species (Pastor and Guarro, 2008). Although usually seen as saprophytic contaminants, *Alternaria* species in particular *A. alternata* and *A. infectoria* are recognised causative agents of subcutaneous phaeohyphomycosis and mycotic keratitis. They are a rare cause of onychomycosis, usually following trauma to the nail.

RG-1 organism

Morphological description: Colonies are fast growing, black to olivaceous-black or greyish, and are suede-like to floccose. Microscopically, branched acropetal chains of multicellular conidia (dictyoconidia) are produced sympodially from simple, sometimes branched, short or elongate conidiophores. Conidia are obclavate, obpyriform, sometimes ovoid or ellipsoidal, often with a short conical or cylindrical beak, pale brown, smooth-walled or verrucose (Fig. 3.1).

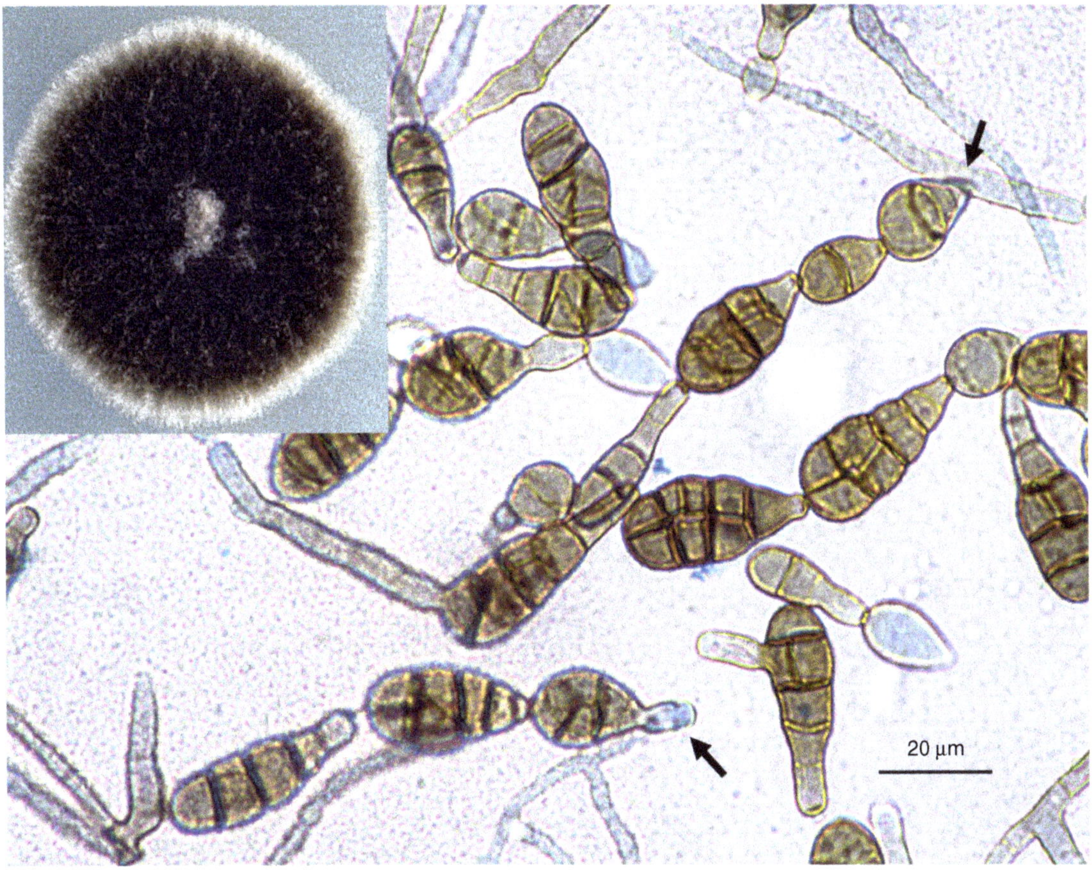

Fig. 3.1. *Alternaria alternata* culture and multi-celled conidia (dictyoconidia) arranged in acropetal chains with the youngest at the tip (arrows).

Key features: Chains of darkly pigmented, ovoid to obclavate dictyoconidia, often with short conical or cylindrical beaks. **Note:** *Alternaria* species soon lose their ability to sporulate in culture. Potato dextrose agar and cornmeal agar are the most suitable media to use, and incubation exposed to ultra-violet light is recommended to maintain sporulation.

Molecular identification: Multilocus genotype studies have shown the *Alternaria* complex currently comprises nine genera and eight *Alternaria* sections (Woudenberg *et al.*, 2013). ITS sequencing is sufficient for genus and usually species level identification and can clearly differentiate *A. alternata* and *A. infectoria* (Pastor and Guarro, 2008). However, it is estimated that >14% of GenBank sequences of *Alternaria* species are misclassified, so unknown sequences should be compared to those of well characterised reference strains (Woudenberg *et al.*, 2013).

Antifungal susceptibility: (Table 3.1).

Table 3.1. *Alternaria* spp. (Australian national data); MIC µg/mL.

Antifungal	No.	≤0.016	0.03	0.06	0.125	0.25	0.5	1	2	4	8	16	32	≥64
AmB	34				4	3	9	11	5	2				
ISAV	7						2		1	3	1			
VORI	34					1	5	13	8	7				
POSA	33		2	3	8	10	9		1					
ITRA	34		1	1	2	14	13	2			1			

References: Simmons (1967, 2007); Ellis (1971); Pryor and Gilberton (2000); de Hoog and Horre (2002); Domsch *et al.* (2007); de Hoog *et al.* (2015).

DOI: 10.1079/9781800622340.0003

4 *Aphanoascus fulvescens* (Cooke) Apinis

Aphanoascus fulvescens is a soil keratinolytic ascomycetous fungus, which occasionally causes dermatomycosis in man and animals. It has a worldwide distribution and is more commonly found in keratin enriched soils from the degradation of animal hair and skin.

RG-2 organism

Morphological description: Colonies are moderately fast growing, white to tan with the production of numerous spherical, pseudoparenchymatous, buff to light brown cleistothecia (non-ostiolate ascocarps) (Fig. 4.1). Asci are subspherical to ellipsoidal and eight-spored. Ascospores light brown, yellowish to pale brown in mass, irregularly reticulate, lens-shaped, 3.5-4.7 x 2.5-3.5 µm. *Aphanoascus fulvescens* has a *Chrysosporium* anamorph showing typical pyriform to clavate-shaped conidia with a truncated base, 15.0-17.5 x 3.7-6.0 µm, which are formed either intercalary, laterally or terminally (Fig. 4.1).

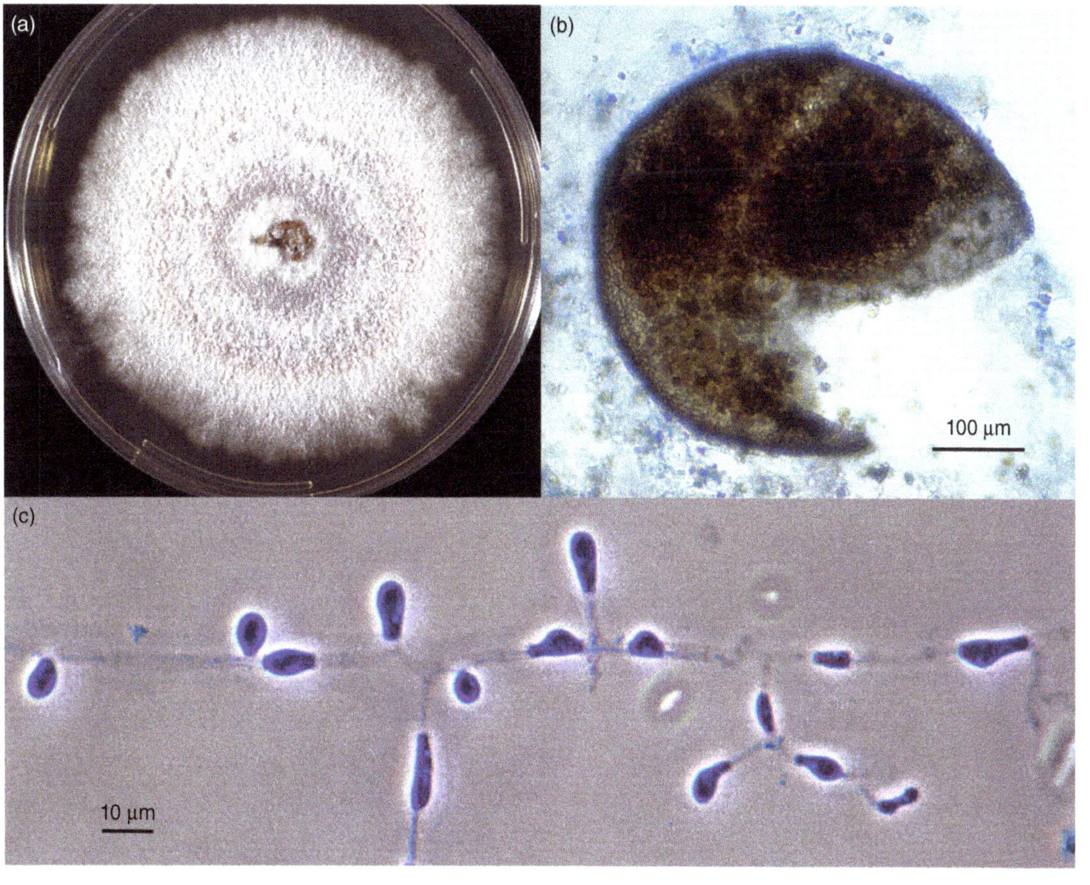

Fig. 4.1. *Aphanoascus fulvescens* showing (a) culture, (b) a ruptured cleistothecium and (c) conidia.
Note: Cleistothecia are formed following colonisation of hair by the fungus.

Key features: Keratinolytic ascomycetous fungus with a *Chrysosporium* anamorph.

Molecular identification: ITS sequencing will differentiate most species. The *Cal* gene may also be useful (Cano *et al.*, 2002; Halliday *et al.*, 2015).

References: McGinnis (1980); Cano and Guarro (1990); Cano *et al.* (2002); Domsch *et al.* (2007); de Hoog *et al.* (2020).

DOI: 10.1079/9781800622340.0004

5 *Apiotrichum* Stautz

In a taxonomic revision of *Trichosporon*, Liu *et al.* (2015) re-defined the genus *Apiotrichum* to include the pathogenic species *Apiotrichum domesticum, A. loubieri* and *A. mycotoxinovorans*. *Apiotrichum* are urease-positive, non-encapsulated basidiomycetous yeasts characterised by the development of hyaline, septate hyphae that fragment into oval or rectangular arthroconidia. This makes morphological identification difficult and molecular identification is often required. Importantly, all species are resistant to echinocandins and other classes of antifungal agents (Arendrup *et al.*, 2014).

Apiotrichum species are often isolated from human and animal sources as well as clinical specimens, some of them being associated with opportunistic infections. All the pathogenic species have the ability to grow at 37°C. Fermentation is absent. Identification to genus level is critical for clinical management and should be performed and provided in a timely manner. Species identification remains difficult and requires molecular analysis or MALDI-ToF MS with an extensive database (Arendrup *et al.*, 2014).

Molecular identification: Species identification requires sequence analysis of the D1/D2 domain and ITS region (Sugita *et al.*, 1999; Arendrup *et al.*, 2014).

MALDI-ToF MS: Appears to be a promising identification tool but requires an extensive database (Kolecka *et al.*, 2013).

References: Gueho *et al.* (1992); Rodriguez-Tudela *et al.* (2005); Chagas-Neto *et al.* (2008); Guo *et al.* (2011); Sugita (2011); Xiao *et al.* (2013); de Hoog *et al.* (2015); Liu *et al.* (2015).

5.1. *Apiotrichum domesticum* (Sugita *et al.*) Yurkov & Boekhout

Synonymy: *Trichosporon domesticum* Sugita *et al.*

Apiotrichum domesticum may occur widely in the environment, including the extreme habitat of soils in the Antarctica dry valleys (Fell *et al.*, 2006). The species is frequently isolated from the houses of patients suffering from summer-type hypersensitivity pneumonitis (Sugita *et al.*, 2004). *A. domesticum* has also been isolated from the urine of a cat with chronic cystitis (Sakamoto *et al.*, 2001), but the pathogenicity of this species remains unknown.

RG-1 organism

Morphological description: Colonies are cream to pale yellow, semi-shiny, pustular to cerebriform, with the margin fringed with hyphae. Yeast cells are globose or ovoid and occur singly or in pairs. Septate hyphae with arthroconidia are present. Does not assimilate D-gluconate but tolerates 0.1% cycloheximide. Growth at 37°C.

Physiological tests: (+ Positive, - Negative, v Variable, w Weak, s Slow, n not done).

Fermentation: Absent.

Growth reactions:

Glucose	+	Cellobiose	+	D-Mannitol	s		
Sucrose	+	L-Sorbose	v	D-Glucitol	v		
Raffinose	–	L-Rhamnose	–	*myo*-Inositol	+		
Melibiose	–	D-Xylose	+	DL-Lactate	+		
Galactose	+	L-Arabinose	+	D-Gluconate	–		
Lactose	+	D-Arabinose	–	2-Keto-D-gluconate	+		
Trehalose	+	D-Ribose	+	D-Glucosamine	+		
Maltose	+	Glycerol	+	*N*-Acetyl-D-glucosamine	n		
Melezitose	+	Erythritol	–	D-Glucuronate	–		
Methyl-α-D-glucoside	+	Ribitol	v	0.1% Cycloheximide	+		
Soluble Starch	+	Galactitol	–	Growth at 37°C	+		

5.2. *Apiotrichum loubieri* (Morenz) Yurkov & Boekhout

Synonymy: *Trichosporon loubieri* Morenz.

Apiotrichum loubieri was originally described by Morenz (1964) from a strain isolated from cow mastitis and has also been reported from the extreme environment of soils in the Antarctic dry valleys (Fell *et al.*, 2006). Although the pathogenicity of this species is not known, two cases of infection caused by *A. loubieri* have been reported (Marty *et al.*, 2003; Padhye *et al.*, 2003).

RG-1 organism

Morphological description: Colonies are whitish, semi-shiny to dull-shiny, and have a fringed margin. Yeast cells are globose or ovoid and occur singly or in pairs. Septate hyphae with arthroconidia are present. Arthroconidia are cubic and short-cylindrical. Tolerates 0.1% cycloheximide and growth at 42°C.

Note: *Apiotrichum loubieri* and *Trichosporon inkin* are both able to grow at 42°C. However, *A. loubieri* can be distinguished from the latter by its ability to use salicin, melibiose or D-glucono-1,5-lactone, which cannot be assimilated by *T. inkin*.

Physiological tests: (+ Positive, - Negative, v Variable, w Weak, s Slow, n not done).

Fermentation: Absent.

Growth reactions:

Glucose	+	Cellobiose	+	D-Mannitol	s
Sucrose	+	L-Sorbose	–	D-Glucitol	v
Raffinose	+	L-Rhamnose	+	*myo*-Inositol	+
Melibiose	+	D-Xylose	+	DL-Lactate	+
Galactose	+	L-Arabinose	+	D-Gluconate	+
Lactose	+	D-Arabinose	–	2-Keto-D-gluconate	+
Trehalose	+	D-Ribose	+	D-Glucosamine	+
Maltose	+	Glycerol	+	*N*-Acetyl-D-glucosamine	n
Melezitose	–	Erythritol	–	D-Glucuronate	+
Methyl-α-D-glucoside	+	Ribitol	v	0.1% Cycloheximide	+
Soluble Starch	+	Galactitol	–	Growth at 42°C	+

5.3. *Apiotrichum mycotoxinovorans* (Molnar *et al.*) Yurkov & Boekhout

Synonymy: *Trichosporon mycotoxinovorans* Molnar *et al.*

Apiotrichum mycotoxinovorans was described by Molnar *et al.* (2004) from a strain isolated from cultivated termites. The species can be distinguished from other species by the ability to assimilate a combination of sucrose, melibiose, galactitol and propane 1,2 diol. In addition, it can detoxify mycotoxins such as ochratoxin A and zearalenone (Molnar *et al.*, 2004).

RG-2 organism

Morphological description: Culture is whitish, dull, smooth, flat, and velutinous, with a margin fringed with hyphae. Yeast cells are globose or ovoid and occur singly or in pairs. Septate hyphae with arthroconidia are present. Arthroconidia are cubic and cylindrical. Tolerates 0.1% cycloheximide and growth at 37°C.

Physiological tests: (+ Positive, - Negative, v Variable, w Weak, s Slow, n not done).

Fermentation: Absent.

Growth reactions:

Glucose	+	Cellobiose	+	D-Mannitol	+
Sucrose	+	L-Sorbose	–	D-Glucitol	v
Raffinose	+	L-Rhamnose	+	*myo*-Inositol	+
Melibiose	+	D-Xylose	+	DL-Lactate	+
Galactose	+	L-Arabinose	+	D-Gluconate	+
Lactose	+	D-Arabinose	+	2-Keto-D-gluconate	+
Trehalose	+	D-Ribose	+	D-Glucosamine	+
Maltose	+	Glycerol	+	*N*-Acetyl-D-glucosamine	n
Melezitose	–	Erythritol	w	D-Glucuronate	+
Methyl-α-D-glucoside	+	Ribitol	-	0.1% Cycloheximide	+
Soluble starch	+	Galactitol	+	Growth at 37°C	+

Antifungal susceptibility: (Table 5.1).

Table 5.1. *Apiotrichum mycotoxinovorans* (Australian national data); MIC μg/mL.

Antifungal	No.	≤0.016	0.03	0.06	0.125	0.25	0.5	1	2	4	8	16	32	≥64
AmB	16						3	5	5	3				
FLU	16							2	4	2	5	2	1	
ISAV	2				1			1						
VORI	16			4	4	1	4	2	1					
POSA	16				2	5	6	3						
ITRA	16				4	7	4	1						

DOI: 10.1079/9781800622340.0005

6 *Apophysomyces* complex

Historically the genus *Apophysomyces* was considered monotypic and *A. elegans* was reported to be an important human pathogen in immunocompetent patients following traumatic implantation. A phylogenetic revision of the genus has now identified additional species, *A. mexicanus*, *A. ossiformis*, *A. trapeziformis* and *A. variabilis*. Interestingly, most of the human clinical isolates tested appear to be *A. variabilis* and not *A. elegans* as previously reported (Alvarez *et al.*, 2010a; Walther *et al.*, 2019). Molecular identification is required to accurately separate these species. Morphological characters overlap so identification to *Apophysomyces* complex level is recommended for most diagnostic laboratories.

Molecular identification: The ITS region and D1/D2 domain may allow accurate species identification (Halliday *et al.*, 2015).

6.1. *Apophysomyces elegans* Misra, Srivastava & Lata

RG-2 organism

Morphological description: Colonies are fast growing, white, becoming brownish grey with age, downy with no reverse pigment, and are composed of broad, sparsely septate (coenocytic) hyphae (Fig. 6.1). Sporangiophores are unbranched, straight or curved, slightly tapering towards the apex, up to 540 µm long, 3-6 µm in width near the apophysis, and hyaline when young but developing a light to dark brown pigmentation and a conspicuous sub-apical thickening 10-16 µm below the apophysis with age (Fig. 6.1). Sporangiophores arise at right angles from the aerial hyphae and often have a septate basal segment resembling the "foot cell" commonly seen in *Aspergillus*. Rhizoids are thin-walled, subhyaline and predominantly unbranched. Sporangia are multispored, small (20-58 µm diameter), typically pyriform in shape, hyaline at first, sepia coloured when mature, columellate and strongly apophysate (Fig. 6.1). Columellae are hemispherical in shape and the apophyses are distinctively funnel or bell-shaped. Sporangiospores are smooth-walled, subspherical to cylindrical, (5-8 x 4-6 µm), subhyaline to sepia in mass. Good growth at 26°C, 37°C and 42°C.

Antifungal susceptibility: (Table 6.1).

Table 6.1. *Apophysomyces elegans* (Chakrabarti *et al.*, 2010; and Australian national data); MIC µg/mL.

Antifungal	No.	≤0.016	0.03	0.06	0.125	0.25	0.5	1	2	4	8	16	32	≥64
AmB	24					3	6	1	12					
ISAV	18							1	9	8				
VORI	24								2		3	19		
POSA	22				4	6	9	3						
ITRA	24					1	6	11	6					

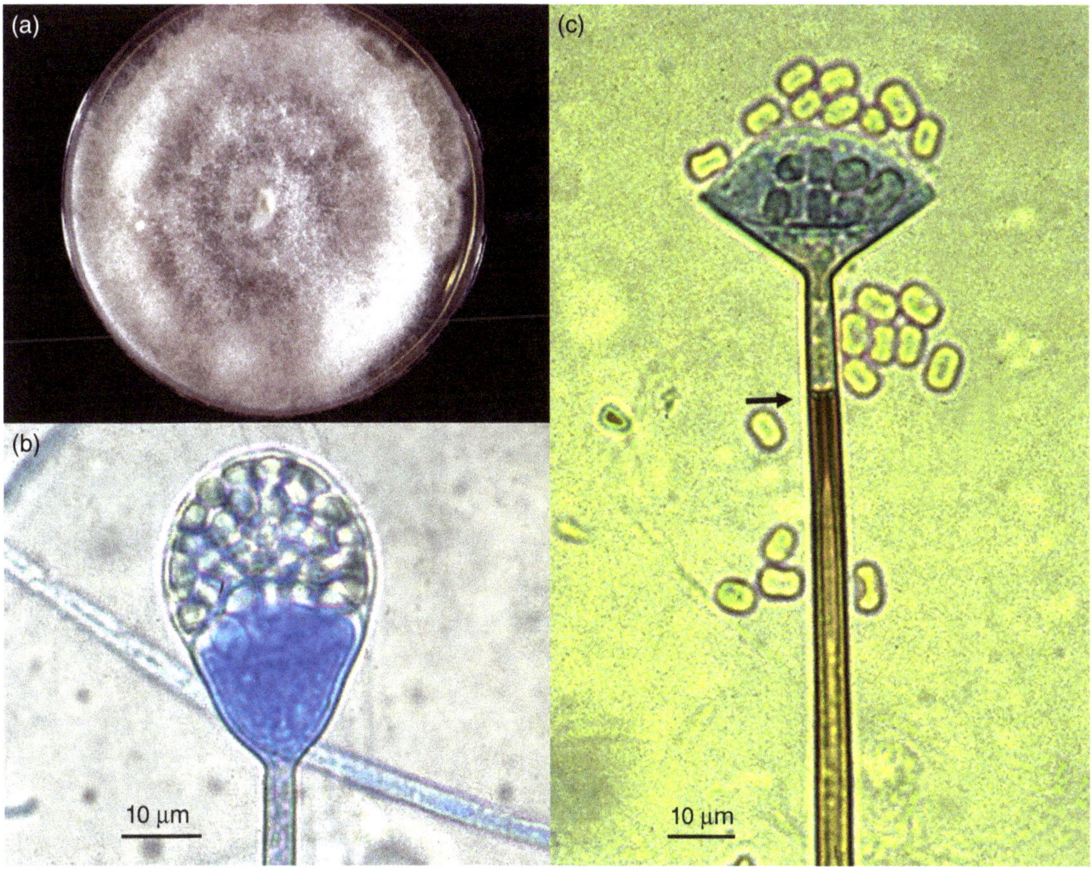

Fig. 6.1. *Apophysomyces elegans* showing (a) culture, (b) young sporangium with a typical funnel-shaped apophysis, and (c) mature sporangium with distinctive funnel-shaped apophyses, columellae, and a conspicuous pigmented sub-apical thickening of the sporangiophore below the apophysis (arrow). Sporangiospores are smooth-walled, oblong and subhyaline.

6.2. *Apophysomyces variabilis* Alvarez *et al.*

RG-2 organism

Morphological description: Colonies are fast growing, whitish with scarce aerial mycelium and no reverse pigment. Sporangiophores are erect, generally arising singly, unbranched, slightly tapering towards the apex, up to 100-400 µm long, 2-3.5 µm in width near the apophysis, hyaline when young but developing a light greyish brown pigmentation with age. Sporangia are multispored, small (15-50 µm diameter), typically pyriform in shape, hyaline at first, sepia coloured when mature, columellate and strongly apophysate. Columellae are hemispherical in shape and the apophyses are short and distinctively funnel shaped. Sporangiospores are smooth-walled, variable in shape trapezoid, ellipsoidal, subtriangular or claviform, (5-14 x 3-6 µm), subhyaline to sepia in mass. Good growth at 26°C, 37°C and 42°C.

Antifungal susceptibility: (Tables 6.2, 6.3).

Table 6.2. *Apophysomyces variabilis* (Espinel-Ingroff *et al.,* 2015a; and Australian national data); MIC µg/mL.

Antifungal	No.	≤0.016	0.03	0.06	0.125	0.25	0.5	1	2	4	8	16	32	≥64
AmB	13			1		1	1	5	5					
ISAV	3								2	1				
VORI	3										3			
POSA	13					2	2	8	1					
ITRA	3							3						

Table 6.3. *Apophysomyces variabilis*, data from 106 isolates (Alvarez *et al.*, 2010a; Prakash *et al.*, 2016); MIC µg/mL.

Antifungal	Range	MIC_{90}	Antifungal	Range	MIC_{90}
AmB	0.125-4	4	ITRA	0.03-2	1
POSA	0.03-2	1	VORI	16-≥16	>16

Key features: Soil fungus with a tropical to subtropical distribution. Characteristic "cocktail glass" apophysate sporangial morphology with conspicuous subapical thickening of the sporangiophore. Resistance to cycloheximide. Rapid growth at 42°C, no growth at 50°C.

Comments: *Apophysomyces* complex is readily distinguished from other mucormycetous fungi, especially the morphologically similar, strongly apophysate pathogen *Lichtheimia corymbifera*, by having sporangiophores with distinctive funnel or bell-shaped apophyses and hemispherical-shaped columellae. In addition, there is a conspicuous pigmented subapical thickening, which constricts the lumen of the sporangiophore below the apophysis, and distinctive foot cells. Laboratory identification of this fungus may be difficult or delayed because of the mould's failure to sporulate on primary isolation media or on subsequent subculture onto potato dextrose agar. Sporulation may be stimulated by the use of nutrient deficient media, like cornmeal-glucose-sucrose-yeast extract agar, Czapek Dox agar, or by using the agar block method described by Ellis and Ajello (1982) and Ellis and Kaminski (1985). Molecular-based identification is particularly helpful for the rapid and definitive identification of poorly sporulating cultures.

References: Misra *et al.* (1979); Wieden *et al.* (1985); Lawrence *et al.* (1986); Padhye and Ajello (1988); Cooter *et al.* (1990); Holland (1997); Ellis (2005b); Guarro *et al.* (2011); Chakrabarti *et al.* (2003, 2010); Walther *et al.* (2019).

DOI: 10.1079/9781800622340.0006

7 *Arthroderma* Curr.

The genus *Arthroderma* currently encompasses 27 species (Hainsworth *et al.*, 2021), mostly occurring in soil, cave environments and infrequently from human and animal cutaneous infections. They are all geophilic dermatophyte species which play an important ecological role in the decomposition of keratin from animal hair and skin.

The molecular taxonomy of this genus has also been evolving with de Hoog *et al.* (2017), transferring *Trichophyton terrestre* and *T. ajelloi* to *Arthroderma* as *A. insingulare* and *A. uncinatum* respectively. However, a more recent multi-gene phylogenetic study by Hainsworth *et al.* (2021) has validated *Arthroderma terrestre* as the correct name for *Trichophyton terrestre*.

Morphological genus description: Colonies mostly granular to cottony, yellowish to brownish, with a cream coloured or brown colony reverse. Macroconidia are multi-celled, (sub)hyaline, clavate, fusiform or cigar-shaped and are thick- and rough-walled. Microconidia are hyaline, one-celled, clavate and smooth-walled.

7.1. *Arthroderma terrestre* (Durie & D. Frey) Kucerova & Hubka

Synonymy: *Trichophyton terrestre* Durie and Fry; *Arthroderma insingulare* Padhye and Carmichael.

Arthroderma terrestre is a geophilic fungus of worldwide distribution which may occur as a saprophytic contaminant on humans and animals. Durie and Frey (1957) first described this soil fungus as *Trichophyton terrestre* from New South Wales, Australia. Since then, *T. terrestre* has been described as an anamorph of three different species of *Arthroderma: A. insingulare, A. lenticulare* and *A. quadrifidum* (Padhye and Carmichael, 1972). However, ITS and D1/D2 sequencing of the original isolates has now identified this fungus as *Arthroderma terrestre* (Hainsworth *et al.*, 2021). *Arthroderma terrestre* is the most frequently encountered *Arthroderma* species recovered from human skin and hair specimens.

RG-1 organism

Morphological description: Colonies are usually flat to downy with a suede-like to granular texture resembling *Trichophyton mentagrophytes*. The surface colour may range from white to cream, buff to yellow, or greenish-yellow. Reverse pigmentation is usually yellowish-brown although some variants have a deep rose red reverse. Microconidia are large, clavate or pedicellate, usually exhibiting transition forms to more or less abundant lateral macroconidia. Macroconidia are clavate to cylindrical with rounded ends, smooth and thin-walled, and are two- to six-celled (Fig. 7.1). Chlamydospores, hyphal spirals, racquet mycelium and antler hyphae may also be present. No growth at 37°C.

Molecular identification: ITS and D1/D2 sequencing is recommended for definitive identification of isolates.

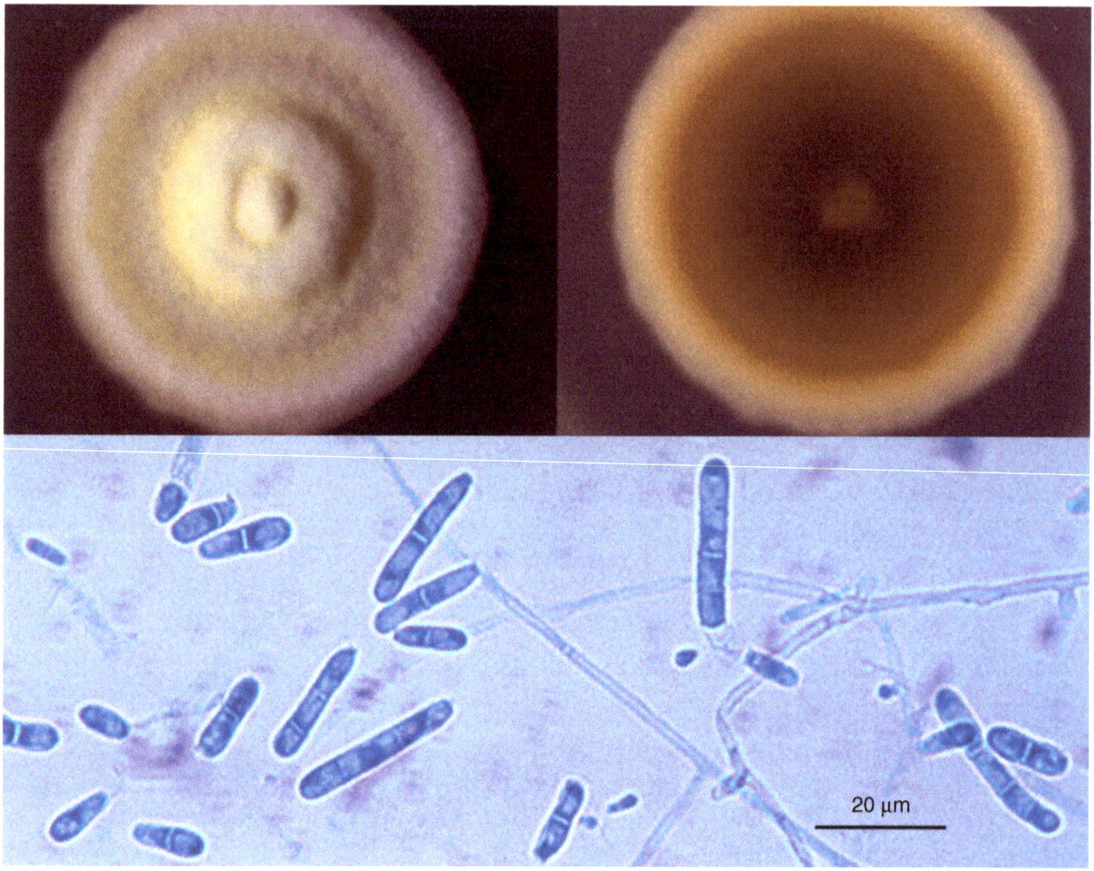

Fig. 7.1. *Arthroderma terrestre* culture and macroconidia.

7.2. *Arthroderma uncinatum* Dawson & Gentles

Synonymy: *Trichophyton ajelloi* (Vanbreus.) Ajello; *Keratinomyces ajelloi* Vanbreusegham.

Arthroderma uncinatum is a geophilic fungus with a worldwide distribution which may occur as a saprophytic contaminant on humans and animals, but infections are doubtful. Not known to invade hair *in vivo* but produces hair perforations *in vitro*.

RG-1 organism

Morphological description: Colonies are usually flat, powdery, cream, tan to orange-tan in colour, with a blackish-purple submerged fringe and reverse. Macroconidia are numerous, smooth, thick-walled, elongate, cigar-shaped, 29-65 x 5-10 µm, and multiseptate with up to nine or ten septa (Fig. 7.2). Microconidia are usually absent, but when present are ovate to pyriform in shape.

Key features: Culture characteristics, macroconidial morphology, urease positive and good growth on Sabouraud's 5% salt agar.

Molecular identification: ITS and D1/D2 sequencing is recommended for definitive identification of isolates.

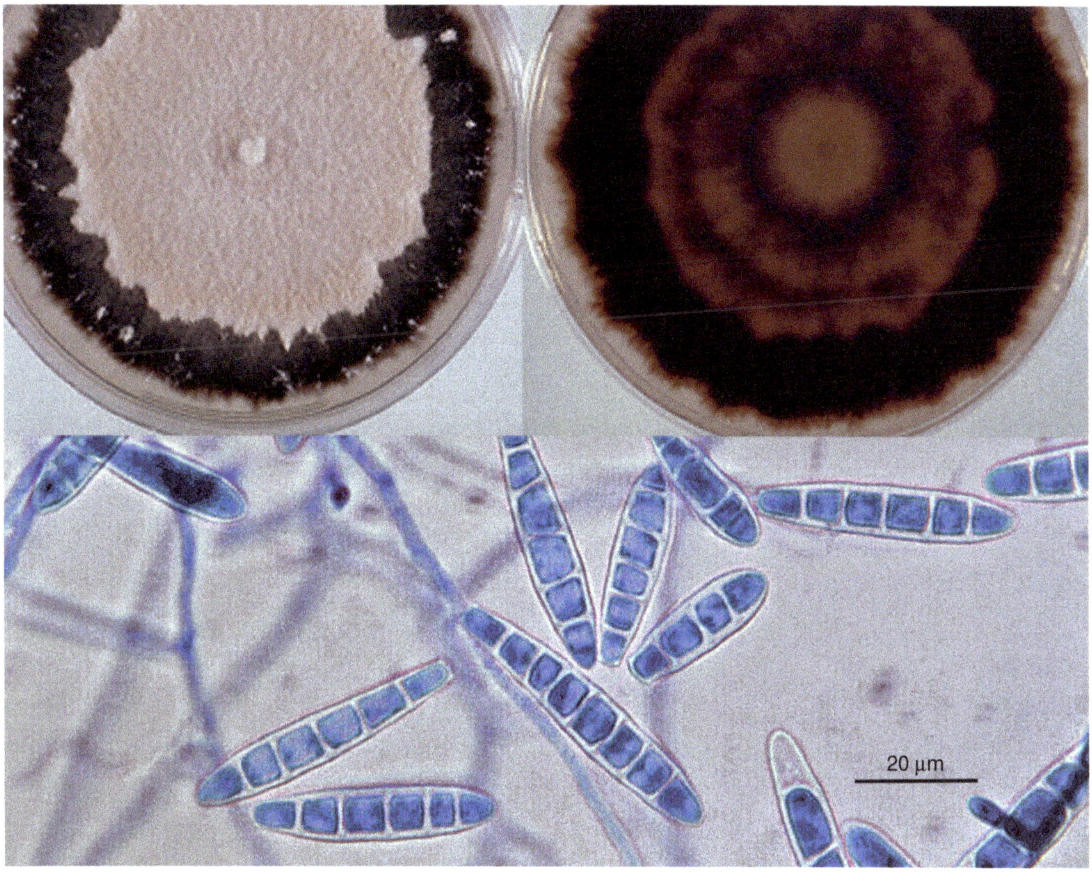

Fig. 7.2. *Arthroderma uncinatum* culture and macroconidia.

References: Rebell and Taplin (1970); Rippon (1988); Zheng *et al.* (2020); Hainsworth *et al.* (2021).

DOI: 10.1079/9781800622340.0007

8 *Arthrographis kalrae* (Tewari & Macpherson) Sigler & Carmichael

Arthrographis kalrae is an arthroconidial mould that is commonly found in environmental samples (soil, wood, air and water), but less frequently in clinical samples, predominantly from nail infections (Sandoval-Denis *et al.*, 2014a).

RG-1 organism

Morphological description: Colonies with slow to moderate growth, creamy white to tan coloured. Initially, yeast-like then developing hyphal growth and conidiophores subhyaline, narrow, branched, often in bundles, occasionally forming whitish, 0.5 cm large, linear synnemata. Arthroconidia are one-celled, hyaline, smooth-walled, oblong to cylindrical, with truncate ends, 2.5-9.0 x 1-2 µm (Fig. 8.1). Spherical blastoconidia 2-4 x 2-3 µm, may also be formed laterally and sessile on undifferentiated hyphae. Chlamydospores may also be present. Very rarely immature ascomata submerged in the agar are produced. Growth at 42°C, and on media containing cycloheximide.

Molecular identification: ITS and D1/D2 sequencing may be used for accurate species identification (Sugiura and Hironaga, 2010; Halliday *et al.*, 2015).

Key features: Keratinolytic; *in vitro* hair perforation positive; growth at 37°C and tolerance to cycloheximide.

Antifungal susceptibility: (Table 8.1).

Table 8.1. *Arthrographis kalrae* (Australian national data); MIC µg/mL.

Antifungal	No.	≤0.016	0.03	0.06	0.125	0.25	0.5	1	2	4	8	16	32	≥64
AmB	13					1	6	2	4					
ISAV	1						1							
VORI	13		1	4	5	2	1							
POSA	13		1	1	4	4	3							
ITRA	13		1		2	7	3							

References: Giraldo *et al.* (2014a); de Hoog *et al.* (2015, 2020).

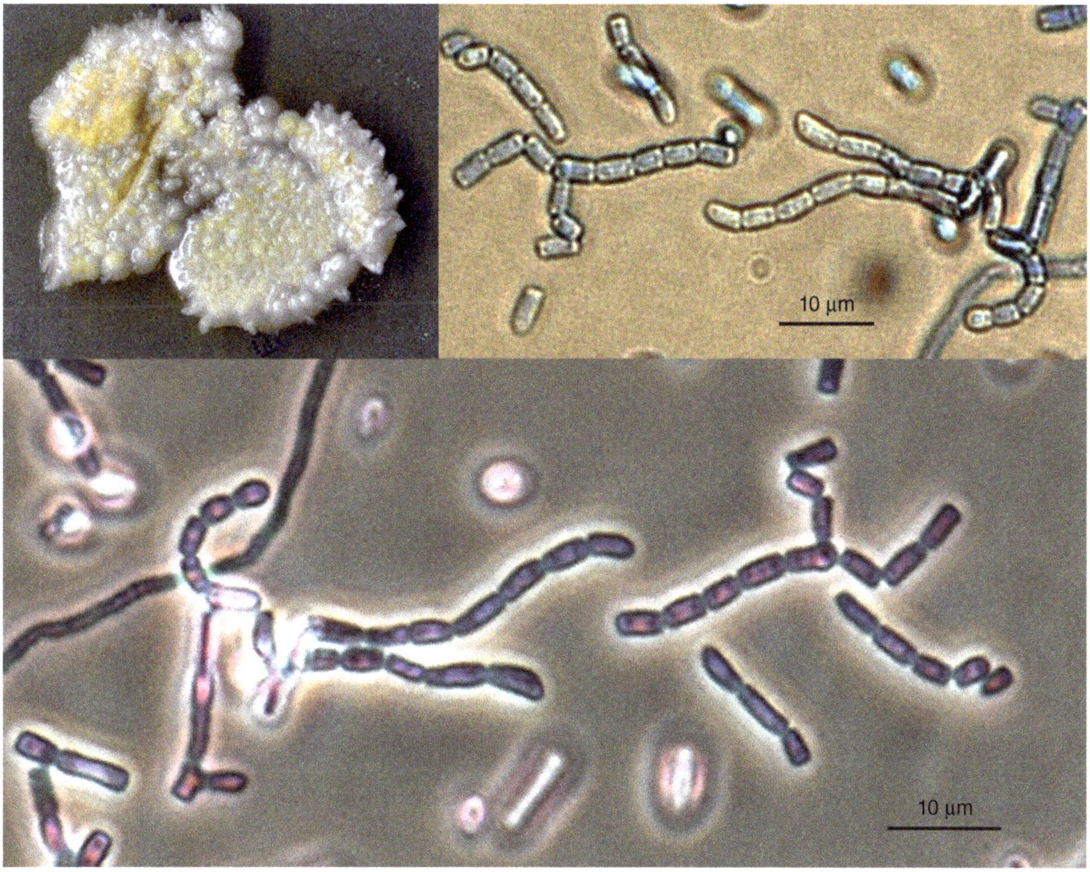

Fig. 8.1. *Arthrographis kalrae* culture and arthroconidia.

DOI: 10.1079/9781800622340.0008

9 *Aspergillus* Micheli ex Link

Aspergillus is a very large genus containing about 250 species, which are currently classified into seven sub-genera that are in turn sub-divided into several sections comprised of related species (Raper and Fennell, 1965; Gams *et al.*, 1985; Geiser *et al.*, 2007; Samson *et al.*, 2014). Traditionally, clinical microbiology laboratories have relied heavily on morphology-based identification methods to separate *Aspergillus* species. However, many species, especially members of the section *Fumigati* have overlapping morphological characteristics (Fig. 9.1), which has allowed several genetically distinct species with very similar morphologies to be misidentified (Balajee *et al.*, 2005a, 2007). This has led to the clustering of species with overlapping morphologies into "species complexes", so that laboratories may better report morphology-based identifications.

Identification of clinical isolates of *Aspergillus* to species level may be clinically important given that different species have variable susceptibilities to multiple antifungal drugs. For example, *in vitro* and *in vivo* studies have demonstrated that *A. terreus* isolates are largely resistant to the antifungal drug amphotericin B, *A. ustus* isolates appear to be refractory to azoles, and *A. lentulus* and *A. alliaceus* have reduced *in vitro* susceptibilities to a wide range of antifungals including amphotericin B, azoles, and echinocandins (Balajee *et al.*, 2005b, 2007).

Molecular identification: Recommended barcoding gene: β-tubulin or *Cal*. General criteria for identification were outlined by Balajee *et al.* (2007). Phylogenetic relationships of the entire genus were presented by Wang *et al.* (1998) and Peterson (2000, 2008).

MALDI-ToF MS: A comprehensive "in-house" database of reference spectra allows accurate identification of species of *Aspergillus* even within complexes e.g. *A. fumigatus sensu stricto* and *A. lentulus* (Lau *et al.*, 2013, 2019).

Morphological genus description: Colonies are usually fast growing, white, yellow, yellow-brown, brown to black or shades of green, mostly consisting of a dense felt of erect conidiophores. Conidiophores terminate in a vesicle covered with either a single palisade-like layer of phialides (uniseriate) or a layer of subtending cells (metulae), which bear small whorls of phialides (the so called biseriate structure) (Fig. 9.2). The vesicle, phialides, metulae (if present) and conidia form the conidial head. Conidia are one-celled, smooth or rough-walled, hyaline or pigmented and produced in long dry chains which may be divergent (radiate) or aggregated in compact columns (columnar). Some species may produce Hülle cells or sclerotia.

For morphologic identification, isolates are usually inoculated at three points on Czapek Dox agar and 2% malt extract agar and incubated at 25°C. Most species sporulate within 7 days. Descriptions are primarily based on colony pigmentation and morphology of the conidial head. Microscopic mounts are best made using cellotape flag or slide culture preparations mounted in lactophenol cotton blue. A drop of alcohol is usually needed to remove bubbles and excess conidia.

Key features: Hyaline ascomycetous mould showing distinctive conidial heads with flask-shaped phialides arranged in whorls on a vesicle.

References: McGinnis (1980); Onions *et al.* (1981); van den Bossche *et al.* (1988); Samson and Pitt (1990, 2000); Samson *et al.* (1995, 2011a, 2014); Klich (2002a,b); Steinbach *et al.* (2005); Domsch *et al.* (2007); de Hoog *et al.* (2020).

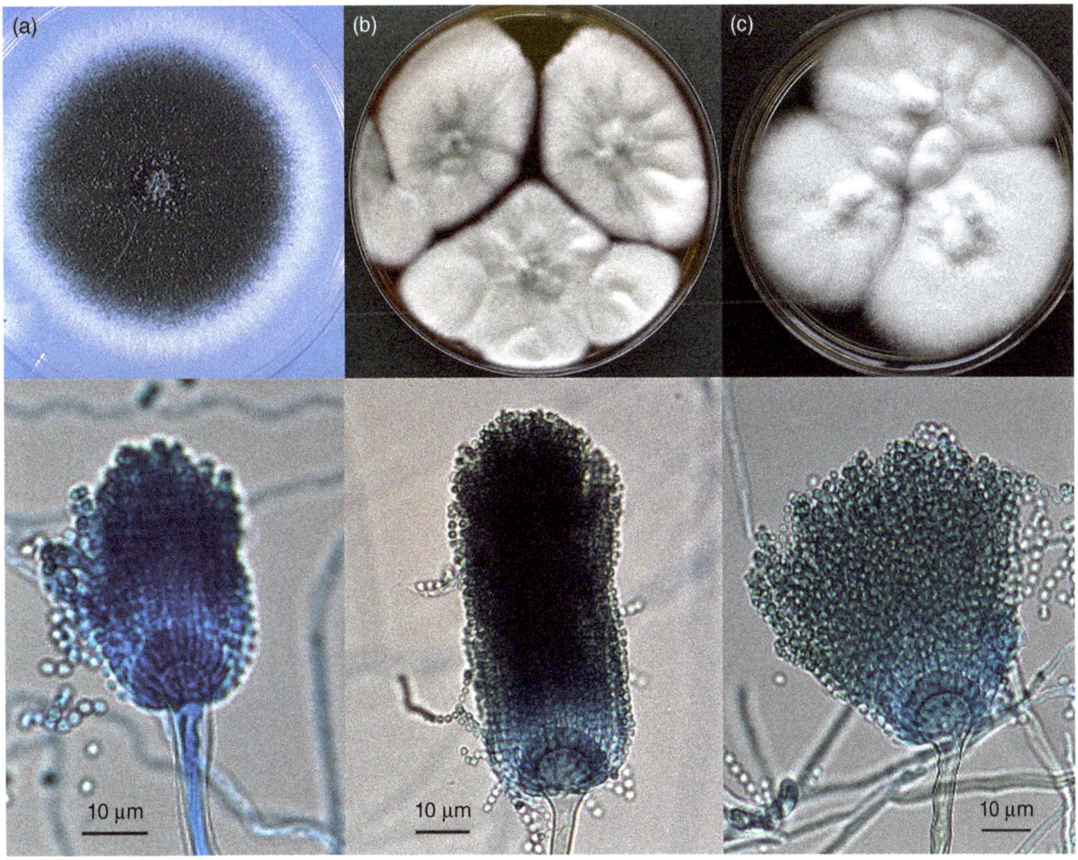

Fig. 9.1 Three species in the *Aspergillus fumigatus* complex showing overlapping morphological characteristics; (a) *A. fumigatus*, (b) *A. lentulus*, (c) *A. fischeri.*

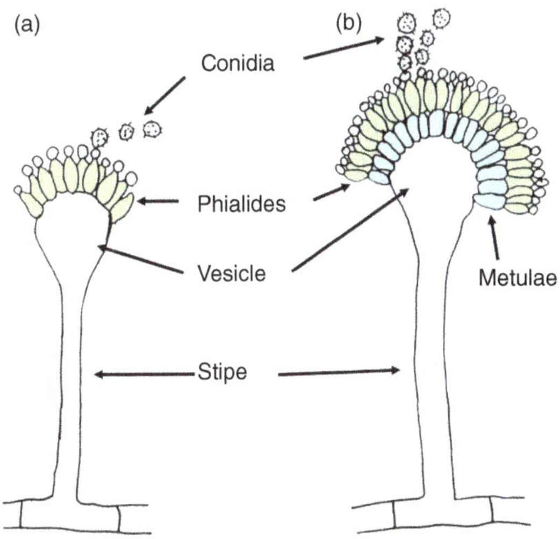

Fig. 9.2. Conidial head morphology in *Aspergillus* (a) uniserate, (b) biseriate.

9.1. *Aspergillus felis* Barrs *et al.*

Aspergillus felis is a member of the *Aspergillus viridinutans* complex within section *Fumigati* (Novakova *et al.*, 2013) and has been reported as a causative agent of invasive aspergillosis and rhinosinusitis in humans, dogs and cats (Barrs *et al.*, 2013). Disease in all host species is often refractory to aggressive antifungal therapeutic regimens.

RG-1 organism

Morphological description: Colonies are suede-like to floccose, white with interspersed grey-green patches of conidia (conidiation is slow to poor) (Fig. 9.3). Conidial heads of *A. felis* are short columnar and uniseriate. Conidiophore stipes are smooth-walled, and vesicles are usually subglobose in shape. Conidia globose (2-3 µm in diameter), smooth to finely roughened (Fig. 9.3).

Molecular identification: *Aspergillus felis* is a molecular sibling of *A. viridinutans* (Novakova *et al.*, 2013) and can be distinguished from other aspergilli by sequence analysis of β-tubulin, *Cal* and *ACT* genes (Barrs *et al.*, 2013).

Comment: *Aspergillus felis* is phenotypically similar to *A. viridinutans* but differs by its ability to grow at 45°C. This species is phylogenetically related to *A. aureoluteus* and *A. udagawae* and differs from *A. aureoluteus* by having a heterothallic mode of reproduction.

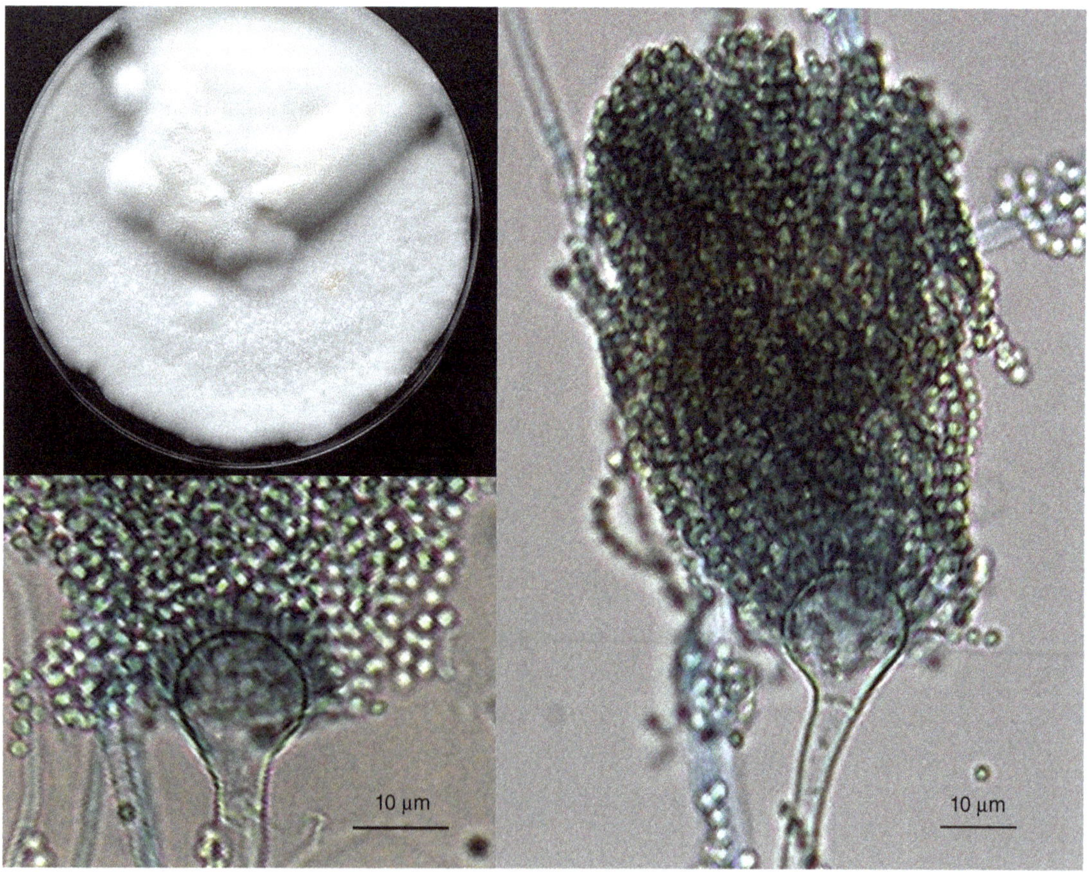

Fig. 9.3. *Aspergillus felis* culture and conidial head morphology.

Antifungal susceptibility: (Table 9.1).

Table 9.1. *Aspergillus felis* (Australian national data); MIC, *MEC μg/mL.

Antifungal	No.	≤0.016	0.03	0.06	0.125	0.25	0.5	1	2	4	8	16	32	≥64
AmB	21					1	11	3	5	1				
ISAV	3									3				
VORI	21					1	3	3	6	8				
POSA	21		4	3	2	1	9	2						
ITRA	21		1	2	2	4	2	9	1					
ANID*	5	4	1											
MICA*	5	3	2											

9.2. *Aspergillus fischeri* Wehmer

Synonymy: *Neosartorya fischeri* (Wehmer) Malloch & Cain.

Aspergillus fischeri is a member of the *A. fumigatus* complex and is mostly found in canned foodstuffs and is now a documented causative agent of invasive aspergillosis in immunosuppressed patients (Chim *et al.*, 1998; Gori *et al.*, 1998).

RG-1 organism

Morphological description: Colonies are suede-like to floccose, white to pale yellow with slow to poor conidiation (Fig. 9.4). Conidial heads are short columnar and uniserate. Conidiophore stipes are smooth-walled, and vesicles are usually subglobose to flask shaped. Conidia are globose to subglobose (2-2.5 μm in diameter), smooth to finely roughened (Fig. 9.4). Good growth at 37°C.

Molecular identification: *A. fischeri* can be distinguished from other members of the section *Fumigati* by sequence analysis of β-tubulin, *Cal* and *ACT* genes (Balajee *et al.*, 2005a; Samson *et al.*, 2007). ITS sequencing is not recommended.

Antifungal susceptibility: (Table 9.2).

Table 9.2. *Aspergillus fischeri* (Australian national data); MIC, *MEC μg/mL.

Antifungal	No	≤0.016	0.03	0.06	0.125	0.25	0.5	1	2	4	8	16	32	≥64
AmB	10					1		6	3					
ISAV	1							1						
VORI	10						2	3	4	1				
POSA	10			1	4	1	2	2						
ITRA	10				2	2	3	3						
ANID*	2	2												
MICA*	2	2												

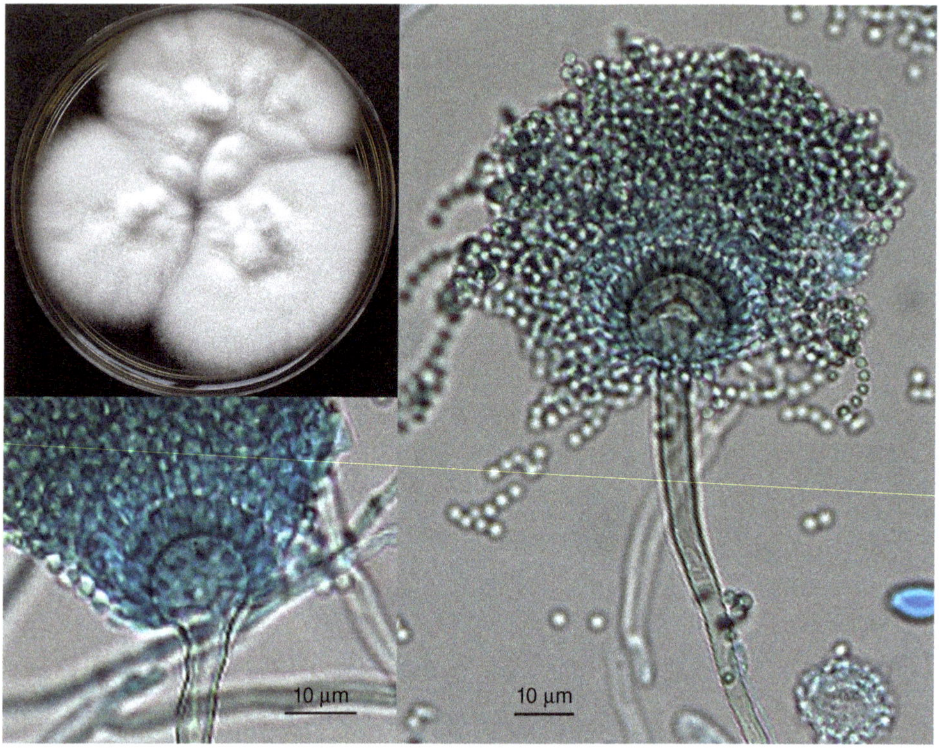

Fig. 9.4. *Aspergillus fischeri* culture and conidial head morphology.

9.3. *Aspergillus flavus* Link ex Grey

Aspergillus section *Flavi* historically includes species with conidial heads in shades of yellow green to brown and dark sclerotia. Hedayati *et al.* (2007) reviewed the *A. flavus* complex and included 23 species or varieties. Several species produce aflatoxins, among which aflatoxin B1 is the most toxic of the many naturally occurring secondary metabolites produced by fungi. Aflatoxins are mainly produced by *A. flavus* and *A. parasiticus*, which coexist with and grow on almost any crop or food (Varga *et al.*, 2011).

Aspergillus flavus has a worldwide distribution and normally occurs as a saprophyte in soil and on many kinds of decaying organic matter, however, it is also a recognised pathogen of humans and animals. It is an agent of otitis, keratitis, acute and chronic invasive sinusitis, and pulmonary and systemic infections in immunocompromised patients. *A. flavus* is second only to *A. fumigatus* as the cause of human invasive aspergillosis (Hedayati *et al.*, 2007). Although, *A. flavus* is the principle medically important pathogen of both humans and animals, some other species in the *A. flavus* complex notably *A. oryzae, A. avenaceus, A. tamari, A. alliaceus* and *A. nomius* may cause rare mostly superficial infections (Hedayati *et al.*, 2007; de Hoog *et al.*, 2020).

RG-2 organism

Morphological description: Colonies are granular, flat, often with radial grooves, yellow at first but quickly becoming bright to dark yellow-green with age (Fig. 9.5). Conidial heads are typically radiate, later splitting to form loose columns (mostly 300-400 µm in diameter), biseriate but having some heads with phialides borne directly on the vesicle (uniseriate). Conidiophore stipes are hyaline and coarsely roughened, often more noticeable near the vesicle. Conidia are globose to subglobose (3-6 µm in diameter), pale green and conspicuously echinulate (Fig. 9.5). Some strains produce brownish sclerotia.

Key features: Spreading yellow-green colonies, rough-walled stipes, mature vesicles bearing phialides over their entire surface and conspicuously echinulate conidia. **Note:** Accurate species identification within *Aspergillus flavus* complex remains difficult due to overlapping morphological and biochemical characteristics. For morphological identifications, it is recommended to report as *Aspergillus flavus* complex.

Molecular identification: ITS sequence analysis is sufficient to identify to species complex level only. Definitive identification requires analysis of β-tubulin, *Cal* and *ACT* genes (Balajee *et al.*, 2005a; Samson *et al.*, 2007).

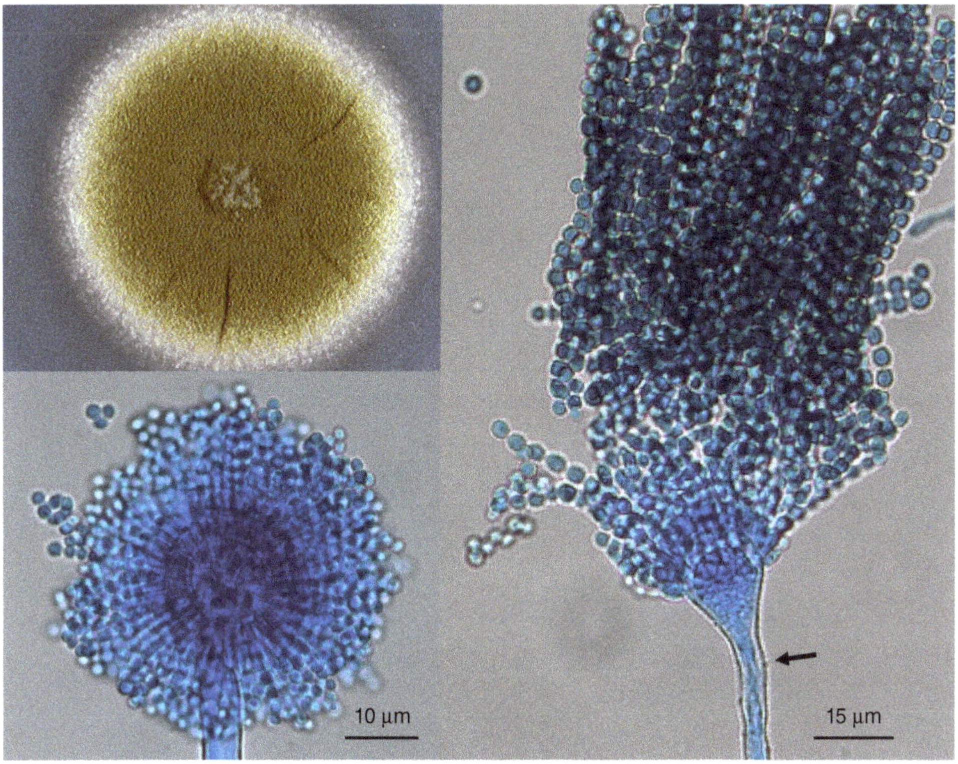

Fig. 9.5. *Aspergillus flavus* culture and conidial head morphology. **Note**: Rough-walled stipe near vesicle (arrow) and that both uniseriate and biseriate conidial heads may be present.

Antifungal susceptibility: (Table 9.3).

Table 9.3. *Aspergillus flavus* complex (Pfaller *et al.,* 2013a; and Australian national data); MIC, *MEC µg/mL.

Antifungal	No.	≤0.016	0.03	0.06	0.125	0.25	0.5	1	2	4	8	16	32	≥64
AmB	184				1	5	7	41	93	36	1			
ISAV	49					6	24	8	8	1	2			
VORI	183		1	1	7	50	97	26			1			
POSA	172	2	1	12	29	98	28	2						
ITRA	184		1	7	19	116	39	1			1			
ANID*	146	110	24	1	1		1				9			
MICA*	146	90	46		1						9			

9.4. *Aspergillus fumigatus* Fresenius

Aspergillus section *Fumigati* includes species characterised by uniseriate aspergilla with columnar conidial heads in shades of blue-green and flask-shaped vesicles (Raper and Fennell, 1965). Teleomorphic species belonging to the "*Aspergillus fischeri* series" of the *A. fumigatus* complex were originally placed in the genus *Neosartorya* (family Trichocomaceae) by Malloch and Cain (1972). These have now been transferred to *Aspergillus* and the section *Fumigati* now includes more than 60 species (Samson *et al.*, 2007).

Although *A. fumigatus sensu stricto* is recognised as the major human pathogen within the complex, recent phylogenetic studies have demonstrated that some human and animal infections may be caused by *A. lentulus*, *A. fumigatiaffinis*, *A. fumisynnematus*, *A. fischeri*, *A. udagawae* and *A. hiratsukae* (Coriglione *et al.*, 1990; Summerbell *et al.*, 1992; Padhye *et al.*, 1994; Lonial *et al.*, 1997; Jarv *et al.*, 2004; Balajee *et al.*, 2005a, 2006; Barrs *et al.*, 2013).

RG-2 organism

Morphological description: Colonies show typical blue-green surface pigmentation with a suede-like surface consisting of a dense felt of conidiophores (Fig. 9.6). Conidial heads are typically columnar (up to 400 x 50 µm but often much shorter and smaller) and uniseriate. Conidiophore stipes are short, smooth-walled and have conical-shaped terminal vesicles which support a single row of phialides on the upper two thirds of the vesicle. Conidia are produced in basipetal succession forming long chains and are globose to subglobose (2.5-3.0 µm in diameter), green and rough-walled to echinulate (Fig. 9.6). **Note:** This species is thermotolerant and grows at temperatures up to 55°C.

Key features: Uniseriate and columnar conidial heads with the phialides limited to the upper two thirds of the vesicle and curving to be roughly parallel to each other. Thermotolerant grows at temperatures up to 55°C. **Note:** *Aspergillus lentulus* differs by slower growth and absence of growth at 48°C (Yu *et al.*, 2020).

Molecular identification: Sequence analysis of ITS is sufficient to identify to species complex level only. Definitive identification requires sequence analysis of the β-tubulin, *Cal* and *ACT* genes (Balajee *et al.*, 2005a; Samson *et al.*, 2007).

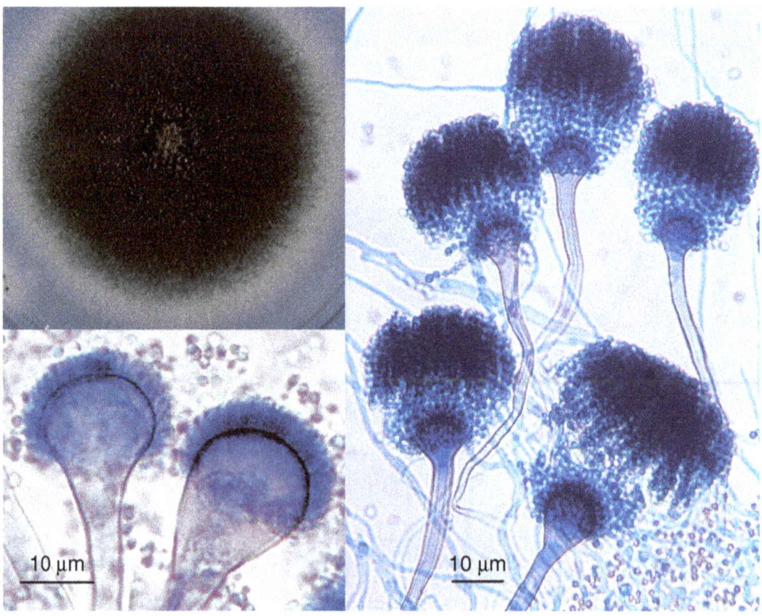

Fig. 9.6. *Aspergillus fumigatus* culture and conidial head morphology. **Note:** Uniseriate row of phialides on the upper two thirds of the vesicle.

Antifungal susceptibility: (Table 9.4).

Table 9.4. *Aspergillus fumigatus* complex (Pfaller *et al.,* 2013a; and Australian national data); MIC, *MEC μg/mL.

Antifungal	No.	≤0.016	0.03	0.06	0.12	0.25	0.5	1	2	4	8	16	32	≥64
AmB	1110			2	40	121	179	290	433	44	1			
ISAV	344	1		1	6	42	188	83	20	1	2			
VORI	1074	1	2	29	138	553	274	42	19	10	6			
POSA	1005	44	88	147	419	258	34	13	1		1			
ITRA	1110	8	20	53	149	542	298	28	4	1	1	6		
ANID*	761	616	119	21	2		1		2					
MICA*	761	653	89	13	4		2							

9.5. *Aspergillus lentulus* Balajee & Marr

Aspergillus lentulus appears to be widely distributed in soil and is now well documented as a causative agent of invasive aspergillosis in immunosuppressed patients. *A. lentulus* is part of the *A. fumigatus* complex (Balajee *et al.*, 2005b).

RG-2 organism

Morphological description: Colonies are suede-like to floccose, white with interspersed grey-green patches of conidia (conidiation is slow to poor in most strains) (Fig. 9.7). Conidial heads are short columnar and uniserate. Conidiophore stipes are smooth-walled, sometimes sinuous and are often constricted at the neck. Vesicles are usually subglobose in shape. Conidia are globose to broadly ellipsoidal (2-3.2 μm in diameter), smooth to finely roughened (Fig. 9.7).

Key features: Morphologically like *A. fumigatus*, but slower growing with poor conidiation and absence of growth at 48°C (Yu *et al.*, 2020).

Molecular identification: *Aspergillus lentulus* can be distinguished from other members of the section *Fumigati* by sequence analysis of β-tubulin, *Cal* and *ACT* genes (Balajee *et al.*, 2005a, 2005b; Samson *et al.*, 2007).

Antifungal susceptibility: (Table 9.5).

Table 9.5. *Aspergillus lentulus* (Australian national data); MIC, *MEC μg/mL.

Antifungal	No.	≤0.016	0.03	0.06	0.125	0.25	0.5	1	2	4	8	16	32	≥64
AmB	17					1	2	2	7	4	1			
ISAV	7							1	3	3				
VORI	16							4	4	5	3			
POSA	15				4	3	7	1						
ITRA	17					4	2	7	2			2		
ANID*	9	7	1	1										
MICA*	9	8	1											

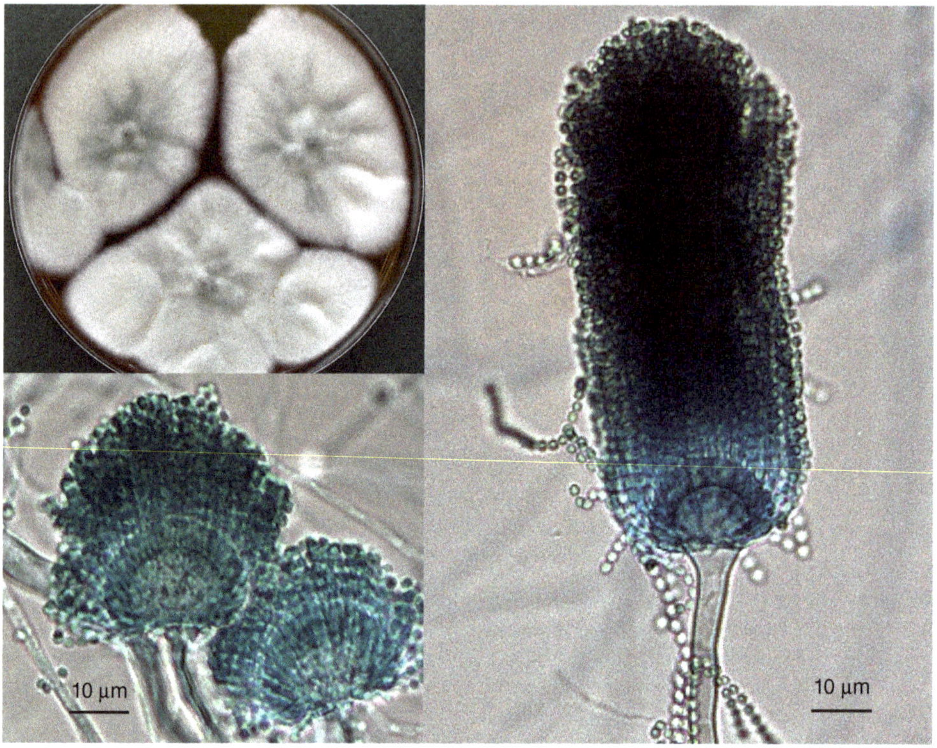

Fig. 9.7. *Aspergillus lentulus* culture and conidial head morphology.

9.6. *Aspergillus nidulans* (Eidam) Winter

Aspergillus section *Nidulantes*; Gams *et al.* (1985) includes species with biserate conidial heads, brown pigmented often short stipes and green conidia, cleistothecia are soft-walled, surrounded by Hülle cells, ascospores are red to purple in colour. Section *Nidulantes* is one of the largest subgenera of the genus *Aspergillus* and includes about 80 species. *A. nidulans* is a typical soil fungus with a worldwide distribution and has been reported to cause disease in human and animals (Bastos *et al.*, 2020). Several other species within the section have also been reported as medical pathogens, *A. sydowii, A. unguis, A. rugulovalvus and A. quadrilineatus.*

RG-1 organism

Morphological description: Colonies are typically plain green in colour with dark red-brown cleistothecia developing within and upon the conidial layer. Reverse may be olive to drab grey or purple-brown (Fig. 9.8). Conidial heads are short columnar (up to 70 x 30 µm in diameter) and biseriate. Conidiophore stipes are usually short, brownish and smooth-walled. Conidia are globose (3-3.5 µm in diameter) and rough-walled (Fig. 9.8). Ascomata are spherical, purple, 100-200 µm in diameter, surrounded by a yellowish to cinnamon layer of scattered hyphae bearing a dense aggregation of pale yellow, thick-walled, spherical to subspherical Hülle cells (Fig. 9.9). Asci are 8-spored, spherical to subspherical, 7-12 µm in diameter. Ascospores are purple-red, smooth-walled, lenticular with two equatorial crests, 3.5-4.5 x 3.5-4.0 µm.

Key features: Conidial heads are short columnar and biseriate. Stipes are usually short, brownish and smooth-walled. Conidia are globose and rough-walled.

Molecular identification: ITS sequencing is sufficient to identify to species complex only. *A. nidulans* can be distinguished from other members of the section *Nidulantes* by sequence analysis of β-tubulin, *Cal* and *ACT* genes.

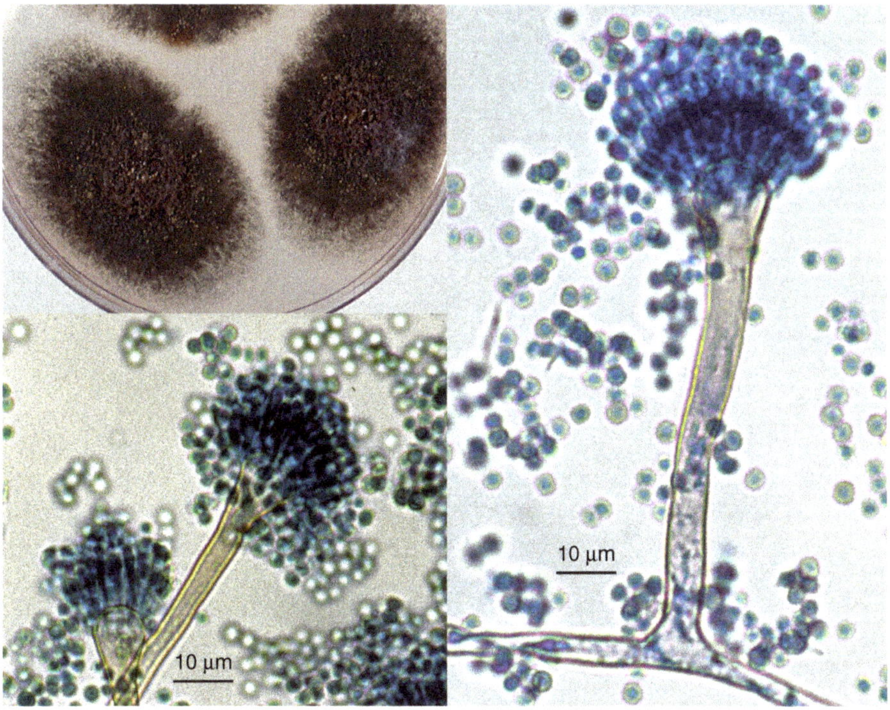

Fig. 9.8. *Aspergillus nidulans* culture and biseriate conidial head morphology. **Note:** The short, brownish conidiophore stipes.

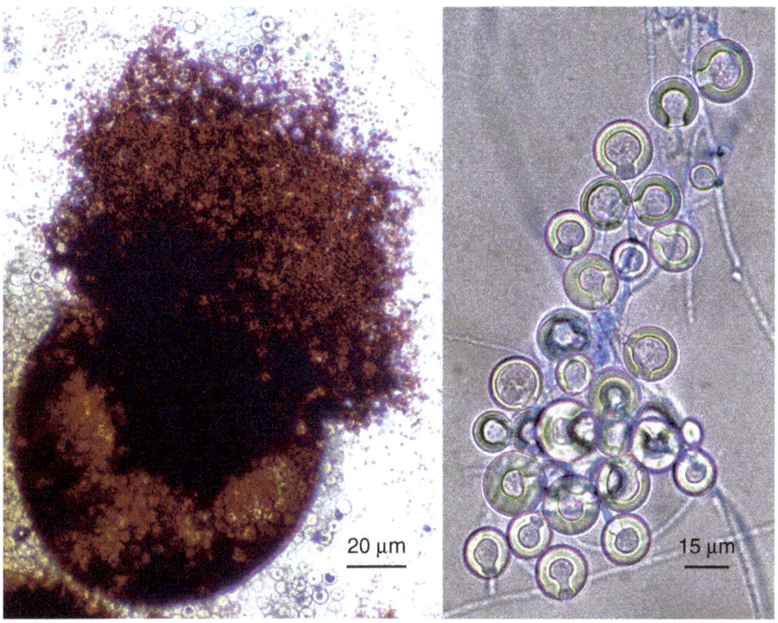

Fig. 9.9. *Aspergillus nidulans* cleistothecium showing numerous reddish-brown ascospores and aggregations of thick-walled, spherical to subspherical thick-walled Hülle cells.

Antifungal susceptibility: (Table 9.6).

Table 9.6. *Aspergillus nidulans* complex (Australian national data); MIC, *MEC μg/mL.

Antifungal	No.	≤0.016	0.03	0.06	0.125	0.25	0.5	1	2	4	8	16	32	≥64
AmB	57				5	10	3	18	16	5				
ISAV	12			3	8	1								
VORI	56	1	3	12	29	9	1			1				
POSA	53	3	8	10	17	13	1		1					
ITRA	57	2	4	9	15	25	1	1						
ANID*	35	18	10	5	2									
MICA*	35	22	11	2										

9.7. *Aspergillus niger* van Tieghem

The black aspergilli, *Aspergillus* section *Nigri* (Gams *et al.*, 1985) includes species with uniseriate or biseriate conidial heads, spherical to pyriform vesicles, smooth-walled stipes and black or near black coloured conidia. This group contains about 26 species with *Aspergillus niger* being the most common species isolated. *A. niger* can be isolated from all continents and is not very selective with respect to environmental conditions. *A. niger* is one of the most common and easily identifiable species of the genus *Aspergillus*, with its white to yellow mycelial culture surface later bearing black conidia. This species is very commonly found in aspergillomas and is the most frequently encountered agent of otomycosis. It is also a common laboratory contaminant. Other species within this group that have been linked to human and animal infection include *A. acidus, A. aculeatus, A. brasiliensis* and *A. tubingensis*.

RG-1 organism

Morphological identification: Colonies consist of a compact white or yellow basal felt covered by a dense layer of dark brown to black conidial heads (Fig. 9.10). Conidial heads are large (up to 3 mm by 15 to 20 μm in diameter), globose, dark brown, becoming radiate and tending to split into several loose columns with age. Conidiophore stipes are smooth-walled, hyaline or turning dark towards the vesicle. Conidial heads are biseriate with the phialides borne on brown coloured metulae, that are twice as long as the phialides (Fig. 9.10). Conidia are globose to subglobose (3.5-5 μm in diameter), dark brown to black and rough-walled.

Key features: Conidial heads are dark brown to black, radiate and biseriate with metulae twice as long as the phialides. Conidia brown and rough-walled.

Molecular identification: In *Aspergillus* section *Nigri,* all species can be distinguished from each other using *Cal* sequence data, and all except one can be distinguished using β-tubulin sequence data. ITS sequencing is not recommended for definitive species identification (Samson *et al.*, 2007).

Antifungal susceptibility: (Table 9.7).

Table 9.7. *Aspergillus niger* complex (Pfaller *et al.,* 2013a; and Australian national data); MIC, *MEC μg/mL.

Antifungal	No.	≤0.016	0.03	0.06	0.125	0.25	0.5	1	2	4	8	16	32	≥64
AmB	185				16	31	33	81	21	3				
ISAV	46			1	3		2	17	18	3	2			
VORI	181		4	9	10	24	79	51	3		1			
POSA	170	2	8	19	28	61	50	2						
ITRA	185		1	4	17	26	85	47	3		1	1		
ANID*	128	89	24	6	2	1								
MICA*	128	58	45	18	2						5			

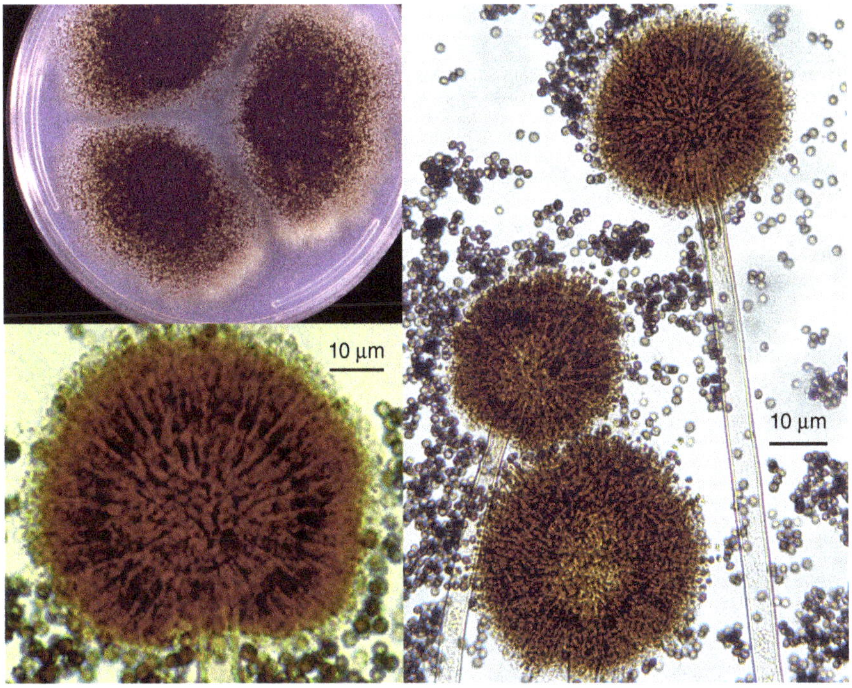

Fig. 9.10. *Aspergillus niger* culture and conidial head morphology. **Note:** Conidial heads are biseriate, large, globose, dark brown, becoming radiate with the phialides borne on metulae.

9.8. *Aspergillus terreus* Thom

Aspergillus section *Terrei* (Gams *et al.*, 1985); *Aspergillus terreus* complex includes species with biseriate, columnar conidial heads in shades of buff to brown (Raper and Fennell, 1965). The most important species of this section is *A. terreus,* which is ubiquitous in the environment (Samson *et al.*, 2011). *Aspergillus terreus* occurs commonly in soil and is occasionally reported as a pathogen of humans and animals. Two other species have been reported as medical pathogens, *A. alabamensis* and *A. niveus.*

RG-2 organism

Morphological identification: Colonies are typically suede-like and cinnamon-buff to sand-brown in colour with a yellow to deep dirty brown reverse (Fig. 9.11). Conidial heads are compact, columnar (up to 500 x 30-50 µm in diameter) and biseriate. Conidiophore stipes are hyaline and smooth-walled (Fig. 9.11). Conidia are globose to ellipsoidal (1.5-2.5 µm in diameter), hyaline to slightly yellow and smooth-walled.

Key features: Cinnamon-brown cultures, conidial heads biseriate with metulae as long as the phialides.

Molecular identification: *Aspergillus terreus* can be distinguished from other members of the section *Terrei* by sequence analysis of β-tubulin, *Cal* and *ACT* genes. ITS sequencing is sufficient to identify to species complex level only.

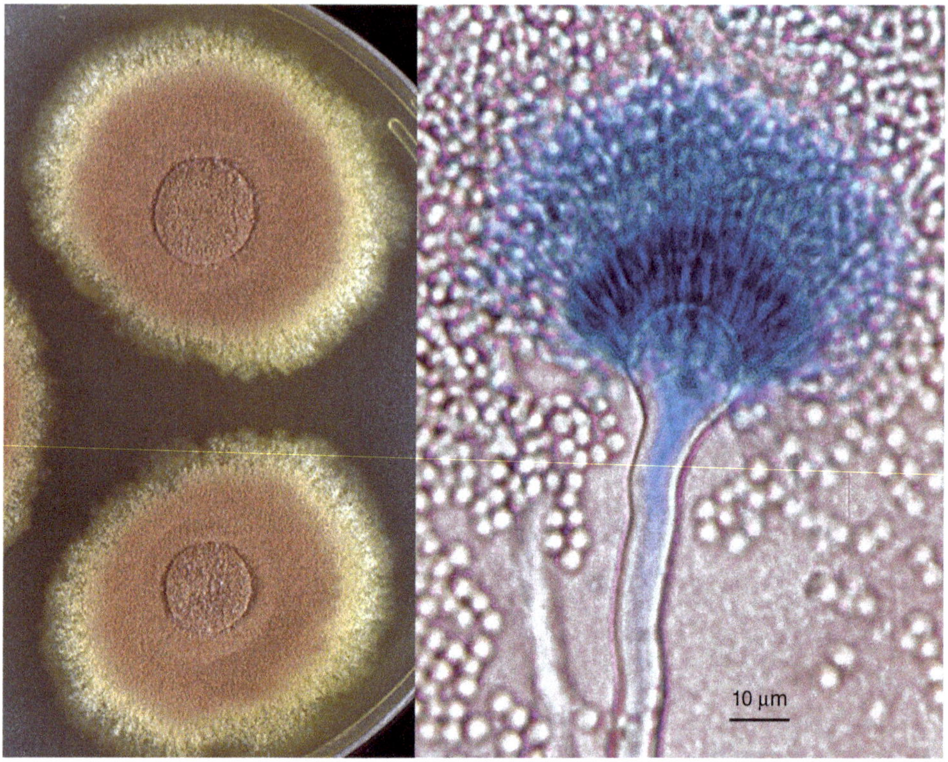

Fig. 9.11. *Aspergillus terreus* culture and conidial head morphology. **Note:** Conidial heads are biseriate and tightly columnar.

Antifungal susceptibility: (Table 9.8).

Table 9.8. *Aspergillus terreus* complex (Pfaller *et al.*, 2013a; and Australian national data); MIC, *MEC µg/mL.

Antifungal	No.	≤0.016	0.03	0.06	0.125	0.25	0.5	1	2	4	8	16	32	≥64
AmB	168				1	3	11	37	78	36	2			
ISAV	43			1	7	20	8	6	1					
VORI	166			5	25	86	38	7	4		1			
POSA	156	2	14	23	70	36	7	4						
ITRA	169	5	3	19	53	78	10		1					
ANID*	128	89	24	6	2	1								
MICA*	128	58	45	8	2						5			

DOI: 10.1079/9781800622340.0009

10 *Aureobasidium pullulans* (de Bary) Arnaud

Aureobasidium pullulans has a worldwide distribution and is usually isolated as a saprophyte, occasionally from skin and nails. However, it has also been reported as a rare causative agent of phaeohyphomycosis, mycotic keratitis and peritonitis in patients on continuous ambulatory peritoneal dialysis (CAPD).

RG-1 organism

Morphological description: Colonies are fast growing, smooth, soon covered with slimy masses of conidia, cream or pink becoming brown or black (Fig. 10.1). Hyphae are hyaline and septate, frequently becoming dark brown with age and forming chains of one- to two-celled, thick-walled, darkly pigmented arthroconidia. The arthroconidia are only of secondary importance in recognising members of this genus. Conidia are produced synchronously in dense groups from indistinct scars or from short denticles on undifferentiated, hyaline to sub-hyaline hyphae. Conidia are hyaline, smooth-walled, single-celled, ellipsoidal but of very variable shape and size (8-12 x 4-6 µm), often with an indistinct hilum (a mark or scar at the point of attachment) (Fig. 10.1). Temperature: optimum 25°C; maximum 35-37°C.

Key features: Slimy masses of conidia produced simultaneously from the vegetative hyphae, which may also form chains of darkly pigmented, thick-walled arthroconidia.

Molecular identification: Recommended barcoding genes are *TEF-1* and D1/D2 domains (de Hoog *et al.*, 2015).

Antifungal susceptibility: (Table 10.1).

Table 10.1. *Aureobasidium pullulans* (Australian national data); MIC µg/mL.

Antifungal	No.	≤0.016	0.03	0.06	0.125	0.25	0.5	1	2	4	8	16	32	≥64
AmB	15				1		6	4	4					
ISAV	6					3	2		1					
VORI	15				2	6	6	1						
POSA	14				3	8	3							
ITRA	15				3	9	3							

References: Hermanides-Nijhof (1977); McGinnis (1980); Domsch *et al.* (2007); Zalar *et al.* (2008); Najafzadeh *et al.* (2014); de Hoog *et al.* (2015, 2020).

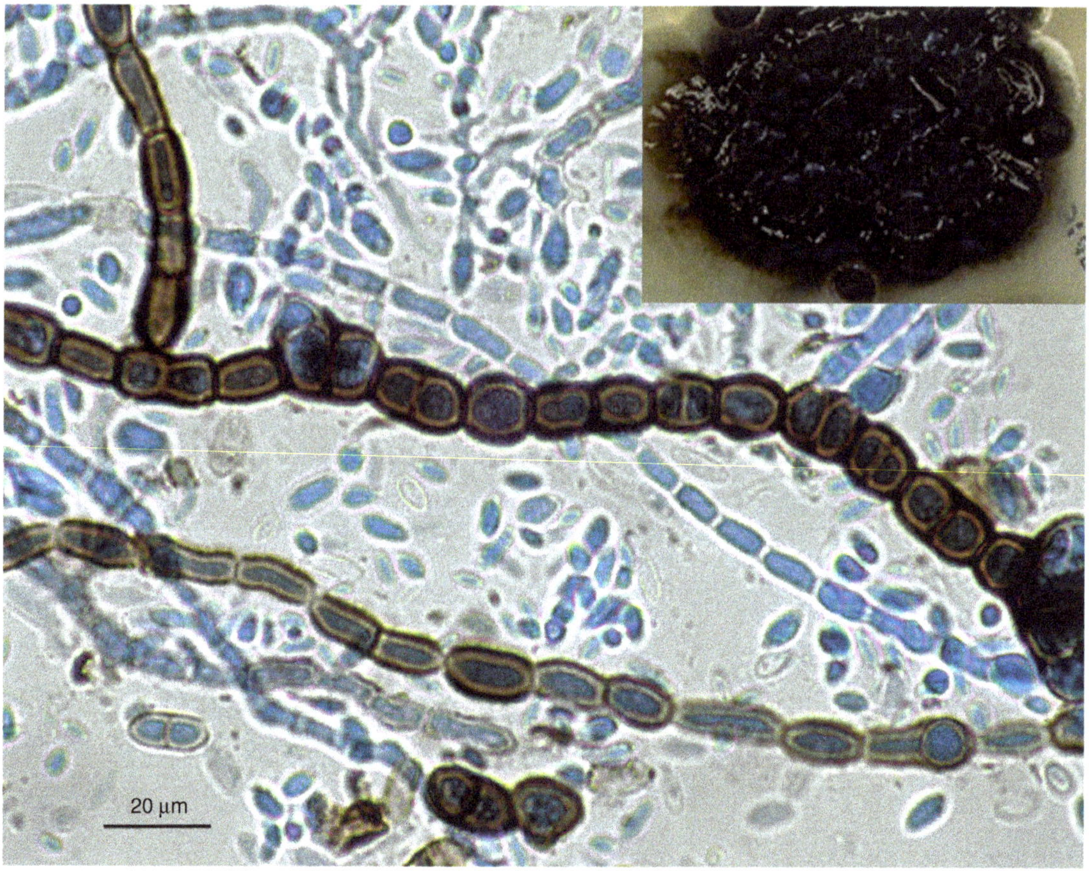

Fig. 10.1. *Aureobasidium pullulans* culture and microscopy showing darkly pigmented arthroconidia and the presence of numerous hyaline conidia which are produced on short denticles.

DOI: 10.1079/9781800622340.0010

11 *Basidiobolus ranarum* Eidem

Synonymy: *Basidiobolus meristosporus* Drechsler; *Basidiobolus heterosporus* Srinivasan & Thirumalachar; *Basidiobolus haptosporus* Drechsler.

Basidiobolus ranarum is commonly present in decaying fruit and vegetable matter, and as a commensal in the intestinal tract of frogs, toads and lizards. It has been reported from tropical regions of Africa and Asia including India, Indonesia and Australia. *B. ranarum* is a known cause of subcutaneous and gastrointestinal entomophthoromycosis often as a result of traumatic inoculation or a barrier break (Ellis, 2005a; Vilela and Mendoza, 2018).

RG-2 organism

Morphological description: Colonies are moderately fast growing at 30°C, flat, yellowish-grey to creamy-grey, glabrous, becoming radially folded and covered by a fine, powdery, white surface mycelium. Satellite colonies are often formed by germinating conidia ejected from the primary colony (Fig. 11.1). Microscopic examination usually shows the presence of large vegetative hyphae (8-20 µm in diameter) forming numerous round (20-50 µm in diameter), smooth, thick-walled zygospores that have two closely appressed beak-like appendages. The production of "beaked" zygospores is characteristic of the genus (Fig. 11.1). Two types of asexual conidia are formed, although isolates often lose their ability to sporulate with subculture. Special media incorporating glucosamine hydrochloride and casein hydrolysate may be needed to stimulate sporulation (Shipton and Zahari, 1987). Primary conidia are globose, one-celled, solitary and are forcibly discharged from a sporophore. The sporophore has a distinct swollen area just below the conidium that actively participates in the discharge of the conidium (Fig. 11.1). Secondary (replicative) conidia are clavate, one-celled and are passively released from a sporophore. These sporophores are not swollen at their bases. The apex of the passively released spore has a knob-like adhesive tip. These spores may function as sporangia, producing several sporangiospores.

Molecular identification: ITS sequencing is useful for identification of most clinical isolates. In the past, pathogenic isolates of *Basidiobolus* were identified as either *B. ranarum, B. meristosporus* or *B. haptosporus,* until all three species were found to be identical by rDNA RFLP analysis (Kwon-Chung and Bennett, 1992). However, recent sequence data may suggest the need to re-establish several species (Bshabshe *et al.*, 2020; Al-Hatmi *et al.,* 2021). Many sequences in GenBank are misidentified, so comparison to sequences from Type cultures is important.

Key features: The genus *Basidiobolus* is distinguished primarily by the morphology and development of forcibly discharged conidia and zygospores, both of which must be observed for reliable identification to genus and species level (King, 1983).

References: Strinivasan and Thirumalachar (1965); Greer and Friedman (1966); Dworzack *et al.* (1978); Rippon (1988); Kwon-Chung and Bennett (1992); Davis *et al.* (1994); Vikram *et al.* (2012); Shaikh *et al.* (2016); Hung *et al.* (2020b).

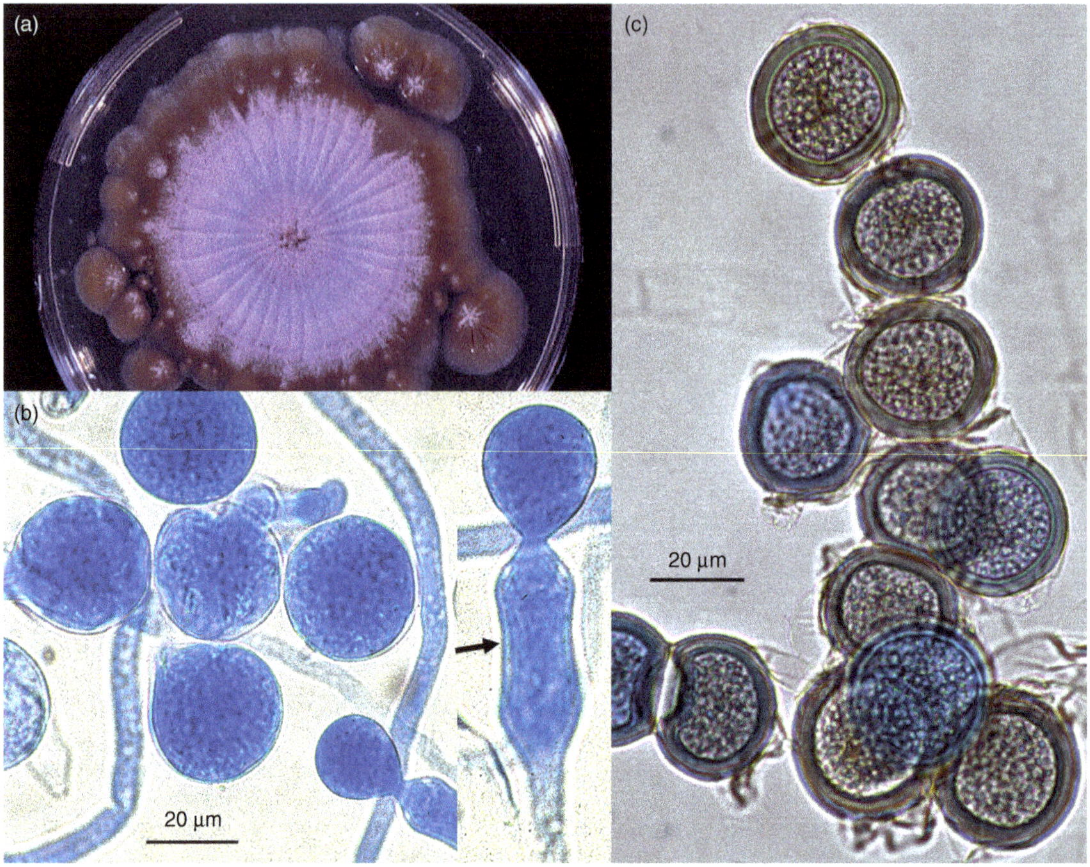

Fig. 11.1. *Basidiobolus ranarum* showing (a) culture with satellite colonies formed by germinating conidia ejected from the primary colony, (b) conidia and a sporophore with a distinct swollen area just below the conidium (arrow), and (c) thick-walled "beaked" zygospores.

DOI: 10.1079/9781800622340.0011

12 *Beauveria bassiana* (Balsamo) Vuillemin

Three species of *Beauveria* are recognised, two of which are well known parasites of insects. *B. bassiana* is the most common species and is best known as the causal agent of the disastrous muscardine in silkworms. *Beauveria* species are occasionally isolated in the clinical laboratory as saprophytic contaminants, although rare human cases of keratitis have been reported (Lara Oya *et al.*, 2016).

RG-1 organism

Morphological description: Colonies are usually slow growing, mostly not exceeding 2 cm in ten days at 20°C, downy, at first white but later often becoming yellow to pinkish with age. The genus *Beauveria* is characterised by the sympodial development of single-celled conidia (ameroconidia) on a geniculate or zig-zag rachis. Conidiogenous cells are flask-shaped, rachiform, proliferating sympodially and are often aggregated into sporodochia or synnemata. Conidia are hyaline and globose or ovoid in shape (Fig. 12.1).

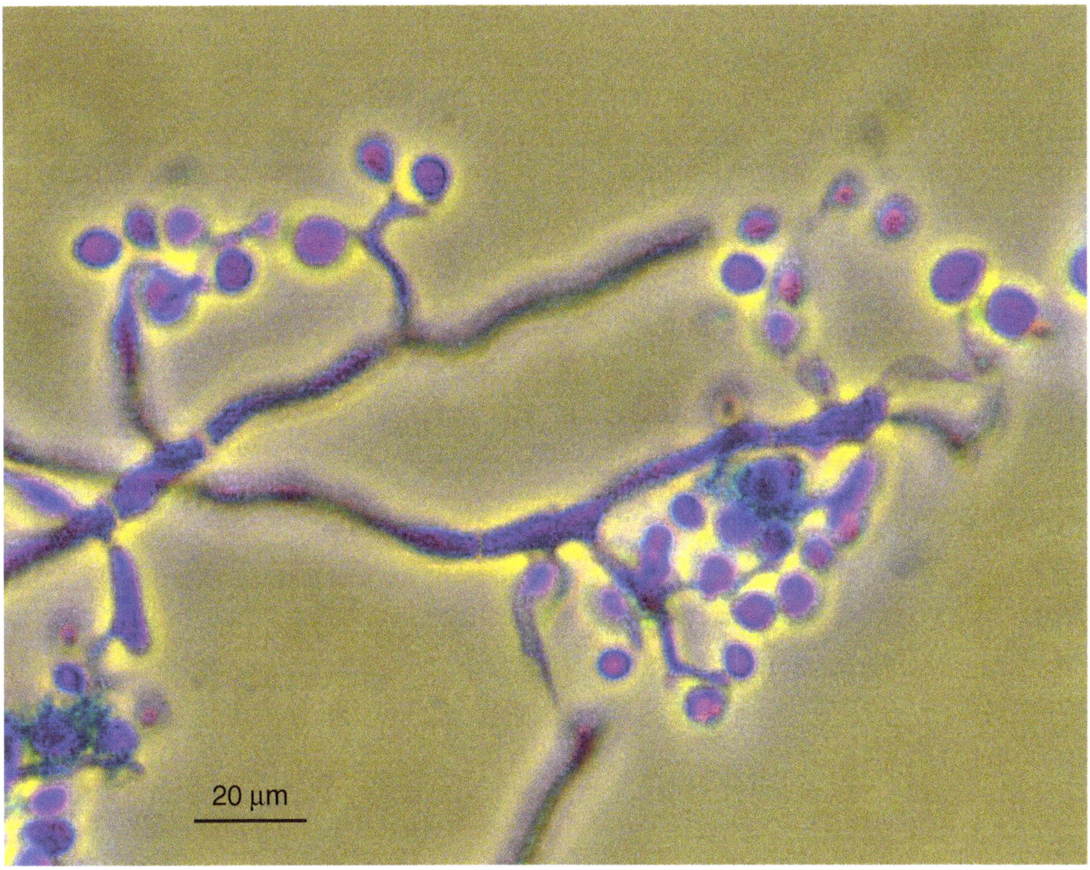

Fig. 12.1. *Beauveria bassiana* showing sympodial development of conidia on a geniculate or zig-zag rachis.

Key features: Sympodial development of single-celled conidia on a geniculate or zig-zag rachis emanating from a flask-shaped conidiophore.

Molecular identification: Specific primers were developed by Hegedus and Khachatourians (1996). Full phylogeny of the genus was provided by Rehner and Buckley (2005). Biogeography of molecular types was elaborated by Ghikas *et al.* (2010).

MALDI-ToF MS: Cassagne *et al.* (2011) published a standardised procedure for mould identification in the clinical laboratory.

References: de Hoog (1972); McGinnis (1980); Domsch *et al.* (2007); de Hoog *et al.* (2015).

DOI: 10.1079/9781800622340.0012

13 *Bipolaris* Shoemaker

The genus *Bipolaris* contains about 47 species, which are mostly subtropical and tropical plant parasites. *Bipolaris cynodontis* and *B. papendorfii* have been isolated from clinical samples (da Cunha *et al.*, 2012a). However recent phylogenetic studies have transferred all the well-documented human pathogens, notably *B. australiensis, B. hawaiiensis* and *B. spicifera* to the genus *Curvularia* (Manamgoda *et al.*, 2012). Other reported clinical isolates include *Curvularia micropus* (formerly *Bipolaris micropus*) and *Cochliobolus setariae* (formerly *Bipolaris setariae*).

RG-1 organisms

Morphological description: Colonies are moderately fast growing, effuse, grey to blackish-brown, suede-like to floccose with a black reverse. Microscopic morphology shows sympodial development of pale brown pigmented, distoseptate conidia on a geniculate or zig-zag rachis. Conidia mostly curved, canoe-shaped, fusoid or obclavate, rarely straight, 2-14 distoseptate (usually more than 6), germinating only from the ends (bipolar).

Key features: Darkly pigmented mould producing sympodial, distoseptate, pale brown, straight, fusiform to ellipsoidal conidia, which are rounded at both ends.

Comments: The genera *Drechslera, Bipolaris*, *Curvularia* and *Exserohilum* are all closely related and morphological differentiation of the genera relies upon a combination of characters including conidial shape, the presence or absence of a protruding hilum, the contour of the basal portion of the conidium and its hilum, the point at which the germ tube originates from the basal cell and, to a lesser degree, the sequence and location of the first three conidial septa. However, Manamgoda *et al.* (2012) reported no clear morphological distinction between the genera *Bipolaris* and *Curvularia* and some species have intermediate morphologies between the two genera. These authors recommend using a combined ITS and *GAPDH* sequence analysis for definitive identification of species (Manamgoda *et al.*, 2012, 2014).

Molecular identification: ITS sequencing may be used to identify clinical species (da Cunha *et al.*, 2012a). *GAPDH* (Glycerol-3-phosphate dehydrogenase) has been determined to be the optimal phylogenetic marker of *Bipolaris* species (Manamgoda *et al.*, 2012, 2014).

References: Ellis (1971, 1976); Alcorn (1983); McGinnis *et al.* (1986b); Sivanesan (1987); Rippon (1988); Domsch *et al.* (2007); de Hoog *et al.* (2015).

DOI: 10.1079/9781800622340.0013

14 *Blastomyces* Gilchrist & Stokes

Blastomyces contains a number of pathogenic species, most notably *Blastomyces dermatitidis* and *Blastomyces gilchristii*, which are morphologically identical but distinguishable by sequence analysis of the ITS region (Brown *et al.*, 2013). These are associated with soil and decaying organic matter such as leaves and wood in North America. *Blastomyces percursus* and *B. emzantsi* have been identified from a culture repository in South Africa (Dukik *et al.*, 2017; Jiang *et al.*, 2018; Maphanga *et al.*, 2020). In addition, *Blastomyces helices* and *B. parvus* have been transferred to this genus from *Emmonsia* (Peterson and Sigler, 1998; Jiang *et al.*, 2018). These are the causative agents of blastomycosis a chronic granulomatous and suppurative disease, having a primary pulmonary stage that is frequently followed by dissemination to other body sites, typically the skin and bone.

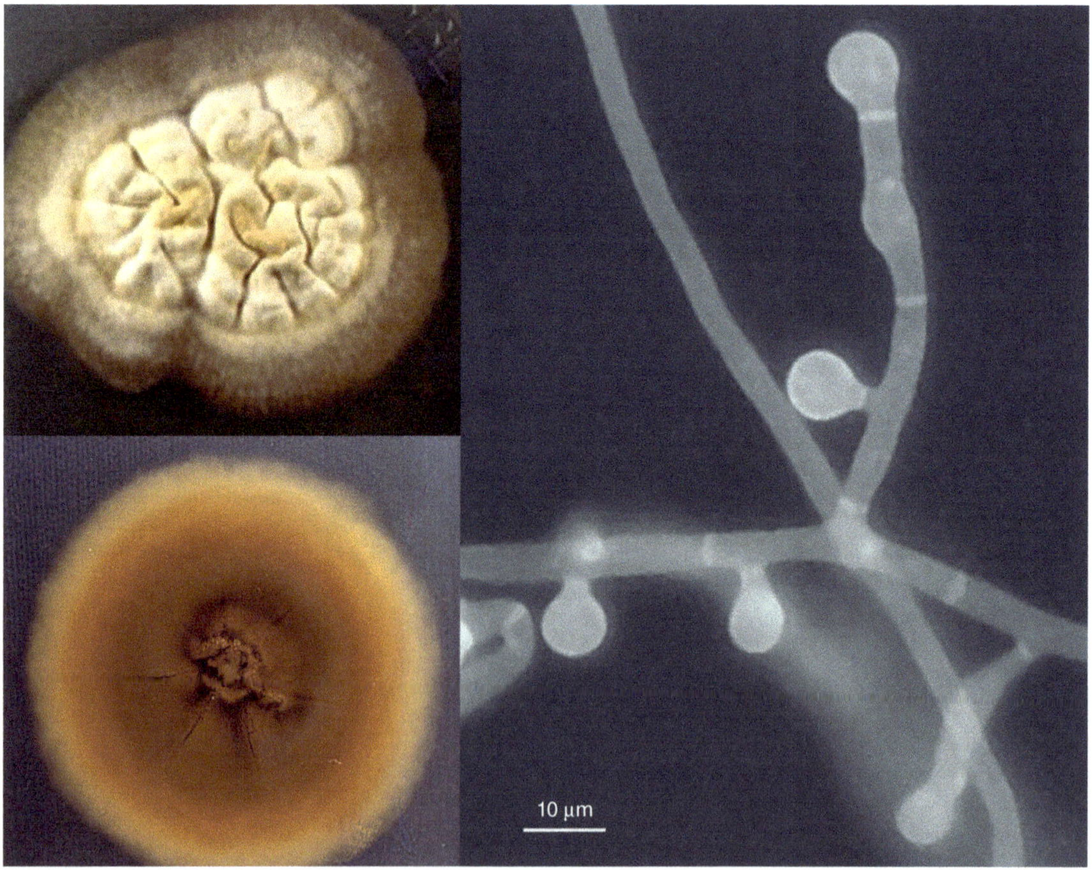

Fig. 14.1. *Blastomyces dermatitidis* culture and microscopy showing one-celled, smooth-walled conidia borne on short lateral or terminal hyphal branches (photographs courtesy of Jon Biehle, Department of Pathology, Creighton University, Omaha, Nebraska).

WARNING: RG-3 organism

Cultures of *Blastomyces* species represent a biohazard to laboratory personnel and must be handled in a Class II Biological safety cabinet (BSCII).

14.1. *Blastomyces dermatitidis* Gilchrist & Stokes

Morphological description: Colonies grown at 25°C are variable in morphology and rate of growth. They may grow rapidly, producing a fluffy white mycelium or slowly as glabrous, tan, non-sporulating colonies. Growth and sporulation may be enhanced by yeast extract. Most strains become pleomorphic with age. Microscopically, hyaline, ovoid to pyriform, one-celled, smooth-walled conidia (2-10 µm in diameter) of the *Chrysosporium* type, are borne on short lateral or terminal hyphal branches (Fig. 14.1).

Blastomyces species are dimorphic fungi and colonies on blood agar at 37°C are wrinkled and folded, glabrous and yeast-like. Microscopically, the organism produces the characteristic yeast phase as seen in tissue pathology.

Comment: In the past, conversion from the mould to the yeast form was necessary to positively identify this dimorphic pathogen from species of *Chrysosporium* or *Sepedonium*. However, identification by molecular methods is preferred to minimise manipulation of the fungus.

Histopathology: *Blastomyces dermatitidis* tissue sections show large, thick-walled, broad-based, unipolar budding yeast-like cells, which may vary in size from 8-15 µm, with some larger forms up to 30 µm in diameter. Tissue sections need to be stained by Grocott-Gomori's methenamine silver method to clearly see the yeast-like cells, which are often difficult to observe in haematoxylin and eosin preparations (Fig. 14.2).

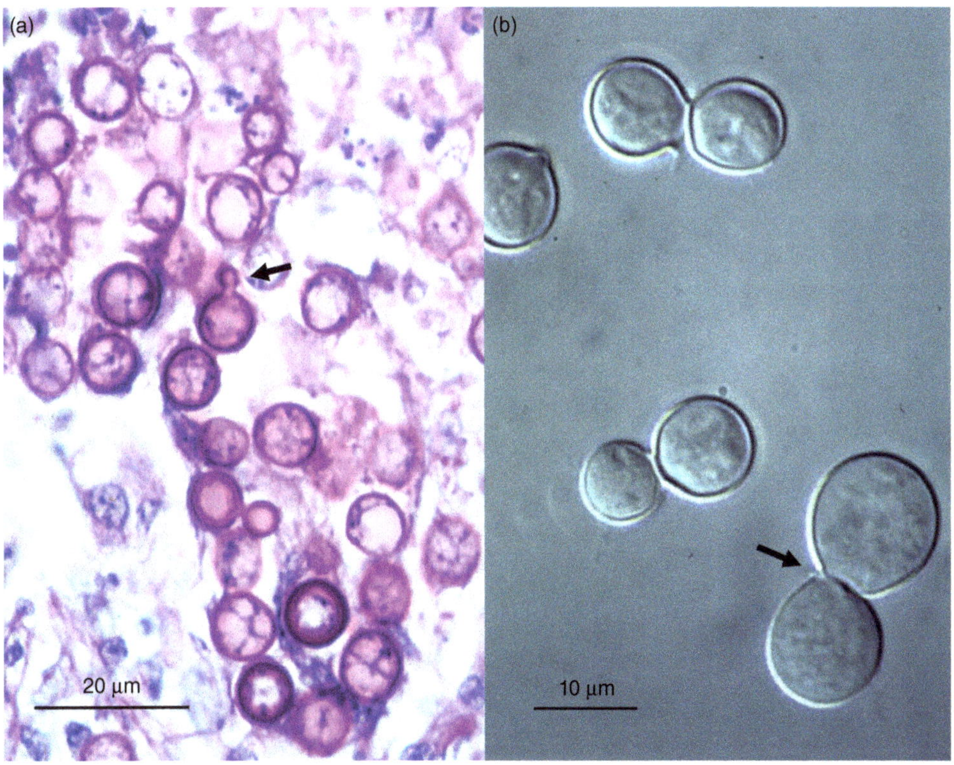

Fig. 14.2. *Blastomyces dermatitidis* showing large, broad-base, unipolar budding yeast-like cells (arrows) in (a) haematoxylin and eosin-stained section of lung tissue and (b) calcofluor white stain of the yeast phase grown at 36°C (photographs courtesy of Jon Biehle, Department of Pathology, Creighton University, Omaha, Nebraska).

Molecular identification: Several conventional PCR assays have been developed for the identification of *B. dermatitidis* from clinical specimens (Bialek *et al.*, 2003) and soil (Burgess *et al.*, 2006). Sidamonidze *et al.* (2012) developed a real-time PCR targeting the *BAD1* (formerly known as *WI-1*) gene for the identification of *B. dermatitidis* in culture and tissue and Morjaria *et al.* (2015) used ITS and D1/D2 sequencing for identification from paraffin embedded tissue.

Key features: Clinical history, tissue pathology, culture identification by positive exoantigen test.

Antifungal susceptibility: (Table 14.1).

Table 14.1. *Blastomyces dermatitidis* limited data available (Sugar and Liu, 1996; Espinel-Ingroff *et al.*, 2001; Espinel-Ingroff, 2003; Gonzalez *et al.*, 2005; Sabatelli *et al.*, 2006); MIC µg/mL. Antifungal susceptibility testing is not recommended.

Antifungal	Range	MIC_{90}	Antifungal	Range	MIC_{90}
FLU	0.125-64	4-16	AmB	0.03-1	0.5
ITRA	0.03->16	0.125-2	VORI	0.03-16	0.25
POSA	0.03-2	0.125			

References: Chandler *et al.* (1980); McGinnis (1980); Kaufman and Standard (1987); Rippon (1988); Jiang *et al.* (2018).

DOI: 10.1079/9781800622340.0014

15 *Candida* Berkhout

The genus *Candida* is characterised by globose to elongate yeast-like cells or blastoconidia that reproduce by narrow-based multilateral budding. Pseudohyphae and occasionally true hyphae may also be present. Colony pigmentation is usually absent. Ballistoconidia are not formed. Arthroconidia may be formed, but not extensively. Sexual reproduction is absent. Glucose may be fermented. Nitrate may be assimilated. Starch-like compounds are not produced. The diazonium blue B reaction is negative.

The genus is highly polyphyletic, as it comprises mitosporic species that are devoid of special distinguishing features (Lachance *et al.*, 2011). Accordingly, several taxonomic rearrangements have been made and many well-known *Candida* species have been moved to other genera, notably *Pichia kudriavzevii* (formerly *Candida krusei*), *Meyerozyma guilliermondii* (formerly *Candida guilliermondii*), *Clavispora lusitaniae* (formerly *Candida lusitaniae*), *Kluyveromyces marxianus* (formerly *Candida kefyr*), *Diutina catenulata* (formerly *Candida catenulata*), *Diutina rugosa* (formerly *Candida rugosa*) and *Wickerhamomyces anomalus* (formerly *Candida pelliculosa*). *Candida glabrata* and *C. parapsilosis* are recognised as species complexes (Alcoba-Florez *et al.*, 2005; Tavanti *et al.*, 2005; Correia *et al.*, 2006).

Several species may be aetiological agents, most commonly *Candida albicans*, followed by *C. glabrata, C. parapsilosis, C. tropicalis* and *Pichia kudriavzevii.* Altogether, these five species account for >95% of human infections. However, a number of other species may also be isolated. All are ubiquitous and occur naturally on humans.

Morphological identification: Ensure that you start with a fresh growing pure culture, streak for single colony isolation if necessary (Fig. 15.1) (Kurtzman *et al.*, 2011).

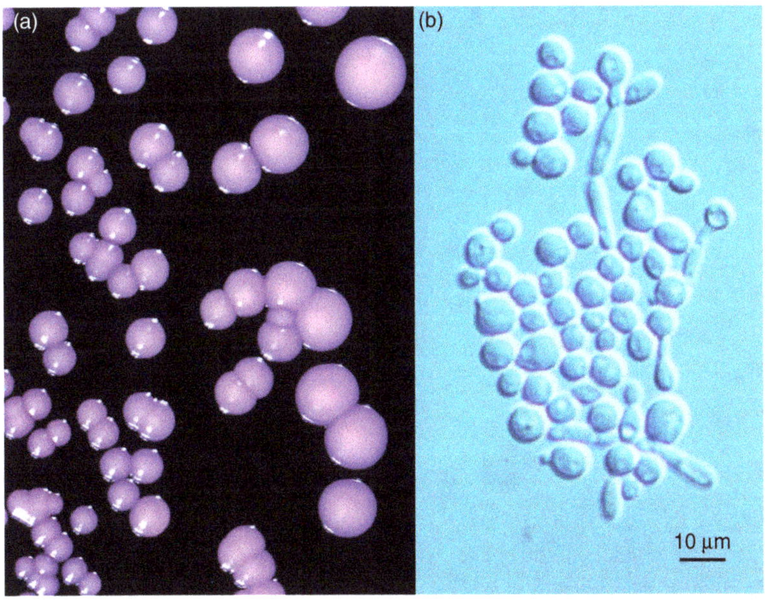

Fig. 15.1. *Candida albicans* showing (a) typical cream coloured, smooth-surfaced, waxy colonies and (b) narrow based budding spherical to ovoid blastoconidia.

Chromogenic agars: Are now being used for primary isolation and for the detection of mixed flora, especially from non-sterile sites. Depending on the brand of chromogenic media presumptive identification of *C. albicans, C. tropicalis, C. auris* and *P. kudriavzevii* is possible. CHROMagar™ Candida (and Candida Plus) (CHROMagar, Paris, France) is particularly useful for the detection of mixed infections (Fig. 15.2).

Fig. 15.2. CHROMagar™ Candida plate showing chromogenic colour change for *Candida albicans* (green), *C. tropicalis* (blue), *C. parapsilosis* (white) and *C. glabrata* (mauve).

Germ tube test: A rapid screening test for *Candida albicans* and *Candida dubliniensis*. 0.5 mL of serum, containing 0.5% glucose, is lightly inoculated with the test organism and incubated at 35°C for 2-3 hours. On microscopy, the production of germ tubes by the cells is presumptive for *Candida albicans* and *Candida dubliniensis* (Fig. 15.3).

Dalmau plate culture: To set up a yeast morphology plate, dip a flamed sterilised straight wire into a culture to make a light inoculum and then lightly scratch the wire into the surface of a cornmeal/Tween 80, rice/Tween 80 or yeast morphology agar plate, then place a flamed coverslip onto the agar surface covering the scratches. Dalmau morphology plates are examined *in situ* directly under the lower power of a microscope for the presence of pseudohyphae which may take up to 4-5 days at 26°C to develop. *Candida albicans* also produces characteristic large, round, terminal, thick-walled vesicles (chlamydospores). For best results a light inoculum should be scratched into the agar surface using a wire (Fig. 15.4).

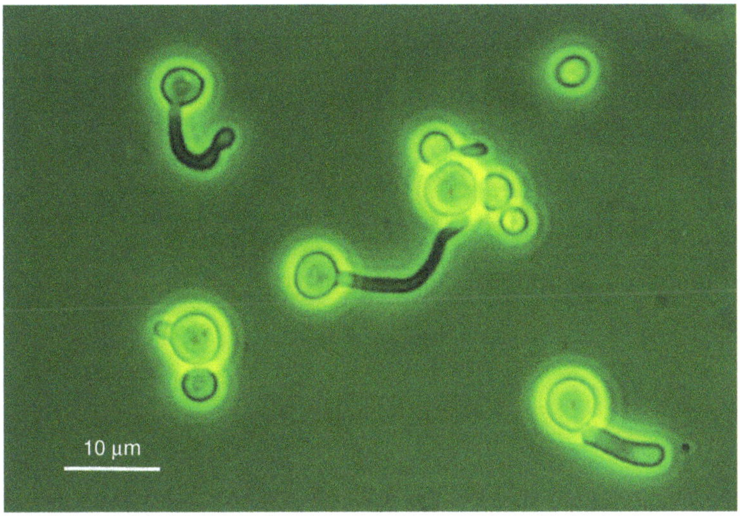

Fig. 15.3. *Candida albicans* showing production of germ tubes.

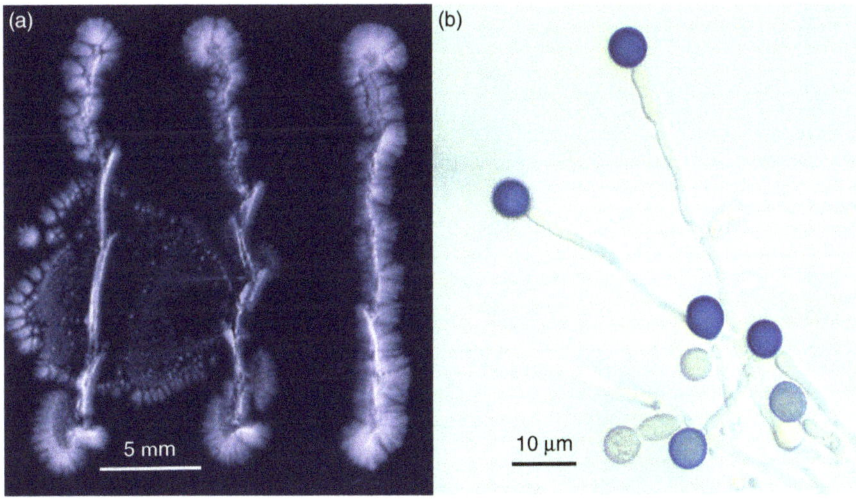

Fig. 15.4. *Candida albicans* Dalmau plate culture showing (a) colonies growing out from scratches on the surface of a cornmeal/Tween 80 agar plate, and (b) the production of large round, thick-walled chlamydospores. **Note:** A coverslip has been placed onto the agar surface covering the scratches.

Physiological and biochemical tests: Fermentation and assimilation studies should be performed based on those used at the Westerdijk Biodiversity Institute (formerly known as the Centraalbureau voor Schimmelcultures) (Kurtzman *et al.*, 2011). Several yeast identification systems are commercially available, for example the API 20C AUX (bioMerieux, Marcy-l'Etoile, France, ID 32C (bioMerieux), YT MicroPlate (Biolog, Hayward, California), AuxaColor 2 (Bio-Rad, Hercules, California) and Vitek 2 ID-YST (bioMerieux). However, they can only be used to identify those species in their respective databases and may misidentify many uncommon or new yeasts, including *Candida auris*. Other supplementary tests include growth at 37°C, cycloheximide resistance and hydrolysis of urea.

MALDI-ToF MS: Matrix Assisted Laser Desorption Ionization-Time of Flight mass spectrometry (MALDI-ToF MS) has emerged as the method of choice for rapid, highly accurate and cost-effective routine identification of yeast species. There are two commercially available MALDI-ToF MS systems, the Bruker Biotyper (Bruker Daltonics, Bremen, Germany) and Vitek MS (bioMerieux). The identification rate of most common and uncommon yeast isolates is comparable using both systems (Lee *et al.*, 2017; Patel, 2019), although studies have reported more misidentifications with the VITEK MS when the microorganism analysed was not present in the database compared to Bruker Biotyper (Levesque *et al.*, 2015). The Vitek MS has been reported to misidentify some yeasts, notably *Candida metapsilosis* as *Candida parapsilosis* (Nobrega de Almeida *et al.*, 2014).

Molecular identification: ITS sequencing is useful for the identification of most clinical yeasts.

References: Barnett *et al.* (1983); Kurtzman *et al.* (2011); de Hoog *et al.* (2015).

15.1. *Candida albicans* (Robin) Berkhout

Candida albicans is a commensal of mucous membranes and the gastrointestinal tract. Environmental isolations have been made from sources contaminated by human or animal excreta, such as polluted water, soil, air and plants. *C. albicans* is the most common cause of candidiasis in humans.

RG-2 organism

Culture: Colonies (SDA) white to cream coloured, smooth, glabrous, yeast-like.

Microscopy: Spherical to subspherical budding blastoconidia, 2-7 × 3-8 µm in size.

India ink preparation: Negative - no capsules present.

Dalmau plate culture: Branched pseudohyphae with dense verticils of blastoconidia. Spherical chlamydospores, mostly terminal, often on a slightly swollen subtending cell are formed near the edge of the cover slip.

Germ tube formation: Positive.

Physiological tests: (+ Positive, - Negative, v Variable, w Weak, s Slow).

Fermentation:

Glucose	+	Sucrose	v	Lactose	–
Galactose	v	Maltose	+	Trehalose	v

Growth reactions:

Glucose	+	L-Sorbose	v	*myo*-Inositol	–
Sucrose	v	L-Rhamnose	–	DL-Lactate	+
Raffinose	–	D-Xylose	+	D-Gluconate	v
Melibiose	–	L-Arabinose	v	2-Keto-D-gluconate	+
Galactose	+	D-Arabinose	v	D-Glucosamine	v
Lactose	–	D-Ribose	v	*N*-Acetyl-D-glucosamine	v
Trehalose	v	Glycerol	v	D-Glucuronate	–
Maltose	+	Erythritol	–	Nitrate	–
Melezitose	v	Ribitol	v	Urease	–
Methyl-α-D-glucoside	v	Galactitol	–	0.1% Cycloheximide	+
Soluble starch	+	D-Mannitol	+	Growth at 37°C	+
Cellobiose	–	D-Glucitol	v	Growth at 42°C	+

Key features: Germ tube positive, production of chlamydospores on Dalmau plate culture, fermentation of glucose, sugar assimilation profile and a distinctive green colour on CHROMagar™ Candida. **Note:** Germ tube negative variants (previously known as *C. claussenii*), and sucrose-negative variants (previously described as *C. stellatoidea*) may occur.

Antifungal susceptibility: (Table 15.1).

Table 15.1. *Candida albicans* (Australian national data with additional ISAV data from Pfaller *et al*., 2013a); MIC µg/mL.

Antifungal	No.	≤0.016	0.03	0.06	0.125	0.25	0.5	1	2	4	8	16	32	≥64
AmB	2635	2	16	164	648	645	875	283	2					
FLU	2638	2	2	3	100	840	1104	378	82	22	17	31	14	43
ISAV	800	733	51	5	5	2	2				1	1		
VORI	2354	1936	217	84	30	28	26	12	10	2	3	6		
POSA	2005	876	809	193	63	30	18	13	1		2			
ITRA	2638	165	749	1114	467	60	46	15	3		1	18		
ANID	1731	526	656	403	137	6	1	1	1					
MICA	1730	1614	96	12	5		2		1					
5FC	2638	4	207	1196	541	273	260	88	23	12	8	3	3	20

15.2. *Candida auris* Satoh and Makimura

Candida auris is an emerging multidrug resistant yeast that causes invasive infections, that was first described in 2009 in Japan and has since been reported from numerous countries (Lockhart *et al.*, 2017a,b; Jeffery-Smith *et al.*, 2018; Spivak and Hanson, 2018; Heath *et al.*, 2019; Zhu *et al.*, 2020). Infections and outbreaks caused by *C. auris* in hospital settings have been rising over the past several years. Difficulty in its identification, multidrug resistance properties, associated high mortality rates, and long-term survival on surfaces in the environment make *C. auris* particularly problematic in clinical settings (Du *et al.*, 2020). *Candida auris* is now a notifiable infection under public health regulations in many countries including Australia.

Candida auris may be misidentified as *C. haemuloni* or other yeast species using conventional phenotypic and biochemical methods (Kathuria *et al.,* 2015; Jeffery-Smith *et al.*, 2018). However, unlike these other *Candida* species, *C. auris* grows at 42°C, and this has become a useful differential characteristic (Casadevall *et al.*, 2019). *Candida auris* is best identified by MALDI-ToF MS or by molecular methods (Du *et al.*, 2020; Zhu *et al.*, 2020).

MALDI-ToF MS: Can accurately differentiate *Candida auris* from other fungal species; however, accurate identification is dependent on the reference database being used (Kathuria *et al.*, 2015; Jeffery-Smith *et al.*, 2018; Zhu *et al.*, 2020).

Molecular identification: ITS and D1/D2 sequencing provide accurate species identification (Kathuria *et al.*, 2015; Kordalewska *et al.*, 2017; Leach *et al.*, 2018; Du *et al.*, 2020; Zhu *et al.*, 2020).

RG-2 organism

Culture: Colonies (SDA) white to cream coloured, smooth, glabrous, yeast-like.

Microscopy: Spherical to subspherical budding blastoconidia, 2-3 x 2-5 µm in size.

India ink preparation: Negative - no capsules present.

Dalmau plate culture: Ovoid budding yeast cells only. No pseudohyphae produced.

Germ tube formation: Negative.

Physiological tests: (+ Positive, – Negative, v Variable, w Weak, s Slow).

Fermentation:

Glucose	+	Sucrose	w,+	Lactose	–
Galactose	–	Maltose	–	Trehalose	w,+

Growth reactions:

Glucose	+	L-Sorbose	–	*myo*-Inositol	–
Sucrose	+	L-Rhamnose	–	DL-Lactate	–
Raffinose	+	D-Xylose	–	D-Gluconate	–
Melibiose	–	L-Arabinose	–	2-Keto-D-gluconate	–
Galactose	–	D-Arabinose	–	D-Glucosamine	–
Lactose	–	D-Ribose	–	*N*-Acetyl-D-glucosamine	+
Trehalose	+	Glycerol	–	D-Glucuronate	–
Maltose	+	Erythritol	–	Nitrate	–
Melezitose	+	Ribitol	w,+	Urease	–
Methyl-α-D-glucoside	–	Galactitol	+	0.1% Cycloheximide	–
Soluble starch	+	D-Mannitol	+	Growth at 37°C	+
Cellobiose	–	D-Glucitol	+	Growth at 42°C	+,w,s

Key features: *Candida auris* is difficult to identify using conventional phenotypic and biochemical methods and has often been misidentified as *Candida haemuloni* or *Saccharomyces cerevisiae*. Growth at 42°C on CHROMagar™ Candida produces white, pink, or dark purple colonies (Kumar *et al.*, 2017), however on CHROMagar™ Candida Plus, *C. auris* colonies display a differential light blue colour with a blue halo (Mulet Bayona *et al.*, 2022). Reliable identification methods are MALDI-ToF MS or ITS and D1/D2 sequencing.

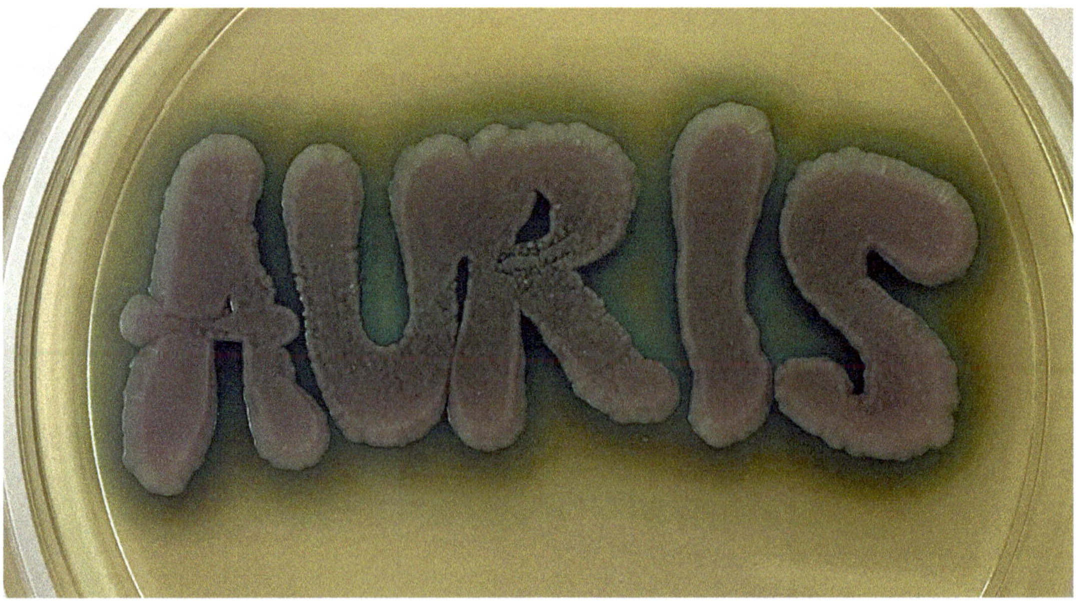

Fig. 15.5. *Candida auris* grown on CHROMagar™ Candida Plus agar showing characteristic blue halo.

Antifungal susceptibility: (Table 15.2).

Table 15.2. *Candida auris* (Chowdhary *et al.*, 2018; and Australian national data); MIC µg/mL.

Antifungal	No.	≤0.016	0.03	0.06	0.125	0.25	0.5	1	2	4	8	16	32	≥64
AmB	369				4	27	129	174	22	11	2			
FLU	369							5	1	5	2	25	62	269
ISAV	353	44	138	53	35	37	23	10	8	5				
VORI	369		21	23	94	91	43	36	28	19	4	10		
POSA	369	97	90	85	48	31	8	3	3	2	2			
ITRA	369	1	66	70	104	72	28	16	8		1	3		
ANID	369	3	28	99	75	112	37	5	1	1	8			
MICA	369	26	109	140	61	17	5	1	1		5	4		
5FC	369				213	44	5	24	2	11	6	5	9	50

15.3. *Candida bracarensis* Correia et al.

Candida bracarensis is recognised as part of the *Candida glabrata* species complex (Correia *et al.*, 2006). It is phenotypically indistinguishable from *C. glabrata* and is best identified by MALDI-ToF or molecular methods. *C. bracarensis* was described based on PCR-fingerprints and sequence divergence in the D1/D2 domains (Correia *et al.*, 2006).

RG-2 organism

Culture: Colonies (SDA) white to cream coloured, smooth, glabrous, yeast-like.

Microscopy: Ellipsoidal budding blastoconidia, 3.9-6 x 2-4 µm in size. No pseudohyphae or chlamydospores produced.

India ink preparation: Negative - no capsules present.

Dalmau plate culture: No pseudohyphae produced.

Germ tube formation: Negative.

Physiological tests: (+ Positive, – Negative, v Variable, w Weak, s Slow, n not done).

Fermentation:

Glucose	+	Sucrose	–	Lactose	–
Galactose	–	Maltose	–	Trehalose	s

Growth reactions:

Glucose	+	L-Sorbose	–	*myo*-Inositol	–
Sucrose	–	L-Rhamnose	–	DL-Lactate	–
Raffinose	–	D-Xylose	–	D-Gluconate	+
Melibiose	–	L-Arabinose	–	2-Keto-D-gluconate	–
Galactose	–	D-Arabinose	–	D-Glucosamine	–
Lactose	–	D-Ribose	–	*N*-Acetyl-D-glucosamine	n
Trehalose	+	Glycerol	v	D-Glucuronate	–
Maltose	–	Erythritol	–	Nitrate	–
Melezitose	–	Ribitol	–	Urease	–
Methyl-α-D-glucoside	–	Galactitol	–	0.1% Cycloheximide	–
Soluble starch	–	D-Mannitol	–	Growth at 37°C	+
Cellobiose	–	D-Glucitol	–	Growth at 42°C	+

Key features: *Candida bracarensis* has variable API 20C patterns that overlap with *C. nivariensis* and some *C. glabrata* isolates and has variable results with a rapid trehalose assay. **Note:** *C. glabrata* produces mauve coloured colonies on CHROMagar™ Candida, whereas isolates of *C. bracarensis* and *C. nivariensis* produce white colonies on CHROMagar™ Candida (Alcoba-Florez *et al.*, 2005; Bishop *et al.*, 2008).

Antifungal susceptibility: (Table 15.3).

Table 15.3. *Candida bracarensis* (Australian national data); MIC µg/mL.

Antifungal	No.	≤0.016	0.03	0.06	0.125	0.25	0.5	1	2	4	8	16	32	≥64
AmB	12					1	3	4	3	1				
FLU	12							1	4	4	2	1		
ISAV	3		3		1									
VORI	12	1	1	4	5	1								
POSA	12			1	3	6	2							
ITRA	12		1		1	7	3							
ANID	12	3	3	5	1									
MICA	12	11	1											
5FC	12			2	2	4	3	1						

15.4. *Candida dubliniensis* Sullivan *et al.*

Candida dubliniensis is an occasional cause of candidaemia and mucosal infection, especially in HIV patients.

RG-2 organism

Culture: Colonies (SDA) white to cream coloured, smooth, glabrous, yeast-like.

Microscopy: Spherical to subspherical budding blastoconidia, 3-8 x 2-7 µm in size.

India ink preparation: Negative - no capsules present.

Dalmau plate culture: Branched pseudohyphae with dense verticils of blastoconidia and spherical, mostly terminal chlamydospores.

Germ tube formation: Positive.

Physiological tests: (+ Positive, – Negative, v Variable, w Weak, s Slow).

Fermentation:

Glucose	+	Sucrose	–	Lactose	–
Galactose	+,s	Maltose	+	Trehalose	v

Growth reactions:

Glucose	+	L-Sorbose	–	*myo*-Inositol	–
Sucrose	+	L-Rhamnose	–	DL-Lactate	+
Raffinose	–	D-Xylose	s,+	D-Gluconate	–
Melibiose	–	L-Arabinose	–	2-Keto-D-gluconate	+
Galactose	+	D-Arabinose	–	D-Glucosamine	v
Lactose	–	D-Ribose	–	*N*-Acetyl-D-glucosamine	+
Trehalose	s,+	Glycerol	w,s,+	D-Glucuronate	–
Maltose	+	Erythritol	–	Nitrate	–

Melezitose	w,+	Ribitol	+	Urease	–
Methyl-α-D-glucoside	+,s	Galactitol	–	0.1% Cycloheximide	s
Soluble starch	w,+	D-Mannitol	+	Growth at 37°C	+
Cellobiose	–	D-Glucitol	+	Growth at 42°C	v

Key features: Germ tube positive, similar to *C. albicans*, except for absence of growth at 45°C (Pinjon *et al.*, 1998); glycerol (mostly +), methyl-α-D-glucoside (-), trehalose (-), and D-xylose (-). Initial colonies dark green colour on CHROMagar™ Candida and producing rough colonies on bird seed agar. ITS sequencing and MALDI-ToF can reliably distinguish *Candida dubliniensis* from *C. albicans*.

Antifungal susceptibility: (Table 15.4).

Table 15.4. *Candida dubliniensis* (Australian national data with additional ISAV data from Pfaller *et al.*, 2013a); MIC µg/mL.

Antifungal	No.	≤0.016	0.03	0.06	0.125	0.25	0.5	1	2	4	8	16	32	≥64
AmB	196	1	1	3	37	82	66	6						
FLU	196				90	75	26	5						
ISAV	54	49	4	1										
VORI	193	193												
POSA	185	99	63	20	3									
ITRA	196	77	49	54	12	4								
ANID	170	24	36	45	56	3					6			
MICA	170	81	70	11	2						6			
5FC	196		12	160	17	4	2		1					

15.5. *Candida glabrata* (Anderson) S.A. Meyer & Yarrow

Synonymy: *Torulopsis glabrata* (Anderson) Lodder & de Vries.

Recently *Candida glabrata* has been recognised as a species complex consisting of *C. glabrata, C. bracarensis* (Correia *et al.*, 2006) and *C. nivariensis* (Alcoba-Florez *et al.*, 2005). These three species are phenotypically indistinguishable and are best identified by molecular methods or MALDI-ToF MS. *C. bracarensis* was described based on PCR-fingerprints and sequence divergence in the D1/D2 domains (Correia *et al.*, 2006). *C. nivariensis* was differentiated from other yeasts based on ITS sequences (Borman *et al.*, 2008).

Candida glabrata is one of the most common yeast species to be found on the body surface and is often isolated as an incidental finding from skin and urine. It has been implicated as an "opportunistic" cause of both superficial and systemic infections, especially in immunocompromised patients, and it has been isolated from patients with septicaemia, pyelonephritis, pulmonary infections, endocarditis and hyperalimentation. Approximately 10% of clinical isolates exhibit azole cross-resistance.

RG-2 organism

Culture: Colonies (SDA) white to cream coloured, smooth, glabrous, yeast-like.

Microscopy: Ovoid to ellipsoidal budding blastoconidia, 3.4 × 2.0 µm in size. No pseudohyphae or chlamydospores produced.

India ink preparation: Negative - no capsules present.

Dalmau plate culture: Ovoid budding yeast cells only. No pseudohyphae produced.

Germ tube formation: Negative.

Physiological tests: (+ Positive, – Negative, v Variable, w Weak, s Slow).

Fermentation:

Glucose	+	Sucrose	–	Lactose	–
Galactose	–	Maltose	–	Trehalose	s

Growth reactions:

Glucose	+	L-Sorbose	–	*myo*-Inositol	–
Sucrose	–	L-Rhamnose	–	DL-Lactate	v
Raffinose	–	D-Xylose	–	D-Gluconate	+
Melibiose	–	L-Arabinose	–	2-Keto-D-gluconate	v
Galactose	–	D-Arabinose	–	D-Glucosamine	–
Lactose	–	D-Ribose	–	*N*-Acetyl-D-glucosamine	–
Trehalose	v	Glycerol	v	D-Glucuronate	–
Maltose	–	Erythritol	–	Nitrate	–
Melezitose	–	Ribitol	–	Urease	–
Methyl-α-D-glucoside	–	Galactitol	–	0.1% Cycloheximide	–
Soluble starch	–	D-Mannitol	–	Growth at 37°C	+
Cellobiose	–	D-Glucitol	–		

Key features: Germ tube negative yeast and sugar assimilation pattern. *Candida bracarensis* (Correia *et al.*, 2006) and *C. nivariensis* (Alcoba-Florez *et al.*, 2005) are recently identified cryptic species.

Antifungal susceptibility: (Table 15.5).

Table 15.5. *Candida glabrata* complex (Australian national data); MIC µg/mL.

Antifungal	No.	≤0.016	0.03	0.06	0.125	0.25	0.5	1	2	4	8	16	32	≥64
AmB	1763	2	8	25	280	393	594	441	19	1				
FLU	1766				1	4	8	29	183	271	236	417	343	274
ISAV	503	9	26	61	74	105	129	62	16	13	8			
VORI	1658	20	68	181	305	390	359	157	59	87	31	1		
POSA	1563	2	11	17	71	178	361	454	289	7	146			
ITRA	1766		6	19	126	358	485	395	123	33	7	214		
ANID	1395	103	566	462	225	9	5	8	15	1	1			
MICA	1394	1214	132	13	5	8	3	7	4	3	5			
5FC	1764	3	308	1362	37	11	10	8	11	3	1			10

15.6. *Candida haemuloni* (Uden & Kolip.) Meyer & Yarrow

Synonymy: *Candida haemuloni* var. *vulneris* Cendejas-Bueno *et al.*

Note: *Candida haemuloni* and *C. haemulonis* are orthographic variants, but *C. haemuloni* is considered the correct name.

Candida haemuloni has recently been reclassified as a complex of three phenotypically identical but genotypically distinct entities: *C. haemuloni, C. duobushaemulonis* and *C. haemuloni* var. *vulneris*, based on ITS and D1/D2 sequencing (Cendejas-Bueno *et al.*, 2012; Ramos *et al.*, 2015). *Candida haemuloni* has been reported from a few cases of fungaemia but clinical isolations remain rare.

RG-1 organism

Culture: Colonies (SDA) white to cream coloured, smooth, glabrous, yeast-like.

Microscopy: Ovoid to globose, budding yeast-like cells or blastoconidia, 2-7 x 2-7 µm. No pseudohyphae produced.

India ink preparation: Negative - no capsules present.

Dalmau plate culture: No pseudohyphae produced.

Germ tube formation: Negative.

Physiological tests: (+ Positive, – Negative, v Variable, w Weak, s Slow).

Fermentation:

Glucose	+	Sucrose	+	Lactose	–
Galactose	–	Maltose	–	Trehalose	+,s

Growth reactions:

Glucose	+	L-Sorbose	–	*myo*-Inositol	–
Sucrose	+	L-Rhamnose	+,w	DL-Lactate	–
Raffinose	+,s	D-Xylose	–	D-Gluconate	+
Melibiose	–	L-Arabinose	–	2-Keto-D-gluconate	+
Galactose	+,w	D-Arabinose	–	D-Glucosamine	+,s
Lactose	–	D-Ribose	–	*N*-Acetyl-D-glucosamine	+
Trehalose	+	Glycerol	+,s	D-Glucuronate	–
Maltose	+	Erythritol	–	Nitrate	–
Melezitose	+,w	Ribitol	+,s	Urease	–
Methyl-α-D-glucoside	–	Galactitol	–	0.1% Cycloheximide	–
Soluble starch	–	D-Mannitol	+	Growth at 30°C	+
Cellobiose	–	D-Glucitol	+	Growth at 37°C	–

Key features: Germ tube negative yeast and sugar assimilation pattern. *C. haemuloni* may be difficult to distinguish from *Debaryomyces hansenii* (previously *C. famata*) using some commercial yeast identification systems due to database limitations. Molecular identification may be required.

Antifungal susceptibility: (Table. 15.6).

Table 15.6. *Candida haemuloni* complex (Australian national data); MIC µg/mL.

Antifungal	No.	≤0.016	0.03	0.06	0.125	0.25	0.5	1	2	4	8	16	32	≥64
AmB	30					2	10	7	7	1	2			
FLU	30						1		3	1				25
VORI	28		1	1		1	1				20	4		
POSA	23	1	2		1					1	3	16		
ITRA	30		2	1	2			1				24		
ANID	21		13	1	3	1				1		2		
MICA	21		2	10	3	2						4		
5FC	25		2	4	7	6	2	3	1					

15.7. *Candida inconspicua* (Lodder & Kreger-van Rij) Meyer & Yarrow

Synonymy: *Torulopsis inconspicua* Lodder & Kreger-van Rij.

Candida inconspicua is a rare cause of candidaemia.

RG-1 organism

Culture: Colonies (SDA) white to cream coloured, smooth, glabrous, yeast-like.

Microscopy: Ovoidal budding blastoconidia, 2.0-5 × 5.0-11.0 µm.

India ink preparation: Negative - no capsules present.

Dalmau plate culture: Spherical to ovoid budding yeast cells only. Primitive pseudohyphae may be produced after 14 days.

Germ tube formation: Negative.

Physiological tests: (+ Positive, – Negative, v Variable, w Weak, s Slow).

Fermentation:

Glucose	+	Sucrose	–	Lactose	–
Galactose	–	Maltose	–	Trehalose	–

Growth reactions:

Glucose	+	L-Sorbose	–	*myo*-Inositol	–
Sucrose	–	L-Rhamnose	–	DL-Lactate	+
Raffinose	–	D-Xylose	–	D-Gluconate	–
Melibiose	–	L-Arabinose	–	2-Keto-D-gluconate	–
Galactose	–	D-Arabinose	–	D-Glucosamine	+
Lactose	–	D-Ribose	–	*N*-Acetyl-D-glucosamine	v
Trehalose	–	Glycerol	+	D-Glucuronate	–
Maltose	–	Erythritol	–	Nitrate	–
Melezitose	–	Ribitol	–	Urease	–
Methyl-α-D-glucoside	–	Galactitol	–	0.1% Cycloheximide	–
Soluble starch	–	D-Mannitol	–	Growth at 37°C	+
Cellobiose	–	D-Glucitol	–		

Key features: Germ tube negative yeast, sugar assimilation pattern, colonies are white on *Candida* CHROMagar™ Candida.

Antifungal susceptibility: (Table 15.7).

Table 15.7. *Candida inconspicua* (Australian national data); MIC µg/mL.

Antifungal	No.	≤0.016	0.03	0.06	0.125	0.25	0.5	1	2	4	8	16	32	≥64
AmB	16	4	1	1	3	6	1							
FLU	16											2	4	10
VORI	13			2	4	3	2	1						
POSA	13			2	3	5	2				1			
ITRA	15				2	6	6	1						
ANID	2	2												
MICA	2	2												
5FC	14				1		1	2	4	3	1	1	1	

15.8. *Candida metapsilosis* Tavanti et al.

Candida metapsilosis is recognised as part of the *Candida parapsilosis* species complex (Tavanti *et al.*, 2005). It is phenotypically indistinguishable from *C. parapsilosis, C. orthopsilosis* and *Lodderomyces elongisporus* and is best identified by ITS sequencing or MALDI-ToF MS.

RG-2 organism

Culture: Colonies (SDA) white to cream coloured, smooth, glabrous, yeast-like.

Microscopy: Ellipsoid, subglobose to fusiform budding blastoconidia, 4 x 3-6 µm, with some larger subglobose forms present.

India ink preparation: Negative - no capsules present.

Dalmau plate culture: Abundant, much branched pseudohyphae produced.

Germ tube formation: Negative.

Physiological tests: (+ Positive, – Negative, v Variable, w Weak, s Slow).

Fermentation:

Glucose	+	Sucrose	-	Lactose	–
Galactose	+	Maltose	-	Trehalose	–

Growth reactions:

Glucose	+	L-Sorbose	+	*myo*-Inositol	–
Sucrose	+	L-Rhamnose	–	DL-Lactate	–
Raffinose	–	D-Xylose	+	D-Gluconate	+
Melibiose	–	L-Arabinose	+	2-Keto-D-gluconate	+
Galactose	+	D-Arabinose	–	D-Glucosamine	–
Lactose	–	D-Ribose	+	*N*-Acetyl-D-glucosamine	+
Trehalose	+	Glycerol	+	D-Glucuronate	–
Maltose	+	Erythritol	–	Nitrate	–
Melezitose	+	Ribitol	+	Urease	–
Methyl-α-D-glucoside	+	Galactitol	–	0.1% Cycloheximide	–
Soluble starch	v	D-Mannitol	+	Growth at 37°C	+
Cellobiose	–	D-Glucitol	+	Growth at 42°C	+

Key features: *Candida metapsilosis* cannot be distinguished phenotypically from *C. parapsilosis* and *C. orthopsilosis*, but can be reliably identified by ITS sequencing (Tavanti *et al.*, 2005; Borman *et al.*, 2008; Asadzadeh *et al.*, 2009) and MALDI-ToF MS.

Antifungal susceptibility: (Table 15.8).

Table 15.8. *Candida metapsilosis* (Australian national data); MIC µg/mL.

Antifungal	No.	≤0.016	0.03	0.06	0.125	0.25	0.5	1	2	4	8	16	32	≥64
AmB	58				2	9	21	22	3	1				
FLU	58					1	4	17	22	4				
ISAV	7	5	1	1										
VORI	58	13	31	10	4									
POSA	58	6	3	15	27	5	1		1					
ITRA	28	2	3	4	9	9	1							
ANID	39			1	11	12	12	2	1					
MICA	39				6	15	17	1						
5FC	27			16	6	4	1							

15.9. *Candida nivariensis* Alcoba-Florez *et al.*

Candida nivariensis is recognised as part of the *Candida glabrata* species complex (Alcoba-Florez *et al.*, 2005). It is closely related to *C. glabrata* and *C. bracarensis* and is best identified by molecular methods (Alcoba-Florez *et al.*, 2005; Correia *et al.*, 2006; Wahyuningsih *et al.*, 2008) and MALDI-ToF MS (Pinto *et al.*, 2011).

RG-2 organism

Culture: Colonies (SDA) white to cream coloured, smooth, glabrous, yeast-like.

Microscopy: Ellipsoidal budding blastoconidia, 3-5 x 1.8-3 µm in size. No pseudohyphae or chlamydospores produced.

India ink preparation: Negative - no capsules present.

Dalmau plate culture: No pseudohyphae produced.

Germ tube formation: Negative.

Physiological tests: (+ Positive, – Negative, v Variable, w Weak, s Slow).

Fermentation:

Glucose	+	Sucrose	–	Lactose	–
Galactose	–	Maltose	–	Trehalose	+

Growth reactions:

Glucose	+	L-Sorbose	–	*myo*-Inositol	–
Sucrose	–	L-Rhamnose	–	DL-Lactate	–
Raffinose	–	D-Xylose	–	D-Gluconate	+
Melibiose	–	L-Arabinose	–	2-Keto-D-gluconate	–
Galactose	–	D-Arabinose	–	D-Glucosamine	–
Lactose	–	D-Ribose	–	*N*-Acetyl-D-glucosamine	–
Trehalose	–	Glycerol	+	D-Glucuronate	–
Maltose	–	Erythritol	–	Nitrate	–
Melezitose	–	Ribitol	–	Urease	–
Methyl-α-D-glucoside	–	Galactitol	–	0.1% Cycloheximide	–
Soluble starch	v	D-Mannitol	–	Growth at 37°C	+
Cellobiose	–	D-Glucitol	–	Growth at 42°C	+

Key features: *Candida nivariensis* is closely related to *C. glabrata* and *C. bracarensis*. These three species were found to differ by DNA-DNA reassociation experiments, RAPD-typing, AFLP-typing and D1/D2 and ITS sequences (Alcoba-Florez *et al.*, 2005; Wahyuningsih *et al.*, 2008).

Antifungal susceptibility: (Table 15.9).

Table 15.9. *Candida nivariensis* (Australian national data); MIC µg/mL.

Antifungal	No.	≤0.016	0.03	0.06	0.125	0.25	0.5	1	2	4	8	16	32	≥64
AmB	20					1	4	13	2					
FLU	20						2	6	4	3	1	3		1
ISAV	6	1	1	4										
VORI	20	2	7	6	1	1	2	1						
POSA	20				5	9	2	2	2					
ITRA	20				5	9	4	2						
ANID	20		10	6	4									
MICA	20	19	1											
5FC	20			1	12	3	3	1						

15.10. *Candida orthopsilosis* Tavanti *et al.*

Candida orthopsilosis is recognised as part of the *Candida parapsilosis* species complex (Tavanti *et al.*, 2005). It is phenotypically indistinguishable from *C. parapsilosis*, *C. metapsilosis* and *Lodderomyces elongisporus* and is best identified by ITS sequencing or MALDI-ToF MS.

RG-2 organism

Culture : Colonies (SDA) white to cream coloured, smooth, glabrous, yeast-like.

Microscopy: Ellipsoid to subglobose budding blastoconidia, 2-5 × 3-7 µm, with some larger elongated forms present.

India ink preparation: Negative - no capsules present.

Dalmau plate culture: Abundant, much-branched pseudohyphae produced.

Germ tube formation: Negative.

Physiological tests: (+ Positive, – Negative, v Variable, w Weak, s Slow).

Fermentation:

Glucose	+	Sucrose	–	Lactose	–
Galactose	–	Maltose	–	Trehalose	–,s

Growth reactions:

Glucose	+	L-Sorbose	+	*myo*-Inositol	–
Sucrose	+	L-Rhamnose	–	DL-Lactate	–
Raffinose	–	D-Xylose	+	D-Gluconate	+
Melibiose	–	L-Arabinose	+	2-Keto-D-gluconate	+
Galactose	+	D-Arabinose	–	D-Glucosamine	–
Lactose	–	D-Ribose	–	*N*-Acetyl-D-glucosamine	+
Trehalose	+	Glycerol	+	D-Glucuronate	–
Maltose	+	Erythritol	–	Nitrate	–
Melezitose	+	Ribitol	+	Urease	–
Methyl-α-D-glucoside	+	Galactitol	–	0.1% Cycloheximide	–
Soluble starch	v	D-Mannitol	+	Growth at 37°C	+
Cellobiose	–	D-Glucitol	+	Growth at 42°C	+

Key features: *Candida orthopsilosis* cannot be distinguished morphologically from *C. parapsilosis* and *C. metapsilosis* but can be separated based on ITS sequencing (Tavanti *et al.*, 2005; Borman *et al.*, 2008; Asadzadeh *et al.*, 2009) and MALDI-ToF MS.

Antifungal susceptibility: (Table 15.10).

Table 15.10. *Candida orthopsilosis* (Australian national data); MIC µg/mL.

Antifungal	No.	≤0.016	0.03	0.06	0.125	0.25	0.5	1	2	4	8	16	32	≥64
AmB	225		3	3	30	46	60	48	23	9				
FLU	197				1	9	45	86	24	16	9	4		3
VORI	177	68	51	36	6	13	2	2						
POSA	137	5	21	49	42	9	11							
ITRA	95	3	14	30	32	16								
ANID	87		1		2	9	20	45	10					
MICA	86	1	1		3	36	38	6	1					
5FC	93			56	22	9	3	1		1				1

15.11. *Candida parapsilosis* (Ashford) Langeron & Talice

Recently *Candida parapsilosis* has been recognised as four species: *C. parapsilosis, C. orthopsilosis, C. metapsilosis* and *Lodderomyces elongisporus* (Tavanti *et al.*, 2005). These four species are phenotypically indistinguishable and are best distinguished by ITS sequencing or MALDI-ToF MS analysis. Antifungal susceptibility data from the Australian candidemia study also shows no significant differences between the species (Chen *et al.*, 2009; Chapman *et al.*, 2017).

RG-2 organism

Culture: Colonies (SDA) white to cream coloured, smooth, glabrous, yeast-like.

Microscopy: Predominantly small, globose to ovoid budding blastoconidia, 3-4 x 5-8 µm, with some larger elongated forms present.

India ink preparation: Negative - no capsules present.

Dalmau plate culture: Abundant, much-branched pseudohyphae in a delicate tree-like pattern with 2-3 blastoconidia in small clusters at intervals along the pseudohyphae.

Germ tube formation: Negative.

Physiological tests: (+ Positive, – Negative, v Variable, w Weak, s Slow).

Fermentation:

Glucose	+	Sucrose	–,s	Lactose	–
Galactose	v	Maltose	–,s	Trehalose	–,s

Growth reactions:

Glucose	+	L-Sorbose	v	*myo*-Inositol	–
Sucrose	+	L-Rhamnose	–	DL-Lactate	–
Raffinose	–	D-Xylose	+	D-Gluconate	v
Melibiose	–	L-Arabinose	+	2-Keto-D-gluconate	+
Galactose	+	D-Arabinose	–	D-Glucosamine	v
Lactose	–	D-Ribose	v	*N*-Acetyl-D-glucosamine	+
Trehalose	+	Glycerol	+	D-Glucuronate	–
Maltose	+	Erythritol	–	Nitrate	–
Melezitose	+	Ribitol	v	Urease	–
Methyl-α-D-glucoside	+	Galactitol	–	0.1% Cycloheximide	–
Soluble starch	–	D-Mannitol	+	Growth at 37°C	+
Cellobiose	–	D-Glucitol	+		

Key features: Germ tube negative yeast and sugar assimilation pattern. *Candida parapsilosis* is commonly found on the skin and is a causative agent of candidaemia.

Antifungal susceptibility: (Table 15.11).

Table 15.11. *Candida parapsilosis* complex (Australian national data; additional ISAV data from Pfaller *et al.*, 2013a); MIC µg/mL.

Antifungal	No.	≤0.016	0.03	0.06	0.125	0.25	0.5	1	2	4	8	16	32	≥64
AmB	935	2	8	28	132	153	265	269	75	3				
FLU	935				15	91	237	299	169	80	29	6	7	2
ISAV	347	72	71	113	70	11	4	4				2		
VORI	858	406	235	132	57	20	5	1	2					
POSA	766	76	190	346	127	25	2							
ITRA	937	30	100	226	454	112	12	3						
ANID	671			3	11	12	54	232	279	57	23			
MICA	671	1		3	2	8	67	263	239	77	11			
5FC	937	3	31	261	285	294	51	7	3					2

15.12. *Candida tropicalis* (Castellani) Berkhout

Candida tropicalis is a major cause of septicaemia and disseminated candidiasis. It is also found as part of the normal human mucocutaneous flora and environmental isolations have been made from faeces, shrimp, kefir and soil.

RG-2 organism

Culture: Colonies (SDA) white to cream coloured, smooth, glabrous, yeast-like.

Microscopy: Spherical to subspherical budding yeast-like cells or blastoconidia, 3.5-7 x 5.5-10 µm.

India ink preparation: Negative - no capsules present.

Dalmau plate culture: Abundant, long, wavy, branched pseudohyphae with numerous ovoid blastoconidia budding off. Chlamydospores are not produced.

Germ tube formation: Negative.

Physiological tests: (+ Positive, – Negative, v Variable, w Weak, s Slow).

Fermentation:

Glucose	+	Sucrose	v	Lactose	–
Galactose	+	Maltose	+	Trehalose	+,s

Growth reactions:

Glucose	+	L-Sorbose	v	*myo*-Inositol	–
Sucrose	v	L-Rhamnose	–	DL-Lactate	v
Raffinose	–	D-Xylose	+	D-Gluconate	v
Melibiose	–	L-Arabinose	–	2-Keto-D-gluconate	+
Galactose	+	D-Arabinose	–	D-Glucosamine	v
Lactose	–	D-Ribose	v	*N*-Acetyl-D-glucosamine	+
Trehalose	+	Glycerol	v	D-Glucuronate	–
Maltose	+	Erythritol	–	Nitrate	–
Melezitose	v	Ribitol	v	Urease	–
Methyl-α-D-glucoside	v	Galactitol	–	0.1% Cycloheximide	+
Soluble starch	+	D-Mannitol	+	Growth at 37°C	+
Cellobiose	v	D-Glucitol	+		

Key features: Germ tube negative yeast and sugar assimilation pattern. Colonies are dark blue on *Candida* CHROMagar™ Candida.

Antifungal susceptibility: (Table 15.12).

Table 15.12. *Candida tropicalis* (Australian national data with additional ISAV data from Pfaller *et al*., 2013a); MIC µg/mL.

Antifungal	No.	≤0.016	0.03	0.06	0.125	0.25	0.5	1	2	4	8	16	32	≥64
AmB	380			5	41	74	114	134	11	1				
FLU	381				1	9	53	137	98	41	15	6	5	16
ISAV	180	11	46	56	36	15	9	5	1	1				
VORI	348	22	42	88	101	54	16	7	9	1	7	1		
POSA	329	8	29	58	111	83	22	10	2		6			
ITRA	381	2	12	22	144	143	42	4	2	1	1	8		
ANID	275	13	14	47	175	23	2				1			
MICA	275	55	185	29	3	1	1				1			
5FC	381		31	237	69	22	5	2	1	1	2	1	2	8

DOI: 10.1079/9781800622340.0015

16 *Chaetomium* Kunze ex Fries

Chaetomium was formerly a large genus with over 250 phenotypic species described. Species were identified by the size and shape of the perithecia, the setae or hairs covering the perithecia, and ascospore morphology. Several species were thermophilic and could grow at temperatures above 37°C. However recent multigene phylogenetic studies have recognised only 44 *Chaetomium* species, and many others have been reclassified in new genera (Wang *et al.*, 2016, 2022). Molecular identification using the β-tubulin and *RPB2* genes is now required to accurately identify species. *Chaetomium* species are important agents for the decomposition of cellulose waste and plant materials and are only rarely isolated in medical mycology laboratories.

RG-1 organism

Morphological description: *Chaetomium* is a common ascomycetous fungus characterised by the formation of darkly pigmented, globose, ovoid, barrel to flask-shaped, ostiolate ascocarps (perithecia) beset with dark coloured terminal hairs (setae) which are straight, branched or curved. Asci are clavate to cylindrical, typically eight-spored and evanescent. Ascospores are one-celled, darkly pigmented, smooth-walled, of varying shape, mostly ovoid, ellipsoidal or lemon-shaped. Chlamydospores and solitary conidia may also be produced (Fig. 16.1).

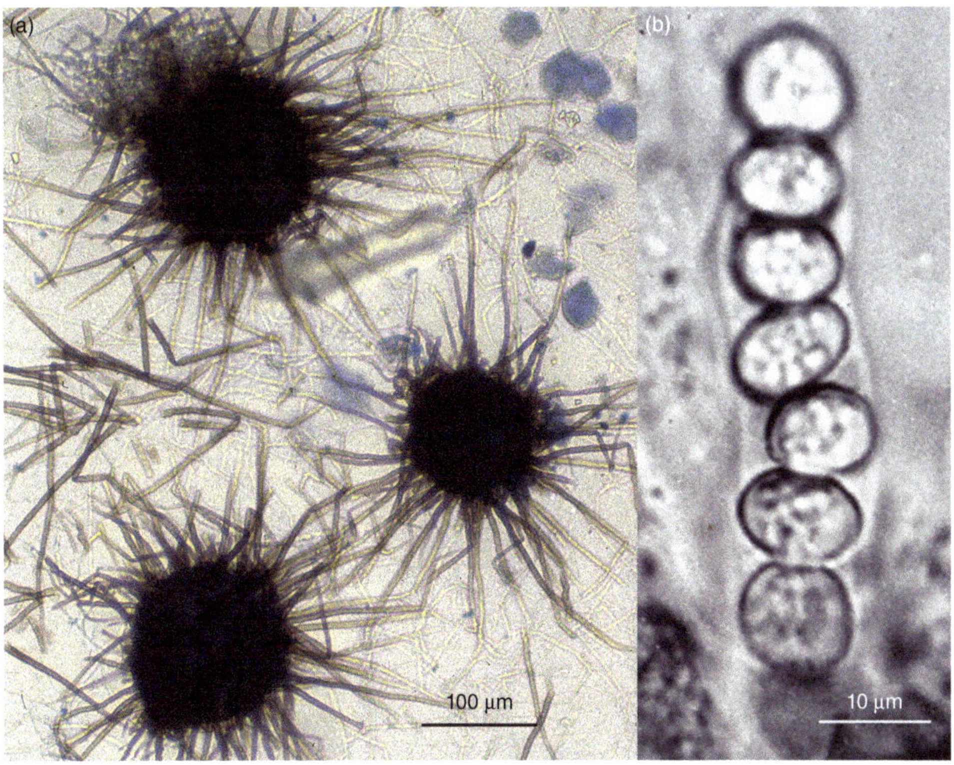

Fig. 16.1. *Chaetomium* spp. showing (a) ascocarps (perithecia) with dark coloured terminal hairs (setae), and (b) an ascus containing eight ascospores.

Molecular identification: Lee and Hanlin (1999) established the phylogenetic relationships of *Chaetomium* based on ribosomal DNA sequences. Sequencing of the β-tubulin and *RPB2* genes is recommended for routine identification (Wang *et al.*, 2016, 2022).

Key features: Ascomycetous fungus producing darkly pigmented ostiolate perithecia beset with long dark terminal setae.

Antifungal susceptibility: (Table 16.1).

Table 16.1. *Chaetomium* spp. (Australian national data); MIC μg/mL.

Antifungal	No.	≤0.016	0.03	0.06	0.125	0.25	0.5	1	2	4	8	16	32	≥64
AmB	8							5	2	1				
ISAV	3						1	1	1					
VORI	8				1	3	2	1	1					
POSA	8					1	1	4	1	1				
ITRA	8						2	5	1					

References: Ames (1963); Seth (1970); Millner (1977); Ellis (1981); Ellis and Keane (1981); von Arx *et al.* (1986); Domsch *et al.* (2007); Wang *et al.* (2016, 2022).

DOI: 10.1079/9781800622340.0016

17 *Chrysosporium* Corda

Species of *Chrysosporium* are occasionally isolated from skin and nail scrapings, especially from feet, but because they are common soil saprophytes they are usually considered as contaminants. There are about 70 species of *Chrysosporium*, several are keratinolytic with some also being thermotolerant, and cultures may closely resemble some dermatophytes, especially *Trichophyton mentagrophytes*, and other strains may also resemble cultures of *Histoplasma* and *Blastomyces*. Fungi previously identified as the *Chrysosporium* anamorph of *Nannizziopsis vriesii* (CANV) have been moved to the *Nannizziopsis* genus and have emerged as the leading cause of fungal dermatitis in both captive and wild reptiles (Sigler *et al.*, 2013).

Morphological genus description: Colonies are moderately fast growing, flat, white to tan to beige in colour, often with a powdery or granular surface texture. Reverse pigment absent or pale brownish-yellow with age. Hyaline, one-celled conidia are produced directly on vegetative hyphae by non-specialised conidiogenous cells. Conidia are typically pyriform to clavate with truncate bases and are formed either intercalary (arthroconidia), laterally (often on pedicels) or terminally.

Molecular identification: *Chrysosporium* is phylogenetically heterogeneous; the polyphyletic origin of the genus was first demonstrated by Vidal *et al.* (2000) based on ITS sequences, and further elaborated by Stchigel *et al.* (2013). ITS sequencing can assist in identification of clinical isolates.

17.1. *Chrysosporium tropicum* Carmichael

RG-2 organism

Morphological description: Colonies are flat, white to cream coloured with a very granular surface (Fig. 17.1). Reverse pigment absent or pale brownish-yellow with age. Microscopically, conidia are numerous, hyaline, single-celled, clavate to pyriform, smooth, slightly thick-walled (6-7 x 3.5-4 µm), and have broad truncate bases and pronounced basal scars. Conidia are formed at the tips of the hyphae, on short or long lateral branches, or sessile along the hyphae (intercalary). No macroconidia or hyphal spirals are seen (Fig. 17.1).

References: Carmichael (1962); Rebell and Taplin (1970); Sigler and Carmichael (1976); van Oorschot (1980); Domsch *et al.* (2007); de Hoog *et al.* (2015).

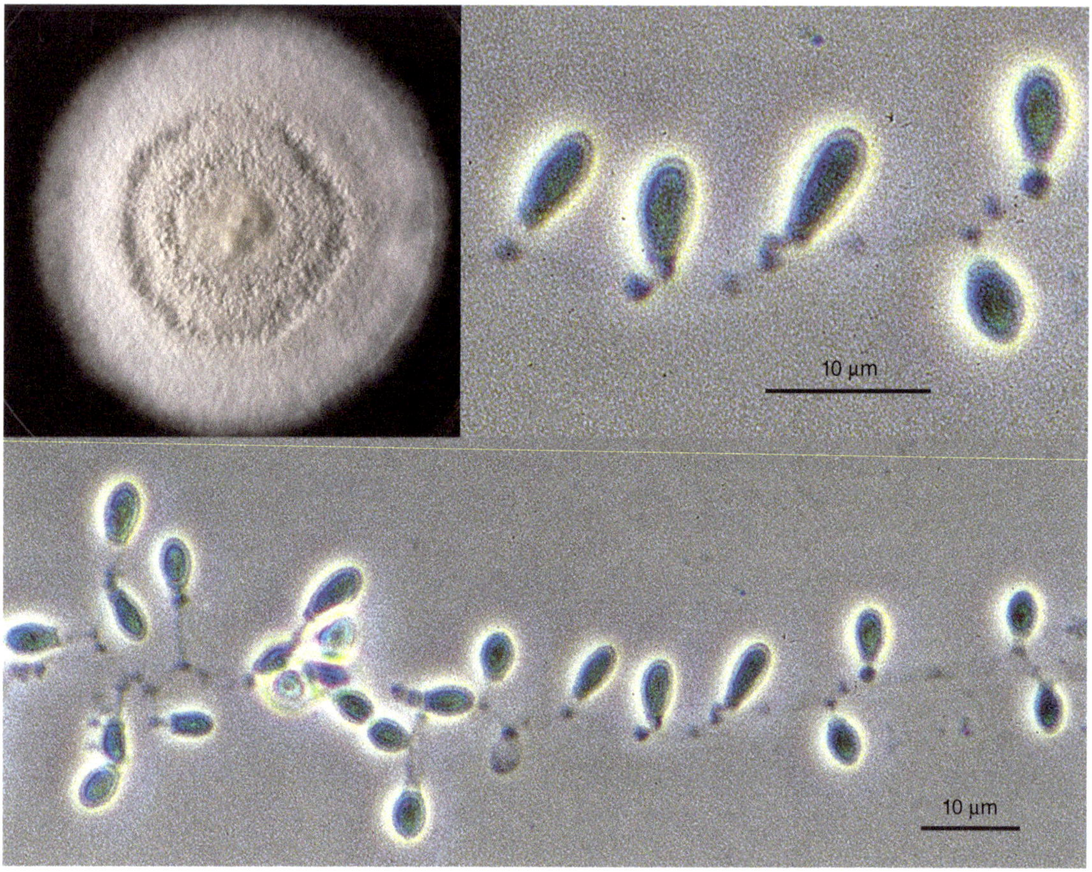

Fig. 17.1. *Chrysosporium tropicum* culture and typical pyriform to clavate shaped conidia.

DOI: 10.1079/9781800622340.0017

18 *Cladophialophora* Borelli

The genus *Cladophialophora* is characterised by: (1) the absence of conidiophores, "shield cells," or prominent hila (attachment points); (2) the ability to grow on media containing cycloheximide; and (3) having dry, non-fragile chains of conidia (Revankar and Sutton, 2010). It has recently been re-evaluated by multilocus sequencing and currently contains nine species associated with human infection (Badali *et al.*, 2008).

Cladophialophora bantiana is the causative agent of numerous cases of cerebral phaeohyphomycosis many of which occur in immunocompetent individuals and most of which are fatal (Chakrabarti *et al.*, 2016). *C.carrionii* and the recently described *C. samoensis* are agents of chromoblastomycosis. Less common species occasionally implicated in deep and superficial mycoses include, *C. arxii, C. boppii, C. devriesii, C. emmonsii, C. modesta* and *C. saturnica. Cladophialophora yegresii* is a closely related environmental sister species to *C. carrionii* (de Hoog *et al.*, 1995; Revankar and Sutton, 2010; de Hoog *et al.*, 2020).

18.1 *Cladophialophora bantiana* (Saccardo) de Hoog *et al.*

Synonymy: *Xylohypha bantiana* (Saccardo) McGinnis, Borelli and Ajello; *Cladosporium bantianum* (Sacc.) Borelli; *Cladosporium trichoides* Emmons.

Cladophialophora bantiana has been isolated from soil and is a recognised agent of cerebral phaeohyphomycosis. The fungus is neurotropic and causes brain abscess in both immunocompetent and immunosuppressed patients and is usually fatal. The fungus is likely introduced via inhalation and direct transfer to the brain via the paranasal sinuses, or traumatic head injury (Chakrabarti *et al.*, 2016; Ozgun *et al.*, 2019).

WARNING: RG-3 organism

Cultures of *C. bantiana* represent a potential biohazard to laboratory personnel and must be handled with extreme caution in a Class II Biological Safety Cabinet (BSCII).

Morphological description: Colonies are moderately fast growing, olivaceous-grey, suede-like to floccose and grow at temperatures up to 42-43°C (Fig. 18.1). Conidia are formed in long, sparsely branched, flexuose, acropetal chains from undifferentiated conidiophores. Conidia are one-celled (very occasionally two-celled), pale brown, smooth-walled, ellipsoid to oblong-ellipsoid and are 5-11 x 2-5 μm in size (Fig. 18.1).

Key features: *Cladophialophora bantiana* may be distinguished from *Cladosporium* species by the absence of conidia with distinctly pigmented hila, the absence of characteristic shield cells and by growth at 42°C (compared with *Cladophialophora carrionii* which has a maximum growth temperature of 35-36°C, and *Cladosporium* species which have a maximum of less than 35°C). *C. bantiana* may be further distinguished from *C. carrionii* by the formation of very long, sparsely branched chains of conidia.

Molecular identification. ITS sequencing is recommended (Gerrits van den Ende and de Hoog, 1999; Badali *et al.*, 2010a).

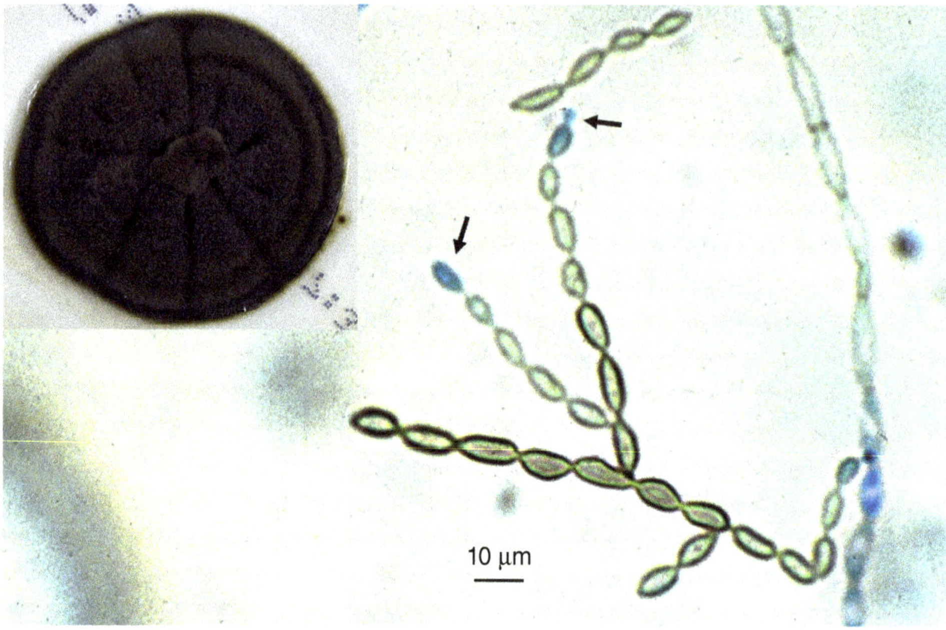

Fig. 18.1. *Cladophialophora bantiana* culture, conidiophores and conidia. **Note:** Conidiophores produce branched acropetal chains of conidia, where the youngest conidium (arrows) is at the tip of the chain.

Antifungal susceptibility: (Table 18.1).

Table 18.1. *Cladophialophora bantiana* (Australian national data); MIC µg/mL.

Antifungal	No.	≤0.016	0.03	0.06	0.125	0.25	0.5	1	2	4	8	16	32	≥64
AmB	27			1		3	2	6	6	9				
ISAV	3					2	1							
VORI	27		3	3	7	11	3							
POSA	25	7	5	12	1									
ITRA	27	6	3	5	11	2								

References: McGinnis (1980); McGinnis and Borelli (1981); McGinnis *et al.* (1986a); Rippon (1988); Kwon-Chung and Bennett (1992); Kantarcioglu *et al.* (2017); de Hoog *et al.* (2020).

18.2. *Cladophialophora carrionii* (Trejos) de Hoog *et al.*

Synonymy: *Cladosporium carrionii* Trejos.

Cladophialophora carrionii is a recognised agent of chromoblastomycosis and has been isolated from soil and fence posts made from *Eucalyptus* wood. Cases of chromoblastomycosis caused by *C. carrionii* are commonly found in Australia, Venezuela, Madagascar and South America. Isolates from phaeomycotic cysts and opportunistic infections have also been reported.

RG-2 organism

Morphological description: Colonies are slow growing, reaching 3-4 cm in diameter after one month, with a compact suede-like to downy surface and are olivaceous-black in colour (Fig. 18.2). Microscopy shows

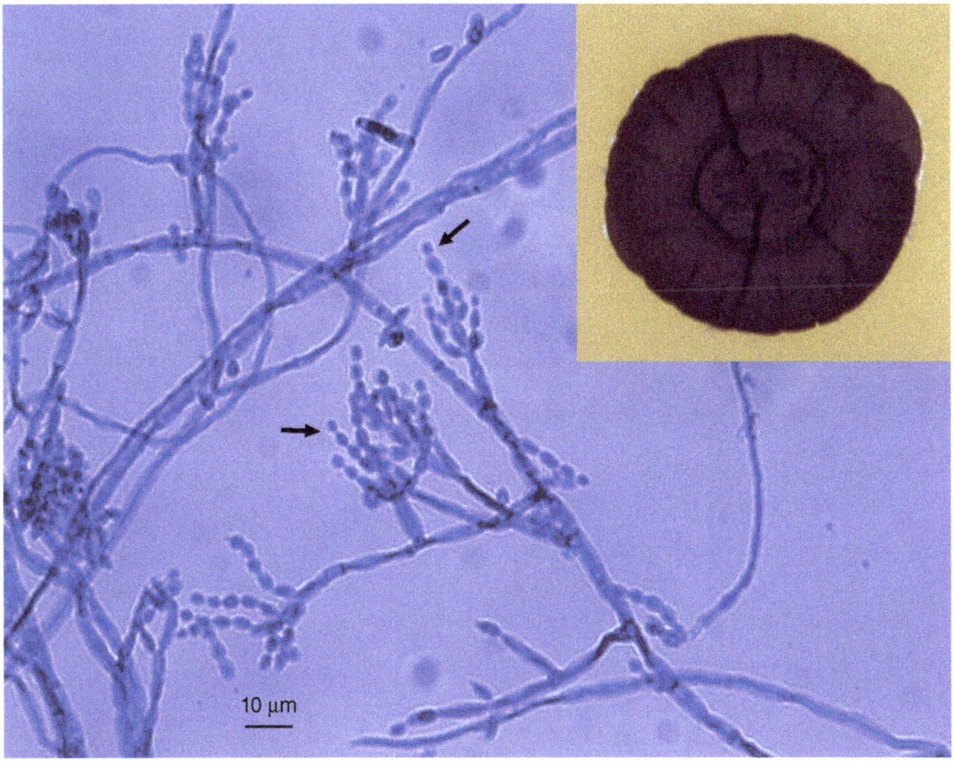

Fig. 18.2. *Cladophialophora carrionii* culture, conidiophores and conidia. **Note:** Conidiophores produce branched acropetal chains of conidia (arrows).

ascending to erect, olivaceous-green, apically branched, elongate conidiophores producing branched acropetal chains of smooth-walled conidia. Conidia are pale olivaceous, smooth-walled or slightly verrucose, limoniform to fusiform, 1.5-3.0 x 2.0-7.0 µm in size (Fig. 18.2). Bulbous phialides with large collarettes and minute, hyaline conidia are occasionally formed on nutritionally poor media. Maximum growth temperature 35-37°C.

Molecular identification: ITS sequencing is recommended (Abliz *et al.*, 2004; de Hoog *et al.*, 2007).

Key features: Conidia are smaller and comprise heavily branched systems which fall apart much more easily than in the other *Cladophialophora* species.

Antifungal susceptibility: (Table 18.2).

Table 18.2. *Cladophialophora carrionii* limited data available (Australian national data); MIC µg/mL.

Antifungal	No.	≤0.016	0.03	0.06	0.125	0.25	0.5	1	2	4	8	16	32	≥64
AmB	2							1	1					
VORI	2	1	1											
POSA	2	1	1											
ITRA	2	1	1											

References: McGinnis (1980); Rippon (1988); de Hoog *et al.* (1995, 2015).

DOI: 10.1079/9781800622340.0018

19 *Cladosporium* Link ex Fries

Cladosporium species are ubiquitous worldwide, and commonly isolated from soil and organic matter. They represent the most frequently isolated airborne fungi. The genus has undergone a number of revisions. The well-known thermotolerant "true" human-pathogenic species, formerly known as *C. bantiana*, *C. carrionii* and *C. devriesii*, characterised by the absence of conidiophores, and unpigmented conidial scars, were reclassified to *Cladophialophora* (de Hoog *et al.*, 1995; Bensch *et al.*, 2012). The remaining species of medical interest were *C. cladosporioides*, *C. herbarum*, *C. oxysporum* and *C. sphaerospermum*. More recently, extensive revisions based on polyphasic approaches have recognised 169 species, and demonstrated that *C. cladosporioides*, *C. herbarum* and *C. sphaerospermum* are species complexes encompassing several sibling species that can only be distinguished by phylogenetic analyses (Crous *et al.*, 2007; Schubert *et al.*, 2007; Zalar *et al.*, 2007; Bensch *et al.*, 2010, 2012).

Sandoval-Denis *et al.* (2015b) analysed 92 clinical isolates from the United States using phenotypic and molecular methods, which included sequence analysis of the ITS and D1/D2 regions, partial *TEF-1* and *ACT* genes. Surprisingly, the most frequently isolated species was *Cladosporium halotolerans* (15%) a member of *C. sphaerospermum* species complex, followed by *C. tenuissimum* (10%), *C. subuliforme* (6%) and *C. pseudocladosporioides* (5%), all members of the *C. cladosporioides* species complex. However, 40% of the isolates did not correspond to any known species and were deemed to represent at least 17 new lineages for *Cladosporium*. The most frequent anatomic site of isolation was the respiratory tract (55%), followed by superficial (28%) and deep tissues and fluids (15%). Species of the two recently described *Cladosporium*-like genera *Toxicocladosporium* and *Penidiella* were also reported for the first time from clinical samples (Sandoval-Denis *et al.*, 2015b).

RG-1 organisms

Morphological description: Colonies are slow growing, mostly olivaceous-brown to blackish-brown but also sometimes grey, buff or brown, suede-like to floccose, often becoming powdery due to the production of abundant conidia. The reverse is olivaceous-black (Fig. 19.1). Vegetative hyphae, conidiophores and conidia are equally pigmented. Conidiophores are more or less distinct from the vegetative hyphae, being erect, straight or flexuose, unbranched or branched only in the apical region, with geniculate sympodial elongation in some species. Conidia are produced in branched acropetal chains, being smooth, verrucose or echinulate, one to four-celled, and have a distinct dark hilum (Fig. 19.1). The term blastocatenate is often used to describe chains of conidia where the youngest conidium is at the apical or distal end of the chain. **Note:** The conidia closest to the conidiophore, and where the chains branch, are usually "shield-shaped". The presence of shield-shaped conidia, a distinct hilum, and chains of conidia that readily disarticulate, are characteristic of the genus *Cladosporium*.

Key features: Darkly pigmented mould forming branched acropetal chains of conidia, each with a distinct hilum.

Molecular identification: Genus level identification is usually sufficient and morphological identification can be confirmed by ITS and D1/D2 sequence analysis. Multilocus gene analysis of the ITS, D1/D2, *TEF-1* and *ACT* gene loci is necessary for accurate species identification (Bensch *et al.*, 2012).

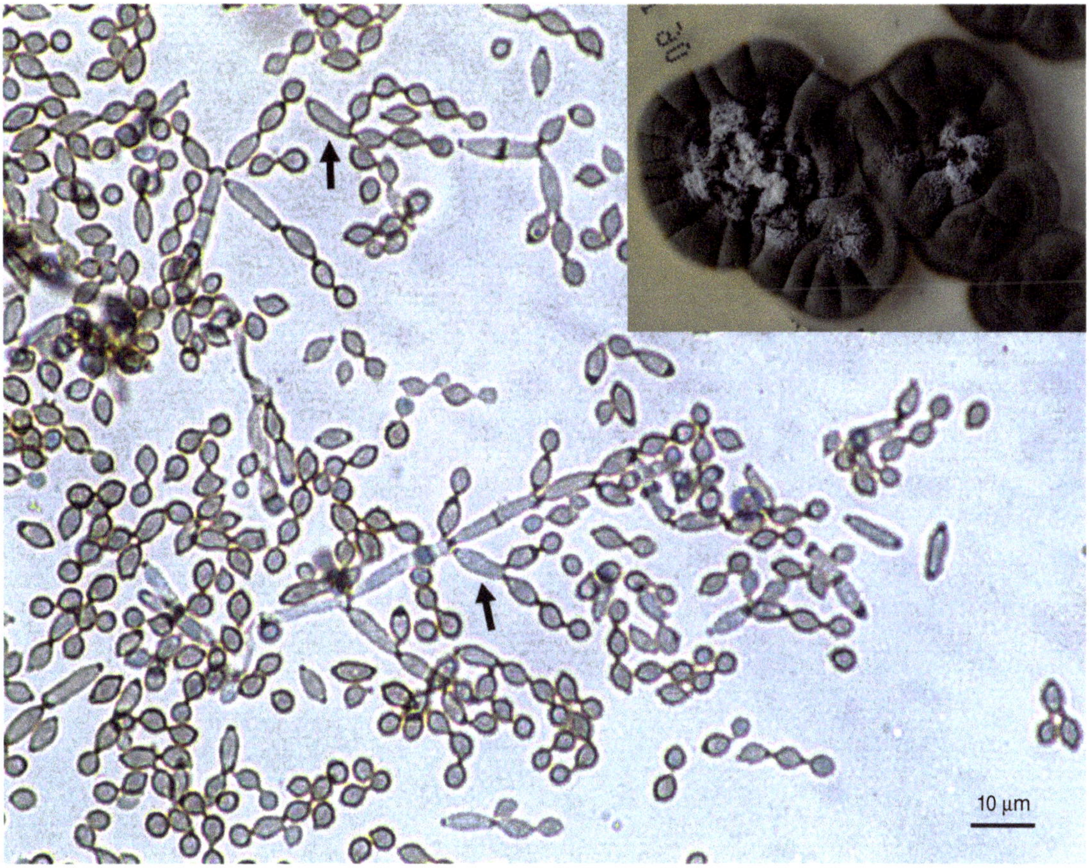

Fig. 19.1. *Cladosporium cladosporioides* culture and conidiophores producing acropetal chains of conidia each with a distinct darkly pigmented hilum. **Note:** The conidia closest to the conidiophore, and where the chains branch, are usually "shield-shaped" (arrows).

Antifungal susceptibility: (Table 19.1).

Table 19.1. *Cladosporium* spp. (Australian national data); MIC µg/mL.

Antifungal	No.	≤0.016	0.03	0.06	0.125	0.25	0.5	1	2	4	8	16	32	≥64
AmB	15							2	4	6	3			
ISAV	9					1	1	3		2	2			
VORI	15						3	1	4	3	4			
POSA	15		1		1	6	4	2			1			
ITRA	15			1	1	2	9	1				1		

References: Ellis (1971,1976); McGinnis (1980); Domsch *et al.* (2007); Crous *et al.* (2007); de Hoog *et al.* (2015).

DOI: 10.1079/9781800622340.0019

20 *Clavispora lusitaniae* Rodr. Mir.

Synonymy: *Candida lusitaniae* van Uden & do Carmo-Sousa.

Clavispora lusitaniae is a known cause of disseminated candidiasis, including septicaemia and pyelonephritis. *C. lusitaniae* was first isolated from the digestive tract of warm-blooded animals and environmental isolations have been made from cornmeal, citrus peel, fruit juices, and milk from cows with mastitis.

RG-2 organism

Culture: Colonies (SDA) white to cream coloured, smooth, glabrous, yeast-like.

Microscopy: Ovoid to ellipsoidal budding blastoconidia, 1.5-6.0 x 2.5-10.0 µm.

India ink preparation: Negative - no capsules present.

Dalmau plate culture: Abundant pseudohyphae with short chains of blastoconidia.

Germ tube formation: Negative.

Physiological tests: (+ Positive, - Negative, v Variable, w Weak, s Slow, n not done).

Fermentation:

Glucose	+	Sucrose	v	Lactose	–
Galactose	v	Maltose	v	Trehalose	v

Growth reactions:

Glucose	+	L-Sorbose	+	*myo*-Inositol	–
Sucrose	+	L-Rhamnose	+	DL-Lactate	v
Raffinose	–	D-Xylose	+	D-Gluconate	v
Melibiose	–	L-Arabinose	–	2-Keto-D-gluconate	+
Galactose	v	D-Arabinose	v	D-Glucosamine	v
Lactose	–	D-Ribose	v	*N*-Acetyl-D-glucosamine	+
Trehalose	+	Glycerol	+	D-Glucuronate	–
Maltose	+	Erythritol	–	Nitrate	–
Melezitose	+	Ribitol	+	Urease	–
Methyl-α-D-glucoside	v	Galactitol	–	0.1% Cycloheximide	–
Soluble starch	–	D-Mannitol	+	Growth at 37°C	+
Cellobiose	v	D-Glucitol	+		

Molecular identification: ITS sequencing recommended.

MALDI-ToF MS: Able to accurately identify this species.

Key features: Germ tube negative yeast and sugar assimilation pattern. **Note:** *Clavispora lusitaniae* may be difficult to distinguish from *Candida tropicalis* using some yeast identification systems.

Antifungal susceptibility: (Table 20.1).

Table 20.1. *Clavispora lusitaniae* (Diekema *et al.*, 2009, Pfaller *et al.*, 2013a,b; and Australian national data); MIC µg/mL.

Antifungal	No.	≤0.016	0.03	0.06	0.125	0.25	0.5	1	2	4	8	16	32	≥64
AmB	273		1	9	34	99	102	20	6		2			
FLU	355			13	38	86	106	52	32	11	4	2	5	6
ISAV	50	17	20	10	3									
VORI	349	291	25	16	4	4	6	2		1				
POSA	326	59	108	111	26	16	2	4						
ITRA	169	3	6	47	61	28	19	4	1					
ANID	263		2	15	65	85	78	16	2					
MICA	247	3	22	60	78	60	16	7	1					
5FC	275		15	77	22	54	67	20	5	3	2	1	5	4

References: Lachance (2011a); de Hoog *et al.* (2015).

DOI: 10.1079/9781800622340.0020

21 *Coccidioides immitis/posadasii* complex

Coccidioides immitis has been separated into two distinct species: *C. immitis* and *C. posadasii* (Fisher *et al.*, 2002). The two species are morphologically identical and can be distinguished only by molecular analysis and different rates of growth in the presence of high salt concentrations (*C. posadasii* grows more slowly). *C. immitis* is geographically limited to California's San Joaquin Valley region and Mexico, whereas *C. posadasii* is found in California, Arizona, Texas, Mexico and South America.

WARNING: RG-3 organism

Cultures of *Coccidioides immitis/posadasii* represent a severe biohazard to laboratory personnel and must be handled with extreme caution in a Class II Biological safety cabinet (BSCII). Cultures can grow rapidly (within 48 hours on blood agar) and are often first encountered in routine microbiology laboratories, rather than specialised and contained mycology laboratories. Therefore, routine bacteriology staff should be made aware of the risks in handling this organism and steps should be taken to limit these risks.

Morphological description: Colonies of *Coccidioides immitis* and *C. posadasii* grown at 25°C may initially be moist and glabrous, but rapidly become suede-like to downy, greyish-white with a tan to brown reverse (Fig. 21.1). However, considerable variation in growth rate and culture morphology has been noted. Microscopy shows typical single-celled, hyaline, rectangular to barrel-shaped, alternate arthroconidia, 2.5-4 x 3-6 μm in size, separated from each other by a disjunctor cell (Fig. 21.1). This arthroconidial state is also seen in the genus *Malbranchea* and is similar to that produced by many non-pathogenic soil fungi, e.g. *Gymnoascus* species.

Histopathology: Tissue morphology shows typical endosporulating spherules. Young spherules have a clear centre with peripheral cytoplasm and a prominent thick wall. Endospores (sporangiospores) are later formed within the spherule by repeated cytoplasmic cleavage. Rupture of the spherule releases endospores into the surrounding tissue where they re-initiate the cycle of spherule development (Fig. 21.2).

Comment: *Coccidioides immitis* and *C. posadasii* are dimorphic fungi, existing in living tissue as spherules and endospores, and in soil or culture in a mycelial form. Despite its dimorphism, the "spherule phase" will not be observed using routine laboratory procedures and inducing this phase should not be attempted. Culture identification by DNA sequencing is preferred to minimise exposure to the infectious propagule.

Key features: Clinical history, tissue pathology, culture identification by ITS sequence analysis.

Molecular identification: ITS sequencing is recommended for differentiation of species (Tintelnot *et al.*, 2007; Binnicker *et al.*, 2010).

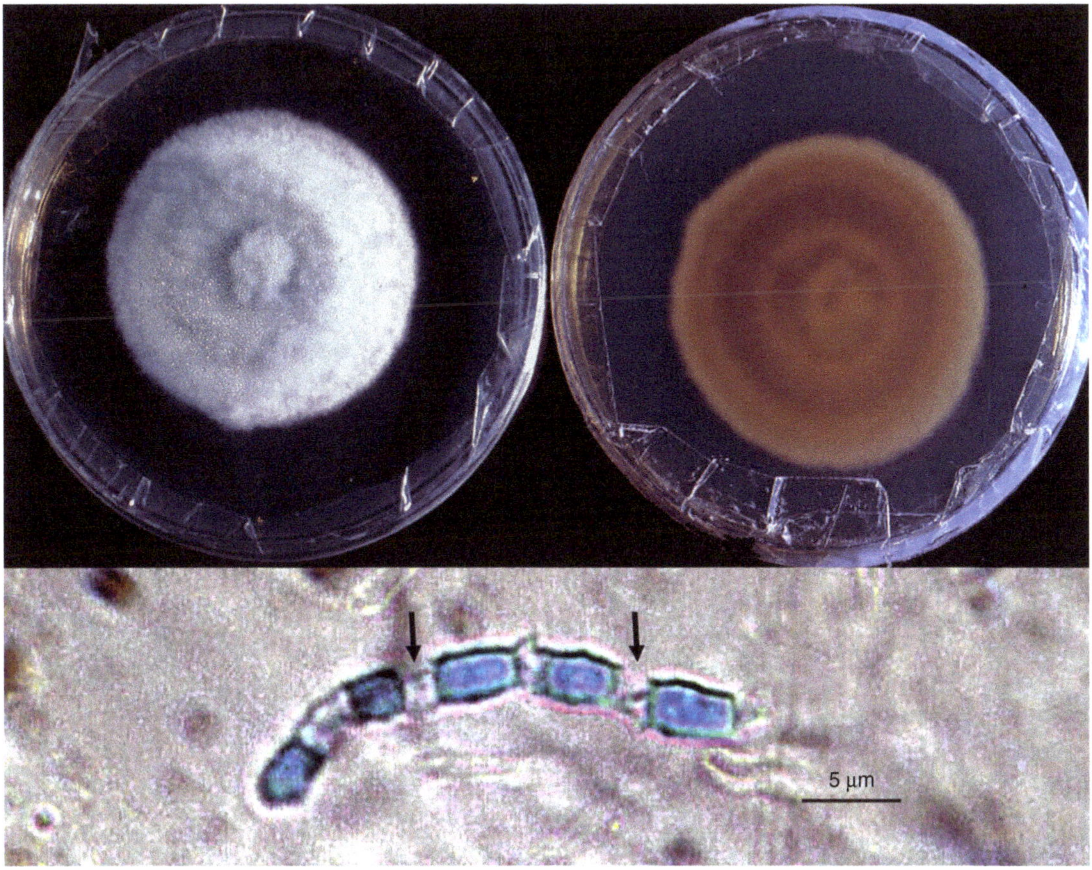

Fig. 21.1. *Coccidioides immitis* culture and arthroconidia with disjunctor cells (arrows).

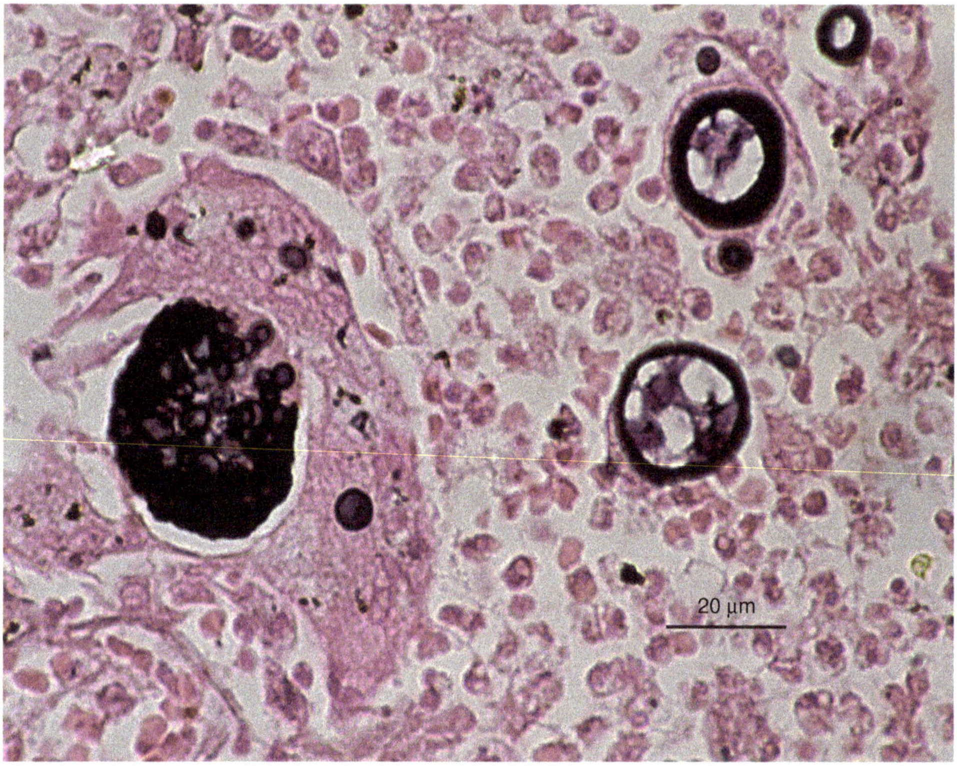

Fig. 21.2. *Coccidioides immitis* showing typical endosporulating spherules in a periodic acid-Schiff stained tissue section.

Antifungal susceptibility: (Table 21.1).

Table 21.1. *Coccidioides immitis* (Ramani and Chaturvedi, 2007); MIC µg/mL. Antifungal susceptibility testing is not recommended.

Antifungal	No.	≤0.016	0.03	0.06	0.125	0.25	0.5	1	2	4	8	16	32	≥64
AmB	45		7	25	8	4	1							
FLU	45							1	4	10	27	3		
VORI	45		13	9	15	8								
POSA	45		25	13	3	4								
ITRA	45		23	22										

References: Sigler and Carmichael (1976); Chandler *et al.* (1980); McGinnis (1980); Stevens (1980); Rippon (1988); de Hoog *et al.* (2015); Kirkland and Fierer (2018); Barker *et al.* (2019).

DOI: 10.1079/9781800622340.0021

22 *Colletotrichum coccodes* (Wallroth) S. Hughes

Over 500 *Colletotrichum* species have been reported. *C. coccodes* is a common soil and plant pathogen widely distributed in Africa, Asia, Australasia, Europe, and the Americas. *Colletotrichum* species are a rare yet important cause of human mycotic keratitis (Hung *et al.*, 2020a).

RG-1 organism

Morphological description: Colonies are usually darkly pigmented with white aerial mycelium, consisting of numerous black sclerotia and light, brown coloured conidial masses, reverse is dark brown (Fig. 22.1). Sclerotia are usually abundant, setose, spherical and are often confluent (Fig. 22.2). Conidia are straight, fusiform, attenuated at the ends, 16-22 x 3-4 µm. Appressoria are common, clavate, brown, 11-16.5 x 6-9.5 µm, variable in shape (Fig. 22.1).

Molecular identification: *GAPDH* gene sequencing is the recommended barcoding gene for species identification. ITS sequencing is not able to distinguish most species (Vieira *et al.*, 2020).

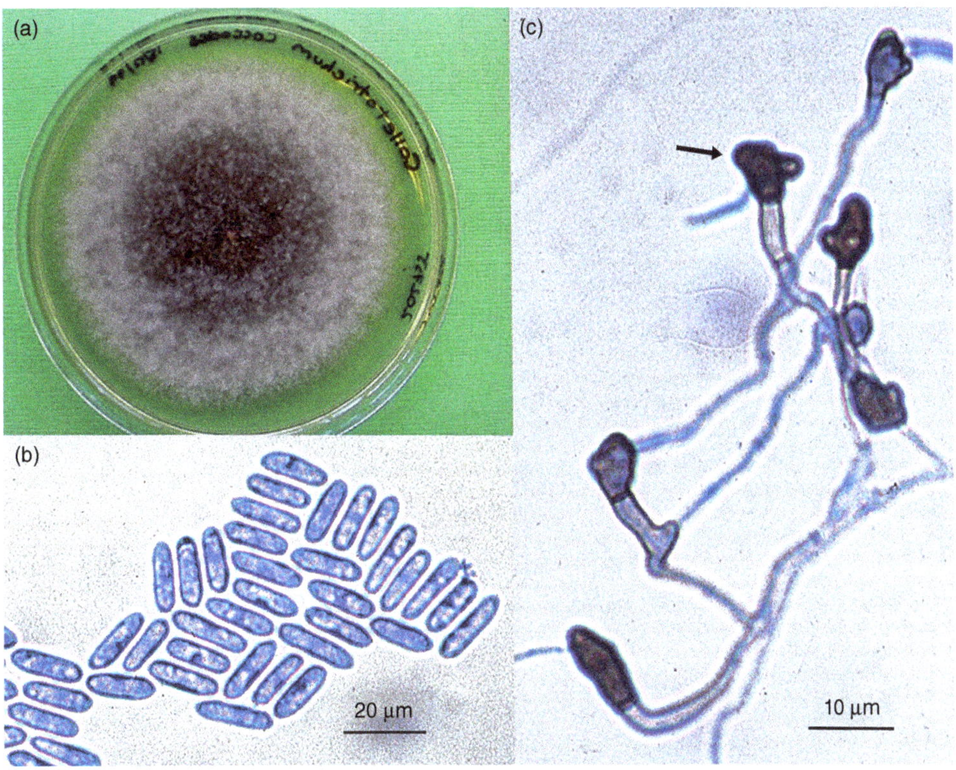

Fig. 22.1. *Colletotrichum coccodes* showing (a) culture, (b) conidia and (c) appressoria (arrow).

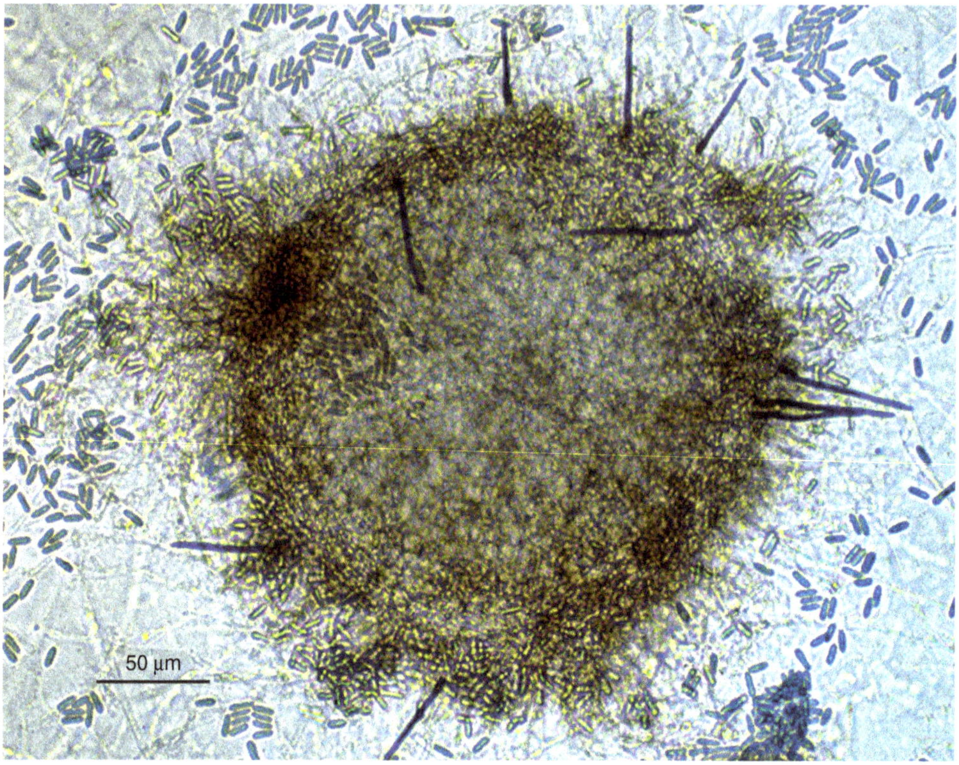

Fig. 22.2. *Colletotrichum coccodes* sclerotia with setae and conidia.

Antifungal susceptibility: (Table 22.1).

Table 22.1. *Colletotrichum* spp. (Australian national data); MIC µg/mL.

Antifungal	No.	≤0.016	0.03	0.06	0.125	0.25	0.5	1	2	4	8	16	32	≥64
AmB	13			1	3	4	3	1	1					
ISAV	4					2	2							
VORI	13				1	1	4	5		2				
POSA	13			1		2	5	4	1					
ITRA	13					1	2	2	2	1		5		

References: McGinnis (1980); Domsch *et al.* (2007); de Hoog *et al.* (2015).

DOI: 10.1079/9781800622340.0022

23 *Conidiobolus coronatus* (Costantin) Batko

Synonym: *Entomophthora coronata* (Costantin) Kevorkian

The species of the genus *Conidiobolus* produce characteristic multinucleate primary and secondary (replicative) conidia on top of unbranched conidiophores. Each subspherical conidium is discharged as a result of the pressure developed within the conidium, and each bears a more or less prominent papilla after discharge (Ellis, 2005a). The genus contains 27 species, however only *C. coronatus, C. incongruus* and *C. lamprauges* have been reported as causative agents of human and animal infection. A morphological identification key for clinical isolates was given by Vilela *et al.* (2010), and an overview of the taxonomy of the genus by Nie *et al.* (2020).

Conidiobolus coronatus is commonly present in soil and decaying leaves. It has a worldwide distribution especially tropical rain forests of Africa. Human infections are usually restricted to the rhinofacial area. However, there are reports of disseminated infection in immunocompromised patients. All human infections have been confined to the tropics (Shaikh *et al.*, 2016).

RG-2 organism

Morphological description: Colonies grow rapidly and are flat, cream coloured, glabrous, becoming radially folded and covered by a fine, powdery, white surface mycelium and conidiophores (Fig. 23.1). The lid of the petri dish soon becomes covered with conidia, which are forcibly discharged by the conidiophores. The colour of the colony may become tan to brown with age. Conidiophores are simple forming solitary, terminal conidia, which are spherical, 10-25 µm in diameter, single-celled and have a prominent papilla (Fig. 23.1). Conidia may also produce hair-like appendages, called villae. Conidia germinate to produce either, (1) single or multiple hyphal tubes that may also become conidiophores which bear secondary conidia, or (2) replicate by producing multiple short conidiophores, each bearing a small secondary conidium.

Molecular identification: ITS sequencing is useful for identification of most clinical isolates.

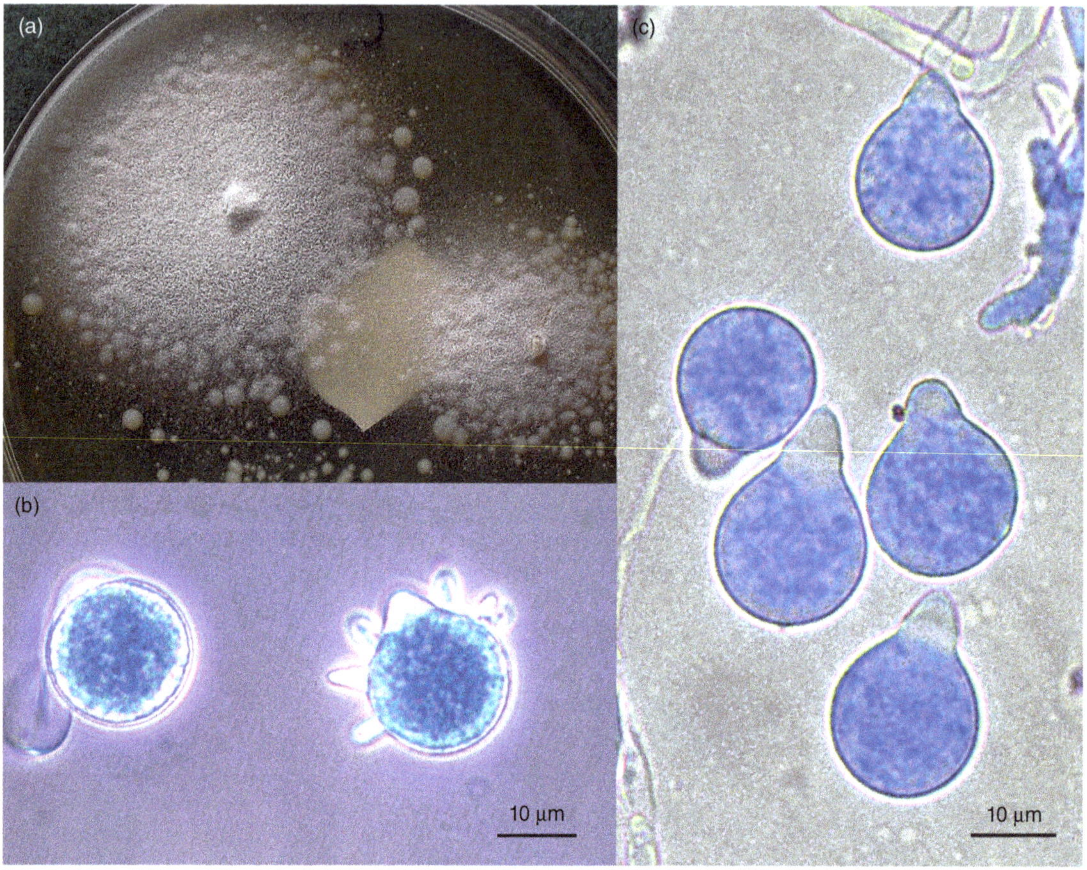

Fig. 23.1. *Conidiobolus coronatus* showing (a) culture with satellite colonies from forcibly discharged conidia, (b) spherical conidia with hair-like appendages (villae) and (c) conidia with prominent papillae.

References: McGinnis (1980); King (1983); Rippon (1988); Kwon-Chung and Bennett (1992); de Hoog *et al.* (2015); Vilela and Mendoza (2018).

DOI: 10.1079/9781800622340.0023

24 *Coniochaeta hoffmannii* (J.F.H. Beyma) Z.U. Khan, Gene & Guarro

Synonymy: *Lecythophora hoffmannii* (J.F.H. Beyma) W. Gams & McGinnis; *Phialophora hoffmannii* (J.F.H. Beyma) Schol-Schwarz; *Margarinomyces hoffmannii* J.F.H. Beyma.

In response to recent changes in the nomenclature for pleomorphic fungi, Khan *et al.* (2013) have transferred all *Lecythophora* species to *Coniochaeta*. The genus *Coniochaeta* is morphologically similar to *Phialemonium*. It also forms adelophialides, but in *Coniochaeta*, these conidiogenous cells show conspicuous collarettes, and the colonies are usually pink salmon to dark brown, although discrete phialides like those of *Acremonium* may be also present (Perdomo *et al.*, 2011b). *Coniochaeta hoffmannii* has been associated with cases of subcutaneous infections, keratitis, sinusitis, peritonitis and canine osteomyelitis. *Coniochaeta mutabilis* has been described to be the causative agent of human peritonitis, endocarditis, endophthalmitis and keratitis (Perdomo *et al.*, 2011b).

RG-1 organism

Morphological description: Colonies are flat, smooth, moist, pink to orange, with regular and sharp margin; reverse pink (Fig. 24.1). Hyphae are narrow, hyaline, producing conidia laterally from small collarettes directly on the hyphae, or from lateral cells which are sometimes arranged in dense groups; lateral cells flask-shaped or nearly cylindrical. Collarettes are unpigmented, about 1.5 µm wide. Conidia are hyaline, smooth, and thin walled, broadly ellipsoidal to cylindrical or allantoid, 3.0-3.5 x 1.5-2.5 µm, produced in slimy heads (Fig. 24.1).

Comment: *Coniochaeta* and *Phialemonium* species are poorly differentiated morphologically and are difficult to identify. They may also be confused with poorly sporulating *Fusarium* or *Acremonium* species.

Molecular identification: Khan *et al.* (2013) used a combined sequence dataset of the ITS region, D1/D2, *ACT* and β-tubulin genes to resolve the unique phylogenetic status of this species.

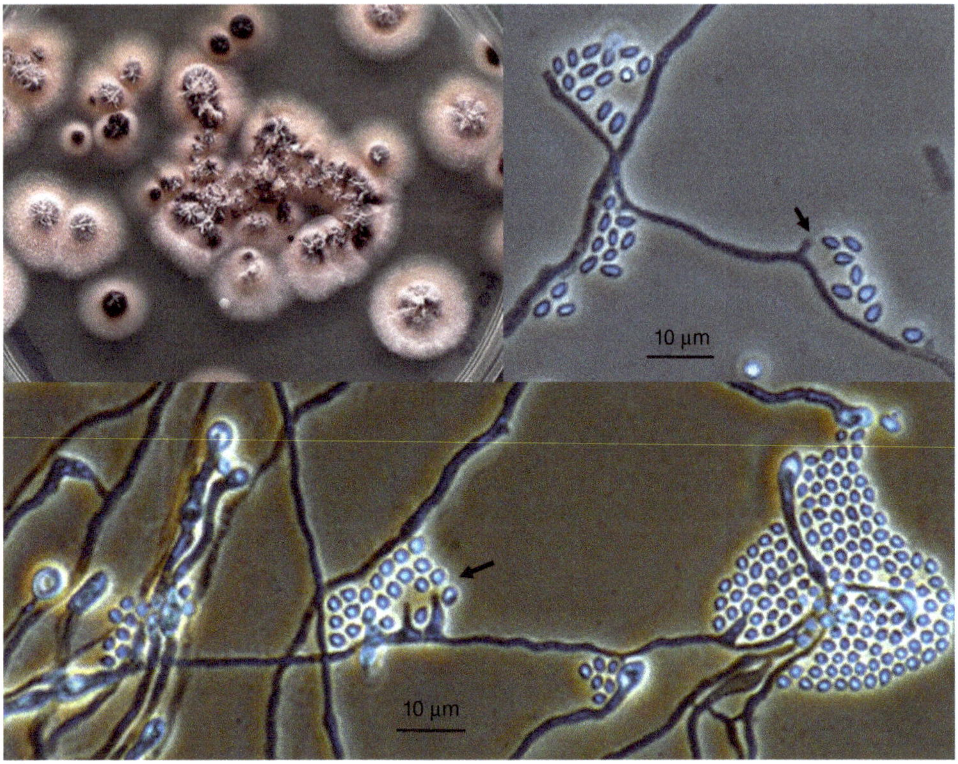

Fig. 24.1. *Coniochaeta hoffmannii* culture and hyphae with small collarettes (arrows) producing conidia.

Antifungal susceptibility: (Table 24.1).

Table 24.1. *Coniochaeta hoffmannii* very limited data (Australian national data); MIC µg/mL.

Antifungal	No.	≤0.016	0.03	0.06	0.125	0.25	0.5	1	2	4	8	16	32	≥64
AmB	3						1	1	1					
ISAV	2							1		1				
VORI	3						1	1		1				
POSA	3					1	1		1					
ITRA	3					1	1	1						

References: McGinnis (1978b); Perdomo *et al.* (2011b); Khan *et al.* (2013); de Hoog *et al.* (2015).

DOI: 10.1079/9781800622340.0024

25 *Cryptococcus* Vuill.

The genus *Cryptococcus* is characterised by globose to elongate yeast-like cells or blastoconidia that reproduce by narrow-necked budding. Pseudohyphae are absent or rudimentary. Most species are encapsulated, although the extent of capsule formation depends on the medium (Fig. 25.1). Under certain conditions of growth, the capsule may contain starch-like compounds, which are released into the medium by many strains. Within tissue sections, mucicarmine or Alcian blue stains the capsule of *Cryptococcus* species to distinguish it from other yeasts with similar morphologies. Historically the *Cryptococcus* genus was large, but many species have been reclassified into other genera including *Naganishia albida* (*Cryptococcus albidus*) and *Papiliotrema laurentii* (*Cryptococcus laurentii*).

On solid media the cultures are generally mucoid or slimy in appearance; red, orange or yellow carotenoid pigments may be produced, but young colonies of most species are usually non-pigmented, and cream in colour (Fig. 25.2). All *Cryptococcus* species produce urease and are non-fermentative. Nitrate may be assimilated or not; inositol assimilated. The genus *Cryptococcus* differs from the genus *Rhodotorula* in its inositol assimilation (Kurtzman *et al.*, 2011).

Cryptococcosis is a chronic, subacute to acute pulmonary, systemic or meningitic disease, initiated by the inhalation of infectious propagules (basidiospores and/or desiccated yeast cells) from the environment. Primary pulmonary infections have no diagnostic symptoms and are usually subclinical. On dissemination, the fungus usually shows a predilection for the central nervous system, however skin, bones and other visceral organs may also become involved. *Cryptococcus neoformans* and *C. gattii* are the principle pathogenic species. *Naganishia albida* and *Papiliotrema laurentii* have on occasion also been implicated in human infection (Liu *et al.*, 2015).

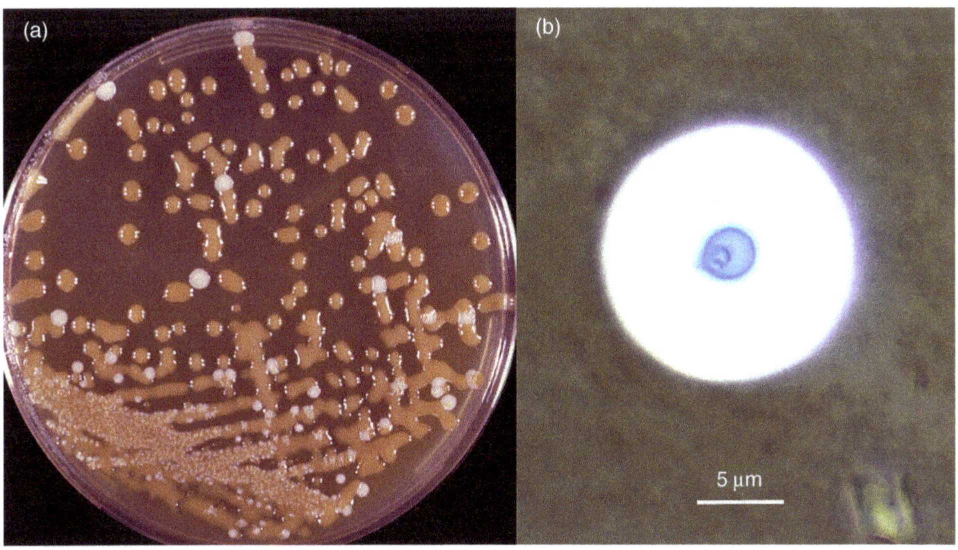

Fig. 25.1. (a) Bird seed agar plate showing brown colonies of *Cryptococcus neoformans* and white colonies of *Candida albicans*, and (b) India ink preparation of a central spinal fluid specimen showing the characteristic wide gelatinous capsule formed by *C. neoformans*.

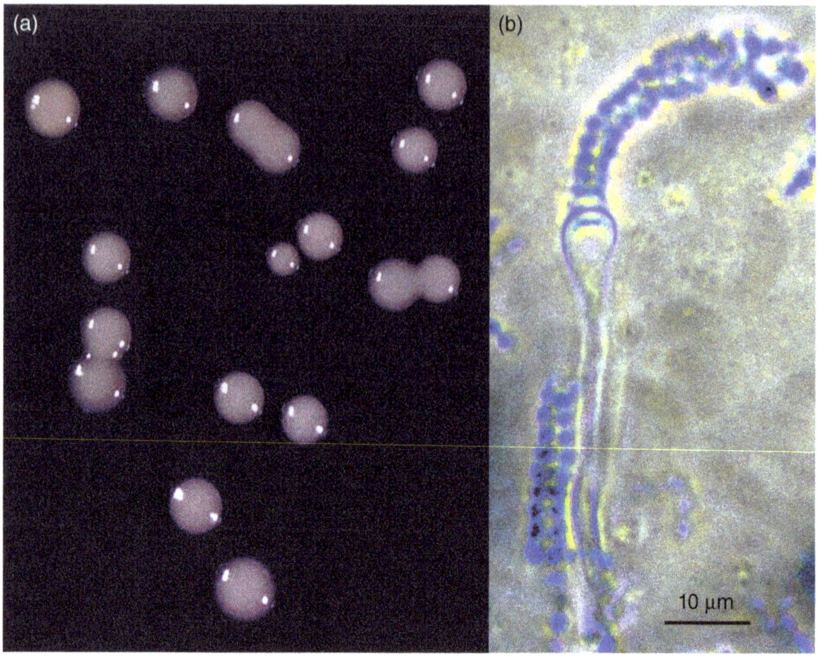

Fig. 25.2. *Cryptococcus neoformans* culture and a basidium with basidiospores.

In a phylogenetic analysis, Hagen *et al.* (2015) proposed to recognise *Cryptococcus neoformans* (formerly *C. neoformans* var. *grubii*) and *Cryptococcus deneoformans* (formerly *C. neoformans* var. *neoformans*) as separate species. In addition, these authors also describe five genotypic clades of *Cryptococcus gattii* as new species; *C. gattii, C. bacillisporus, C. deuterogattii, C. tetragattii* and *C. decagattii* (Hagen *et al.*, 2015). This proposal has not been universally accepted (Kwon-Chung *et al.*, 2017). It has been recommended to continue to use "*Cryptococcus neoformans* species complex" and "*C. gattii* species complex" for practical identification purposes, rather than creating more species (Kwon-Chung *et al.*, 2017).

Molecular identification: Requires ITS and/or D1/D2 sequencing, particularly for differentiating molecular types (proposed species) within the species complex.

MALDI-ToF MS: Can provide reliable species and subspecies level identification, but its accuracy is dependent on database quality (Arendrup *et al.*, 2014).

References: Barnett *et al.* (1983); Rippon (1988); Kwon-Chung and Bennett (1992); Casadevall and Perfect (1998); Kwon-Chung (2011); McTaggart *et al.* (2011); de Hoog *et al.* (2015); Lockhart *et al.* (2016); Sugita *et al.* (2017); Morales-Lopez and Garcia-Effron (2021).

25.1. *Cryptococcus gattii* complex (Vanbreus. & Takashio) Kwon-Chung & Boekhout

Synonymy: *Filobasidiella bacillispora* Kwon-Chung; *Cryptococcus neoformans* var. *gattii* Vanbreus & Takashio.

Cryptococcus gattii complex has two serotypes (B and C) and was reclassified as a separate species from *C. neoformans* in 2002 (Kwon-Chung *et al.*, 2002). *C. gattii* generally has a more restricted geographical distribution than *C. neoformans*, causing human disease in climates ranging from temperate to tropical Australia, Papua New Guinea, parts of Africa, India, Southeast Asia, Mexico, Brazil, Paraguay and Southern

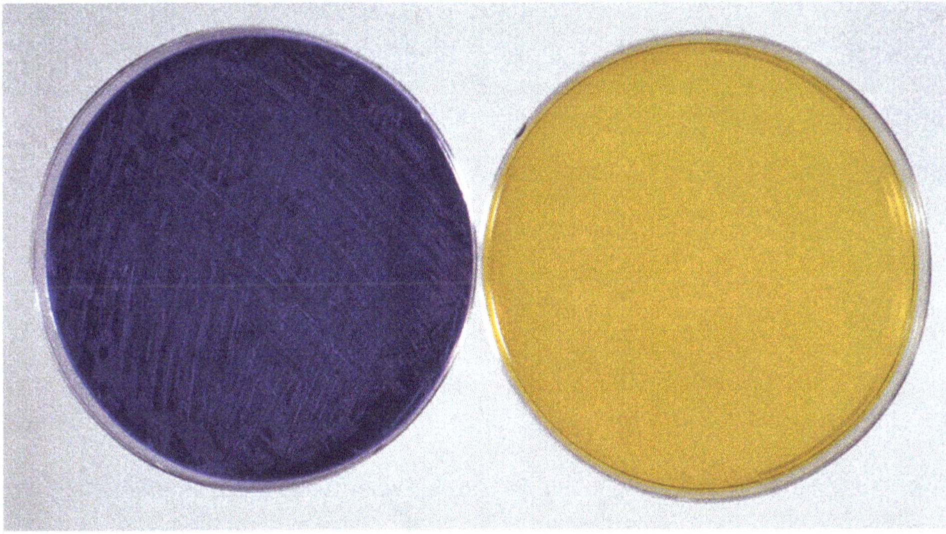

Fig. 25.3. *Cryptococcus gattii* complex turns CGB agar blue within 2-5 days; *C. neoformans* complex does not grow on this medium.

California, although recent infections have also been reported from Vancouver Island, Canada and in the Pacific Northwest, USA (Pfaller and Diekema, 2010; Espinel-Ingroff and Kidd, 2015). *C. gattii* has a specific ecological association with numerous species of *Eucalyptus* trees, although the Canadian isolates are associated with a range of native non-*Eucalyptus* species (Kidd *et al.*, 2007a,b). Historically considered a pathogen in immunocompetent hosts, a recent review in Australia noted an increase in *C. gattii* infections in HIV-negative immunocompromised patients (Chen *et al.*, 2012). Cryptococcosis caused by *C. gattii* is often associated with large mass lesions (cryptococcomas) in the lung and/or brain (Sorrell, 2001).

Canavanine glycine bromothymol blue (CGB) agar (Kwon-Chung *et al.*, 1982) is the media of choice to differentiate *C. gattii* complex from *C. neoformans* complex (Fig. 25.3). This simple biotype test is based on the ability of *C. gattii* isolates to grow in the presence of L-canavanine and to assimilate glycine as a sole carbon source.

RG-2 organism

Culture: Colonies (SDA) cream coloured, smooth, mucoid, yeast-like colonies. Microscopy: Globose to ovoid budding yeast-like cells 3.0-7.0 x 3.3- 7.9 µm.

India ink preparation: Positive - distinct, wide gelatinous capsules are present.

Dalmau plate culture: Budding yeast cells only. No pseudohyphae present.

Bird seed agar: Colonies turn dark brown in colour as colonies selectively absorb a brown pigment from this media. Colonies are often more mucoid when compared to *C. neoformans* (Staib *et al.*, 1989).

Canavanine-glycine-bromothymol blue (CGB) agar: Turns blue within 2-5 days.

Germ tube formation: Negative.

Physiological tests: (+ Positive, - Negative, v Variable, w Weak, s Slow, n not done).

Fermentation: Absent.

Growth reactions:

Glucose	+	L-Sorbose	–	*myo*-Inositol	+
Sucrose	+	L-Rhamnose	+	DL-Lactate	–
Raffinose	+,w	D-Xylose	+	D-Gluconate	+
Melibiose	–	L-Arabinose	+,w	2-Keto-D-gluconate	v
Galactose	+	D-Arabinose	+	D-Glucosamine	v
Lactose	–	D-Ribose	v	*N*-Acetyl-D-glucosamine	v
Trehalose	+	Glycerol	–	D-Glucuronate	+
Maltose	+	Erythritol	–	Nitrate	–
Melezitose	+	Ribitol	v	Urease	+
Methyl-½-D-glucoside	+	Galactitol	+	0.1% Cycloheximide	–
Soluble starch	+	D-Mannitol	+	Growth at 37°C	+
Cellobiose	+,w	D-Glucitol	+		

Key features: Encapsulated yeast; absence of pseudohyphae; growth at 37°C; positive hydrolysis of urea; negative fermentation of sugars and positive assimilation of glucose, maltose, sucrose, galactose, trehalose, raffinose, inositol, cellobiose, rhamnose, arabinose, melezitose and xylose, and negative assimilation of nitrate, lactose, melibiose, erythritol and soluble starch; growth on bird seed (*Guizotia abyssinica* seed) or caffeic acid agar colonies turn a dark brown colour; growth on CGB agar turning it blue within 2-5 days.

Antifungal susceptibility: (Table 25.1).

Table 25.1. *Cryptococcus gattii* complex (Australian national data with additional ISAV data from Espinel-Ingroff *et al.* (2015b); MIC µg/mL. **Note:** All *Cryptococcus* species are intrinsically resistant to echinocandins.

Antifungal	No.	≤0.016	0.03	0.06	0.125	0.25	0.5	1	2	4	8	16	32	≥64
AmB	230	1	3	18	52	57	41	55	3					
FLU	230					1	4	16	42	73	58	31	4	1
ISAV	430	106	109	93	90	25	7							
VORI	230	21	43	72	41	26	5							
POSA	168	20	13	32	57	41	9	1	1					
ITRA	230	11	22	57	92	45	3							
5FC	230				3	12	53	93	54	11	4			

25.2. *Cryptococcus neoformans* complex (Sanfelice) Vuillemin

Synonymy: *Filobasidiella neoformans* Kwon-Chung; *Cryptococcus neoformans* var. *neoformans*.

This species has two varieties: *C. neoformans* var. *grubii* (serotype A) and *C. neoformans* var. *neoformans* (serotype D).

***C. neoformans* var. *grubii*:** Has a worldwide distribution, causing 95% of all *C. neoformans* infections. It has been isolated from various sources in nature and is noted for its association with accumulations of avian guano, especially with pigeon excreta. The fungus has also been isolated from the dung of caged birds including canaries, parrots and budgerigars. Other environmental isolations of *C. neoformans* var. *grubii* include rotting vegetables, fruits and fruit juices, wood, dairy products and soil.

***C. neoformans* var. *neoformans*:** Has a more restricted distribution with infections being more prevalent in Europe, including France, Italy and Denmark, where it accounts for 30% of isolates. Moreover, *C. neoformans* var. *neoformans* infections are more strongly correlated with older patients, the skin, and the use of corticosteroids (Franzot *et al.*, 1999).

RG-2 organism

Culture: Colonies (SDA) cream coloured, smooth, mucoid, yeast-like colonies.

Microscopy: Globose to ovoid budding yeast-like cells 3.0-7.0 x 3.3-7.9 µm.

India ink preparation: Positive - distinct, wide gelatinous capsules are present.

Dalmau plate culture: Budding yeast cells only. No pseudohyphae present.

Bird seed agar: Colonies turn dark brown in colour as colonies selectively absorb a brown pigment from this media (Staib *et al.,* 1989).

Canavanine-glycine-bromothymol blue (CGB) agar: Does not grow on this medium and leaves the colour unchanged.

Creatinine dextrose bromothymol blue thymine (CDBT) agar: *C. neoformans* var. *neoformans* grows as bright red colonies, turning the medium a bright orange after 5 days. No colour change is observed for *C. neoformans* var. *grubii* (Irokanulo *et al.*, 1994).

Germ tube formation: Negative.

Physiological tests: (+ Positive, – Negative, v Variable, w Weak, s Slow, n not done).

Fermentation: Absent.

Growth reactions:

Glucose	+	L-Sorbose	–	*myo*-Inositol	+
Sucrose	+	L-Rhamnose	+	DL-Lactate	–
Raffinose	+,w	D-Xylose	+	D-Gluconate	+
Melibiose	–	L-Arabinose	+,w	2-Keto-D-gluconate	v
Galactose	+	D-Arabinose	+	D-Glucosamine	v
Lactose	–	D-Ribose	v	*N*-Acetyl-D-glucosamine	v
Trehalose	+	Glycerol	–	D-Glucuronate	+
Maltose	+	Erythritol	–	Nitrate	–
Melezitose	+	Ribitol	v	Urease	+
Methyl-½-D-glucoside	+	Galactitol	+	0.1% Cycloheximide	–
Soluble starch	+	D-Mannitol	+	Growth at 37°C	+
Cellobiose	+,w	D-Glucitol	+		

Key features: Encapsulated yeast; absence of pseudohyphae; growth at 37°C; positive hydrolysis of urea; negative fermentation of sugars and positive assimilation of glucose, maltose, sucrose, galactose, trehalose, raffinose, inositol, cellobiose, rhamnose, arabinose, melezitose and xylose, and negative assimilation of nitrate, lactose, melibiose, erythritol and soluble starch; growth on bird seed (*Guizotia abyssinica* seed) or caffeic acid agar colonies turn a dark brown colour; does not grow on CGB agar (no colour change).

Antifungal susceptibility: (Table 25.2).

Table 25.2. *Cryptococcus neoformans* complex (Australian national data with additional ISAV data from Espinel-Ingroff *et al.*, 2015b); MIC µg/mL. **Note:** All *Cryptococcus* species are intrinsically resistant to echinocandins.

Antifungal	No.	≤0.016	0.03	0.06	0.125	0.25	0.5	1	2	4	8	16	32	≥64
AmB	397	2	5	17	85	59	111	109	9					
FLU	398				1	1	17	44	78	102	110	34	9	2
ISAV	494	178	179	106	22	7	2							
VORI	359	85	90	103	59	21	1							
POSA	314	17	46	70	104	66	11							
ITRA	398	18	59	111	157	46	6	1						
5FC	398			1	1	7	31	72	98	120	58	7	1	2

DOI: 10.1079/9781800622340.0025

26 *Cunninghamella bertholletiae* Stadel

Synonymy: *Cunninghamella elegans; Cunninghamella echinulate* var. *elegans.*

The genus *Cunninghamella* is characterised by white to grey, rapidly growing colonies, producing erect, straight, branching sporangiophores. These sporangiophores end in globose or pyriform-shaped vesicles from which several one-celled, globose to ovoid, echinulate or smooth-walled sporangiola develop on swollen denticles. Chlamydospores and zygospores may also be present.

Cunninghamella species are mainly soil fungi of the Mediterranean and subtropical zones; they are only rarely isolated in temperate regions. The genus now contains around 15 species with *C. bertholletiae* responsible for almost all infections in humans and animals, often in association with trauma and immunosuppression.

RG-2 organism

Morphological description: Colonies are very fast growing, white at first, but becoming rather dark grey and powdery with sporangiola development. Sporangiophores to 20 µm wide, straight, with verticillate or solitary branches. Vesicles subglobose to pyriform, the terminal ones up to 40 µm and the lateral ones 10-30 µm in diameter. Sporangiola are globose (7-11 µm diameter), or ellipsoidal (9-13 x 6-10 µm), verrucose or short-echinulate, hyaline singly but brownish in mass (Fig. 26.1). Temperature: optimum 25 to 30°C: maximum up to 50°C.

Molecular identification: ITS sequencing is recommended (Yu *et al.*, 2015).

Key features: Mucorales, clinical isolates grow at 40°C, one-celled, globose to ovoid, echinulate sporangiola borne on swollen terminal or lateral globose to clavate fertile vesicles.

Antifungal susceptibility: (Table 26.1).

Table 26.1. *Cunninghamella bertholletiae* (Espinel-Ingroff *et al.*, 2015a; and Australian national data); MIC µg/mL.

Antifungal	No.	≤0.016	0.03	0.06	0.125	0.25	0.5	1	2	4	8	16	32	≥64
AmB	40					1	1	5	20	11	2			
ISAV	2								1	1				
VORI	8									4	2	2		
POSA	37					8	19	9	1					
ITRA	33				1	7	6	10	9					

References: McGinnis (1980); Weitzman and Crist (1980); Lunn and Shipton (1983); Weitzman (1984); Zheng and Chen (2001); Ellis (2005b); Domsch *et al.* (2007); de Hoog *et al.* (2015); Bragulat *et al.* (2017); Walther *et al.* (2019).

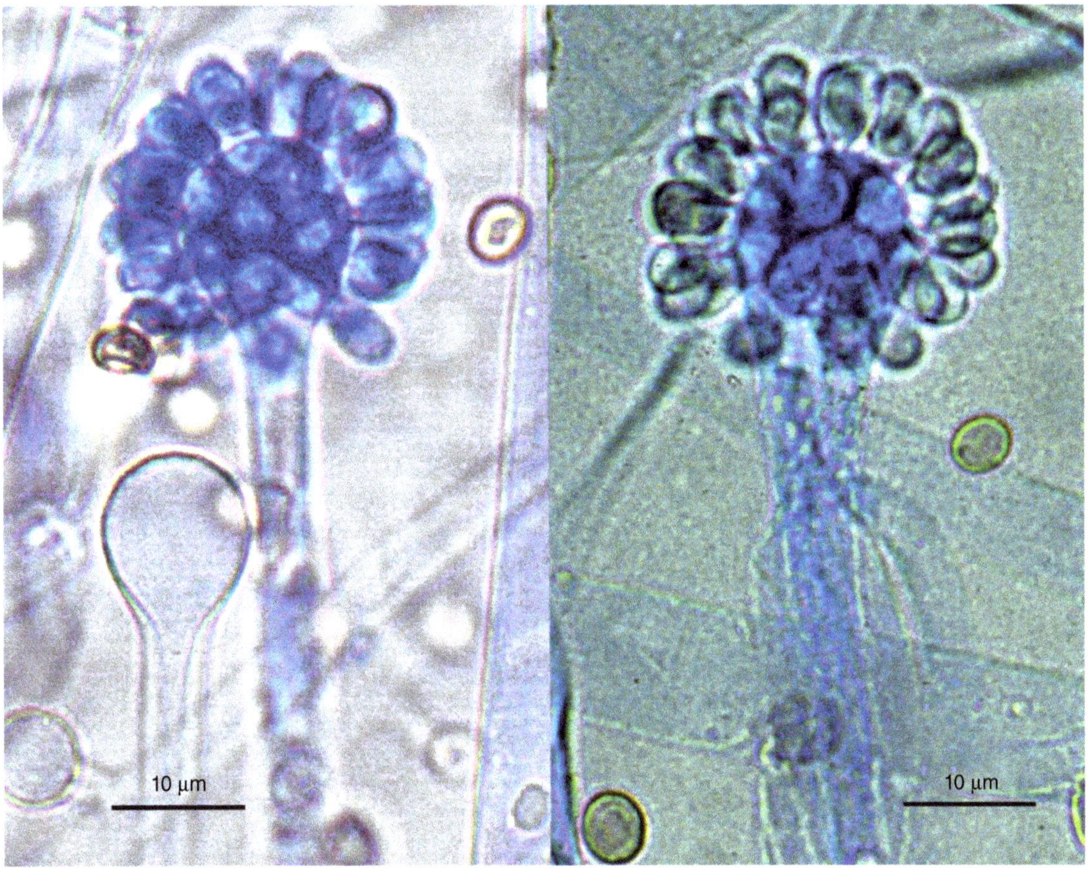

Fig. 26.1. *Cunninghamella bertholletiae* showing simple sporangiophores forming a swollen, terminal vesicle around which single-celled, globose to ovoid sporangiola develop on swollen denticles.

DOI: 10.1079/9781800622340.0026

27 *Curvularia* Boedijn

The genus *Curvularia* contains about 80 species, which are mostly soil or plant pathogens. Recent studies have shown that morphological identification does not correlate with molecular identification (Yanagihara *et al.*, 2010; Manamgoda *et al.*, 2012; da Cunha *et al.*, 2013). A phylogenetic analysis of the genera *Bipolaris* and *Curvularia* has resulted in a re-alignment of several species. In particular, clinical isolates previously identified as *Bipolaris* species, notably *B. australiensis, B. hawaiiensis* and *B. spicifera* have been transferred to *Curvularia* species (Manamgoda *et al.*, 2012).

Previously *Curvularia lunata* was the most frequently reported clinical species, other species, such as *C. americana, C. brachyspora, C. chlamydospora, C. clavata, C. hominis, C. inaequalis, C. muehlenbeckiae, C. pallescens, C. pseudolunata, C. senegalensis* and *C. verruculosa* have now also been reported in clinical cases (Revankar and Sutton, 2010; da Cunha *et al.*, 2013; Madrid *et al.*, 2014).

RG-1 organisms

Morphological genus description: Colonies are fast growing, suede-like to downy, brown to blackish brown with a black reverse. Conidiophores erect, straight to flexuous, septate, often geniculate (producing conidia in sympodial succession) sometimes nodulose. Conidia are ellipsoidal, often curved or lunate, rounded at the ends or sometimes tapering slightly towards the base, pale brown, medium reddish- brown to dark brown, three to ten (usually three to five) septa, conidial wall smooth to verrucose. Hila protuberant in some species.

Key features: Darkly pigmented mould producing sympodial, pale brown, cylindrical or slightly curved phragmoconidia.

Molecular identification: *GAPDH* gene (Manamgoda *et al.*, 2012) and/or ITS sequencing (da Cunha *et al.*, 2013; Irinyi *et al.*, 2015).

References: Ellis (1971); McGinnis (1980); Rippon (1988); Domsch *et al.* (2007); de Hoog *et al.* (2015).

27.1. *Curvularia australiensis* (Bugnic ex M.B. Ellis) Manamgoda *et al.*

Synonymy: *Bipolaris australiensis* (Bugnic. ex M.B. Ellis) Tsuda & Ueyama.

RG-1 organism

Morphological description: Colonies are grey to blackish-brown and suede-like in surface texture. Conidiophores are brown, flexuose or geniculate, smooth-walled, up to 150 µm long, mostly 3-7 µm wide. Conidia are straight, rounded at the ends, pale brown to mid reddish-brown, mostly three (rarely four to five) distoseptate, ellipsoidal or oblong, 14-40 x 6-11 µm, smooth-walled to finely roughened (Fig. 27.1).

Fig. 27.1. *Curvularia australiensis* showing sympodial development of pale brown, ellipsoidal, distoseptate conidia on a geniculate or zig-zag rachis (arrow).

Antifungal susceptibility: (Table 27.1).

Table 27.1. *Curvularia australiensis* (Australian national data); MIC µg/mL.

Antifungal	No.	≤0.016	0.03	0.06	0.125	0.25	0.5	1	2	4	8	16	32	≥64
AmB	27		1		6	3	7	10						
ISAV	4								1	1	2			
VORI	27				3	6	7	4	7					
POSA	25	1	4	5	6	5	2		2					
ITRA	27		1	4	3	13	5	1						

27.2. *Curvularia hawaiiensis* (Bugnic. ex M.B. Ellis) Manamgoda *et al.*

Synonymy: *Bipolaris hawaiiensis* (Bugnic. ex M.B. Ellis) J.Y. Uchida & Aragaki.

RG-1 organism

Morphological description: Colonies powdery to hairy, black. Conidiophores erect, unbranched, septate, apically flexuose with flat conidial scars on the edges, up to 80 µm long. Conidia smooth- and rather thick-walled, brown, with three to seven, mostly five distosepta, cylindrical to cigar-shaped, 18-35 x 6-9 µm.

Antifungal susceptibility: (Table 27.2).

Table 27.2. *Curvularia hawaiiensis* (Australian national data); MIC µg/mL.

Antifungal	No.	≤0.016	0.03	0.06	0.125	0.25	0.5	1	2	4	8	16	32	≥64
AmB	8				1	4	1		2					
ISAV	2							2						
VORI	8				2	3	2		1					
POSA	6	1	1	1	1	1	1							
ITRA	8			1	3	3	1							

27.3. *Curvularia lunata* (Wakker) Boedijn

RG-1 organism

Morphological description: Colonies black, downy. Conidiophores erect, unbranched, septate, flexuose in the apical part, with flat, dark brown scars. Conidia smooth-walled, olivaceous-brown, end cells somewhat paler; conidia obovoidal to broadly clavate, curved at the subterminal cell, 21-31 x 8.5-12.0 µm, three septate, the subterminal cell swollen and distinctly larger than the remaining cells (Fig. 27.2).

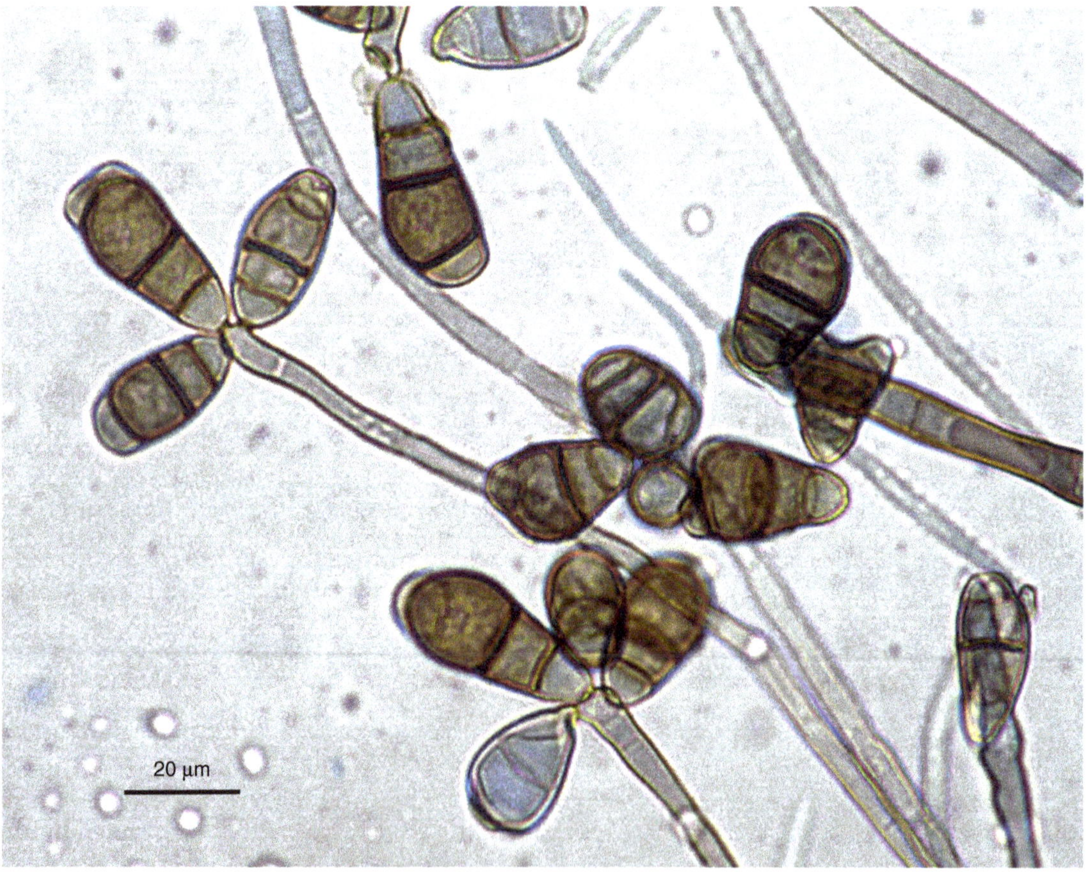

Fig. 27.2. *Curvularia lunata* conidiophores and conidia.

Antifungal susceptibility: (Table 27.3).

Table 27.3. *Curvularia lunata* (Australian national data); MIC µg/mL.

Antifungal	No.	≤0.016	0.03	0.06	0.125	0.25	0.5	1	2	4	8	16	32	≥64
AmB	34			1		5	7	10	8	3				
ISAV	4							2		1	1			
VORI	34					8	7	9	4	6				
POSA	33		2	1	14	10	2	4						
ITRA	34				7	16	3	1	1		6			

27.4. *Curvularia spicifera* (Bainier) Boedijn

Synonymy: *Bipolaris spicifera* Bainier.

RG-1 organism

Morphological description: Colonies appearing glassy with sooty powder of conidia, or hairy if sporulation is poor. Conidiophores erect, unbranched, septate, up to 250 µm long and 4-8 µm wide, regularly zig-zagged in the apical part, with flat, dark brown scars on the edges. Conidia brown, cylindrical with rounded ends, medium brown except for narrow subhyaline spots at the extremities, 20-40 x 9-14 µm, with three distosepta.

Antifungal susceptibility: (Table 27.4).

Table 27.4. *Curvularia spicifera* (Australian national data); MIC µg/mL.

Antifungal	No.	≤0.016	0.03	0.06	0.125	0.25	0.5	1	2	4	8	16	32	≥64
AmB	3					1	1	1						
ISAV	1									1				
VORI	3							1	2					
POSA	3				1	1	1							
ITRA	3					1	2							

DOI: 10.1079/9781800622340.0027

28 *Cutaneotrichosporon* Liu *et al.*

The genus *Cutaneotrichosporon* was proposed based on DNA sequence analysis of seven genes; 18*S*, D1/D2 domain of the 18*S*, ITS, *RBP1* and *RBP2*, *TEF-1*, and cytochrome B (Liu *et al.*, 2015). Fifteen species were initially assigned to the new genus, however four species reassigned from the genus *Trichosporon*, namely *Cutaneotrichosporon cutaneum*, *C. dermatis, C. jirovecii and C. mucoides* are the most common clinical isolates.

Cutaneotrichosporon species are urease-positive, non-encapsulated basidiomycetous yeasts characterised by the development of hyaline, septate hyphae that fragment into oval or rectangular arthroconidia. Some blastoconidia may be seen. The colonies are usually raised and have a waxy appearance, which develop radial furrows and irregular folds.

Cutaneotrichosporon species are often isolated from human and animal sources as well as clinical specimens, some of them being associated with opportunistic infections. It is also worth mentioning that their ability to grow at 37°C is variable. Importantly, all species are resistant to echinocandins and other classes of antifungal agents including amphotericin B and some azoles (Arendrup *et al.*, 2014; do Espirito Santo *et al.*, 2019; Borman *et al.*, 2020).

Note: Identification to genus level is mandatory for clinical management and should be performed and provided in a timely manner. Species identification remains difficult and requires molecular analysis or MALDI-ToF MS (with an extensive database) (Arendrup *et al.*, 2014).

Molecular identification: ITS and D1/D2 sequencing is recommended (Sugita *et al.*, 1999; Arendrup *et al.*, 2014).

MALDI-ToF MS: Appears to be a promising identification tool (with an extensive or supplemented database) (Kolecka *et al.*, 2013; de Almeida *et al.*, 2017).

References: Gueho *et al.* (1992); Rodriguez-Tudela *et al.* (2005); Chagas-Neto *et al.* (2008); Guo *et al.* (2011); Sugita (2011); Xiao *et al.* (2013); Liu *et al.* (2015); de Hoog *et al.* (2020).

28.1. *Cutaneotrichosporon cutaneum* (Beurm. *et al.*) Liu *et al.*

Synonymy: *Trichosporon cutaneum* (Beurm. *et al.*) Ota.

RG-2 organism

Morphological description: Colonies are cream coloured, cerebriform, glabrous, with radial furrows and irregular folds. Budding cells abundant in primary cultures; hyphae developing in older cultures. Arthroconidia are cylindrical to ellipsoidal. Appressoria absent. This species assimilates melibiose and is not tolerant to 0.1% (variable tolerance to 0.01%) cycloheximide. No growth at 37°C. Uncommon species usually associated with superficial infections.

Physiological tests: (+ Positive, - Negative, v Variable, w Weak, s Slow).

Fermentation: Absent.

Growth reactions:

Glucose	+	Cellobiose	+	D-Mannitol	+
Sucrose	+	L-Sorbose	v	D-Glucitol	+
Raffinose	+	L-Rhamnose	+	*myo*-Inositol	+
Melibiose	+	D-Xylose	+	DL-Lactate	+
Galactose	+	L-Arabinose	+	D-Gluconate	+
Lactose	+	D-Arabinose	v	2-Keto-D-gluconate	+
Trehalose	+	D-Ribose	+	D-Glucosamine	v
Maltose	+	Glycerol	+	*N*-Acetyl-D-glucosamine	+
Melezitose	+	Erythritol	+	D-Glucuronate	+
Methyl-α-D-glucoside	+	Ribitol	+	0.1% Cycloheximide	v
Soluble starch	+	Galactitol	–	Growth at 37°C	–

28.2. *Cutaneotrichosporon dermatis* (Sugita *et al.*) Liu *et al.*

Synonymy: *Trichosporon dermatis* Sugita *et al.*

RG-1 organism

Morphological description: Colonies (SDA) are yellowish-white, semi-shiny, butyrous and wrinkled, and have a fringed margin. Yeast cells are ovoid, ellipsoidal or elongate, and occur singly or in pairs. Hyphae disarticulate into cylindrical arthroconidia. This species cannot be distinguished reliably from *C. mucoides* using standard carbon assimilation characteristics. Growth at 37°C. Uncommon species usually associated with human skin and hypersensitivity pneumonitis.

Physiological tests: (+ Positive, - Negative, v Variable, w Weak, s Slow, n not done).

Fermentation: Absent.

Growth reactions:

Glucose	+	Cellobiose	+	D-Mannitol	+
Sucrose	+	L-Sorbose	+	D-Glucitol	+
Raffinose	+	L-Rhamnose	+	*myo*-Inositol	+
Melibiose	+	D-Xylose	+	DL-Lactate	+
Galactose	+	L-Arabinose	+	D-Gluconate	+
Lactose	+	D-Arabinose	+	2-Keto-D-gluconate	+
Trehalose	+	D-Ribose	+	D-Glucosamine	+
Maltose	+	Glycerol	+	*N*-Acetyl-D-glucosamine	n
Melezitose	+	Erythritol	+	D-Glucuronate	+
Methyl-α-D-glucoside	+	Ribitol	+	0.1% Cycloheximide	+
Soluble starch	+	Galactitol	+	Growth at 37°C	+

28.3. *Cutaneotrichosporon jirovecii* (Fragner) Liu *et al.*

Synonymy: *Trichosporon jirovecii* Fragner.

RG-1 organism

Morphological description: Colonies (SDA) are whitish to pale yellow, semi-shiny, smooth, slightly cerebriform near the bottom, with a fringed margin. Yeast cells are ovoid, ellipsoidal or elongate, and occur singly, in pairs or in clusters. Lateral globose conidia develop on hyphae. Arthroconidia are cylindrical and form lateral globose blastoconidia. This species is unable to assimilate ribitol. No growth at 37°C. Originally isolated from human nails, rarely isolated from clinical specimens such as skin or urine.

Physiological tests: (+ Positive, - Negative, v Variable, w Weak, s Slow, n not done).

Fermentation: Absent.

Growth reactions:

Glucose	+	Cellobiose	+	D-Mannitol	+
Sucrose	+	L-Sorbose	+	D-Glucitol	+
Raffinose	+	L-Rhamnose	v	*myo*-Inositol	+
Melibiose	+	D-Xylose	+	DL-Lactate	+
Galactose	+	L-Arabinose	+	D-Gluconate	+
Lactose	+	D-Arabinose	–	2-Keto-D-gluconate	+
Trehalose	+	D-Ribose	+	D-Glucosamine	–
Maltose	+	Glycerol	+	*N*-Acetyl-D-glucosamine	n
Melezitose	+	Erythritol	+	D-Glucuronate	+
Methyl-α-D-glucoside	+	Ribitol	–	0.1% Cycloheximide	v
Soluble starch	+	Galactitol	+	Growth at 37°C	–

28.4. *Cutaneotrichosporon mucoides* (Gueho & M.Th. Smith) Liu *et al.*

Synonymy: *Trichosporon mucoides* Gueho & M.Th. Smith.

RG-2 organism

Morphological description: Colonies are moist and glabrous, white, cerebriform, heaped and folded. Budding cells present in primary cultures. Broadly clavate, terminal or lateral blastoconidia often present, becoming thick-walled with age. Arthroconidia are barrel-shaped. Appressoria absent. This species assimilates melibiose and is tolerant to 0.01% (variable tolerance to 0.1%) cycloheximide. Growth at 37°C. Common species associated with superficial infections, white piedra and onychomycosis.

Physiological tests: (+ Positive, - Negative, v Variable, w Weak, s Slow).

Fermentation: Negative.

Growth reactions:

Glucose	+	Cellobiose	+	D-Mannitol	+
Sucrose	+	L-Sorbose	+	D-Glucitol	+
Raffinose	+	L-Rhamnose	+	*myo*-Inositol	+
Melibiose	+	D-Xylose	+	DL-Lactate	+
Galactose	+	L-Arabinose	+	D-Gluconate	+
Lactose	+	D-Arabinose	+	2-Keto-D-gluconate	+
Trehalose	+	D-Ribose	+	D-Glucosamine	+
Maltose	+	Glycerol	+	*N*-Acetyl-D-glucosamine	+
Melezitose	+	Erythritol	w	D-Glucuronate	+
Methyl-α-D-glucoside	+	Ribitol	+	0.1% Cycloheximide	+
Soluble starch	+	Galactitol	+	Growth at 37°C	+

Antifungal susceptibility: (Table 28.1).

Table 28.1. *Cutaneotrichosporon mucoides* (Australian national data); MIC µg/mL.

Antifungal	No.	≤0.016	0.03	0.06	0.125	0.25	0.5	1	2	4	8	16	32	≥64
AmB	9			1	1	1	4	2						
FLU	9								4	2		3		
VORI	8		2	2		1	1	2						
POSA	8		2	2			3	1						
ITRA	9	1	1	1	2		3	1						

DOI: 10.1079/9781800622340.0028

29 *Cyberlindnera fabianii* (Wick.) Minter

Synonymy: *Candida fabianii* Kodama *et al.*

Cyberlindnera fabianii is a rare opportunistic yeast species. Its ability to cause septicaemia, produce biofilm, and rapid acquisition of resistance to fluconazole and voriconazole, reinforced the need for its identification from its closely related species (Arastehfar *et al.*, 2019; Park *et al.*, 2019).

RG-1 organism

Culture: Colonies (SDA) white to cream coloured, smooth, glabrous, yeast-like.

Microscopy: Spheroidal to ellipsoidal budding blastoconidia, 3.0-6.5 x 2-5.5 µm in size. No pseudohyphae are produced. Asci when present, are spherical, containing one to four spherical, finely roughened ascospores.

India ink preparation: Negative - no capsules present.

Dalmau plate culture: Spherical to ovoid budding yeast cells and occasional pseudohyphae produced.

Molecular identification: ITS sequencing recommended.

MALDI-ToF MS: Able to accurately identify this species.

Germ tube formation: Negative.

Physiological tests: (+ Positive, - Negative, v Variable, w Weak, s Slow, n not done).

Fermentation:

Glucose	+	Sucrose	+	Lactose	–
Galactose	–	Maltose	+,s	Trehalose	–

Growth reactions:

Glucose	+	L-Sorbose	–	*myo*-Inositol	–
Sucrose	+	L-Rhamnose	–	DL-Lactate	+
Raffinose	+	D-Xylose	+	D-Gluconate	+
Melibiose	–	L-Arabinose	–	2-Keto-D-gluconate	–
Galactose	–	D-Arabinose	–	D-Glucosamine	–
Lactose	–	D-Ribose	–	*N*-Acetyl-D-glucosamine	–
Trehalose	+	Glycerol	+	D-Glucuronate	n
Maltose	+	Erythritol	–	Nitrate	+
Melezitose	+	Ribitol	–	Urease	–
Methyl-α-D-glucoside	+	Galactitol	–	0.1% Cycloheximide	n
Soluble starch	+	D-Mannitol	+	Growth at 37°C	+
Cellobiose	+	D-Glucitol	+		

Key features: Germ tube negative yeast and sugar assimilation pattern. Molecular identification may be required.

Antifungal susceptibility: (Table 29.1).

Table 29.1. *Cyberlindnera fabianii* (Pfaller *et al.*, 2015; and Australian national data); MIC µg/mL.

Antifungal	No.	≤0.016	0.03	0.06	0.125	0.25	0.5	1	2	4	8	16	32	≥64
AmB	1				1									
FLU	1										1			
VORI	5		4		1									
POSA	5					3	2							
ITRA	1						1							
ANID	1						1							
MICA	4		2	1	1									
5FC	1		1											

References: Kurtzman (2011a).

DOI: 10.1079/9781800622340.0029

30 *Cylindrocarpon* Wollenw.

The genus contains 35 species, widespread, mostly from soil and as an occasional human and animal pathogen. *Cylindrocarpon* differs from *Fusarium* by lacking an asymmetrical foot-cell on the macroconidia (Summerbell and Schroers, 2002).

RG-2 organism if isolated from humans

Morphological description: Colonies are fast growing, hyaline or bright coloured, suede-like or woolly (Fig. 30.1). Sporodochia may occasionally be present. Conidiophores consist of simple or repeatedly verticillate phialides, arranged in brush-like structures. Phialides are cylindrical to subulate, with small collarettes producing hyaline, smooth-walled conidia, which are arranged in slimy masses. Two types of conidia may be produced; macroconidia which have one to several septate, hyaline, straight or curved, cylindrical to fusiform, with a rounded apex and flat base; and microconidia which are one-celled, and usually clearly distinct from the macroconidia (Fig. 30.1). Chlamydospores may be present or absent, hyaline to brown, spherical, formed singly, in chains or in clumps, intercalary or terminal (Fig. 30.1).

Molecular identification: ITS sequencing recommended.

References: Booth (1966); Domsch *et al.* (2007); de Hoog *et al.* (2015).

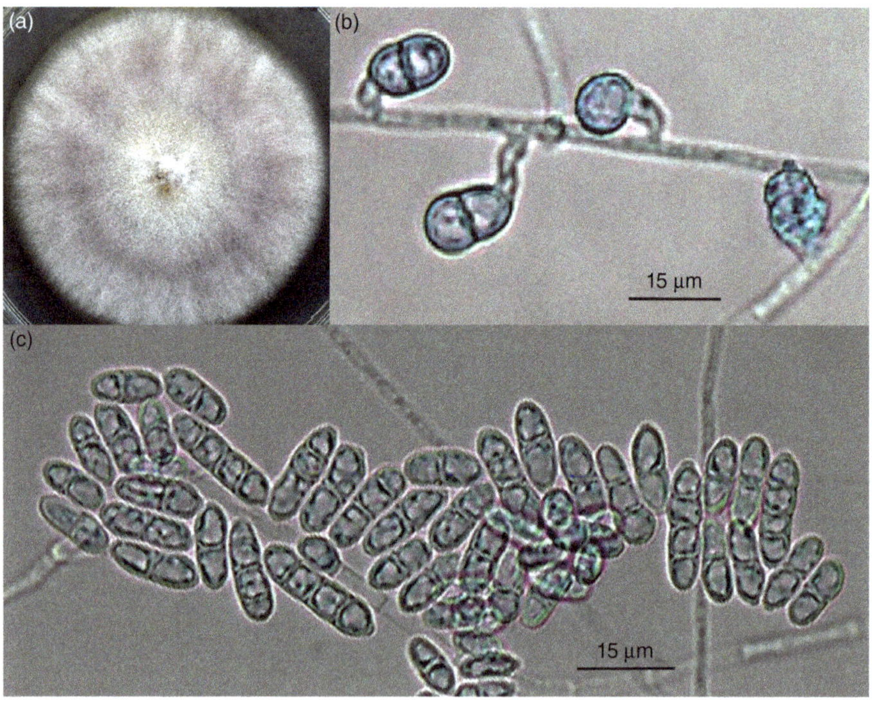

Fig. 30.1. *Cylindrocarpon* spp. showing (a) culture, (b) chlamydospores and (c) macroconidia.

DOI: 10.1079/9781800622340.0030

31 *Debaryomyces hansenii* (Zopf) Lodder & Kreger-van Rij.

Synonymy: *Candida famata* (Harrison) S.A. Meyer & Yarrow.

Debaryomyces hansenii is a common environmental isolate. It may be isolated from human skin and is only rarely recovered from blood stream infections (Desnos-Ollivier *et al.*, 2008).

RG-1 organism

Culture: Colonies (SDA) white to cream coloured, smooth, glabrous, yeast-like.

Microscopy: Ovoid to broadly ellipsoidal budding blastoconidia, 3.5-5 x 2-3.5 μm in size. No pseudohyphae produced. Asci when present are spherical, persistent, containing one to two spherical ascospores with rough walls.

India ink preparation: Negative - no capsules present.

Dalmau plate culture: Spherical to ovoid budding yeast cells only. Pseudohyphae are usually lacking.

Molecular identification: ITS sequencing recommended.

MALDI-ToF MS: Able to accurately identify this species.

Germ tube formation: Negative.

Physiological tests: (+ Positive, – Negative, v Variable, w Weak, s Slow, n not done).

Fermentation:

Glucose	–,+	Sucrose	–,w	Lactose	–
Galactose	–,w	Maltose	–,w	Trehalose	–,w

Growth reactions:

Glucose	+	L-Sorbose	v	*myo*-Inositol	–
Sucrose	+	L-Rhamnose	v	DL-Lactate	v
Raffinose	+	D-Xylose	+	D-Gluconate	+,w
Melibiose	v	L-Arabinose	+,w	2-Keto-D-gluconate	+
Galactose	+	D-Arabinose	v	D-Glucosamine	v
Lactose	v	D-Ribose	v	*N*-Acetyl-D-glucosamine	v
Trehalose	+	Glycerol	+	D-Glucuronate	v
Maltose	+	Erythritol	v	Nitrate	–
Melezitose	v	Ribitol	+	Urease	–
Methyl-α-D-glucoside	+	Galactitol	v	0.1% Cycloheximide	v
Soluble starch	v	D-Mannitol	+	Growth at 37°C	–
Cellobiose	+	D-Glucitol	+,w		

Key features: Germ tube negative yeast and sugar assimilation pattern.

Antifungal susceptibility: (Table 31.1).

Table 31.1. *Debaryomyces hansenii* (Diekema *et al.*, 2009; Beyda *et al.*, 2013; Espinel-Ingroff *et al.*, 2014; and Australian national data); MIC μg/mL.

Antifungal	No.	≤0.016	0.03	0.06	0.125	0.25	0.5	1	2	4	8	16	32	≥64
AmB	29			2	3	7	14	2	1					
FLU	78				9	18	14	9	9	11	2	3		3
VORI	81	20	28	16	9	4		2	1	1				
POSA	75	8	12	22	10	8	10	2	3					
ITRA	14		2	2	1	4	3					2		
ANID	26		1	2	3	1		9	10					
MICA	25	1	4	1		2	8	6	3					
5FC	14		1	7	1	1	2							2

References: Suzuki *et al.* (2011).

DOI: 10.1079/9781800622340.0031

32 *Diutina* Khunnamw. *et al.*

Several species previously classified within the genus *Candida* have now been transferred to the genus *Diutina*, including *Diutina catenulata* (formerly *Candida catenulata*) and *Diutina rugosa* (formerly *Candida rugosa*). *D. rugosa* is a complex of four highly similar species, and is referred to as the *D. rugosa* complex (Khunnamwong *et al.*, 2015; Ming *et al.*, 2019).

32.1. *Diutina catenulata* (Diddens & Lodder) Khunnamwong *et al.*

Synonymy: *Candida catenulata* Diddens & Lodder.

Although most isolates of *Diutina catenulata* originate from human sources, cases of candidaemia are uncommon.

RG-1 organism

Culture: Colonies (SDA) white to cream coloured, smooth, soft and wrinkled, yeast-like.

Microscopy: Ovoid to cylindrical budding blastoconidia, 1.5-4.5 x 4-12 µm.

India ink preparation: Negative - no capsules present.

Dalmau plate culture: Pseudohyphae consisting of chains of ovoid or cylindroid cells, and sometimes small verticils of ovoid blastoconidia.

Germ tube formation: Negative.

Physiological tests: (+ Positive, - Negative, v Variable, w Weak, s Slow).

Fermentation:

Glucose	v	Sucrose	–	Lactose	–
Galactose	–,s	Maltose	–,s	Trehalose	–

Growth reactions:

Glucose	+	L-Sorbose	–	*myo*-Inositol	–
Sucrose	–	L-Rhamnose	–	DL-Lactate	–
Raffinose	–	D-Xylose	v	D-Gluconate	v
Melibiose	–	L-Arabinose	–	2-Keto-D-gluconate	v
Galactose	+	D-Arabinose	–	D-Glucosamine	v
Lactose	–	D-Ribose	v	*N*-Acetyl-D-glucosamine	+
Trehalose	v	Glycerol	+	D-Glucuronate	–
Maltose	v	Erythritol	–	Nitrate	–
Melezitose	–	Ribitol	v	Urease	–
Methyl-α-D-glucoside	–	Galactitol	–	0.1% Cycloheximide	+
Soluble starch	v	D-Mannitol	+	Growth at 30°C	+
Cellobiose	-	D-Glucitol	v	Growth at 37°C	v

Key features: Separation from most physiologically similar species can be accomplished based on positive growth responses on D-mannitol, D-glucitol and resistance to 0.1% cycloheximide, combined with negative responses for sorbose and erythritol utilisation or growth in vitamin-free medium.

Antifungal susceptibility: (Table 32.1).

Table 32.1. *Diutina catenulata* very limited data (Australian national data); MIC µg/mL.

Antifungal	No.	≤0.016	0.03	0.06	0.125	0.25	0.5	1	2	4	8	16	32	≥64
AmB	3				2	1								
FLU	3						1		1		1			
VORI	3		2	1										
POSA	3		2	1										
ITRA	3		1	1		1								
ANID	3			1	2									
MICA	3		3											

32.2. *Diutina rugosa* complex

Synonymy: *Candida rugosa* (Anderson) Diddens & Lodder.

Diutina rugosa is a complex of at least four species, including *D. rugosa, D. pararugosa, D. neorugosa,* and *D. pseudorugosa* (Khunnamwong *et al.*, 2015; Ming *et al.*, 2019). This group has been associated with catheter related fungaemia and has been isolated from human and bovine faeces, sea water and soil (Sugar and Stevens, 1985).

RG-1 organism

Culture: Colonies (SDA) white to cream coloured, smooth, glabrous, yeast-like.

Microscopy: Ellipsoidal to elongate budding blastoconidia, 5-11 x 1.5-2.5 µm. Sometimes short pseudohyphae may be produced.

India ink preparation: Negative - no capsules present.

Dalmau plate culture: Densely branched pseudohyphae produced.

Germ tube formation: Negative.

Physiological tests: (+ Positive, - Negative, v Variable, w Weak, s Slow, n not done).

Fermentation: Absent.

Growth reactions:

Glucose	+	L-Sorbose	v	*myo*-Inositol	–
Sucrose	–	L-Rhamnose	–	DL-Lactate	v
Raffinose	–	D-Xylose	v	D-Gluconate	v
Melibiose	–	L-Arabinose	–	2-Keto-D-gluconate	–
Galactose	v	D-Arabinose	–	D-Glucosamine	–
Lactose	–	D-Ribose	–	*N*-Acetyl-D-glucosamine	+
Trehalose	–	Glycerol	+	D-Glucuronate	–
Maltose	–	Erythritol	–	Nitrate	–
Melezitose	–	Ribitol	v	Urease	–
Methyl-α-D-glucoside	–	Galactitol	–	0.1% Cycloheximide	–
Soluble starch	–	D-Mannitol	v	Growth at 37°C	+
Cellobiose	–	D-Glucitol	v		

Key features: Germ tube negative yeast and sugar assimilation pattern (D-xylose and glycerol +ve; ribitol –ve).

Antifungal susceptibility: (Table 32.2).

Table 32.2. *Diutina rugosa* (Diekema *et al.*, 2009; Espinel-Ingroff *et al.*, 2014; and Australian national data); MIC µg/mL.

Antifungal	No.	≤0.016	0.03	0.06	0.125	0.25	0.5	1	2	4	8	16	32	≥64
AmB	21				1	1	5	5	7	1	1			
FLU	97				1	3	15	24	29	8	8	6	12	2
ISAV	1							1						
VORI	80		50	16	4	7	3							
POSA	17		12	27	9	8	5							
ITRA	5		1	1	1	1		1						
ANID	20			3	4	4	4	3			2			
MICA	20		2	3	5	6	2					2		

References: Lachance *et al.* (2011).

DOI: 10.1079/9781800622340.0032

33 *Drechslera* Ito

McGinnis *et al.* (1986b) reviewed the isolates from human and animal disease purported to be *Drechslera* or *Helminthosporium* and concluded that all pathogenic isolates examined belong to the genera *Bipolaris* (and more recently to *Curvularia*) or *Exserohilum*.

RG-1 organism

Morphological genus description: Colonies are fast growing, suede-like to downy, brown to blackish-brown with a black reverse. Conidia are pale to dark brown, usually cylindrical or subcylindrical, straight, smooth-walled, and are formed apically through a pore (poroconidia) in a sympodially elongating, geniculate conidiophore. Conidia are transversely septate (phragmoconidia), with the septum delimiting the basal cell formed first during conidium maturation. Germinating is from any or all cells and the hilum is not protuberant (Fig. 33.1).

Key features: Darkly pigmented mould producing sympodial, pale brown, cylindrical or subcylindrical, transversely septate poroconidia.

Molecular identification: *GAPDH* gene sequencing recommended.

References: Ellis (1971,1976); Luttrell (1978); McGinnis (1980); Sivanesan (1987); Rippon (1988); Domsch *et al.* (2007); de Hoog *et al.* (2015).

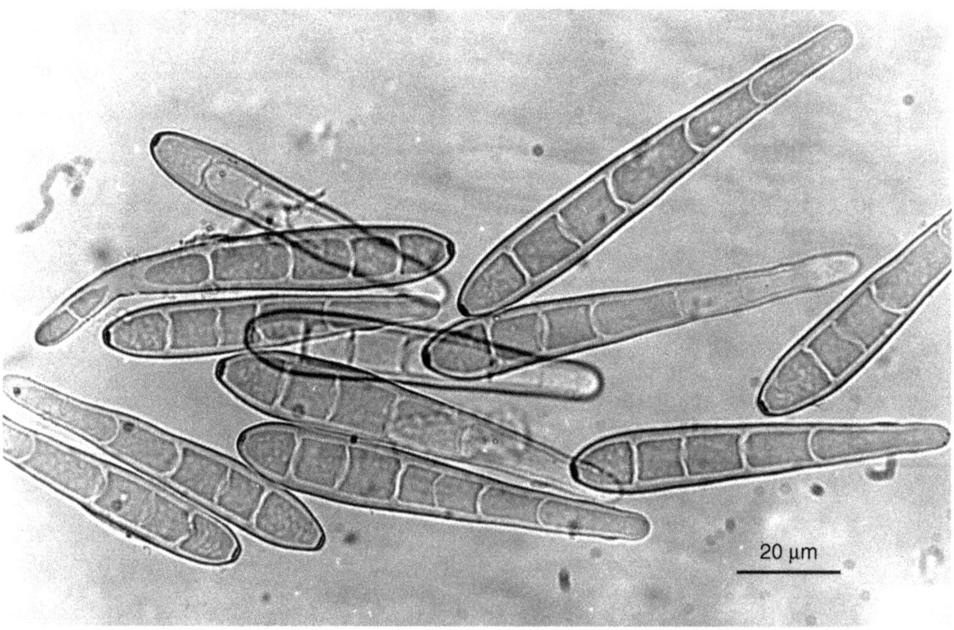

Fig. 33.1. *Drechslera dictyoides* conidia (photograph courtesy John Alcorn, Plant Pathology Herbarium (BRIP), Brisbane, Australia).

DOI: 10.1079/9781800622340.0033

34 *Emergomyces* Dukik, Sigler & de Hoog

The genus, *Emergomyces*, has recently been created to accommodate a number of novel dimorphic human pathogens that had previously been described in the genus *Emmonsia*. Five species are recognised: *Emergomyces pasteurianus*, *E. africanus, E. orientalis, E. canadensis* and *E. europaeus* (Dukik *et al.*, 2017; Wang *et al.*, 2017; Jiang *et al.*, 2018; Schwartz *et al.*, 2019). A review of the genus was provided by Samaddar and Sharma (2021).

Almost all reported cases of *Emergomyces* infection (emergomycosis) have occurred in HIV-infected individuals, often presenting as disseminated disease with cutaneous lesions. **Note:** Clinical features mimic those of histoplasmosis. To date, *Emergomyces* infections have been reported from four continents: Asia, Africa, Europe and North America. The most extensive burden of disease has been observed among HIV-infected patients from South Africa attributed to *E. africanus* (Moodley *et al.*, 2019).

RG-2 organism

Morphological genus description: *Emergomyces* species exhibit thermal dimorphism growing in living tissue or in culture at 37°C as a budding yeast-like fungus and in culture at temperatures below 30°C as a mould. Colonies at 25°C are slow growing, yellowish-white to tan, initially glabrous, becoming powdery, with a pale yellow reverse. Conidia are borne on slender stalks alongside hyphae or in groups on inflated cells. They are small, thin-walled, one-celled, sub-globose, and 2-4 µm in diameter. Colonies at 37°C on brain heart infusion (BHI) agar containing blood are moist, white to cream coloured and yeast-like. Microscopically, numerous small, oval shaped, narrow based budding yeast-like cells, 2-5 µm in size are observed. **Note:** Histopathological features of emergomycosis also mimic histoplasmosis. In fact, these infections are indistinguishable from one another, both showing intracellular narrow based budding yeasts, measuring 2-5 µm in size.

Molecular identification: ITS sequencing is necessary for accurate identification of species (Jiang *et al.*, 2018; Samaddar and Sharma, 2021).

Key features: Clinical history, tissue morphology culture identification by ITS sequencing.

Antifungal susceptibility: (Table 34.1).

Table 34.1. *Emergomyces africanus* (Maphanga *et al.*, 2017); MIC µg/mL.

	Mould form		Yeast form	
Antifungal	Range	MIC_{90}	Range	MIC_{90}
AmB	≤0.008-1	0.25	≤0.008-1	0.25
FLU	0.12-8	0.5	0.12-1	0.5
ITRA	0.008-0.5	0.06	0.008-0.12	0.008
VORI	0.008-0.06	0.03	0.008-0.015	0.015
POSA	0.008-1	0.06	0.008-0.25	0.03

References: Kenyon *et al.* (2013); Dukik *et al.* (2017); Wang *et al.* (2017); Jiang *et al.* (2018); Schwartz *et al.* (2018a,b, 2019); Samaddar and Sharma (2021).

DOI: 10.1079/9781800622340.0034

35 *Epicoccum nigrum* Link

Synonym: *Epicoccum purpurascens* Ehrenb. ex Schlecht.

Epicoccum nigrum is a cosmopolitan saprophyte of worldwide distribution which is occasionally isolated as a contaminant from clinical specimens like skin.

RG-1 organism

Morphological description: Colonies are fast growing, suede-like to downy, with a strong yellow to orange-brown diffusible pigment. When sporulating numerous black sporodochia (aggregates of conidiophores) are visible. Conidia are formed singly on densely compacted, non-specialised, determinant, slightly pigmented conidiophores. Conidia are globose to pyriform, mostly 15-25 μm diameter with a funnel-shaped base and broad attachment scar, often seceding with a protuberant basal cell, i.e. rhexolytic dehiscence of conidia. Conidia become multicellular (dictyoconidia), darkly pigmented and have a verrucose external surface (Fig. 35.1).

Key features: Darkly pigmented mould producing, large globose to pyriform, verrucose dictyoconidia on a sporodochium.

Reference: Ellis (1971); Samson *et al.* (1995); Domsch *et al.* (2007).

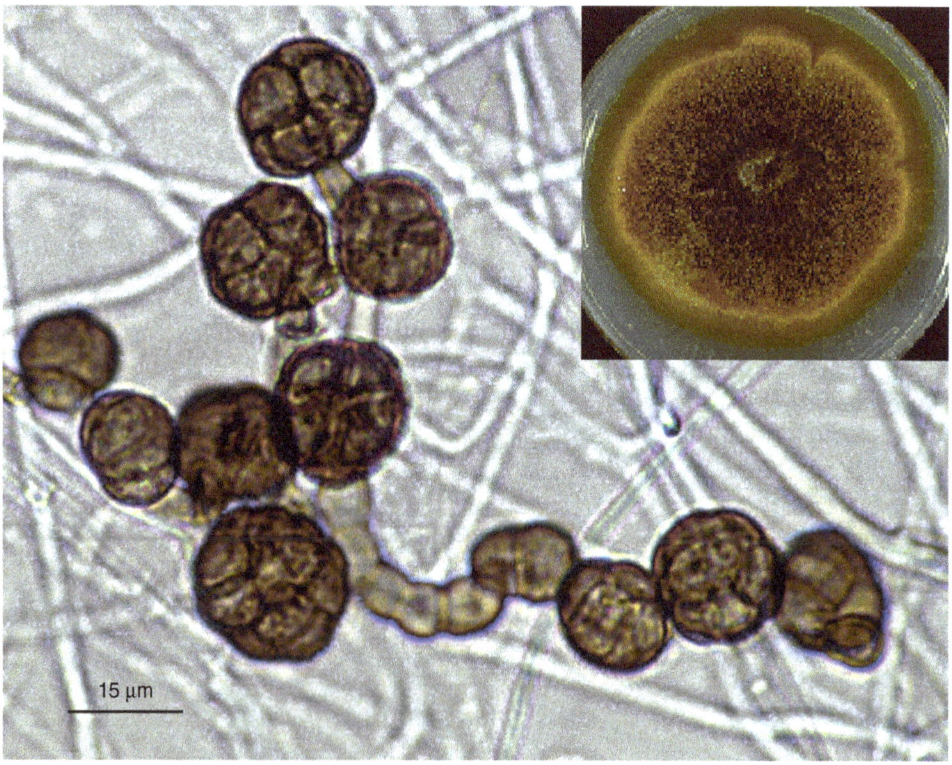

Fig. 35.1. *Epicoccum nigrum* culture and conidia.

DOI: 10.1079/9781800622340.0035

36 *Epidermophyton floccosum* (Harz) Langeron & Milochevitch

Epidermophyton floccosum is an anthropophilic dermatophyte with a worldwide distribution often causing tinea pedis, tinea cruris, tinea corporis and onychomycosis. It is not known to invade hair *in vivo* and no specific growth requirements have been reported.

RG-2 organism

Morphological description: Colonies are usually slow growing, greenish-brown or khaki coloured with a suede-like surface, raised and folded in the centre, with a flat periphery and submerged fringe of growth (Fig. 36.1). Older cultures may develop white pleomorphic tufts of mycelium. A deep yellowish-brown reverse pigment is usually present. Microscopic morphology shows characteristic smooth, thin-walled macroconidia, which are often produced in clusters growing directly from the hyphae (Fig. 36.1). Numerous chlamydospores are formed in older cultures. Microconidia are not formed.

Key features: Culture characteristics, microscopic morphology and clinical disease.

Molecular identification: ITS sequencing is recommended for species identification (de Hoog *et al.*, 2020).

References: Rebell and Taplin (1970); Mackenzie *et al.* (1986); Rippon (1988); de Hoog *et al.* (2015); Liu *et al.* (2021).

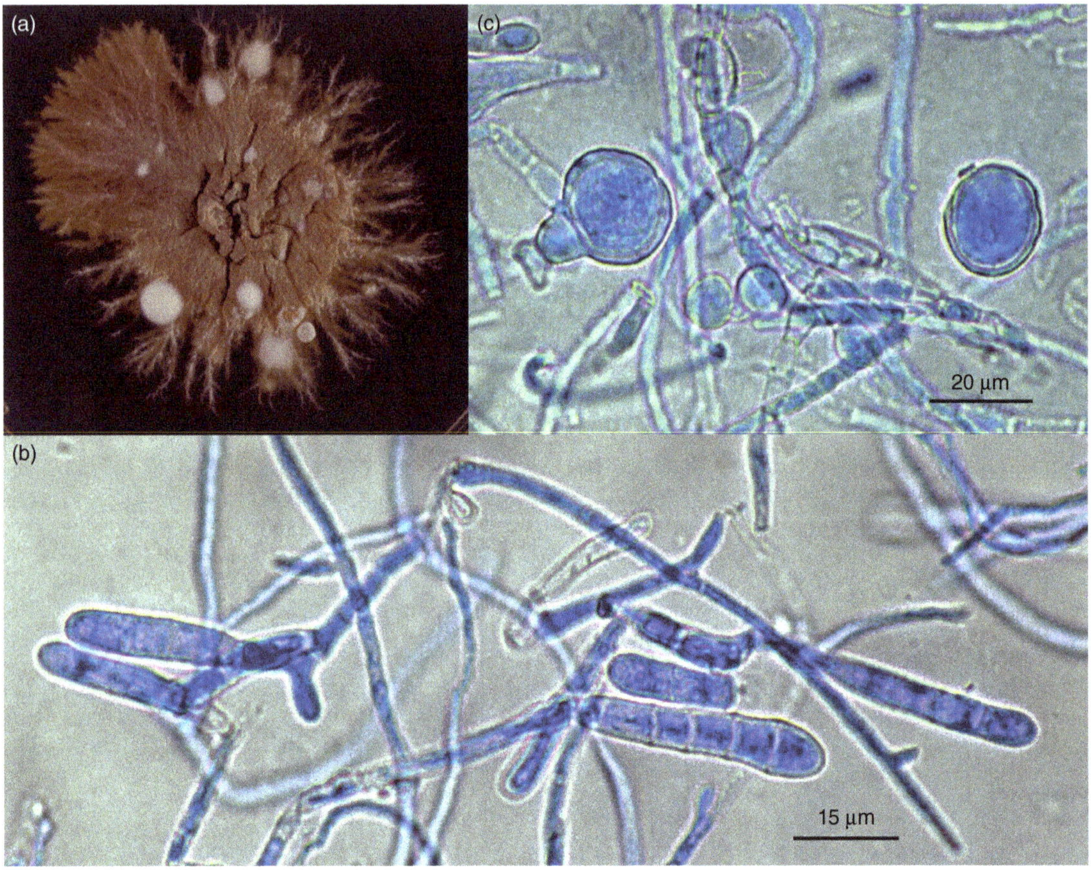

Fig. 36.1. *Epidermophyton floccosum* showing (a) culture, (b) macroconidia and (c) chlamydospores.

DOI: 10.1079/9781800622340.0036

37 *Exophiala* J.W. Carmich.

Exophiala species are common environmental fungi often associated with decaying wood and soil enriched with organic wastes. However, several species, notably *E. jeanselmei* complex and *E. spinifera* complex, are well documented human pathogens. Clinical manifestations include mycetoma (especially for *E. jeanselmei* complex), localised cutaneous infections, subcutaneous cysts, endocarditis and cerebral and disseminated infections. Phaeohyphomycosis caused by *Exophiala* species has been reported in both normal and immunosuppressed patients.

37.1. *Exophiala dermatitidis* (Kano) de Hoog

Synonymy: *Wangiella dermatitidis* (Kano) McGinnis.

Exophiala dermatitidis has been isolated from plant debris and soil and is a recognised causative agent of mycetoma and phaeohyphomycosis in humans (Zeng *et al.*, 2007; Kirchhoff *et al.*, 2019).

RG-2 organism

Morphological description: Colonies are slow growing, initially yeast-like and black, becoming suede-like, olivaceous-grey with the development of aerial mycelium with age. A brown pigment is often produced in the agar. The initial yeast-like phase is characterised by unicellular, ovoid to elliptical, budding yeast-like cells. The yeast-like cells are hyaline and thin-walled when young becoming darkly pigmented (dematiaceous) and thick-walled when mature. With the development of mycelium, flask-shaped to cylindrical annellides are produced. Conidia are hyaline to pale brown, one-celled, round to obovoid, 2-4 x 2.5-6 µm, smooth-walled and accumulate in slimy balls at the apices of the annellides or down their sides (Fig. 37.1). Cultures grow at 42°C and on media containing 0.1% cycloheximide.

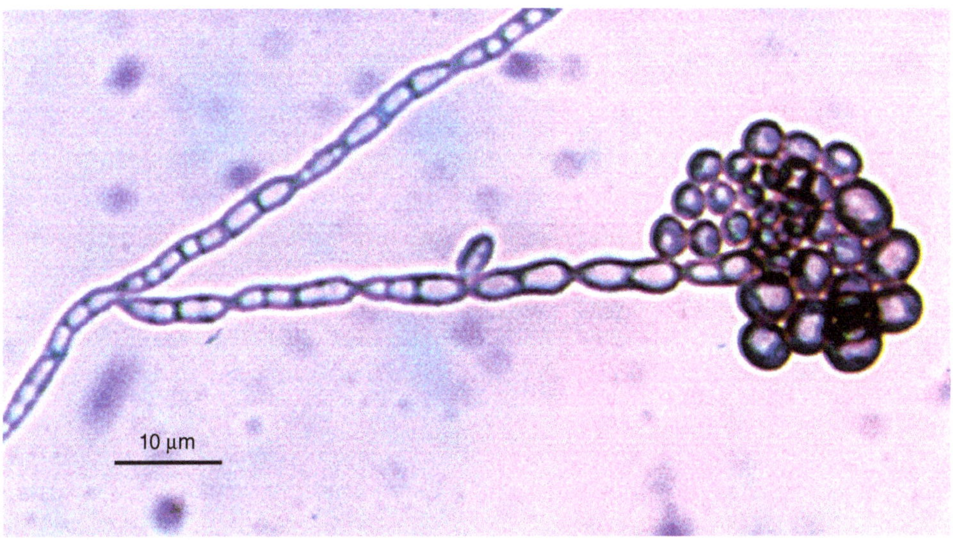

Fig. 37.1. *Exophiala dermatitidis* showing an annellide producing a ball of conidia.

Molecular identification: ITS or D1/D2 sequencing is recommended for species identification (Halliday *et al.*, 2015).

MALDI-ToF MS: Appears to be a promising identification tool with an extensive database (Kondori *et al.*, 2015; Ozhak-Baysan *et al.*, 2015).

Antifungal susceptibility: (Table 37.1).

Table 37.1. *Exophiala dermatitidis* (Australian national data); MIC µg/mL.

Antifungal	No.	≤0.016	0.03	0.06	0.125	0.25	0.5	1	2	4	8	16	32	≥64
AmB	25				1	4	9	11						
ISAV	4					2	2							
VORI	25			3	14	7		1						
POSA	25		1	4	7	12		1						
ITRA	25				5	14	5	1						

References: de Hoog and Hermanides-Nijhof (1977); McGinnis (1980); Hohl *et al.* (1983); Nishimura and Miyaji (1983); Matsumoto *et al.* (1984); Dixon and Polak-Wyss (1991); de Hoog *et al.* (2015).

37.2. *Exophiala jeanselmei* complex McGinnis & Padhye

Exophiala jeanselmei has a worldwide distribution and is a recognised causative agent of mycetoma and phaeohyphomycosis in humans. Zeng *et al.* (2007) presented an overview of the medically important *Exophiala* species.

Exophiala jeanselmei has long been recognised as heterogeneous (de Hoog and Hermanides-Nijhof, 1977). Recent molecular studies have redefined *Exophiala jeanselmei* and three additional species have been identified: *E. oligosperma, E. nishimurae* and *E. xenobiotica* (Vitale and de Hoog, 2002; de Hoog *et al.*, 2003, 2006). These species are morphologically very similar and can best be distinguished by molecular methods (Zeng *et al.,* 2014).

Molecular identification: ITS and/or D1/D2 sequencing is recommended for species identification (Halliday *et al.*, 2015).

Morphological description: Conidiogenous cells are predominantly annellidic and erect, multicellular conidiophores are absent (Table 37.2.). No growth at 40°C.

Table 37.2. Morphological characteristics of *Exophiala jeanselmei* complex species.

E. jeanselmei	Mature conidiogenous cells rocket-shaped, slightly darker than the supporting hyphae, with regular tapering annellated zones.
E. oligosperma	Mature conidiogenous cells remain concolorous with supporting hyphae and may be intercalary and lateral, the latter being flask or rocket shaped. Annellated zones have the appearance of inconspicuous flat scars. Chlamydospores are absent.
E. nishimurae	Similar morphology to *E. oligosperma*, however large chlamydospore-like cells are present.
E. xenobiotica	A segregant genotype of the *E. jeanselmei* complex with less melanised conidiogenous cells.

RG-2 organism

Morphological description of *Exophiala jeanselmei* **McGinnis & Padhye:** Colonies are initially smooth, greenish-grey to black, mucoid and yeast-like, becoming raised and developing tufts of aerial mycelium with age, often becoming dome-shaped and suede-like in texture. Reverse is olivaceous-black. Numerous ellipsoidal, yeast-like, budding cells are usually present, especially in young cultures. Scattered amongst these yeast-like cells are larger, inflated, subglobose to broadly ellipsoidal cells (germinating cells) which give rise to short torulose hyphae that gradually change into unswollen hyphae. Conidia are formed on lateral pegs either arising apically or laterally at right or acute angles from essentially undifferentiated hyphae or from strongly inflated detached conidia. Conidiogenous pegs are 1-3 µm long, slightly tapering and imperceptibly annellate. Conidia are hyaline, smooth, thin-walled, broadly ellipsoidal, 3.2-4.4 x 1.2-2.2 µm, and with inconspicuous basal scars (Fig. 37.2). Cultures grow at 37°C but not at 40°C.

Antifungal susceptibility: *Exophiala jeanselmei* (Table 37.3).

Antifungal susceptibility: *Exophiala oligosperma* (Table 37.4).

Antifungal susceptibility: *Exophiala xenobiotica* (Table 37.5).

References: de Hoog and Hermanides-Nijhof (1977); McGinnis and Padhye (1977); McGinnis (1978a, 1980); Nishimura and Miyaji (1983); Matsumoto *et al.* (1984); Dixon and Polak-Wyss (1991); Domsch *et al.* (2007); Badali *et al.* (2010b); de Hoog *et al.* (2015).

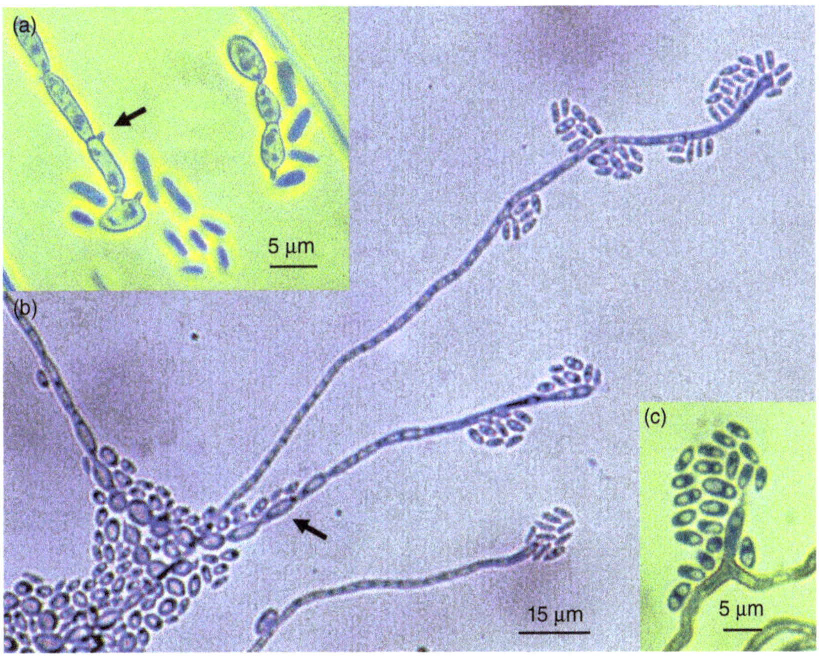

Fig. 37.2. *Exophiala jeanselmei* showing (a) conidiogenous pegs or annellides (arrow) on yeast-like cells, (b) yeast-like cells giving rise to torulose hyphae (arrow) bearing annellides, and (c) an annellide and conidia.

Table 37.3. *Exophiala jeanselmei* (Australian national data); MIC µg/mL.

Antifungal	No.	≤0.016	0.03	0.06	0.125	0.25	0.5	1	2	4	8	16	32	≥64
AmB	14			1	3	2	3		3	1	1			
ISAV	3						2	1						
VORI	14			2	3	3	3	1	2					
POSA	11	1	2	3	4	1								
ITRA	14	2		3	3	4	2							

Table 37.4. *Exophiala oligosperma* (Australian national data); MIC µg/mL.

Antifungal	No.	≤0.016	0.03	0.06	0.125	0.25	0.5	1	2	4	8	16	32	≥64
AmB	6							3		3				
VORI	6					2	3	1						
POSA	6		1	2	1	2								
ITRA	6			1	2	2	1							

Table 37.5. *Exophiala xenobiotica* (Australian national data); MIC µg/mL.

Antifungal	No.	≤0.016	0.03	0.06	0.125	0.25	0.5	1	2	4	8	16	32	≥64
AmB	9				1	1	2	1	4					
ISAV	4					1	1	2						
VORI	9					3	6							
POSA	9	1		3	4				1					
ITRA	9			2	4	1	2							

37.3. *Exophiala spinifera* complex

Synonymy: *Phialophora spinifera* Nielsen & Conant; *Rhinocladiella spinifera* (Nielsen & Conant) de Hoog.

Exophiala spinifera has a worldwide distribution and is a recognised causative agent of mycetoma and phaeohyphomycosis in humans. Zeng *et al.* (2007) presented an overview of the medically important *Exophiala* species.

Recent molecular studies have re-examined *Exophiala spinifera* and have recognised two species: *E. spinifera* and *E. attenuata* (Vitale and de Hoog, 2002). These two species are morphologically very similar and can best be distinguished by genetic analysis.

Molecular identification: ITS sequencing is recommended for accurate species identification (Zeng and de Hoog, 2008).

Morphological description: Conidiogenous cells are predominantly annellidic and erect, multicellular conidiophores are darker than the supporting hyphae present (Table 37.6). No growth at 40°C.

Table 37.6. Morphological characteristics of *Exophiala spinifera* complex species.

E. spinifera	Annellated zones are long with clearly visible, frilled annellations.
E. attenuata	Annellated zones are inconspicuous and degenerate.

RG-2 organism

Morphological description of ***Exophiala spinifera*** **(Nielsen & Conant) McGinnis:** Colonies are initially mucoid and yeast-like, black, becoming raised and developing tufts of aerial mycelium with age, finally becoming suede-like to downy in texture (Fig. 37.3). Reverse is olivaceous-black. Conidiophores are simple or branched, erect or sub-erect, spine-like with rather thick brown pigmented walls. Conidia are formed in basipetal succession on lateral pegs either arising apically or laterally at right or acute angles from the spine-like conidiophores or from undifferentiated hyphae. Conidiogenous pegs are 1-3 µm long, slightly tapering and imperceptibly annellate. Conidia are one-celled, subhyaline, smooth, thin-walled, sub-globose to ellipsoidal, 1.0-2.9 x 1.8-2.5 µm, and aggregate in clusters at the tip of each annellide. Toruloid hyphae and yeast-like cells with secondary conidia are typically present (Fig. 37.3). **Note:** Yeast cells show the presence of capsules in India ink-stained mounts and cultures will grow on media containing 0.1% cycloheximide. No growth at 40°C.

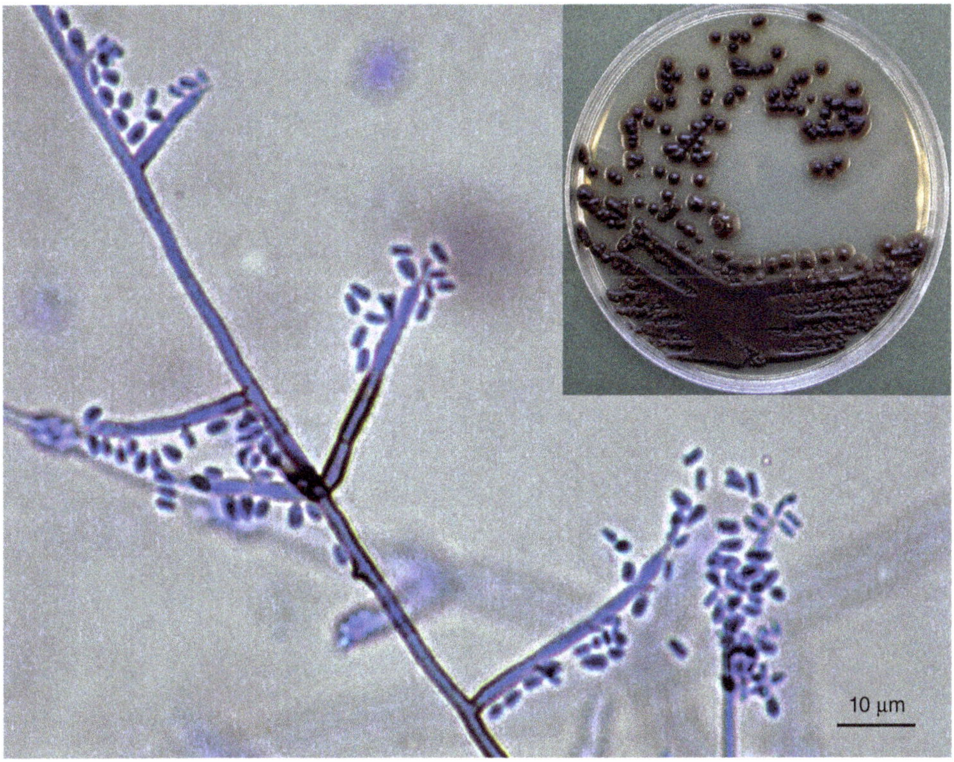

Fig. 37.3. *Exophiala spinifera* culture showing typical black, mucoid yeast-like growth and conidia being formed on erect, multiseptate conidiophores that are darker than the supporting hyphae, with long annellated zones.

Antifungal susceptibility: (Table 37.7).

Table 37.7. *Exophiala spinifera* (Australian national data); MIC µg/mL.

Antifungal	No.	≤0.016	0.03	0.06	0.125	0.25	0.5	1	2	4	8	16	32	≥64
AmB	7					1	2	2	2					
VORI	7				1	2	2	2						
POSA	6				1	4	1							
ITRA	7				1	4	1	1						

References: de Hoog and Hermanides-Nijhof (1977); McGinnis and Padhye (1977); McGinnis (1980); Nishimura and Miyaji (1983); Matsumoto *et al.* (1984); Dixon and Polak-Wyss (1991); de Hoog *et al.* (2003, 2006, 2015); Domsch *et al.* (2007).

DOI: 10.1079/9781800622340.0037

38 *Exserohilum* Leonard and Suggs

The genus *Exserohilum* contains about 35 species and may be differentiated from the closely related genera *Bipolaris, Curvularia* and *Drechslera* by forming conidia with a strongly protruding truncate hilum (i.e. exserted hilum).

In *Drechslera* species, the hilum does not protrude; in *Bipolaris* species the hilum protrudes only slightly. Several species of *Exserohilum* have been reported as agents of phaeohyphomycosis, notably *E. rostratum, E. mcginnisii* and *E. longirostratum*. Although recent molecular studies have demonstrated that the latter two species are probable synonyms of *E. rostratum* (da Cunha *et al.,* 2012b; Katragkou *et al.*, 2014).

38.1. *Exserohilum rostratum* (Drechsler) Leonard and Suggs

Exserohilum rostratum has been reported as causing skin and corneal infection, invasive disease, and allergic fungal sinusitis. It was recently implicated in an outbreak of fungal meningitis associated with contaminated methylprednisolone in the United States (Smith *et al.*, 2013). Human infections mainly come from warm, tropical and subtropical locations such as southern United States, India and Israel (Adler *et al.*, 2006).

RG-1 organism

Morphological description: Colonies are grey to blackish-brown, suede-like to floccose in texture and have an olivaceous black reverse. Conidia are straight, curved or slightly bent, ellipsoidal to fusiform, 30-128 x 9-23 μm and are formed apically through a pore (poroconidia) on a sympodially elongating geniculate conidiophore (Fig. 38.1). Conidia have a strongly protruding, truncate hilum and the septum above the hilum is usually thickened and dark, with the end cells often paler than other cells, walls often finely roughed, usually four to nine distoseptate, but some up to 14 (Fig. 38.1). Conidial germination is bipolar.

Key features: Darkly pigmented mould producing sympodial, transverse septate, ellipsoidal to fusiform conidia with a strongly protruding, truncate hilum.

Molecular identification: ITS sequencing is recommended for barcoding. A specific real-time PCR was developed by Zhao *et al.,* (2013).

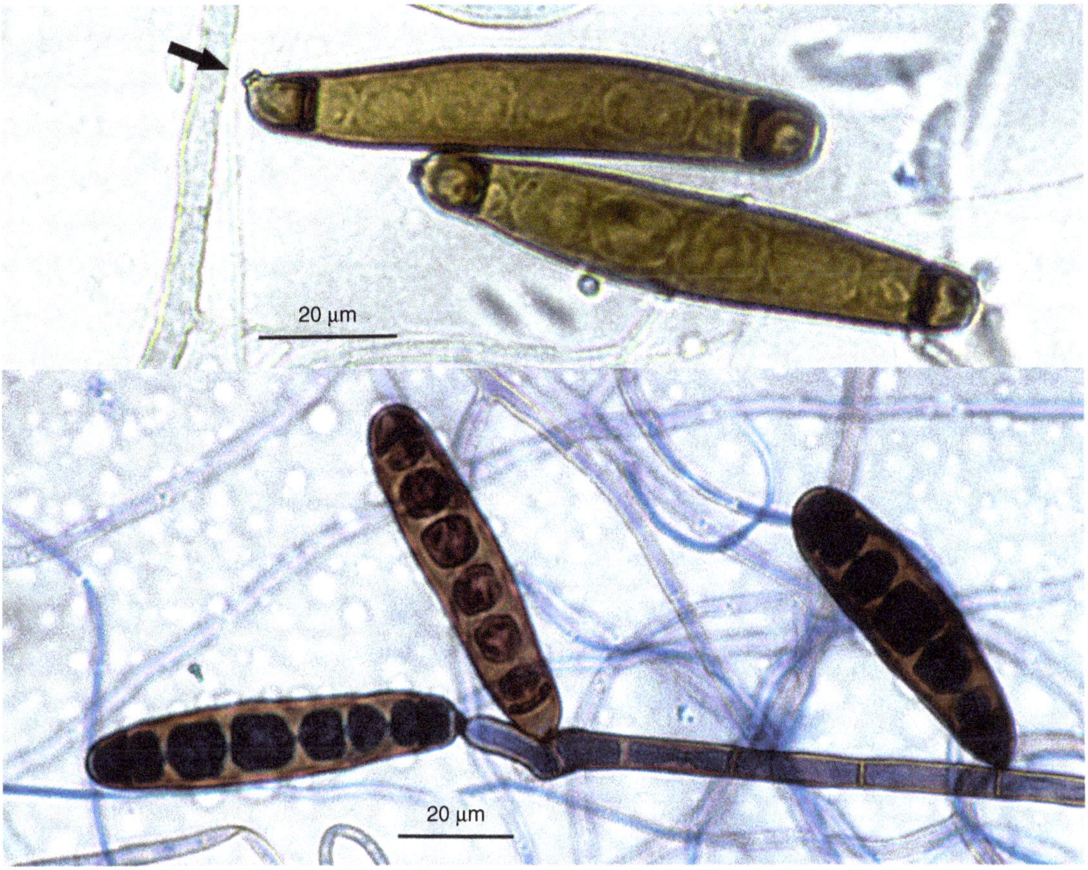

Fig. 38.1. *Exserohilum rostratum* showing a sympodially elongating conidiophore and conidia with a distinctive hilum (arrow).

Antifungal susceptibility: (Table 38.1).

Table 38.1. *Exserohilum rostratum* (Australian national data); MIC µg/mL.

Antifungal	No.	≤0.016	0.03	0.06	0.125	0.25	0.5	1	2	4	8	16	32	≥64
AmB	27				4	5	6	6	3	2	1			
ISAV	4					2			2					
VORI	27			1	2	4	11	7	2					
POSA	25		6	1	5	9	4							
ITRA	27			4	11	10	2							

References: Alcorn (1983); Adam *et al.* (1986); McGinnis *et al.* (1986b); Burges *et al.* (1987); Rippon (1988); Dixon and Polak-Wyss (1991); Domsch *et al.* (2007); da Cunha *et al.* (2012b); Hernandez-Restrepo *et al.* (2018); de Hoog *et al.* (2020).

DOI: 10.1079/9781800622340.0038

39 *Fonsecaea* complex

Morphologically the genus *Fonsecaea* is defined by the presence of indistinct melanised conidiophores with blunt, scattered denticles bearing conidia singly or in short chains that eventually become branched. de Hoog *et al.* (2004b) revised the genus on the basis of ITS sequencing data. Three species from humans are currently recognised, *F. monophora, F. nubica* and *F. pedrosoi*, although they are morphologically indistinguishable (Najafzadeh *et al.* 2009, 2010b; Xi *et al.* 2009; de Hoog *et al.* 2015). All strains grow at 37°C but not at 40°C and the three species are recognised as aetiological agents of chromoblastomycosis.

RG-2 organism

Morphological description: Colonies are slow growing, flat to heaped and folded, suede-like to downy, olivaceous to black with black reverse. Conidiogenous cells pale olivaceous, arranged in loosely branched systems, with prominent denticles. Conidia pale olivaceous, clavate to ellipsoidal, in short chains, subhyaline, smooth and thin-walled, 3.5-5 x 1.5-2 µm (Fig. 39.1).

Molecular identification: ITS sequencing is recommended for species identification (Abliz *et al.* 2003; Najafzadeh *et al.* 2009, 2010b; Xi *et al.* 2009).

Antifungal susceptibility: (Table 39.1).

Table 39.1. *Fonsecaea* spp. (Australian national data); MIC µg/mL.

Antifungal	No.	≤0.016	0.03	0.06	0.125	0.25	0.5	1	2	4	8	16	32	≥64
AmB	7				1			1	5					
VORI	7		1	1	4	1								
POSA	7	1		5	1									
ITRA	7			1	5	1								

References: McGinnis (1980); Dixon and Polak-Wyss (1991); de Hoog *et al.* (2004b, 2015); Abliz *et al.* (2003); Najafzadeh *et al.* (2009, 2010a,b).

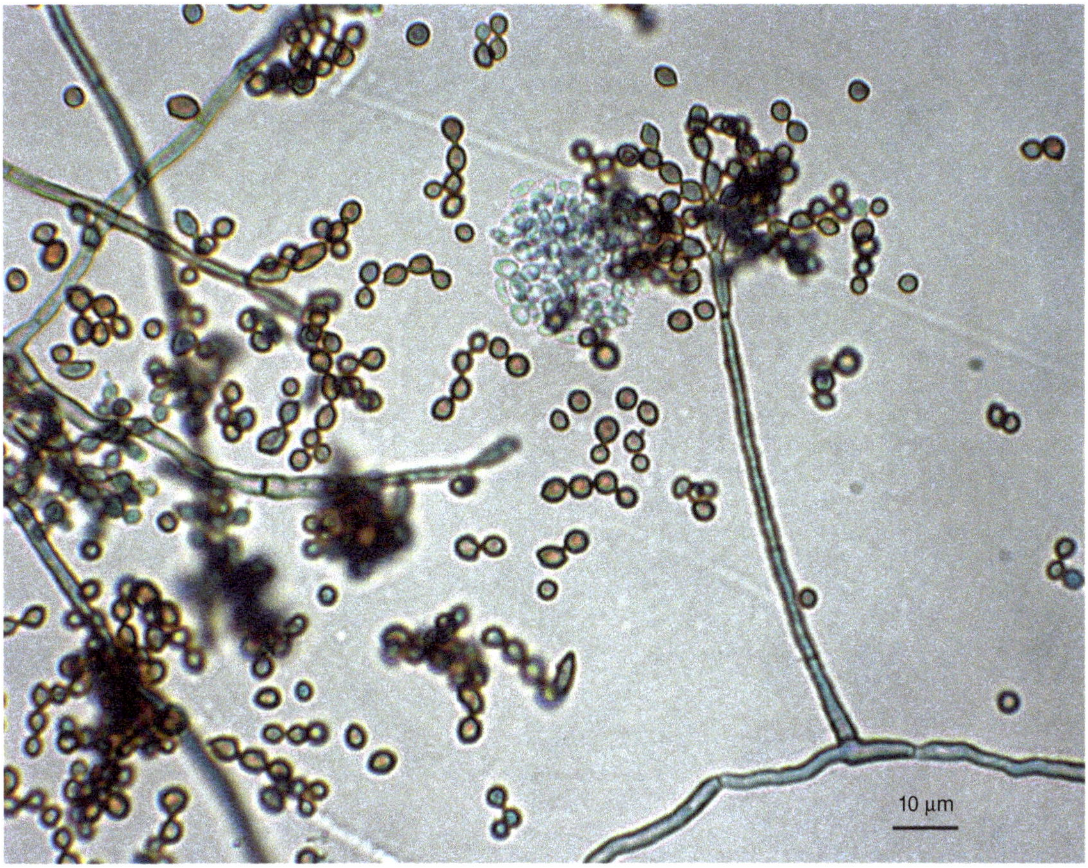

Fig. 39.1. *Fonsecaea* spp. conidiophores and conidia.

DOI: 10.1079/9781800622340.0039

40 *Fusarium* Link ex Fries

Most *Fusarium* species are soil fungi and have a worldwide distribution. Some are plant pathogens, causing root and stem rot, vascular wilt or fruit rot. Several species have emerged as important opportunistic pathogens in humans causing hyalohyphomycosis (especially in burn victims and bone marrow transplant patients), mycotic keratitis and onychomycosis (Guarro, 2013). Other species cause storage rot and are important mycotoxin producers.

Multilocus sequence analysis of *TEF-1*, β-tubulin, *Cal* and *RPB2* have revealed the presence of multiple cryptic species within each "morphospecies" of medically important fusaria (Balajee *et al.* 2009). For instance, *Fusarium solani* represents a complex (i.e. *F. solani* complex) of over 45 phylogenetically distinct species of which at least 20 are associated with human infections. Similarly, members of the *Fusarium oxysporum* complex are phylogenetically diverse, as are members of the *Fusarium incarnatum-equiseti* complex and *Fusarium chlamydosporum* complex (Balajee *et al.* 2009; Tortorano *et al.* 2014; Salah *et al.* 2015).

Currently the genus *Fusarium* comprises at least 300 phylogenetically distinct species, 20 species complexes and nine monotypic lineages (Balajee *et al.* 2009). Most of the identified opportunistic *Fusarium* pathogens belong to the *F. solani* complex, *F. oxysporum* complex and *F. fujikuroi* complex. Less frequently encountered are members of the *F. incarnatum-equiseti*, *F. dimerum* and *F. chlamydosporum* complexes, or species such as *F. sporotrichioides* (O'Donnell *et al.*, 2022; van Diepeningen *et al.* 2015).

Morphological genus description: Colonies are usually fast growing, pale or bright coloured (depending on the species) with or without a cottony aerial mycelium. The colour of the thallus varies from whitish to yellow, pink, red or purple shades (Fig. 40.1). Species of *Fusarium* typically produce both macro- and microconidia from slender phialides. Macroconidia are hyaline, two to several-celled, fusiform to sickle-shaped, mostly with an elongated apical cell and pedicellate basal cell. Microconidia are one or two-celled, hyaline, smaller than macroconidia, pyriform, fusiform to ovoid, straight or curved. Chlamydospores may be present or absent.

Identification of *Fusarium* species is often difficult due to the variability between isolates (e.g. in shape and size of conidia and colony colour) and because not all features required are always well developed (e.g. the absence of macroconidia in some isolates after subculture).

Note: Sporulation may need to be induced in some isolates and a good slide culture is essential. The important characters used in the identification of *Fusarium* species are as follows.

1. Colony growth diameters on potato dextrose agar and/or potato sucrose agar after incubation in the dark for four days at 25°C.
2. Culture pigmentation on potato dextrose agar and/or potato sucrose agar after incubation for 10-14 days with daily exposure to light.
3. Microscopic morphology including shape of the macroconidia; presence or absence of microconidia; shape and mode of formation of microconidia; nature of the conidiogenous cell bearing microconidia; and presence or absence of chlamydospores.

Molecular identification: Current species identification is based on multilocus sequence data (Guarro, 2013; O'Donnell *et al.*, 2022; van Diepeningen *et al.* 2015). FUSARIOID-ID (https://www.fusarium.org) is an internet-accessible validated database dedicated to the identification of fusaria via ITS or multi-locus nucleotide BLAST queries, available through the Westerdijk Fungal Biodiversity Institute.

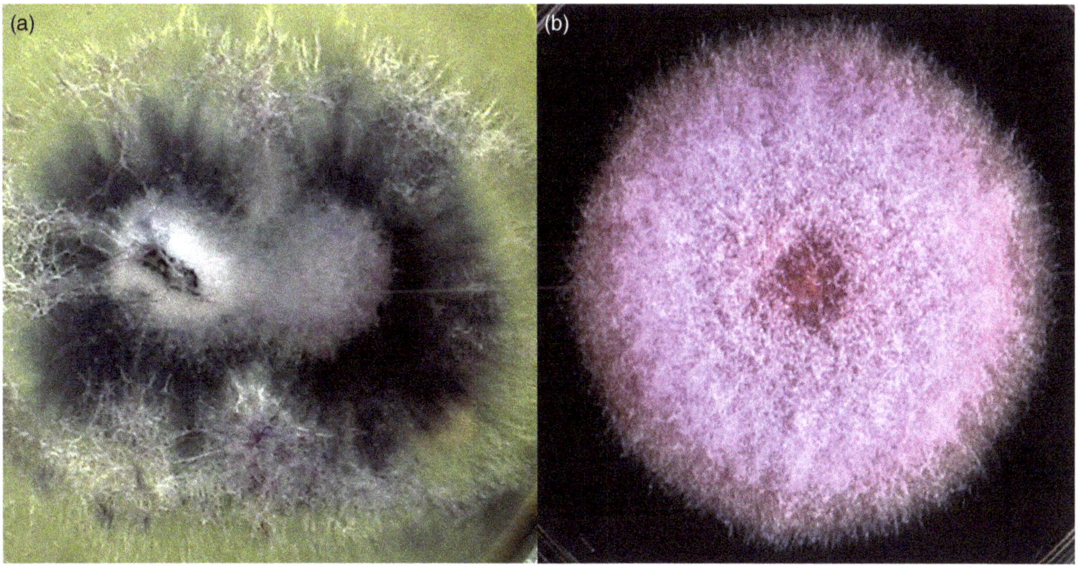

Fig. 40.1. Cultures of (a) *Fusarium oxysporum* complex showing purple pigmentation and (b) *Fusarium fujikuroi* complex showing pink pigmentation.

For sequence-based identification of *Fusarium* species (O'Donnell *et al.*, 2022).

1. FUSARIOID-ID is the recommended sequence database, rather than GenBank.
2. Three loci *TEF-1*, *RBP1* and *RBP2* may be used to accurately identify all species.
3. Ensure sequences are carefully edited and free of ambiguities.
4. Ensure the species names associated with the top BLASTn matches are the same. If multiple species names have similar scores, it may be necessary to sequence additional loci.

Note: ITS and D1/D2 sequences are too conserved to resolve species limits of most fusaria. O'Donnell *et al.*, (2022) recommend avoiding ITS or D1/D2 sequences from an unknown isolate to query GenBank, because >50% of the sequences from *Fusarium* species are misidentified in this database.

Identifications based on morphology and/or ITS and D1/D2 sequences should be reported as species complexes. Sequencing of *TEF-1*, *RPB1* and/or *RPB2* is required for accurate species identification.

MALDI-ToF MS: A comprehensive "in-house" database of reference spectra allows accurate identification of *Fusarium* isolates to species complex level (Lau *et al.* 2013).

References: Booth (1971, 1977); McGinnis (1980); Burgess and Liddell (1983); Rippon (1988); Samson *et al.* (1995); Domsch *et al.* (2007); O'Donnell *et al.* (2008, 2009a,b); Balajee *et al.* (2009); Guarro (2013); Geiser *et al.* (2013); Tortorano *et al.* (2014); van Diepeningen *et al.* (2015); Salah *et al.* (2015); de Hoog *et al.* (2020); Torres-Cruz *et al.* (2021).

40.1. *Fusarium chlamydosporum* complex

Fusarium chlamydosporum complex contains five distinct species and is common in soils and the rhizosphere of numerous vascular plants worldwide. It is occasionally isolated from human and animal infections (O'Donnell *et al.*, 2009b; Guarro, 2013).

RG-1 organisms

Morphological description: Colonies growing rapidly, with abundant aerial mycelium, deep pink, red or ochraceous to brownish; reverse carmine red or tan to brown (Fig. 40.2). Sporodochia orange, flesh coloured or ochraceous. Conidiophores scattered over the aerial mycelium, branched; numerous polyblastic conidiogenous cells are present. Macroconidia rarely produced and appearing only on sporodochial phialides, usually three (some up to five) septate, slightly curved, 30-38 x 3.0-4.5 µm, with no distinct foot-shaped cell. Microconidia and blastoconidia fusiform, rounded apically and tapered towards the base, single-celled to one (some up to three) septate, 6-26 x 2-4 µm. Chlamydospores abundant, intercalary, often roughened.

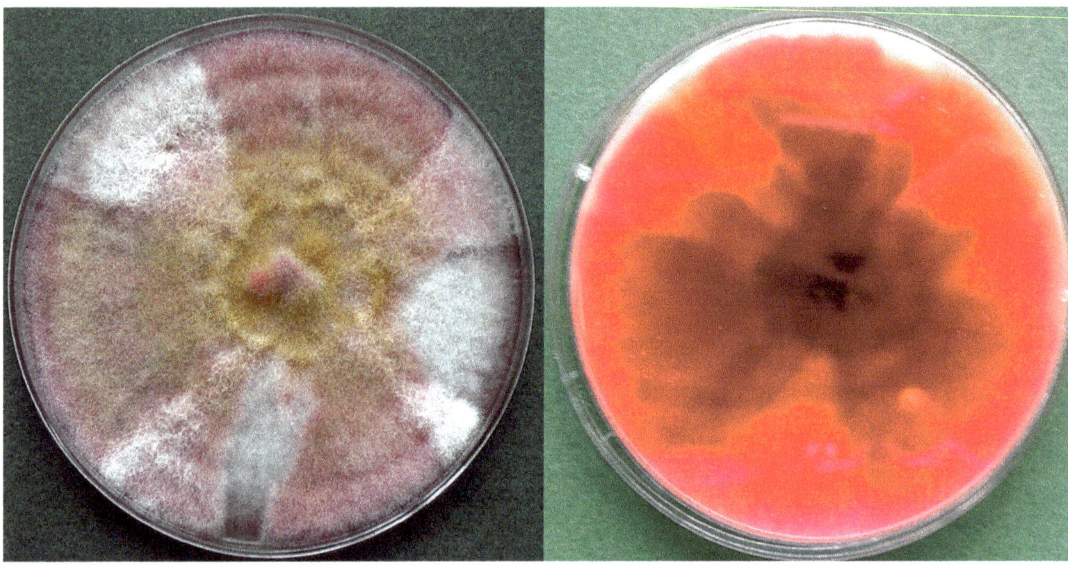

Fig. 40.2. *Fusarium chlamydosporum* complex, culture showing pink to ochraceous to brownish surface and a carmine red reverse.

Antifungal susceptibility: (Table 40.1).

Table 40.1. *Fusarium chlamydosporum* complex (Australian national data); MIC µg/mL.

Antifungal	No.	≤0.016	0.03	0.06	0.125	0.25	0.5	1	2	4	8	16	32	≥64
AmB	10				1	1	4	2	1		1			
ISAV	2									2				
VORI	10						1	1	6	2				
POSA	9						2	4	3					
ITRA	10					1	1	1		1	1	5		

40.2. *Fusarium dimerum* complex

The *Fusarium dimerum* complex contains 12 distinct species including *F. delphinoides, F. penzigii* and *F. dimerum*. These are regarded as cosmopolitan saprotrophs in soil and on plant materials (Domsch *et al.*, 2007). They have also been isolated from human corneal ulcers after trauma and from disseminated or localised infections in immunocompromised patients (Schroers *et al.*, 2009; Guarro, 2013).

RG-1 organisms

Morphological description: Colonies grow slowly; surface usually orange to deep apricot due to confluent conidial slime; aerial mycelium sometimes floccose and whitish (Fig. 40.3). Conidiophores loosely branched, with short, often swollen phialides, 10-18 x 4-5 µm. Macroconidia strongly curved and pointed at the apex, mostly one (some up to three) septate, 5-25 (-32) x 1.5-4.2 µm (Fig. 40.3). Microconidia absent. Chlamydospores mostly intercalary, exceptionally terminal, spherical to ovoidal, 6-12 µm diameter, smooth-walled, single or in chains.

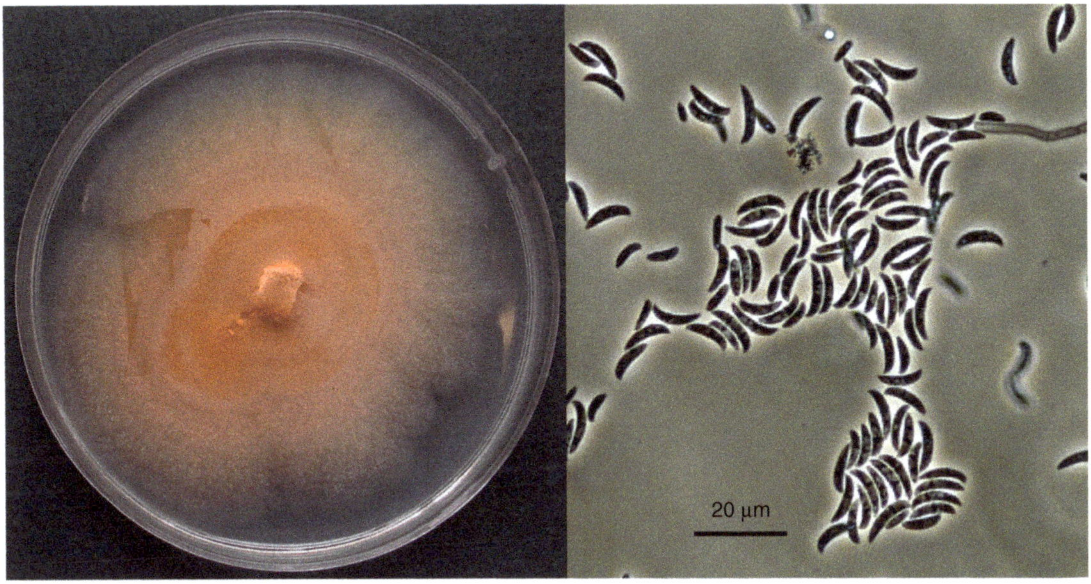

Fig. 40.3. *Fusarium dimerum* complex culture showing orange to deep apricot colour due to confluent conidial slime, and macroconidia.

Antifungal susceptibility: (Table 40.2).

Table 40.2. *Fusarium dimerum* complex (Australian national data); MIC µg/mL.

Antifungal	No.	≤0.016	0.03	0.06	0.125	0.25	0.5	1	2	4	8	16	32	≥64
AmB	37				1	6	9	9	12					
ISAV	3								1	1	1			
VORI	36						6	11	13	5	1			
POSA	34						1	5	6		22			
ITRA	37							2	1			34		

40.3. *Fusarium fujikuroi* complex

Synonymy: *Gibberella fujikuroi* complex.

Fusarium fujikuroi complex consists of 50 distinct species, 13 of which have been reported to cause human infection; *F. acutatum, F. ananatum, F. andiyazi, F. fujikuroi, F. guttiforme, F. napiforme, F. nygamai, F. verticillioides, F. proliferatum, F. sacchari, F. subglutinans, F. temperatum* and *F. thapsinum* (Guarro, 2013; Al-Hatmi *et al.*, 2015).

RG-1 organisms

Morphological description: Colonies grow rapidly, pink or vinaceous to violet; aerial mycelium abundant. Sporodochia present or absent, when present they are tan to orange. Conidiophores usually erect and branched. Macroconidia abundant, falcate to rather straight, three to five-septate, with a distinct foot-cell, 27-73 x 3.4-5.2 µm. Blastoconidia straight or slightly curved, two to three-septate, fusiform to lanceolate, with a somewhat pointed, often slightly asymmetrical apical cell and a truncate basal cell, 16-43 x 3.0-4.5 µm. Microconidia produced on polyphialides and aggregating in heads, usually unicellular, ovoidal, ellipsoidal or allantoid, 4-20 x 1.5-4.5 µm. Chlamydospores absent.

Antifungal susceptibility: (Table 40.3).

Table 40.3. *Fusarium fujikuroi* complex (Castanheira *et al.*, 2012; and Australian national data); MIC µg/mL.

Antifungal	No.	≤0.016	0.03	0.06	0.125	0.25	0.5	1	2	4	8	16	32	≥64
AmB	42							2	21	18	1			
ISAV	2										2			
VORI	42							1	14	16	11			
POSA	42						4	11	6	3	8	10		
ITRA	42							2	2	4	2	32		

40.4. *Fusarium incarnatum-equiseti* complex

Fusarium incarnatum-equiseti complex consists of 40 distinct species. They occasionally cause infections in humans and animals (O'Donnell *et al.*, 2009b; Guarro, 2013).

RG-1 organisms

Morphological description: Colonies grow rapidly; aerial mycelium floccose, at first whitish, later becoming avellaneous to buff-brown; reverse pale, becoming peach coloured. Conidiophores scattered in the aerial mycelium, loosely branched; polyblastic conidiogenous cells abundant. Sporodochial macroconidia slightly curved, with foot-cell, three to seven-septate, 20-46 x 3.0-5.5 µm. Conidia on aerial conidiophores (blastoconidia) usually borne singly on scattered denticles, fusiform to falcate, mostly three to five-septate, 7.5-35 x 2.5-4.0 µm. Microconidia sparse or absent. Chlamydospores sparse, spherical, 10-12 µm diameter, becoming brown, intercalary, single or in chains.

Antifungal susceptibility: (Table 40.4).

Table 40.4. *Fusarium incarnatum-equiseti* complex (Australian national data); MIC µg/mL.

Antifungal	No.	≤0.016	0.03	0.06	0.125	0.25	0.5	1	2	4	8	16	32	≥64
AmB	9					1	1	1	5		1			
ISAV	2								1		1			
VORI	9				1			1		5	2			
POSA	8								3		5			
ITRA	9					1						8		

40.5. *Fusarium oxysporum* complex

This complex contains at least five distinct species and accounts for about 20% of human infections caused by fusaria (Guarro, 2013; Tortorano *et al.*, 2014; Salah *et al.*, 2015). All are ubiquitous soil borne pathogens responsible for vascular wilts, rots, and damping-off diseases of a broad range of plants. A number of these fusaria are also clinically important, causing localised or deeply invasive life-threatening infections in humans and other animals (O'Donnell *et al.*, 2009a). Mortality in patients who are persistently and severely neutropenic is typically 100% (Nucci and Anaissie, 2007).

RG-2 organisms

Morphological description: Colonies grow rapidly, 4.5 cm in four days, aerial mycelium white, becoming purple, with discrete orange sporodochia present in some strains; reverse hyaline to dark blue or dark purple. Conidiophores are short, single, lateral monophialides in the aerial mycelium, later arranged in densely branched clusters. Macroconidia are fusiform, slightly curved, pointed at the tip, mostly three septate, basal cells pedicellate, 23-54 x 3-4.5 µm (Fig. 40.4). Microconidia are abundant, never in chains, mostly non-septate, ellipsoidal to cylindrical, straight or often curved, 5-12 x 2.3-3.5 µm (Fig. 40.4). Chlamydospores are terminal or intercalary, hyaline, smooth or rough-walled, 5-13 µm. In contrast to *F. solani* complex, the phialides are short and mostly non-septate.

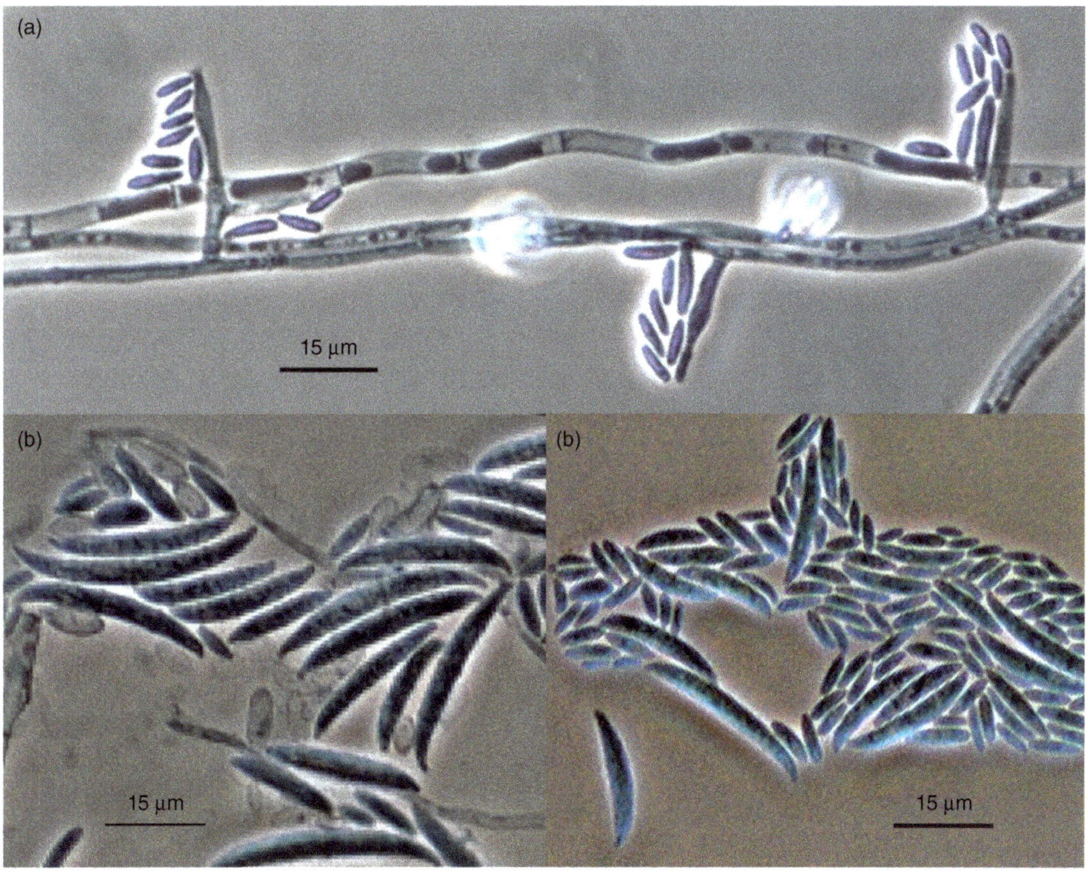

Fig. 40.4. *Fusarium oxysporum* (a) microconidia on short phialides and (b) macroconidia.

Antifungal susceptibility: (Table 40.5).

Table 40.5. *Fusarium oxysporum* complex (Australian national data); MIC µg/mL.

Antifungal	No.	≤0.016	0.03	0.06	0.125	0.25	0.5	1	2	4	8	16	32	≥64
AmB	61				1	4	6	16	19	8	7			
ISAV	3										3			
VORI	57						5	5	13	23	10	1		
POSA	48					1	1	6	9	1	29	1		
ITRA	61						3	1			1	56		

40.6. *Fusarium solani* complex

The *Fusarium solani* complex contains at least 60 species and accounts for about 50% of human infections caused by fusaria (Guarro, 2013; Tortorano *et al.*, 2014; Salah *et al.*, 2015). All are ubiquitous soil borne pathogens responsible for vascular wilts, rots, and damping-off diseases of a broad range of plants. A number of these fusaria, notably *F. keratoplasticum, F. lichenicola* and *F. solani* are clinically important, causing localised or deeply invasive life-threatening infections in humans and other animals (Guarro, 2013; O'Donnell *et al.*, 2008).

RG-2 organisms

Morphological description: Colonies grow rapidly, 4.5 cm in four days, aerial mycelium white to cream, becoming bluish-brown when sporodochia are present. Macroconidia are formed after 4-7 days from short multiple branched conidiophores which may form sporodochia. They are three to five-septate (usually three-septate), fusiform, cylindrical, often moderately curved, with an indistinct pedicellate foot cell and a short blunt apical cell, 28-42 x 4-6 µm (Fig. 40.6). Microconidia are usually abundant, cylindrical to oval, one to two-celled and formed from long lateral phialides, 8-16 x 2-4.5 µm (Fig. 40.5). Chlamydospores are hyaline, globose, smooth to rough-walled, borne singly or in pairs on short lateral hyphal branches or intercalary, 6-10 µm (Fig. 40.6).

Antifungal susceptibility: (Table 40.6).

Table 40.6. *Fusarium solani* complex (Australian national data); MIC µg/mL.

Antifungal	No.	≤0.016	0.03	0.06	0.125	0.25	0.5	1	2	4	8	16	32	≥64
AmB	216				1	2	19	49	91	49	4		1	
ISAV	29										29			
VORI	212							8	25	68	111			
POSA	200					2		2	2	2	192			
ITRA	216					2					2	199	11	1

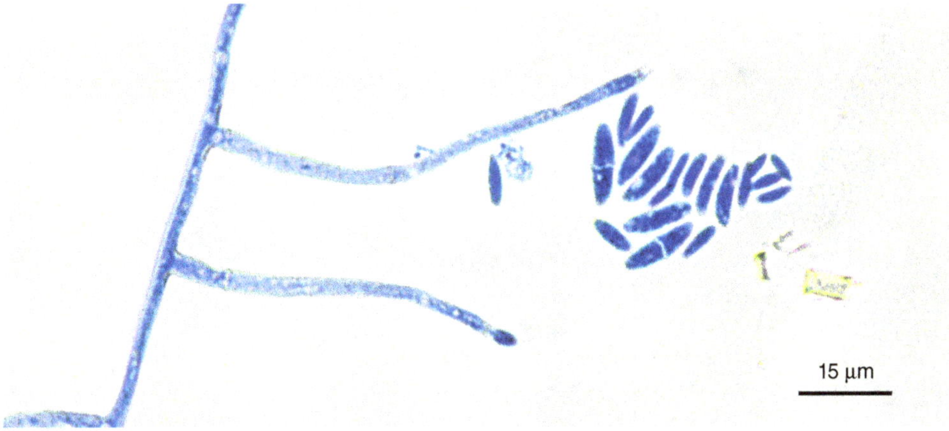

Fig. 40.5. *Fusarium solani* microconidia produced on long phialides.

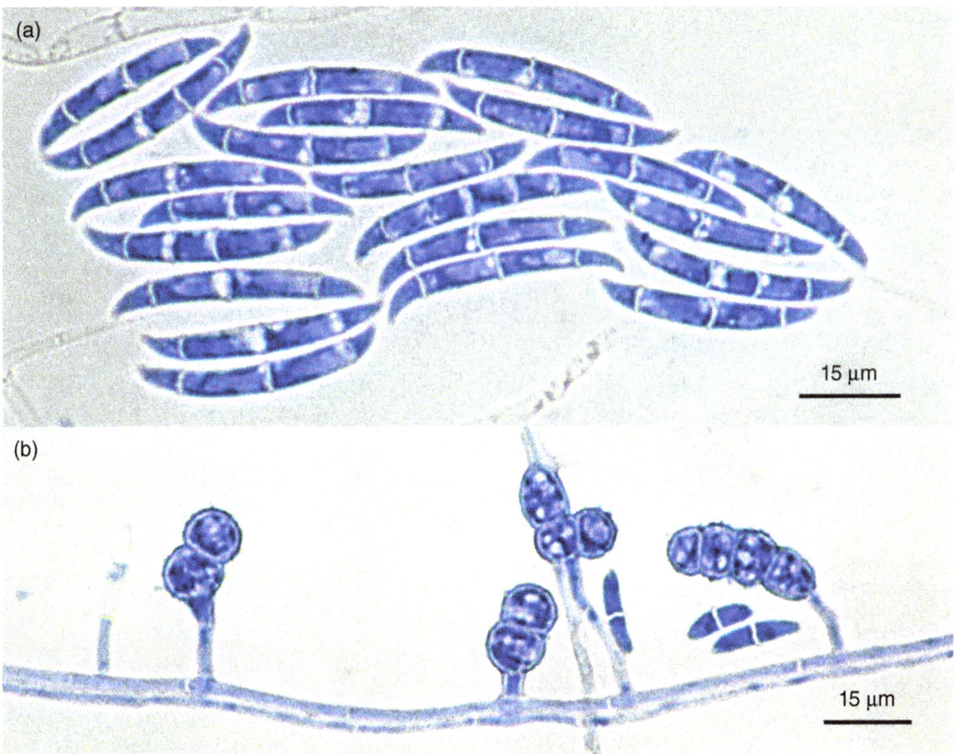

Fig. 40.6. *Fusarium solani* (a) macroconidia and (b) chlamydospores.

DOI: 10.1079/9781800622340.0040

41 *Geotrichum candidum* Link

Synonymy: *Dipodascus geotrichum* (E.E. Butler & L.J. Petersen) Arx., *Galactomyces candidus* de Hoog & M Th. Smith.

Geotrichum and related species have undergone extensive taxonomic revision (de Hoog and Smith, 2004, 2011a,b,c). The three species of prime interest to medical mycology are as *Geotrichum candidum*, *Magnusiomyces capitatus* (previously known as *Geotrichum capitatum*), and *Magnusiomyces clavatus* (previously known as *Saprochaete clavata* or *Geotrichum clavatum*) (Kaplan *et al.*, 2018, 2021).

Geotrichum candidum is an extremely common fungus with a worldwide distribution. It is commonly isolated from soil, air, water, milk, silage, plant tissues, and the digestive tract in humans and other mammals (Pottier *et al.*, 2008). Pulmonary involvement is the most frequently reported form of the disease in humans and animals, but bronchial, oral, vaginal, cutaneous and alimentary infections have also been noted (Arendrup *et al.*, 2014).

RG-1 organism

Morphological description: Colonies are fast growing, flat, white to cream, dry and finely suede-like with no reverse pigment. Hyphae are hyaline, septate, branched and break up into chains of hyaline, smooth, one-celled, subglobose to cylindrical arthroconidia. Arthroconidia are 6-12 x 3-6 μm in size and are released by the separation of a double septum (Fig. 41.1).

Note: True blastoconidia production is not found in this genus. This characteristic distinguishes *Geotrichum candidum* from *Trichosporon*, which usually do produce blastoconidia.

Molecular identification: ITS sequencing recommended for species identification (de Hoog and Smith, 2004).

Physiological tests: (+ Positive, – Negative, v Variable, w Weak, s Slow, n not done).

Fermentation:

Glucose	v	Sucrose	–	Lactose	–
Galactose	–,w	Maltose	–	Trehalose	–

Growth reactions:

Glucose	+	L-Sorbose	+	*myo*-Inositol	–
Sucrose	–	L-Rhamnose	–	DL-Lactate	+
Raffinose	–	D-Xylose	+	D-Gluconate	–
Melibiose	–	L-Arabinose	–	2-Keto-D-gluconate	–
Galactose	+	D-Arabinose	–	D-Glucosamine	–
Lactose	–	D-Ribose	–	*N*-Acetyl-D-glucosamine	n
Trehalose	–	Glycerol	+	D-Glucuronate	n
Maltose	–	Erythritol	–	Nitrate	–
Melezitose	–	Ribitol	v	Urease	–
Methyl-α-D-glucoside	–	Galactitol	–	0.1% Cycloheximide	n
Soluble starch	–	D-Mannitol	+	Growth at 37°C	v
Cellobiose	–	D-Glucitol	+		

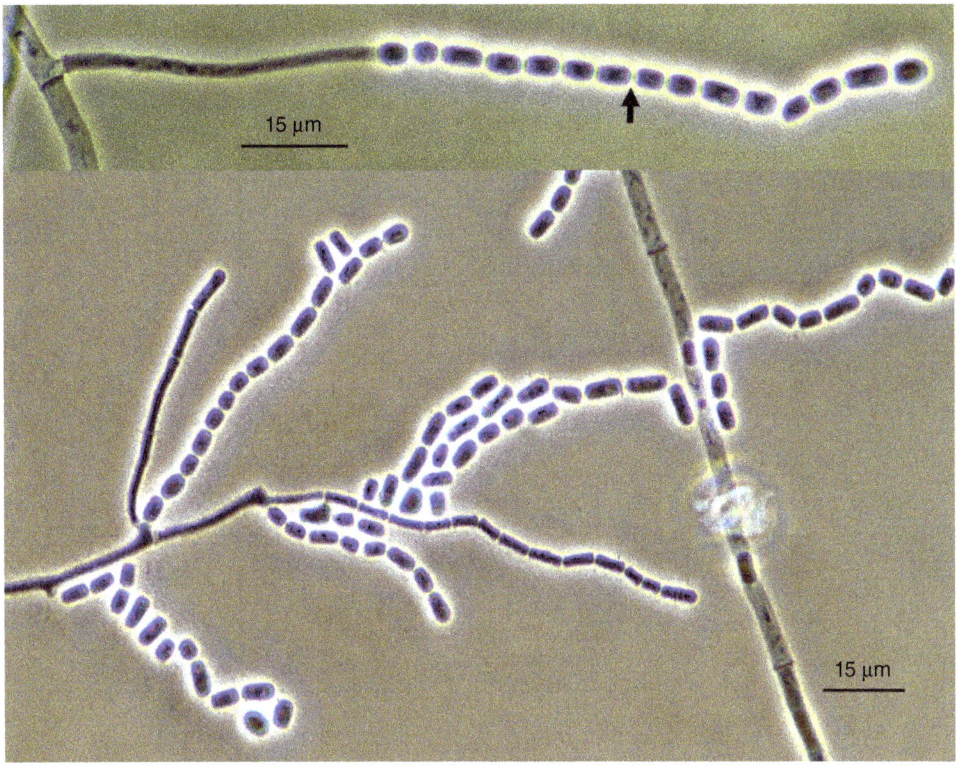

Fig. 41.1. *Geotrichum candidum* arthroconidium formation. Conidial secession is by the centripetal separation (schizolysis) of a so-called double septum (arrow).

Antifungal susceptibility: (Table 41.1).

Table 41.1. *Geotrichum candidum* limited data (Australian national data); MIC µg/mL.

Antifungal	No.	≤0.016	0.03	0.06	0.125	0.25	0.5	1	2	4	8	16	32	≥64
AmB	7						1	2	2	2				
FLU	7										3	3		1
ISAV	3			1		1	1							
VORI	7		1	1		4	1							
POSA	6				1		5							
ITRA	7					4	3							
ANID	5				1						4			
MICA	5				1						4			
5FC	7			6										1

References: Gueho (1979); de Hoog *et al.* (1986); Buchta and Otcenasek (1988); Samson *et al.* (1995); Domsch *et al.* (2007); Bilman and Yetik (2017).

DOI: 10.1079/9781800622340.0041

42 *Gliocladium* Corda

Gliocladium species have a worldwide distribution and are commonly isolated from a wide range of plant debris and soil.

RG-1 organism

Morphological description: The genus *Gliocladium* is often described as a counterpart of *Penicillium* with slimy conidia. Colonies are fast growing, suede-like to downy in texture, white at first, sometimes pink to salmon, becoming pale to dark green with sporulation. The most characteristic feature of the genus is the distinctive erect, often densely penicillate conidiophores with phialides which bear slimy, one-celled hyaline to green, smooth-walled conidia in heads or columns (Fig. 42.1). Although, some penicillate conidiophores are always present, *Gliocladium* species may also produce verticillate branching conidiophores, which can be confused with *Verticillium* or *Trichoderma* species.

Key features: Hyaline mould producing distinctive erect penicillate conidiophores with phialides bearing clusters of single-celled conidia.

References: McGinnis (1980); Onions *et al.* (1981); Rippon (1988); Domsch *et al.* (2007); de Hoog *et al.* (2015).

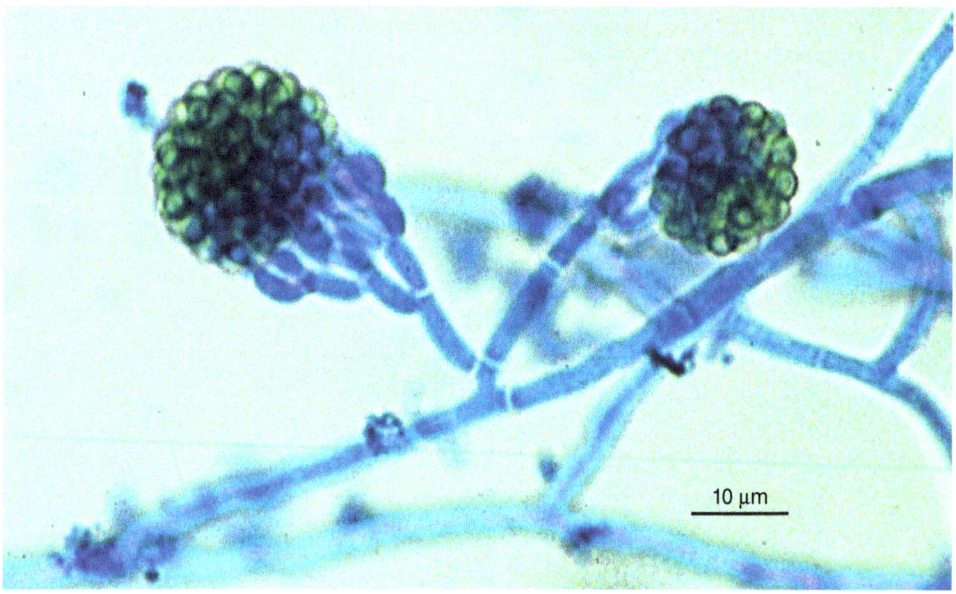

Fig. 42.1. *Gliocladium* spp. conidiophore and conidia.

DOI: 10.1079/9781800622340.0042

43 *Graphium* Corda

The genus *Graphium* is characterised by the formation of synnemata which consist of a more or less compact group of erect conidiophores that are cemented together, usually splaying out and bearing conidia at the apex. *Graphium* species are commonly found on woody plant material. *Graphium basitruncatum* has been reported as causing fungaemia in an immunosuppressed child post stem-cell transplantation (El Feghaly *et al.*, 2012).

Note: Many other fungi such as *Scedosporium* species may also produce synnemata.

RG-1 organism

Morphological description: Colonies are effuse, grey, olivaceous-brown or black. Synnemata are darkly pigmented, erect and occur solitarily or in clusters. Conidia are hyaline, one-celled, smooth, subglobose to ovoid and are usually aggregated in slimy heads at the apex of the synnemata (Fig. 43.1).

Molecular identification: The genus is phylogenetically related to *Scedosporium,* but ITS sequencing can be used to resolve all species (Okada *et al.*, 2000; Lackner and de Hoog, 2011).

Key features: Darkly pigmented mould producing erect synnemata with apical aggregates of single-celled conidia in slimy heads.

References: Barron (1968); Ellis (1971); McGinnis (1980); de Hoog *et al.* (2020).

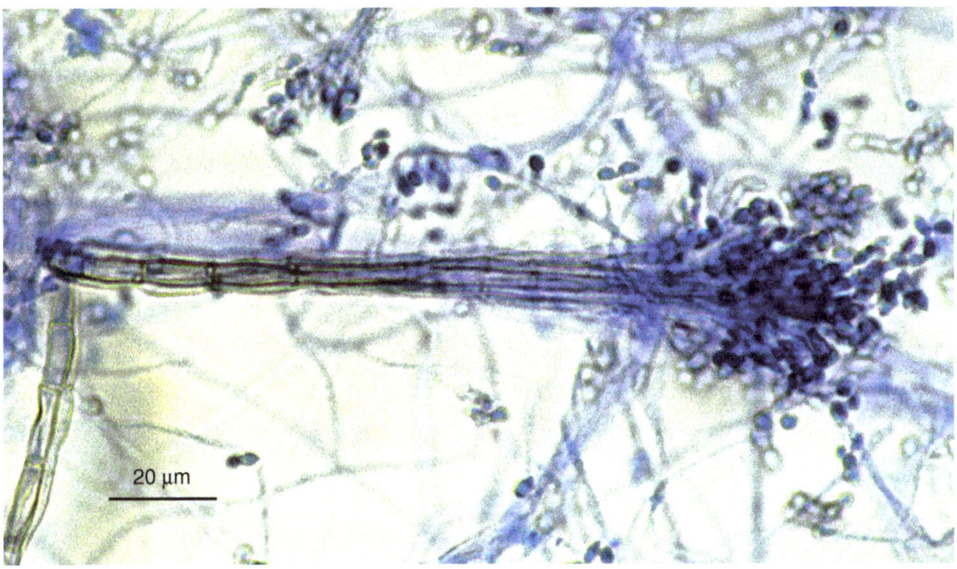

Fig. 43.1. *Graphium* spp. synnemata and conidia.

DOI: 10.1079/9781800622340.0043

44 *Histoplasma capsulatum* complex

Histoplasma capsulatum has a worldwide distribution, however the Mississippi-Ohio River Valley in the USA is recognised as a major endemic region. Environmental isolations of the fungus have been made from soil enriched with excreta from chickens, starlings and bats.

Histoplasma capsulatum has recently been subdivided into a number of molecular siblings (Sepulveda *et al.*, 2017), which in part show extensive hybridization (Maxwell *et al.*, 2018) and which may be difficult to distinguish in practice (de Almeida *et al.*, 2020). On a global scale, and particularly in South America, the *Histoplasma capsulatum* complex shows enormous diversity (Rodrigues *et al.*, 2020a). Patients may be infected by several genotypes (Damasceno *et al.*, 2019).

Histoplasmosis is an intracellular mycotic infection of the reticuloendothelial system acquired by inhalation of the fungus. Approximately 95% of cases of histoplasmosis are inapparent, subclinical or benign. The remaining 5% of cases may develop chronic progressive lung disease, chronic cutaneous or systemic disease or an acute fulminating fatal systemic disease. All stages of this disease may mimic tuberculosis. Sporadic cases have been reported in Australia.

WARNING: RG-3 organism

Cultures of *Histoplasma capsulatum* represent a severe biohazard to laboratory personnel and must be handled with extreme caution in a Class II Biological safety cabinet (BSCII).

Morphological description: *Histoplasma capsulatum* complex exhibits thermal dimorphism growing in living tissue or in culture at 37°C as a budding yeast-like fungus and in soil or culture at temperatures below 30°C as a mould.

Colonies at 25°C are slow growing, white or buff-brown, suede-like to cottony with a pale yellow-brown reverse (Fig. 44.1). Other colony types are glabrous or verrucose, and a red pigmented strain has been noted (Rippon, 1988). Microscopic morphology shows the presence of characteristic large, rounded, single-celled, 8-14 µm in diameter, tuberculate macroconidia formed on short, hyaline, undifferentiated conidiophores. Small, round to pyriform, 2-4 µm in diameter, microconidia borne on short branches or directly on the sides of the hyphae may also be present (Fig. 44.1).

Colonies at 37°C grown on brain heart infusion (BHI) agar blood are smooth, moist, white and yeast-like (Fig. 44.1). Microscopically, numerous small round to oval budding yeast-like cells, 3-4 x 2-3 µm in size are observed.

Three varieties of *Histoplasma capsulatum* are recognised, depending on the clinical disease: var. *capsulatum* is the common cause of histoplasmosis; var. *duboisii* is the African type and var. *farciminosum* causes lymphangitis in horses. *Histoplasma* isolates may also resemble species of *Sepedonium* and *Chrysosporium*. Traditionally, positive identification required conversion of the mould form to the yeast phase by growth at 37°C on enriched media, however for laboratory safety, culture identification by DNA sequencing is preferred.

Key features: Clinical history, tissue morphology, culture morphology and positive exoantigen test or DNA sequence.

Molecular identification: Elias *et al.* (2012) developed a multiplex-PCR for identification from cultures. Scheel *et al.* (2014) developed a loop-mediated isothermal amplification (LAMP) assay for detection directly in clinical samples which is affordable and useful in resource poor facilities. ITS sequencing may be used for accurate identification (Estrada-Barcenas *et al.*, 2014; Irinyi *et al.*, 2015).

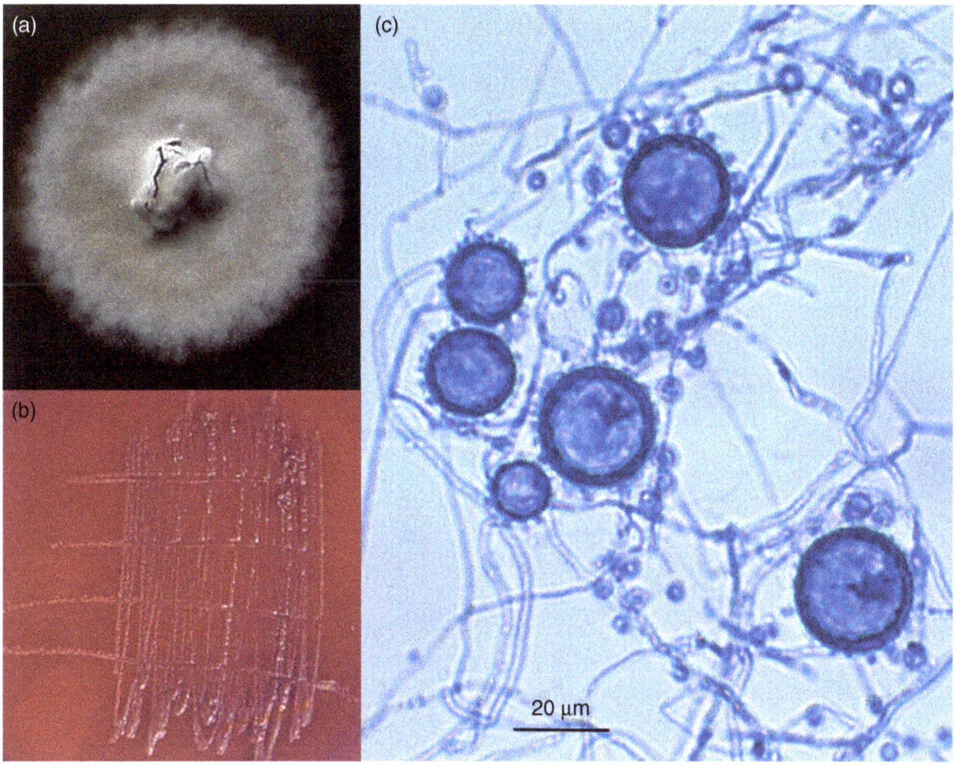

Fig. 44.1. *Histoplasma capsulatum* showing (a) mould-like culture at 25°C, (b) yeast-like culture at 37°C, (c) large, single-celled, tuberculate macroconidia and smaller microconidia.

MALD-ToF MS: Valero *et al.* (2018) developed a reference database for the identification of *Histoplasma capsulatum*.

Antifungal susceptibility: (Table 44.1).

Table 44.1. *Histoplasma capsulatum* complex limited data available (Espinel-Ingroff, 2003; Gonzalez *et al.*, 2005; Sabatelli *et al.*, 2006; Brilhante *et al.*, 2012; Kathuria *et al.*, 2014); MIC µg/mL. Antifungal susceptibility testing not recommended.

	Mould form		Yeast form	
Antifungal	Range	MIC_{90}	Range	MIC_{90}
AmB	≤0.03-0.5	0.5	0.03-0.5	0.25
FLU	1-125	16	2-8	8
ITRA	≤0.03-1	0.125 (1)	≤0.03-0.25	0.125
VORI	≤0.03-2	0.25 (1)	≤0.03-0.5	0.5
POSA	≤0.03-2	0.125 (2)	0.03-0.5	0.25

References: Chandler *et al.* (1980); McGinnis (1980); George and Penn (1986); Rippon (1988); de Hoog *et al.* (2015).

DOI: 10.1079/9781800622340.0044

45 *Hortaea werneckii* (Horta) Nishimura & Miyaji

Synonyms: *Phaeoannellomyces werneckii* (Horta) McGinnis & Schell; *Exophiala werneckii* (Horta) v. Arx; *Cladosporium werneckii* Horta.

Hortaea werneckii is a common saprophytic fungus believed to occur in soil, compost, humus and on wood in humid tropical and sub-tropical regions and is the causative agent of tinea nigra in humans.

RG-1 organism

Morphological description: Colonies are slow growing, initially mucoid, yeast-like and shiny black. However, with age they develop abundant aerial mycelia and become dark olivaceous in colour (Fig. 45.1). Microscopically, colonies consist of brown to dark olivaceous, septate hyphal elements and numerous two-celled, pale brown, cylindrical to spindle-shaped yeast-like cells that taper towards the ends to form an annellide. Most yeast-like cells also have prominent darkly pigmented septa. Annellides may also arise from the hyphae. Conidia are one to two-celled, cylindrical to spindle-shaped, hyaline to pale brown and usually occur in aggregated masses (Fig. 45.1). Chlamydospores also present.

Key features: Darkly pigmented culture, two-celled yeast-like cells producing annelloconidia.

Molecular identification: ITS sequencing can also assist identification.

Antifungal susceptibility: (Table 45.1).

Table 45.1. *Hortaea werneckii* (Ng *et al.*, 2005; Formoso *et al.*, 2015); MIC µg/mL.

Antifungal	No.	≤0.016	0.03	0.06	0.125	0.25	0.5	1	2	4	8	16	32	≥64
AmB	6						1	1	1	1	2			
FLU	6										2	1		3
VORI	6		2	1		3								
POSA	4			2	2									
ITRA	8		2		2	2	2							

References: McGinnis (1980); Mok (1982); McGinnis *et al.* (1985); Rippon (1988); Ng *et al.* (2005); de Hoog *et al.* (2015).

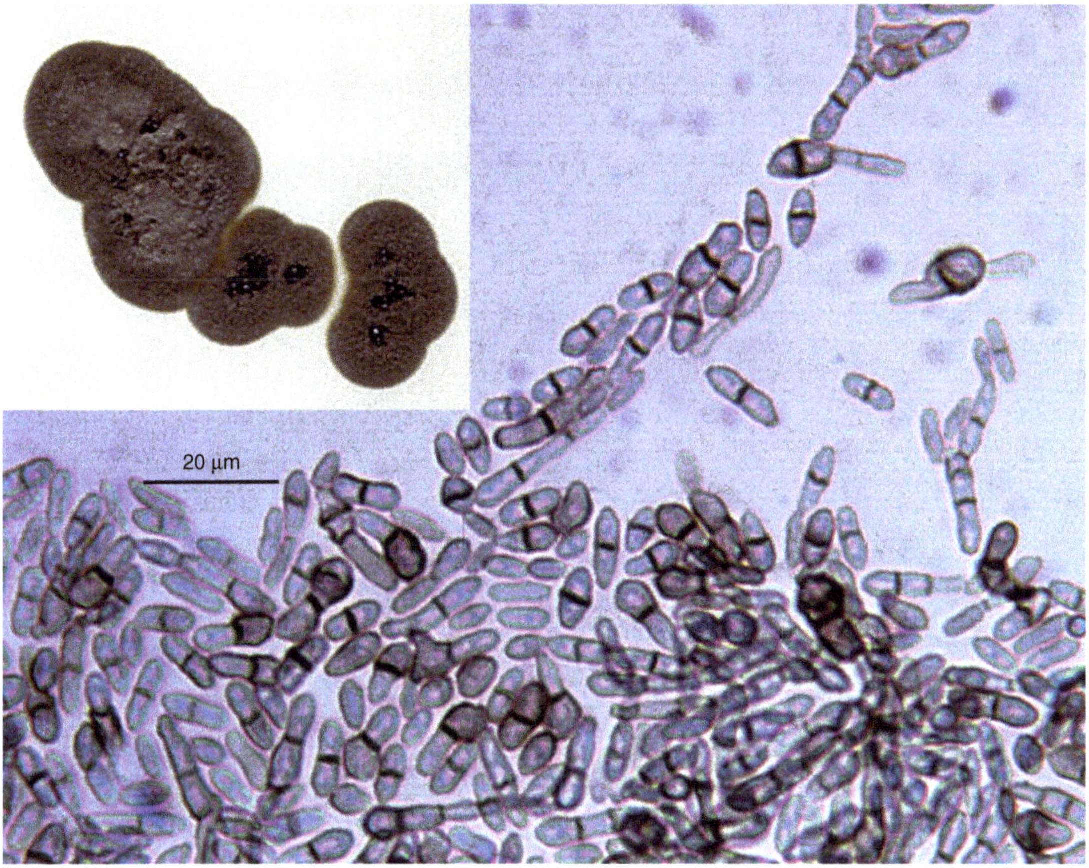

Fig. 45.1. *Hortaea werneckii* culture and conidia.

DOI: 10.1079/9781800622340.0045

46 *Kluyveromyces marxianus* (Hansen) van der Walt

Synonymy: *Candida kefyr* (Beijerineck) van Uden & Buckley; *Candida pseudotropicalis* (Castellani) Basgal.

Kluyveromyces marxianus is a rare cause of candidiasis and is usually associated with superficial cutaneous manifestations rather than systemic disease. Environmental isolations have been made from cheese and dairy products.

RG-1 organism

Culture: Colonies (SDA) white to cream coloured, smooth, glabrous, yeast-like.

Microscopy: Short-ovoid to long-ovoid, budding blastoconidia, 3.0-6.5 x 5.5-11.0 µm, sometimes becoming elongate (up to 16.0 µm).

India ink preparation: Negative - no capsules present.

Dalmau plate culture: Abundant, long, wavy, branched pseudohyphae usually formed, with ovoid blastoconidia, budding off singly, in pairs or chains, often in a verticillated position. **Note:** In some strains pseudohyphae may be scarce or absent.

Molecular identification: ITS sequencing recommended.

MALDI-ToF MS: Able to accurately identify this species.

Germ tube formation: Negative.

Physiological tests: (+ Positive, – Negative, v Variable, w Weak, s Slow, n not done).

Fermentation:

Glucose	+	Sucrose	+	Lactose	v
Galactose	+,s	Maltose	–	Trehalose	–

Growth reactions:

Glucose	+	L-Sorbose	–	*myo*-Inositol	–
Sucrose	+	L-Rhamnose	–	DL-Lactate	+
Raffinose	+	D-Xylose	v	D-Gluconate	–
Melibiose	–	L-Arabinose	v	2-Keto-D-gluconate	–
Galactose	+	D-Arabinose	–	D-Glucosamine	–
Lactose	v	D-Ribose	v	*N*-Acetyl-D-glucosamine	–
Trehalose	–,w	Glycerol	v	D-Glucuronate	–
Maltose	–	Erythritol	–	Nitrate	–
Melezitose	–	Ribitol	v	Urease	–
Methyl-½-D-glucoside	–	Galactitol	–	0.1% Cycloheximide	+
Soluble starch	–	D-Mannitol	v	Growth at 37°C	+
Cellobiose	v	D-Glucitol	v		

Key features: Germ tube negative yeast and sugar assimilation pattern.

Antifungal susceptibility: (Table 46.1).

Table 46.1. *Kluyveromyces marxianus* (Diekema *et al.*, 2009; Pfaller *et al.*, 2013a,b; and Australian national data); MIC μg/mL.

Antifungal	No.	≤0.008	0.016	0.03	0.06	0.125	0.25	0.5	1	2	4	8	16	32	≥64
AmB	123				7	35	66	14				1			
FLU	150				1	31	80	25	6	4	3				
ISAV	36	8	9	12	5	2									
VORI	150	108	32	6	3	1									
POSA	145	2	11	31	51	37	10	3							
ITRA	65		7	12	25	17	4								
ANID	125		3	15	54	42	7	1	1						
MICA	120		1	32	52	30	2	2	1						
5FC	49			4	28	2		5		1	3	4	2		

References: Lachance (2011b).

DOI: 10.1079/9781800622340.0046

47 *Lasiodiplodia theobromae* (Pat.) Griffon & Maublanc

Synonymy: *Botryosphaeria rhodina* (Berk. & Curt.) v. Arx; *Botryodiplodia theobromae* Patouillard.

Lasiodiplodia theobromae is a well-known plant pathogen reported from about 500 host plants, mainly confined to an area 40 degrees north to 40 degrees south of the equator. It has also been associated with ulcerated human cornea, lesions on nail and subcutaneous tissue.

RG-1 organism

Morphological description: Colonies are greyish-sepia to mouse-grey to black, fluffy with abundant aerial mycelium; reverse fuscous black to black. Pycnidia are simple or compound, often aggregated, stromatic, ostiolate, frequently setose, up to 5 mm wide (Fig. 47.1). Conidiophores are hyaline, simple, sometimes septate, rarely branched cylindrical, arising from the inner layers of cells lining the pycnidial cavity. Conidiogenous cells are hyaline, simple, cylindrical to sub-obpyriform, holoblastic, annellidic. Conidia are initially unicellular, hyaline, granulose, sub-ovoid to ellipsoid-oblong, thick-walled, base truncate; mature conidia one-septate, cinnamon to fawn, often longitudinally striate, 20-30 x 10-15 µm (Fig. 47.1). Paraphyses when present are hyaline, cylindrical, sometimes septate, up to 50 µm long.

Key features: Pycnidia producing characteristic two-celled, dark brown, striated conidia.

Molecular identification: Recommended genetic marker: *TEF-1* (de Hoog *et al.*, 2015). ITS sequencing is useful for identifying clinically important species (Bagyalakshmi *et al.* 2008).

Antifungal susceptibility: (Table 47.1)

Table 47.1. *Lasiodiplodia theobromae* (Australian national data); MIC µg/mL.

Antifungal	No.	≤0.016	0.03	0.06	0.125	0.25	0.5	1	2	4	8	16	32	≥64
AmB	4	1				1	1	1						
ISAV	1										1			
VORI	4					1		2		1				
POSA	3								1		2			
ITRA	4							1				3		

References: Liu *et al.* (2012); Phillips *et al.* (2013); de Hoog *et al.* (2015).

Fig. 47.1. *Lasiodiplodia theobromae* showing (a) pycnidia growing on carnation leaf agar and (b) two-celled dark brown conidia with typical longitudinal striations.

DOI: 10.1079/9781800622340.0047

48 *Lichtheimia corymbifera* (Cohn) Vuill.

Synonyms: *Mycocladus corymbifera* (Cohn) J.H. Mirza*; Absidia corymbifera* (Cohn) Saccardo & Trotter; *Mucor corymbifera* Cohn.

Recent phylogenetic based taxonomic revisions of the genus *Absidia* have placed the thermotolerant species into the genus *Lichtheimia* (Hoffmann *et al.*, 2007, 2009). The genus *Lichtheimia* currently contains five mostly saprophytic plant decaying and soil-borne species. *Lichtheimia corymbifera* is the principal pathogen causing human and animal infections, however *L. ramosa* and *L. ornata* have also been reported as human pathogens (often misidentified morphologically as *L. corymbifera*) (Alatruey-Izquierdo *et al.*, 2010; Walther *et al.*, 2019).

RG-2 organism

Morphological description: Colonies are fast growing, floccose, white at first becoming pale grey with age, and up to 1.5 cm high (Fig. 48.1). Sporangiophores are hyaline to faintly pigmented, simple or sometimes branched, arising solitarily or in groups. Sub-sporangial septa are absent or rare. Rhizoids are very sparingly produced and may be difficult to find without the aid of a dissecting microscope to examine the colony on the agar surface. Sporangia are small (10-40 µm in diameter) and are typically pyriform in shape with a characteristic conical-shaped columella and pronounced apophysis, often with a short projection at the top (Fig. 48.1). Sporangiospores vary from subglobose to oblong-ellipsoidal (3-7 x 2.5-4.5 µm), hyaline to light grey and smooth-walled. Intercalary giant cells may also be present. Temperature: optimum 35-37°C; maximum 46-52°C.

Key features: Mucorales, small pyriform-shaped sporangia with a characteristic conical-shaped columellae and pronounced apophysis, rapid growth at 40°C.

Comment: Morphological characteristics alone are not sufficient to reliably differentiate between *Lichtheimia corymbifera, L. ramosa* and *L. ornata*. While *L. ornata* develops large, densely branched giant cells and *L. ramosa* has more ellipsoidal to cylindrical sporangiospores and a faster growth rate, these characters are often difficult to interpret. Molecular methods are needed to accurately separate these species.

Molecular identification. Species recognition in *Lichtheimia* is based on ITS and D1/D2 sequencing (Garcia-Hermoso *et al.*, 2009; Alatruey-Izquierdo *et al.*, 2010).

MALDI-ToF MS: Direct identification of *Lichtheimia* species was described by Schroedl *et al.* (2012).

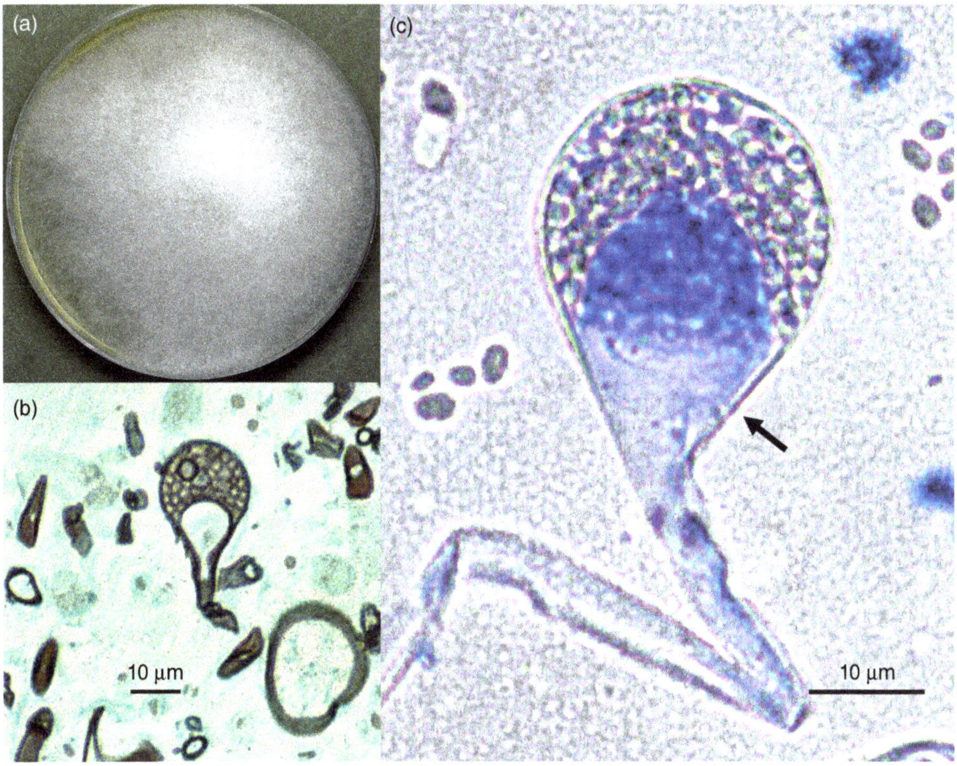

Fig. 48.1. *Lichtheimia corymbifera* showing (a) culture, (b) Grocott-Gomori's methenamine silver-stained tissue section from a lung biopsy showing a typical sporangium, and (c) typical pyriform-shaped sporangium with a conical-shaped columella and pronounced apophysis (arrow).

Antifungal susceptibility: (Table 48.1).

Table 48.1. *Lichtheimia corymbifera* (Espinel-Ingroff *et al.*, 2015a; and Australian national data); MIC µg/mL.

Antifungal	No.	≤0.016	0.03	0.06	0.125	0.25	0.5	1	2	4	8	16	32	≥64
AmB	174			9	31	47	58	23	4		2			
ISAV	6					2	1	2	1					
VORI	36							1	6	7	19	3		
POSA	146			6	14	40	61	22	2	1				
ITRA	131			8	16	41	32	24	6	3	1			

References: Hesseltine and Ellis (1964a, 1966); Ellis and Hesseltine (1965, 1966a); Nottebrock *et al.* (1974); O'Donnell (1979); Samson *et al.* (1995); McGinnis (1980); Ellis (2005b); Domsch *et al.* (2007); de Hoog *et al.* (2015).

DOI: 10.1079/9781800622340.0048

49 *Lodderomyces elongisporus* (Recca & Mrak) van der Walt

Lodderomyces elongisporus has been isolated from soft drinks and juice concentrates, natural fermentations of cocoa, soil, an infected fingernail, human blood infections and from baby cream. Initially isolates appeared to be atypical forms of *Candida parapsilosis*, but D1/D2 sequence analysis showed that they were *L. elongisporus*. In view of these results, *L. elongisporus* may be more common among clinical isolates than initially thought (Lockhart *et al*., 2008; Kurtzman *et al*., 2011).

RG-1 organism

Culture: Colonies (SDA) white to cream coloured, smooth, glabrous, yeast-like.

Microscopy: Ellipsoid to elongate budding blastoconidia, 2.6-6.3 x 4-7.4 µm, with occasional spherical forms present.

India ink preparation: Negative - no capsules present.

Dalmau plate culture: Abundant, much-branched pseudohyphae produced.

Ascospore formation: Asci are unconjugated, persistent, and are transformed from budding cells. Each ascus forms one, rarely two, long-ellipsoid ascospores. Ascospores were observed on V8 agar after 7-10 days at 25°C.

Germ tube formation: Negative.

Physiological tests: (+ Positive, – Negative, v Variable, w Weak, s Slow).

Fermentation:

Glucose	+	Sucrose	–	Lactose	–
Galactose	–	Maltose	–	Trehalose	+

Growth reactions:

Glucose	+	L-Sorbose	+	*myo*-Inositol	–
Sucrose	+	L-Rhamnose	–	DL-Lactate	–
Raffinose	–	D-Xylose	+,w	D-Gluconate	+,w
Melibiose	–	L-Arabinose	–	2-Keto-D-gluconate	+
Galactose	+	D-Arabinose	–	D-Glucosamine	–
Lactose	–	D-Ribose	–	*N*-Acetyl-D-glucosamine	+
Trehalose	+	Glycerol	+	D-Glucuronate	–
Maltose	+	Erythritol	–	Nitrate	–
Melezitose	+	Ribitol	+	Urease	–
Methyl-½-D-glucoside	+	Galactitol	–	0.1% Cycloheximide	–
Soluble starch	–	D-Mannitol	+	Growth at 37°C	+
Cellobiose	–	D-Glucitol	+		

Antifungal susceptibility: (Fig. 49.1).

Table 49.1. *Lodderomyces elongisporus* (Australian national data); MIC µg/mL.

Antifungal	No.	≤0.016	0.03	0.06	0.125	0.25	0.5	1	2	4	8	16	32	≥64
AmB	9				2	3	1	3						
FLU	9				3	3	1	1	1					
ISAV	1	1												
VORI	9	9												
POSA	9	1	1	1	2	3	1							
ITRA	9	1	2	1	2	3								
ANID	9	3	5	1										
MICA	9	5	3	1										
5FC	9		1		2	5		1						

Key features: In the absence of ascospores, *Lodderomyces elongisporus* cannot be distinguished physiologically from *Candida parapsilosis*, *C. orthopsilosis* and *C. metapsilosis* but can be identified based on ITS sequences (Asadzadeh *et al.*, 2009; Borman *et al.*, 2009; Tavanti *et al.*, 2005) and MALDI-ToF MS analysis.

DOI: 10.1079/9781800622340.0049

50 *Lomentospora prolificans* (Hennebert & Desai)

Synonymy: *Scedosporium prolificans* (Hennebert & Desai) Gueho & de Hoog; *Scedosporium inflatum* Malloch & Salkin.

Lomentospora prolificans (formerly *Scedosporium prolificans*) is distinguishable phylogenetically and morphologically from *Scedosporium* species (Lennon *et al.*, 1994; Lackner *et al.*, 2014a). *L. prolificans* appears to occupy a more restricted geographic range, with infections occurring mainly in Australia, Spain, and the United States (Heath *et al.*, 2009; Revankar and Sutton, 2010). *L. prolificans* infections are refractory to antifungal therapy and are associated with high mortality. Major risk factors include malignancy, cystic fibrosis, and solid organ transplantation. The main clinical presentations are disseminated infection and pulmonary mycoses, followed by bone and joint infections (Cortez *et al.*, 2008; Heath *et al.*, 2009; Rodriguez-Tudela *et al.*, 2009; Revankar and Sutton, 2010).

RG-2 organism

Morphological description: Colonies are rapid growing, flat, spreading, olive-grey to black and have a suede-like to downy surface texture (Fig. 50.1). Conidia are borne in small groups on distinctive basally swollen, flask-shaped conidiophores, which occur singly or in clusters along the vegetative hyphae. Conidia are aggregated in slimy heads, single-celled, hyaline to pale brown, ovoid to pyriform, 3-7 x 2-5 µm, and have smooth thick walls (Fig. 50.1). Growth at 45°C.

Key features: Darkly pigmented mould with initial black pasty colony, conidiophores with distinctly swollen bases, and the conidial mass forms apical aggregates of conidia. A *Graphium* synanamorph is absent and there is no growth on media containing cycloheximide (actidione).

Molecular identification: ITS or β-tubulin sequencing recommended.

MALD-ToF MS: A comprehensive "in-house" database of reference spectra allows accurate identification of *Lomentospora prolificans* (Sleiman *et al.* 2016).

Antifungal susceptibility: (Table 50.1).

Antifungal synergy: (Table 50.2).

References: Malloch and Salkin (1984); Salkin *et al.* (1988); Rippon (1988); Wilson *et al.* (1990); Gilgado *et al.* (2005); Rainer and de Hoog (2006); Guarro *et al.* (2006); de Hoog *et al.* (2015).

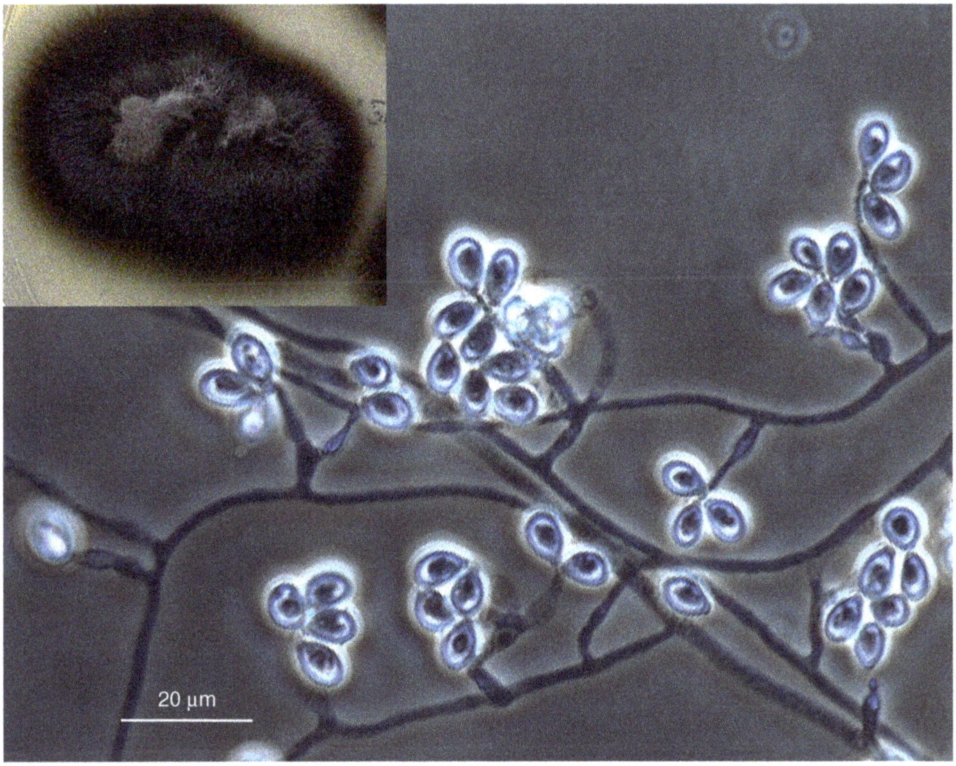

Fig. 50.1. *Lomentospora prolificans* culture and flask-shaped conidiophores bearing small clusters of conidia.

Table 50.1. *Lomentospora prolificans* (Australian national data); MIC μg/mL.

Antifungal	No.	≤0.016	0.03	0.06	0.125	0.25	0.5	1	2	4	8	16	32	≥64
AmB	389					1		2	14	57	315			
ISAV	73								2	3	68			
VORI	384							6	34	68	276			
POSA	305								1	1	303			
ITRA	392										2	387		3

Table 50.2. Synergy testing results for *Lomentospora prolificans* (Australian national data).

Antifungal Combination	No.	ΣFIC < 0.5 (S)	ΣFIC < 0.5-4 (I)	ΣFIC > 4 (A)
VORI/TERB	109	94 (86%)	14 (14%)	0
ITRA/TERB	93	56 (60%)	37 (40%)	0

Abbreviations: A, antagonism; FIC, fractional inhibitory concentration; I, indifferent (synergy or antagonism not demonstrated); S, synergy.

DOI: 10.1079/9781800622340.0050

51 *Lophophyton gallinae* (Ajello) Graser & de Hoog

Synonymy: *Microsporum gallinae* Ajello.

A recent taxonomic revision of the dermatophytes has reduced the genus *Lophophyton* to a single species (de Hoog *et al.*, 2017). *Lophophyton gallinae* is a zoophilic fungus causing fowl favus in chickens and other fowl, affecting the comb and wattles producing "white comb" lesions. A rare cause of tinea in humans. Invaded hairs show a sparse ectothrix infection but do not fluoresce under Wood's ultra-violet light.

RG-2 organism

Morphological description: Colonies are flat with a suede-like texture and are white with a pinkish tinge in colour. Some cultures show radial folding. An orange-pink "strawberry" reverse pigment is usually present (Fig. 51.1). Macroconidia, when present, are usually five to six celled, thin to thick-walled, slightly echinulate, cylindrical to clavate with a narrow base and blunt tip, 15-60 x 6-10 µm (Fig. 51.1). Microconidia are ovoidal to pyriform in shape.

Key features: Macroconidial morphology, culture characteristics and clinical lesions in chickens.

Molecular identification: ITS sequencing is recommended.

References: Rebell and Taplin (1970); Rippon (1988); Graser *et al.* (2008); Cafarchia *et al.*, (2013); de Hoog *et al.* (2017).

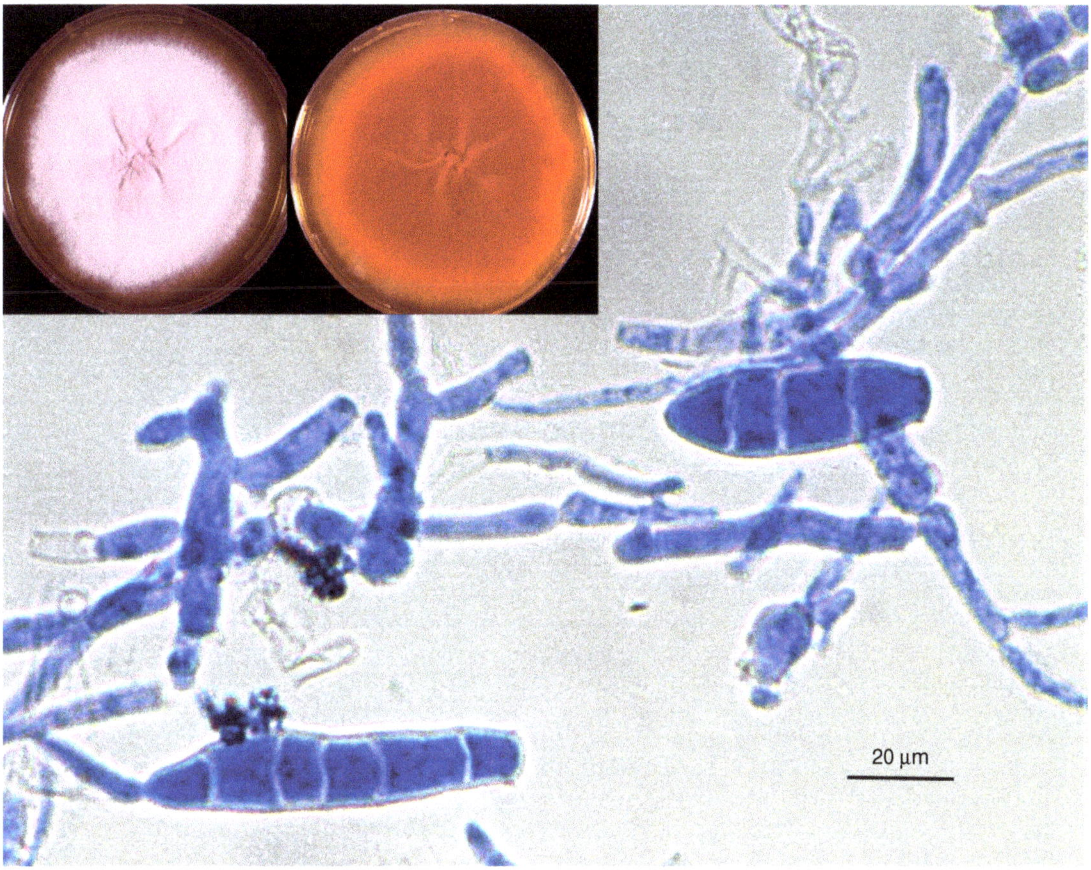

Fig. 51.1. *Lophophyton gallinae* culture and macroconidia.

DOI: 10.1079/9781800622340.0051

52 *Madurella* complex

The genus *Madurella* was originally based on tissue morphology (mycetoma with black grains) and the formation of sterile cultures on mycological media. Initially, two species were described *M. mycetomatis* and *M. grisea.* However recent molecular studies have recognised five species: *Madurella mycetomatis, Trematosphaeria grisea* (formerly *M. grisea*), *M. fahalii, M. pseudomycetomatis* and *M. tropicana* (de Hoog *et al.*, 2004a, 2012; Desnos-Ollivier *et al.*, 2006; Yan *et al.*, 2010; Ahmed *et al.,* 2020). All species have been isolated from soil and are major causative agents of mycetoma.

52.1. *Madurella mycetomatis* (Laveran) Brumpt

RG-2 organism

Morphological description: Colonies are slow growing, flat and leathery at first, white to yellow to yellowish-brown, becoming brownish, folded and heaped with age and the formation of aerial mycelia. A brown diffusible pigment is characteristically produced in primary cultures. Although most cultures are sterile, phialides that bear small rounded conidia in short chains may be occasionally present (Fig. 52.1). Sclerotia are frequently formed. The optimum temperature for growth of this mould is 37°C.

Grains of *Madurella mycetomatis* (tissue microcolonies) are brown or black, 0.5-1.0 mm in size, round or lobed, hard and brittle, composed of hyphae which are 2-5 µm in diameter, with terminal cells expanded to 12-15 (30) µm in diameter.

Madurella mycetomatis can be distinguished from *Trematosphaeria grisea* by growth at 37°C and its ability to assimilate lactose but not sucrose.

Key features: Black grain mycetoma, growth at 37°C, diffusible brown pigment produced on culture and the occasional presence of phialides.

Molecular identification: ITS sequencing is recommended for species identification (Ahmed *et al.*, 2014a; Desnos-Olliver *et al.*, 2006; Irinyi *et al.*, 2015). A five-locus phylogenetic analysis was performed by Ahmed *et al.* (2014b) using the ITS region 18*S*, D1/D2 region of the 28*S*, *RPB2* and *TEF-1* genes.

References: Chandler *et al.* (1980); McGinnis (1980); Rippon (1988); Yan *et al.* (2010); de Hoog *et al.* (2015); Ahmed *et al.* (2020).

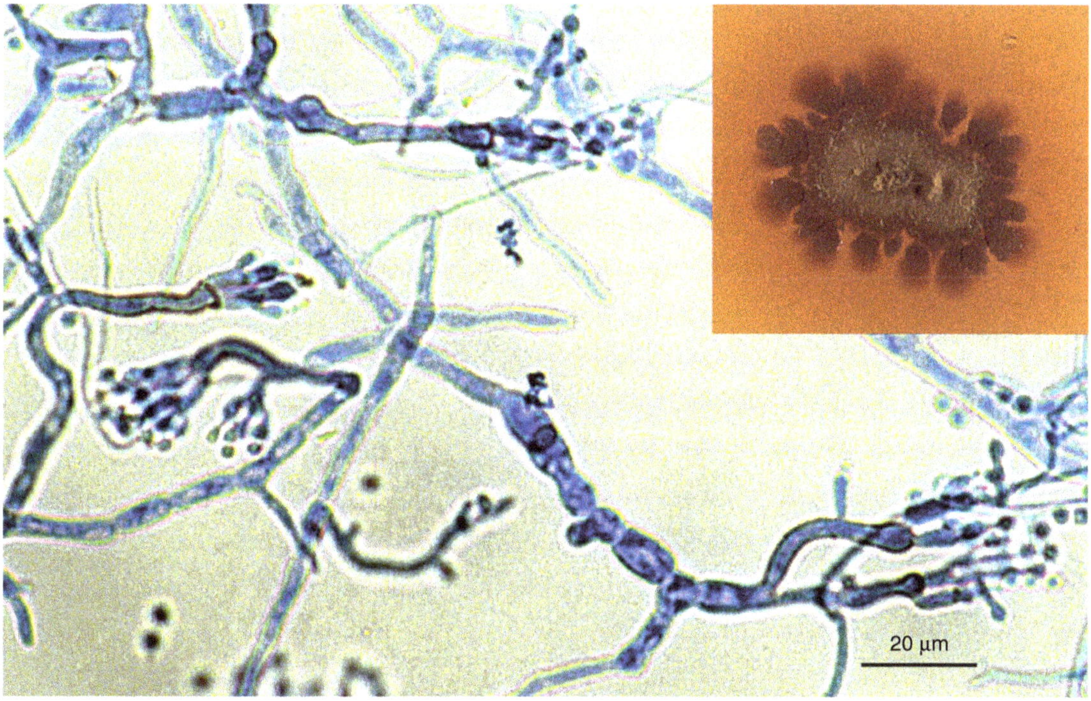

Fig. 52.1. *Madurella mycetomatis* culture showing the typical brown diffusible pigment in the agar and phialides producing small conidia (rarely seen as most isolates are sterile).

DOI: 10.1079/9781800622340.0052

53 *Magnusiomyces* Zender

The name of this genus has changed several times in recent years because of changes in nomenclatural rules (de Hoog *et al.*, 2020). Based on a phylogenetic analysis of rDNA gene sequences, de Hoog and Smith (2004) transferred *Geotrichum capitatum* to the genus *Magnusiomyces*, as *M. capitatus;* and *Geotrichum clavatum* to *Saprochaete clavata*. More recently Kaplan *et al.* (2018, 2021) using multilocus sequence typing of five genes, but principally *RPB2*, have re-defined *Saprochaete clavata* as *Magnusiomyces clavatus*.

Magnusiomyces capitatus and *M. clavatus* are human pathogens that are closely related and are frequently mistaken for each other (Noster *et al.*, 2022). Desnos-Ollivier *et al.* (2014) proposed species-specific carbon assimilation patterns and MALDI-ToF MS profiles to enable the identification of *M. capitatus, M. clavatus,* and *Dipodascus geotrichum* (*Geotrichum candidum*) to the species level.

53.1. *Magnusiomyces capitatus* (de Hoog *et al.*) de Hoog & M Th. Smith

Synonymy: *Saprochaete capitata* (Diddens & Lodder) de Hoog & M.Th. Smith; *Geotrichum capitatum* (Diddens & Lodder) v. Arx; *Trichosporon capitatum* Diddens & Lodder; *Blastoschizomyces capitis* (Diddens & Lodder) Salkin *et al.*

Magnusiomyces capitatus occurs quite commonly in humans, usually as a transient component of normal skin flora and sputum. Systemic infections including pulmonary, fungaemia and endocarditis have been reported in immunosuppressed patients.

RG-1 organism

Morphological description: Colonies are moderately fast growing, flat, whitish, and finely suede-like with no reverse pigment. Hyphae are profusely branched at acute angles, with terminal and intercalary conidiogenous cells which form long, cicatrised rachis on which conidia are borne. Conidia are hyaline, smooth, one-celled, cylindrical to clavate, with a rounded apex and flat base, 7-10 x 2.5-3.5 µm. Rectangular arthroconidia are also often present.

Molecular identification: Molecular detection was developed by Arrieta-Aguirre *et al.* (2017). ITS sequencing may not resolve species; use of *RPB2* is recommended for definitive identification (Kaplan *et al.*, 2018, 2021).

Note: *Magnusiomyces capitatus* can be recognised by its arthroconidia and by absence of growth with D-xylose, sucrose, cellobiose, but growth at 45°C and resistance to 0.1% cycloheximide. The closely related species *M. cla*vatus differs by being cellobiose (+), arbutin (+) and salicin (+) (Desnos-Ollivier *et al.*, 2014).

Physiological tests: (+ Positive, – Negative, v Variable, w Weak, s Slow, n not done).

Fermentation: Absent.

Growth reactions:

Glucose	+	L-Sorbose	v	*myo*-Inositol	–
Sucrose	–	L-Rhamnose	–	DL-Lactate	+
Raffinose	–	D-Xylose	–	D-Gluconate	–
Melibiose	–	L-Arabinose	–	2-Keto-D-gluconate	–
Galactose	+	D-Arabinose	–	D-Glucosamine	–
Lactose	–	D-Ribose	–	*N*-Acetyl-D-glucosamine	n
Trehalose	–	Glycerol	+	D-Glucuronate	n
Maltose	–	Erythritol	–	Nitrate	–
Melezitose	–	Ribitol	–	Urease	–
Methyl-½-D-glucoside	–	Galactitol	–	0.1% Cycloheximide	+
Soluble starch	–	D-Mannitol	+	Growth at 40°C	+
Cellobiose	–	D-Glucitol	+		

Antifungal susceptibility: (Table 53.1).

Table 53.1. *Magnusiomyces capitatus* limited data (Garcia-Ruiz *et al.*, 2013; and Australian national data); MIC µg/mL. **Note:** Isolates of *M. capitatus* are intrinsically resistant to echinocandins (Arendrup *et al.*, 2014).

Antifungal	No.	≤0.016	0.03	0.06	0.125	0.25	0.5	1	2	4	8	16	32	≥64
AmB	11					1	1	5	4					
FLU	11								2	2	2	3	2	
VORI	10		2	3	2	1	1	1						
POSA	10			2	1	3	2	2						
ITRA	11				2	4	5							
5FC	6			4	1		1							

References: de Hoog and Smith (2004, 2011c); Garcia-Ruiz *et al.* (2013); Desnos-Ollivier *et al.* (2014); Arendrup *et al.* (2014); Kaplan *et al.* (2018, 2021); Noster *et al.* (2022).

53.2. *Magnusiomyces clavatus* (de Hoog *et al.*) Kaplan

Synonymy: *Geotrichum clavatum* de Hoog, M.Th. Smith & Gueho; *Saprochaete clavata* (de Hoog, M.Th. Smith & Gueho) de Hoog & M.Th. Smith.

Magnusiomyces clavatus (formerly known as *Geotrichum capitatum* or *Saprochaete clavata*), has only very infrequently been implicated in invasive human infection. However, an outbreak of invasive infections caused by *M. clavatus* in haematology patients has been reported (Vaux *et al.*, 2014).

RG-1 organism

Morphological description: Colonies are moderately fast growing, flat, whitish and butyrous. True hyphae are abundant, soon breaking up into rectangular arthroconidia of variable size, 2.8-4 3 x 6-20 µm. Sympodial conidiogenesis is occasionally present. Terminal parts of hyphae may swell and become thick-walled.

Note: *Magnusiomyces clavatus* and *M. capitatus* are human pathogens that are closely related and are frequently mistaken for each other.

Molecular identification: ITS and/or *RPB2* sequencing is recommended for accurate species identification (Kaplan *et al.*, 2018, 2021).

MALDI-ToF MS: Reliably identifies *M. capitatus, M. clavatus,* and *Dipodascus geotrichum* (*Geotrichum candidum*) to species level (Desnos-Ollivier *et al.*, 2014).

Physiological tests: (+ Positive, - Negative, v Variable, w Weak, s Slow, n not done).

Fermentation: Absent.

Growth reactions:

Glucose	+	L-Sorbose	+	*myo*-Inositol	–
Sucrose	–	L-Rhamnose	–	DL-Lactate	+,w
Raffinose	–	D-Xylose	–	D-Gluconate	–
Melibiose	–	L-Arabinose	–	2-Keto-D-gluconate	–
Galactose	+	D-Arabinose	–	D-Glucosamine	–
Lactose	–	D-Ribose	–	*N*-Acetyl-D-glucosamine	n
Trehalose	–	Glycerol	+	D-Glucuronate	n
Maltose	–	Erythritol	–	Nitrate	–
Melezitose	–	Ribitol	–	Urease	–
Methyl-α-D-glucoside	–	Galactitol	–	0.1% Cycloheximide	–
Soluble starch	–	D-Mannitol	+	Growth at 40°C	+
Cellobiose	+	D-Glucitol	–		

Antifungal susceptibility: (Table 53.2).

Table 53.2. *Magnusiomyces clavatus* very limited data (Australian national data); MIC µg/mL. **Note:** Isolates of *M. clavatus* are intrinsically resistant to echinocandins (Arendrup *et al.*, 2014).

Antifungal	No.	≤0.016	0.03	0.06	0.125	0.25	0.5	1	2	4	8	16	32	≥64
AmB	4					1	1	1	1					
VORI	3			1		1	1							
POSA	3						2	1						
ITRA	4					2	2							
5FC	4			2	1		1							

References: de Hoog and Smith (2004, 2011b); Desnos-Ollivier *et al.* (2014); Arendrup *et al.* (2014); Kaplan *et al.* (2018, 2021); de Hoog *et al.* (2020); Noster *et al.* (2022).

DOI: 10.1079/9781800622340.0053

54 *Malassezia* Baillon

Malassezia species are basidiomycetous yeasts and form part of the normal skin flora of humans and animals. The genus now includes 16 species of which 15 are lipid dependent. These include *M. arunalokei* (human), *M. caprae* (goat, horse), *M. cuniculi* (rabbit), *M. dermatis* (human), *M. equina* (horse, cow), *M. furfur* (human, cow, elephant, pig, monkey, ostrich, pelican), *M. globosa* (human, cheetah, cow), *M. japonica* (human), *M. nana* (cat, cow, dog), *M. obtusa* (human), *M. pachydermatis* (dog, cat, carnivores, birds), *M. restricta* (human), *M. slooffiae* (human, pig, goat, sheep), *M. sympodialis* (human, horse, pig, sheep), *M. vespertilionis* (bats) and *M. yamatoensis* (human) (Cabanes *et al.*, 2011, 2016; Honnavar *et al.*, 2016; Lorch *et al.*, 2018).

Malassezia species may cause various skin manifestations including pityriasis versicolor, seborrhoeic dermatitis, dandruff, atopic eczema and folliculitis. *M. pachydermatis* known to cause external otitis in dogs. Fungaemia due to lipid-dependent *Malassezia* species usually occurs in patients with central line catheters receiving lipid replacement therapy, especially in infants (Tragiannidis *et al.*, 2010; Gaitanis *et al.*, 2012; Arendrup *et al.*, 2014).

Note: With the exception of *M. pachydermatis,* the primary isolation and culture of *Malassezia* species is challenging because *in vitro* growth must be stimulated by natural oils or other fatty substances. The most common method used is to overlay Sabouraud's dextrose agar containing cycloheximide (actidione) with olive oil or alternatively to use a more specialised media like modified Leeming and Notham agar (Kaneko *et al.*, 2007), or modified Dixon's agar (see chapter 111.8).

Comment: For clinical management at the level of the individual patient, species identification is less important, although it is necessary for epidemiological surveillance and outbreak investigation (Arendrup *et al.*, 2014).

RG-1 organisms

Morphological description: On media like modified Dixon's agar, colonies are cream to yellowish, smooth or lightly wrinkled, glistening or dull, and with the margin being either entire or lobate (Fig. 54.1). *Malassezia* is characterised by globose, oblong-ellipsoidal to cylindrical, yeast cells. Reproduction is by budding on a broad base and from the same site at one pole (unipolar) (Fig. 54.1). *Malassezia sympodialis, M. globosa, M. slooffiae* and *M. restricta* are the most frequently found species colonising humans (Table 54.1). CHROMagar™ Malassezia medium is commercially available for the primary isolation and differentiation of the most common *Malassezia* species. Identification and growth criteria for the differentiation of all *Malassezia* species is provided (Table 54.2).

Molecular identification: ITS and D1/D2 sequencing may be used for accurate species identification (de Hoog *et al.*, 2015).

MALDI-ToF MS: Capable of identifying all *Malassezia* species (Kolecka *et al.*, 2014).

Antifungal susceptibility: There is no standardised method for testing *Malassezia* species, special growth conditions are needed, and published results may be variable (Table 54.3). **Note:** Susceptibility testing is not recommended for guiding treatment.

References: Guillot and Gueho (1995); Gueho *et al.* (1996); Guillot *et al.* (1996, 2000); Boekhout *et al.* (2010); Cabanes *et al.* (2011, 2016); Cafarchia *et al.* (2011); Honnavar *et al.* (2016); Lorch *et al.* (2018); de Hoog *et al.* (2020).

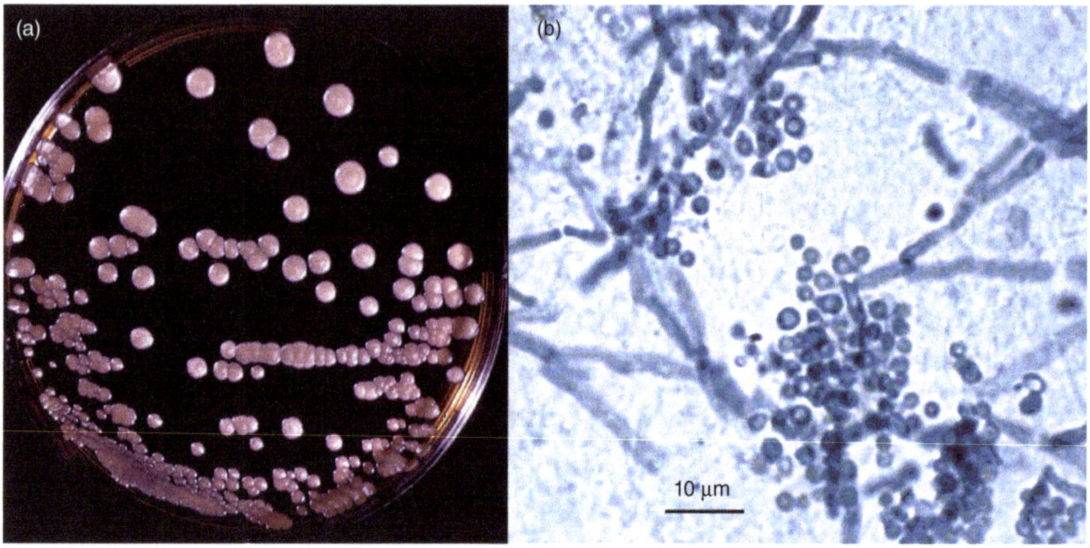

Fig. 54.1. *Malassezia furfur* showing (a) culture on modified Dixon's agar and (b) direct microscopy of skin scrapings showing budding yeast-like cells and short angular hyphal forms (the so-called spaghetti and meatballs appearance) typically seen in pityriasis versicolor.

Table 54.1. Basic phenotypic key to medically important species (Arendrup *et al.*, 2014; de Hoog *et al.*, 2020).

1	Growth on SGA	*M. pachydermatis*
	No growth on SGA	2
2	Catalase reaction positive	3
	No catalase reaction	*M. restricta*
3	Growth on SGA with Tween 40	4
	No growth on SGA with Tween 40	5
4	Growth on SGA with Tween 80	6
	No growth on SGA with Tween 80	*M. slooffiae*
5	Cells long, cylindrical	*M. obtusa*
	Cells spherical	*M. globosa*
6	Growth on SGA with cremophor EL	*M. furfur*
	No growth on SGA with cremophor EL	*M. sympodialis*

Table 54.2. Identification and growth criteria for the differentiation of *Malassezia* species (Cabanes *et al.*, 2011, 2016; Honnavar *et al.*, 2016; Lorch *et al.*, 2018).

+ Positive, – Negative, v Variable, w Weak, s Slow, nd No Data, SDA Sabouraud's dextrose agar, T80 Tween 80, T60 Tween 60, T40 Tween 40, T20 Tween 20.

Species	Buds	SDA	40°C	Cremophor EL	T80	T60	T40	T20	Esculine	Catalase
M. arunalokei	narrow	–	–	–	v	–	–	–	nd	–
M. caprae	narrow	–	–	–	+	+	+	+	+	+
M. cuniculi	narrow	–	+	–	–	–	–	–	nd	+
M. dermatis	wide	–	+	w,+	+	+	+	+	w	+
M. equina	narrow	–	–	–	+	+	+	+	+	+
M. furfur	wide	–	+	+	+	+	+	+	w	+
M. globosa	narrow	–	–	–	–		–	–	–	+
M. japonica	wide	–	–	nd	–	+	w	–	nd	+
M. nana	narrow	–	v	–	w	+	+	v	nd	+
M. obtusa	wide	–	–	–	–		–	–	+	+
M. pachydermatis	wide	+	+	–,w	+	+	+	–,w	v	v
M. restricta	narrow	–	–	–	–		–	–	–	–
M. slooffiae	wide	–	+	–	+,w	+	+	+	–	+
M. sympodialis	narrow	–	+	+–,w	+	+	+	+	+	+
M. vespertilionis	narrow	–	w	–	w	+	+	w	nd	–
M. yamatoensis	wide	–	–	nd	+	+	+	+	nd	+

Table 54.3. *Malassezia* spp., limited data available (Velegraki *et al.*, 2004; Miranda *et al.*, 2007; Rojas *et al.*, 2014); MIC μg/mL.

	M. furfur		*M. sympodialis*		*M. globosa*	
Antifungal	Range	MIC_{90}	Range	MIC_{90}	Range	MIC_{90}
AmB	0.125–16	2	0.06–4	2	0.03–4	1
FLU	0.125–64	16	0.125–16	8	0.125–32	4
KETO	0.03–4	0.25	0.03–0.25	0.125	0.03–0.5	0.06
ITRA	0.03–0.25	0.25	0.03–0.125	0.06	0.03–0.125	0.125
VORI	0.03–16	1	0.03–0.125	0.125	0.3–0. 25	0.125
POS	0.03–32	2	0.03–0.06	0.03	0.03–0.06	0.06

DOI: 10.1079/9781800622340.0054

55 *Malbranchea pulchella* Sacc. & Penz.

Malbranchea species are soil fungi of worldwide distribution, which microscopically may resemble *Coccidioides immitis/posadasii*. **Note:** Culture identification by ITS sequencing is the method of choice if there is any suspicion or possibility of an isolate being *C. immitis/posadasii*.

RG-1 organism

Morphological description: Colonies are white to sulphur-yellow to ochre-brown in colour, suede-like in texture, with a reddish-brown reverse, and often a reddish diffusible pigment. Microscopic morphology shows typical hyaline, one-celled, cylindrical, truncate, alternate arthroconidia produced in terminal fertile portions of the hyphae (Fig. 55.1). Arthroconidia are released by lysis of the disjunctor cells. These arthroconidia may be perceived as a yellow dust when released at maturity.

Key features: Ascomycetous mould producing alternate arthroconidia with disjunctor cells.

References: Cooney and Emerson (1964); Sigler and Carmichael (1976); McGinnis (1980); Rippon (1988); de Hoog *et al.* (2015).

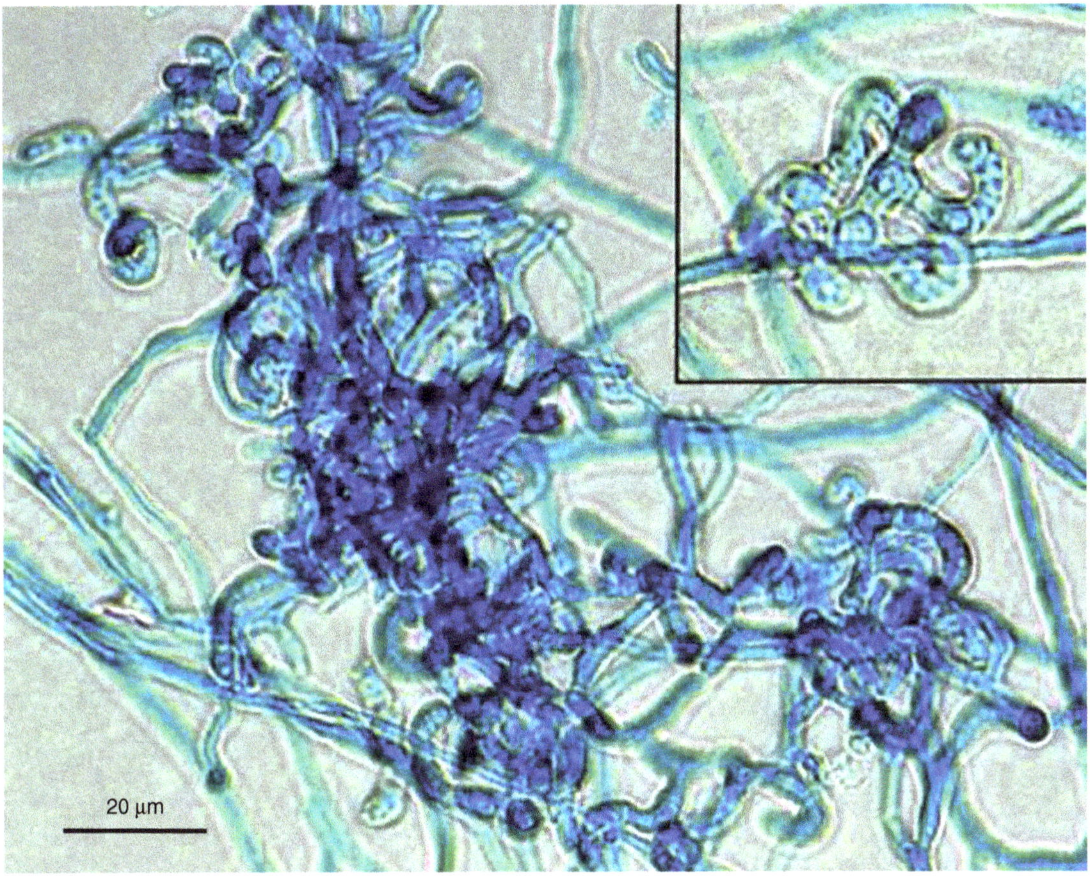

Fig. 55.1. *Malbranchea pulchella* arthroconidia produced in tightly coiled, terminal fertile branches of the hyphae.

DOI: 10.1079/9781800622340.0055

56 *Marquandomyces marquandii* (Massee) Samson, Houbraken & Luangsa-ard

Synonymy: *Paecilomyces marquandii* (Massee) Hughes.

In a review of the molecular phylogeny and classification of the genus *Metarhizium*, Mongkolsamrit *et al.* (2020) have transferred *Paecilomyces marquandii* into a new genus as *Marquandomyces marquandii*. *M. marquandii* is a soil fungus of worldwide distribution from temperate to tropical regions and is seldom associated with human infection.

RG-1 organism

Morphological description: Colonies are fast growing, suede-like to floccose, pale vinaceous to violet coloured, with a yellow to orange yellow reverse (Fig. 56.1). Conidiophores are erect, arising from submerged hyphae, 50-300 µm in length, bearing loose whorls of branches and phialides. Conidiophore stipes are 2.5-3.0 µm wide, hyaline and smooth-walled. Phialides are swollen at their bases, tapering into a thin, distinct neck (Fig. 56.2). Conidia are ellipsoidal to fusiform, smooth-walled to slightly roughened, hyaline to purple in mass, 2.5-3.0 x 2-2.2 µm. Spherical to ellipsoidal chlamydospores, 3-5 µm diameter are present. No growth at 37°C.

Key features: Colony pigmentation with yellow reverse pigment, phialides with swollen bases, smooth conidiophore stipes, presence of chlamydospores, and no growth at 37°C. **Note:** *Purpureocillium lilacinum* has no yellow reverse pigment, rough-walled conidiophore stipes, absence of chlamydospores and growth at 37°C.

Antifungal susceptibility: (Table 56.1).

Table 56.1. *Marquandomyces marquandii* very limited data (Australian national data); MIC µg/mL.

Antifungal	No.	≤0.016	0.03	0.06	0.125	0.25	0.5	1	2	4	8	16	32	≥64
AmB	7					1			1	1	4			
VORI	4			1	2			1						
POSA	2		1		1									
ITRA	7		1			2	2		1			1		

References: Samson (1974); Domsch *et al.* (2007); de Hoog *et al.* (2015); Mongkolsamrit *et al.* (2020).

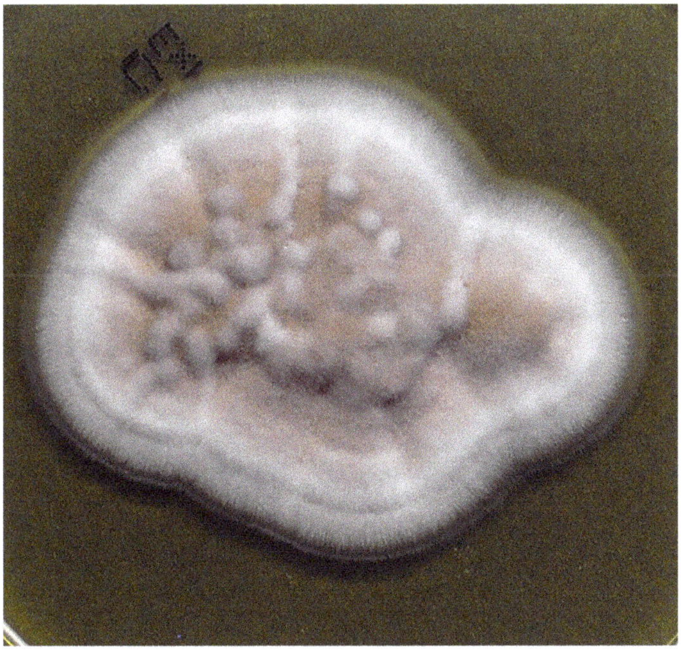

Fig. 56.1. *Marquandomyces marquandii* culture.

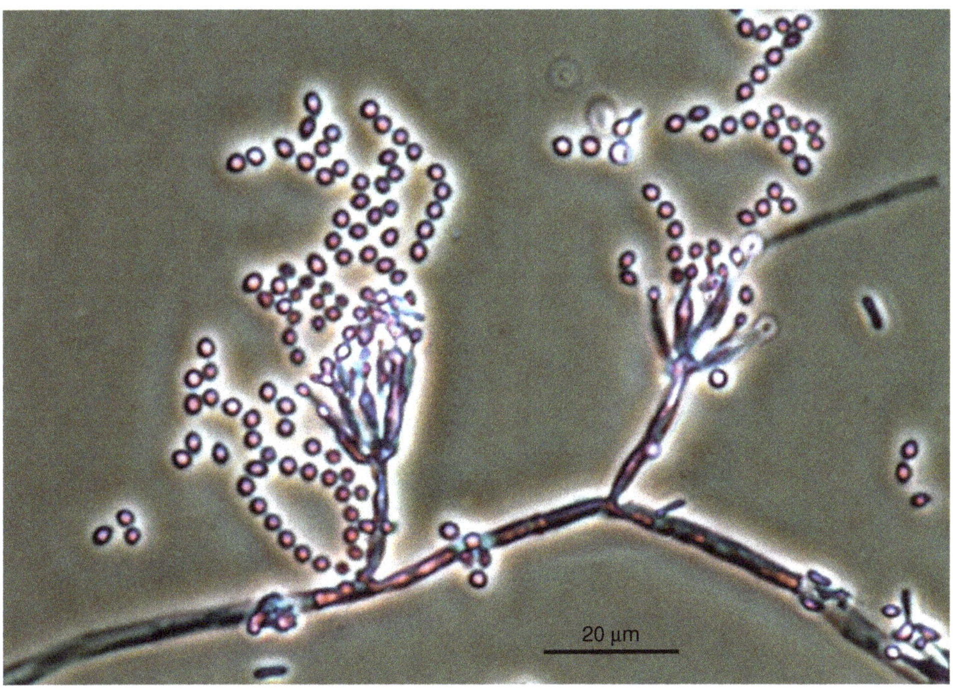

Fig. 56.2. *Marquandomyces marquandii* conidiophores, phialides and conidia.

DOI: 10.1079/9781800622340.0056

57 *Meyerozyma guilliermondii* (Wick.) Kurtzman & M. Suzuki

Synonymy: *Candida guilliermondii* (Castellani) Langeron & Guerra.

Meyerozyma guilliermondii has been isolated from numerous human infections, mostly of cutaneous origin. It is also found on normal skin and in sea water, faeces of animals, fig wasps, buttermilk, leather, fish and beer.

RG-1 organism

Culture: White to cream coloured, smooth, glabrous, yeast-like colonies.

Microscopy: Spherical to subspherical budding yeast-like cells or blastoconidia, 2.0-4.0 × 3.0-6.5 µm.

India ink preparation: Negative - no capsules present.

Dalmau plate culture: Branched pseudohyphae with dense verticils of blastoconidia.

Molecular identification: ITS sequencing recommended.

MALDI-ToF MS: Able to accurately identify this species.

Germ tube formation: Negative.

Physiological tests: (+ Positive, – Negative, v Variable, w Weak, s Slow, n not done).

Fermentation:

Glucose	+	Sucrose	+	Lactose	–
Galactose	v	Maltose	–	Trehalose	+

Growth reactions:

Glucose	+	L-Sorbose	v	*myo*-Inositol	–
Sucrose	+	L-Rhamnose	v	DL-Lactate	v
Raffinose	+	D-Xylose	+	D-Gluconate	v
Melibiose	v	L-Arabinose	+	2-Keto-D-gluconate	+
Galactose	+	D-Arabinose	+	D-Glucosamine	+
Lactose	–	D-Ribose	+	*N*-Acetyl-D-glucosamine	+
Trehalose	+	Glycerol	+	D-Glucuronate	n
Maltose	+	Erythritol	–	Nitrate	–
Melezitose	+	Ribitol	+	Urease	–
Methyl-α-D-glucoside	+	Galactitol	v	0.1% Cycloheximide	v
Soluble starch	–	D-Mannitol	+	Growth at 37°C	+
Cellobiose	+	D-Glucitol	+		

Key features: Germ tube negative yeast and sugar assimilation pattern.

Antifungal susceptibility: (Table 57.1).

Table 57.1. *Meyerozyma guilliermondii* (Diekema *et al.*, 2009; Pfaller *et al.*, 2013a,b; and Australian national data); MIC µg/mL.

Antifungal	No.	≤0.016	0.03	0.06	0.125	0.25	0.5	1	2	4	8	16	32	≥64
AmB	234	2	2	9	72	77	49	14	2	1	6			
FLU	235					1	7	17	93	79	19	9	3	7
ISAV	29		1	1	6	6	9	1	3	2				
VORI	241	14	27	103	67	16	3	2	3			6		
POSA	238	1	11	15	55	93	48	5	4			6		
ITRA	60				10	11	32	5	1			1		
ANID	164	1		4	8	6	13	64	59	8	1			
MICA	153	3	1	4	11	20	62	37	13		1	1		
5FC	60		8	47	2	1	1	1						

Reference: Kurtzman (2011b).

DOI: 10.1079/9781800622340.0057

58 *Microsphaeropsis arundinis* (S. Ahmad) B. Sutton

Microsphaeropsis arundinis is an ascomycetous fungus that produces conidia within a pycnidium (an asexual fruiting body containing conidia). It is ubiquitous in soil and fresh water and typically inhabits terrestrial plant hosts. It has a well-known association with *Aruno donax*, a garden escape weed known as "giant reed" or "elephant grass". *M. arundinis* is an emerging cause of phaeohyphomycosis in cats and immunosuppressed humans.

RG-1 organism

Morphological description: Colonies growing slowly, with dense aerial mycelium, initially greenish-grey, later becoming dark brown to grey-brown (Fig. 58.1). Hyphae are septate, pigmented, and irregularly shaped, with swollen segments up to 4 µm in diameter (Fig. 58.1). Pycnidia are subspherical, 250-350 µm in diameter; with a pseudoparenchymatous wall composed of very densely packed cells that appear angular in cross section (textura angularis). Conidiogenous cells ampulliform, up to 5 µm long. Conidia brown, thick- and smooth-walled, cylindrical, 3.5-4.5 x 1.0-1.5 µm.

Key features: Ascomycetous fungus producing ostiolate pycnidia, ampulliform conidiogenous cells, and small, smooth-walled, brown, cylindrical conidia.

Molecular identification: ITS and D1/D2 sequencing are recommended, especially as it may take many weeks before pycnidia are produced in culture (Reppas *et al.*, 2015). **Note:** ITS and D1/D2 sequencing alone cannot discriminate between *Microsphaeropsis arundinis* and *Paraconiothyrium cyclothyrioides*, additional sequencing of *ACT* and β-tubulin is required (Gordon *et al.*, 2012).

References: Kluger *et al.* (2004); Pendle *et al.* (2004); Krockenberger *et al.* (2010); Hall *et al.* (2013); de Hoog *et al.* (2015); Reppas *et al.* (2015).

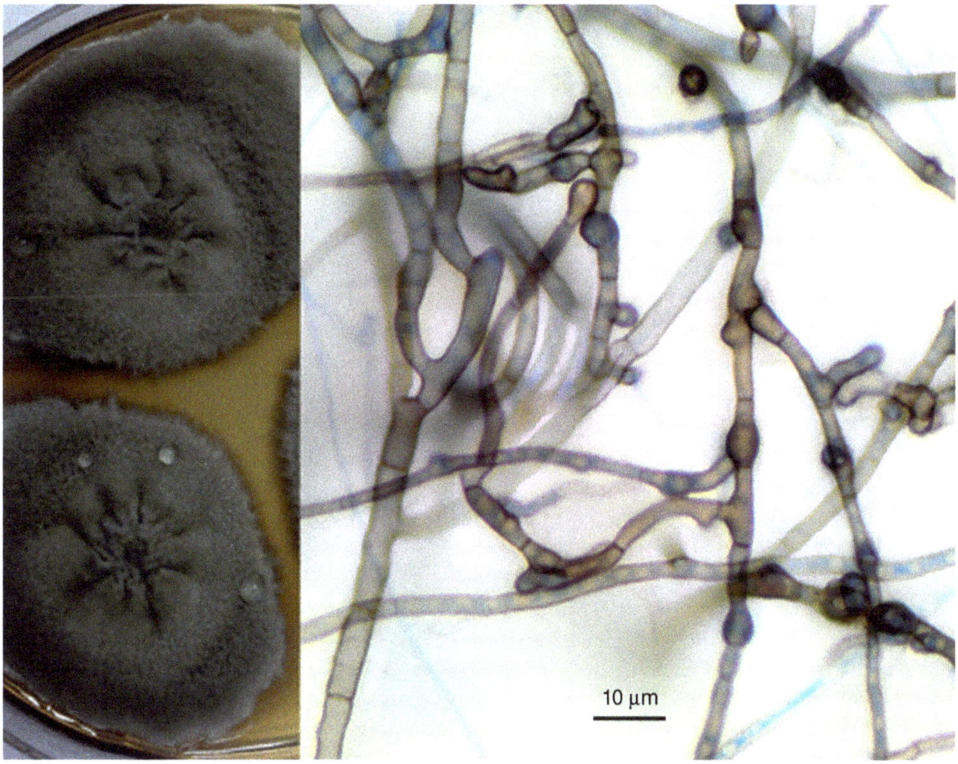

Fig. 58.1. *Microsphaeropsis arundinis* culture and microscopy showing pigmented septate hyphae, with swollen segments.

DOI: 10.1079/9781800622340.0058

59 *Microsporum* Gruby

The genus *Microsporum* is now restricted to just three species: *M. audouinii*, *M. canis* and *M. ferrugineum*. The remaining geophilic and zoophilic species, previously considered *Microsporum* species, have been transferred to the genera *Lophophyton*, *Nannizzia* and *Paraphyton* (de Hoog *et al.*, 2017).

Microsporum species may form both macro- and microconidia, although they are not always present. Cultures are mostly granular to cottony, yellowish to brownish, with a cream coloured or brown colony reverse. Macroconidia are hyaline, multiseptate, with thick, rough cell walls, and are clavate, fusiform or spindle-shaped. Microconidia are single-celled, hyaline, smooth-walled, and are predominantly clavate in shape.

Note: Strains of *Microsporum canis* often do not produce macroconidia and/or microconidia on primary isolation media and subcultures onto polished rice grains or lactritmel agar are recommended to stimulate sporulation. These non-sporulating strains of *M. canis* are often erroneously identified as *M. audouinii*.

Molecular identification: ITS sequencing is recommended (Graser *et al.*, 1998, 2000; Brillowska-Dabrowska *et al.*, 2013).

MALDI-ToF MS: A comprehensive "in-house" database of reference spectra allows accurate identification of *Microsporum* spp. (de Respinis *et al.*, 2013; Calderaro *et al.*, 2014).

References: Rebell and Taplin (1970); Ajello (1977); McGinnis (1980); Mackenzie *et al.* (1986); Weitzman *et al.* (1986); Rippon (1988); Kane *et al.* (1997); Graser *et al.* (1999a, 2008); Domsch *et al.* (2007); Cafarchia *et al.* (2013); de Hoog *et al.* (2015).

59.1. *Microsporum audouinii* Gruby

Microsporum audouinii is an anthropophilic fungus causing non-inflammatory infections of scalp and skin especially in children. Once the cause of epidemics of tinea capitis in Europe and North America, it is now becoming less common. Invaded hairs show an ectothrix infection and usually fluoresce a bright greenish-yellow under Wood's ultra-violet light. Only rarely found in Australasia, typically in newly arrived migrants (McPherson *et al.*, 2008), most reports are thought to be erroneously identified non-sporulating strains of *Microsporum canis*.

RG-2 organism

Morphological description: Colonies are flat, spreading, greyish-white to light tan-white in colour, and have a dense suede-like to downy surface, suggestive of mouse fur in texture. Reverse can be yellow-brown to reddish brown in colour (Fig. 59.1). Some strains may show no reverse pigment. Macroconidia and microconidia are only rarely produced, most cultures are sterile or produce only occasional thick-walled terminal or intercalary chlamydospores (Fig. 59.1). When present macroconidia may resemble those of *M. canis* but are usually longer, smoother, and more irregularly fusiform in shape; microconidia when present are pyriform to clavate in shape and are similar to those seen in other species of *Microsporum*. So called pectinate (comb-like) hyphae and racquet hyphae (a series of hyphal segments swollen at one end) may also be present.

Confirmatory tests

Growth on rice grains: Very poor or absent, usually being visible only as a brown discolouration. This is one of the features, which distinguish *M. audouinii* from *M. canis* (Fig. 59.2).

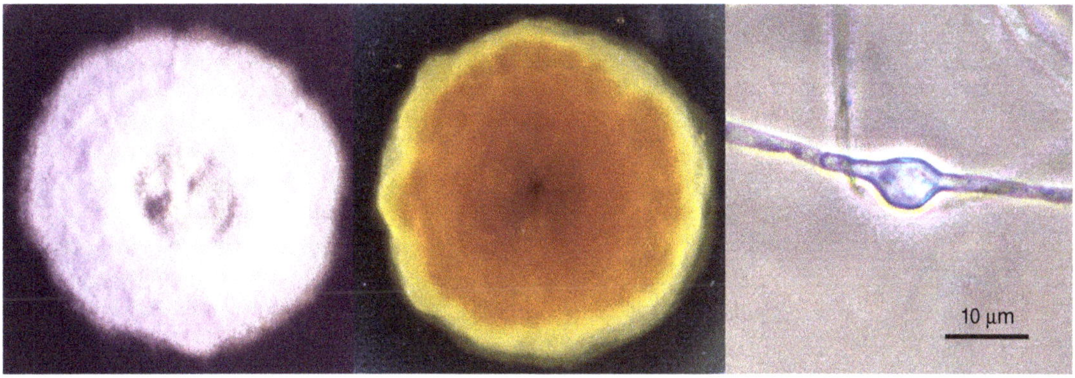

Fig. 59.1. *Microsporum audouinii* culture and microscopy showing a chlamydospore. **Note:** Most cultures are sterile or produce only occasional thick-walled chlamydospores.

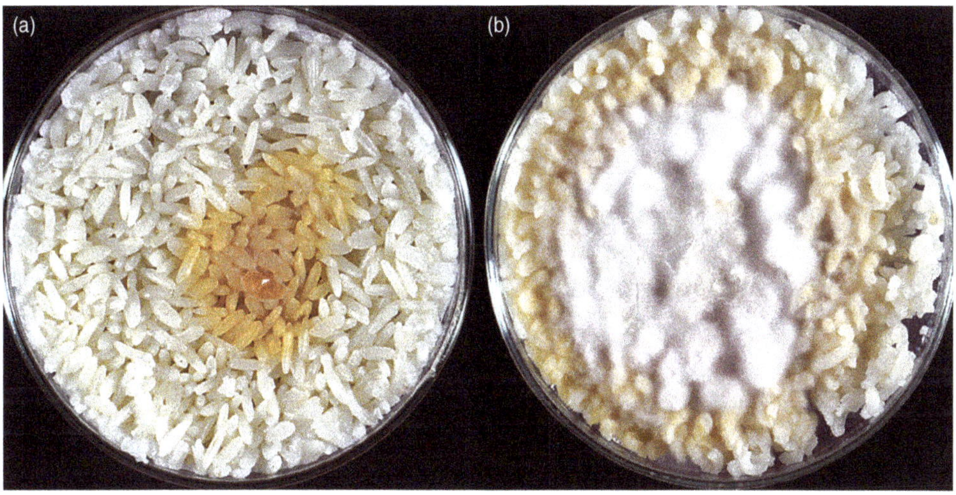

Fig. 59.2. Growth on rice grains (a) *Microsporum audouinii* showing poor growth usually being visible only as a brown discolouration and (b) *M. canis* showing good growth, yellow pigmentation and sporulation.

Reverse pigment on potato dextrose agar: Salmon to pinkish-brown (*M. canis* is bright yellow).

BCP: Both *M. canis* and *M. audounii* demonstrate profuse growth, but only *M. audouinii* shows a rapid pH change to alkaline (purple colour).

Vitamin free agar (Trichophyton agar No.1): Good growth indicating no special nutritional requirements. Cultures are flat, white, suede-like to downy, with a yellow-brown reverse. **Note:** Growth of some strains of *M. audouinii* is enhanced by the presence of thiamine (Trichophyton agar No.4).

Hair perforation test: Negative after 28 days.

Key features: Absence of conidia, poor or no growth on polished rice grains, inability to perforate hair *in vitro*, and culture characteristics.

59.2. *Microsporum canis* Bodin

Microsporum canis is a zoophilic dermatophyte of worldwide distribution and is a frequent cause of ringworm in humans, especially children. Invades hair, skin and rarely nails. Cats and dogs are the main sources of infection. Invaded hairs show an ectothrix infection and fluoresce a bright greenish-yellow under Wood's ultra-violet light.

RG-2 organism

Morphological description: Colonies are flat, spreading, white to cream coloured, with a dense cottony surface, which may show some radial grooves. Colonies usually have a bright golden-yellow to brownish-yellow reverse pigment, but non-pigmented strains may also occur (Fig. 59.3). Macroconidia are typically spindle-shaped with 5-15 cells, verrucose, thick-walled and often have a terminal knob, 35-110 x 12-25 µm (Fig. 59.4). A few pyriform to clavate microconidia are also present. Macroconidia and/or microconidia are

Fig. 59.3. *Microsporum canis* culture.

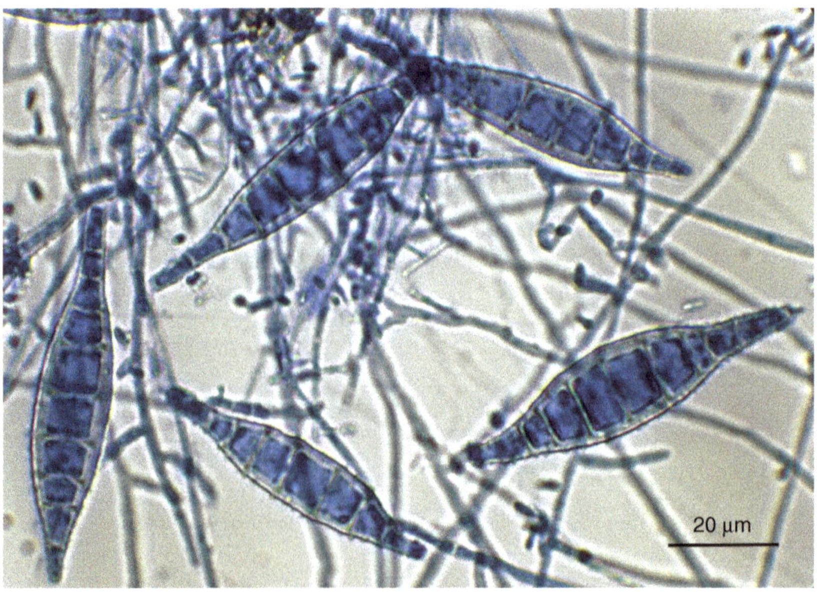

Fig. 59.4. *Microsporum canis* macroconidia.

often not produced on primary isolation media, and it is recommended that sub-cultures be made onto Lactritmel agar and/or boiled polished rice grains to stimulate sporulation.

Confirmatory tests

Growth on rice grains: Good growth of white aerial mycelium with production of yellow pigment. Microscopy reveals numerous macroconidia and microconidia similar to those described above.

Reverse pigment on potato dextrose agar: Bright yellow (both *M. audouinii* and *M. canis* var. *equinum* are salmon to pinkish-brown).

Vitamin free agar (Trichophyton Agar No.1): Good growth indicating no special nutritional requirements. Cultures are flat, white, suede-like to downy, with a yellow to pale yellow-brown reverse.

Hair perforation test: Positive at 14 days.

Key features: Distinctive macroconidia and culture characteristics. Abundant growth and sporulation on polished rice grains and *in vitro* perforation of hair.

***Microsporum canis* dysgonic strains:** Are rare but may also occur. These dysgonic strains have a heaped and folded, yellow-brown thallus and macroconidia are usually absent (Fig. 59.5). However, typical colonies and macroconidia of *M. canis* are usually produced by this variant when subcultured onto polished rice grains. **Note:** The dysgonic type colony of *M. canis* is similar to that of *Microsporum ferrugineum*.

Supplementary description for *Microsporum canis* var. *distortum*: A dysgonic variant of *M. canis* with distinctive distorted macroconidia. Abundant growth and sporulation on rice grains. *M. canis* var. *distortum* is a zoophilic fungus known to cause infections in cats, dogs and other animals. It is a rare cause of tinea capitis in New Zealand, Australia and North America. Clinical disease is similar to *M. canis*. Invaded hairs show an ectothrix infection and fluoresce a bright greenish-yellow under Wood's ultra-violet light (Fig. 59.6).

Supplementary description for *Microsporum canis* var. *equinum*: Now considered to be a genotypic synonym of *M. canis* (de Hoog *et al.*, 2015). This variant of *M. canis* is a rare cause of ringworm of horses. Invaded hairs show an ectothrix infection and fluoresce a bright greenish-yellow under Wood's ultra-violet light. Rarely infects man or other animal species. Reported from Australia, Europe and North America. Colonies are flat, spreading, suede-like, pale buff to pale salmon, usually with some radial furrows. A buff to pinkish-buff to yellow-brown reverse pigment is produced (Fig. 59.7). **Note:** Only a few strains produce conidia on primary isolation, however growth on urea agar usually stimulates the production of macroconidia. Macroconidia are small (especially when compared to those produced by *M. canis*), broad, irregular, spindle-shaped, 18-60 x 5-15 µm with rough, thick walls and few septa (Fig. 59.7). Microconidia are pyriform to clavate in shape, 3-9 x 1.5-3.5 µm, but are rarely produced.

Fig. 59.5. *Microsporum canis* dysgonic strain culture.

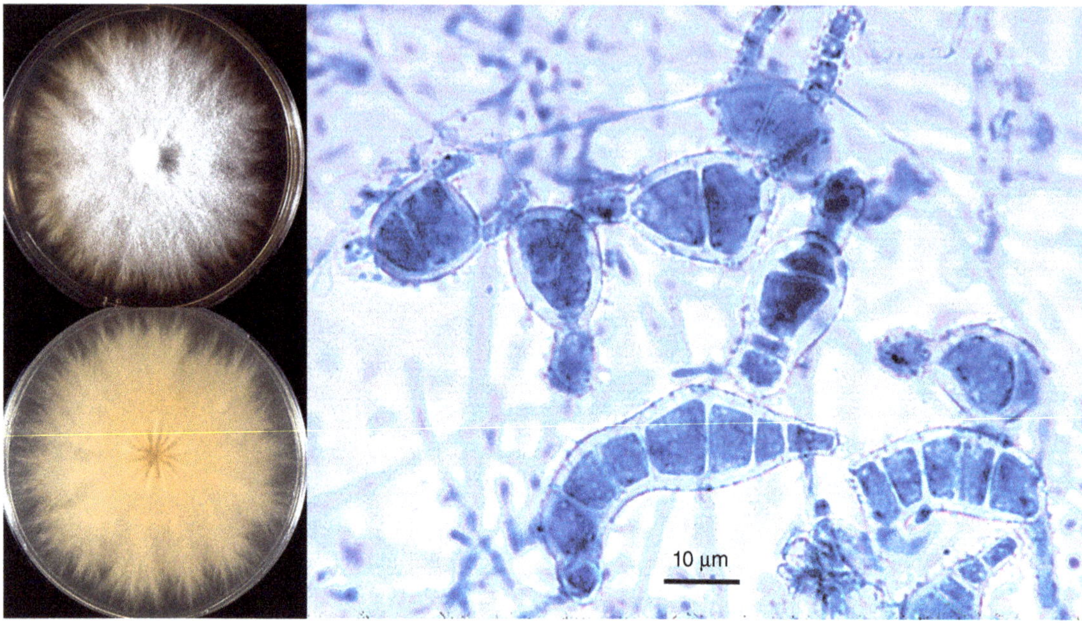

Fig. 59.6. *Microsporum canis* var. *distortum* culture and microscopy showing distorted macroconidia.

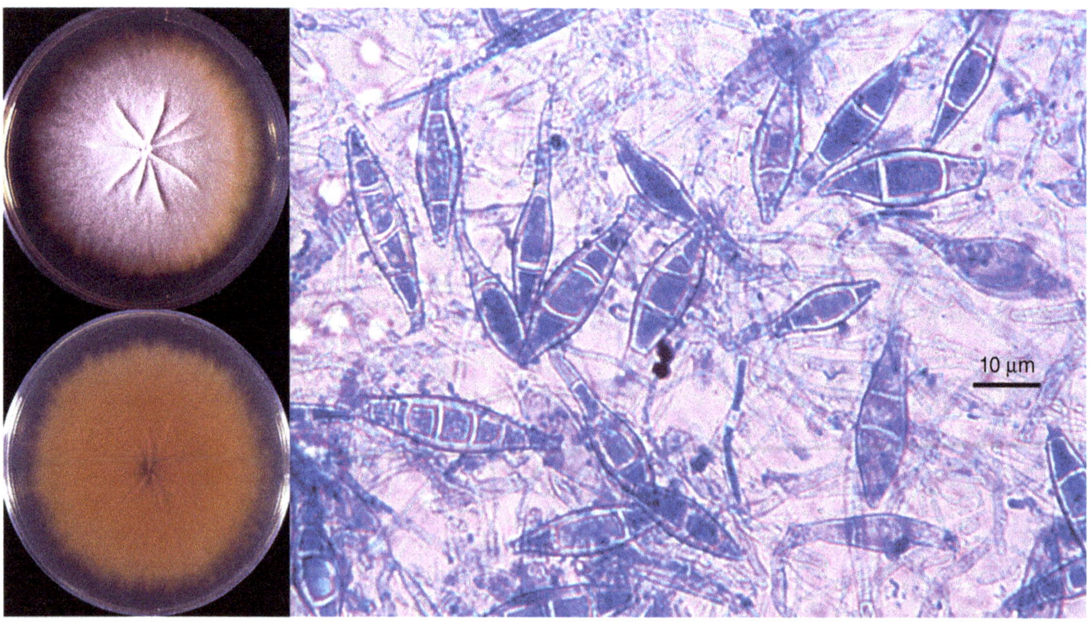

Fig. 59.7. *Microsporum canis* var. *equinum* culture and small spindle-shaped macroconidia.

59.3. *Microsporum ferrugineum* Ota

Microsporum ferrugineum is an anthropophilic fungus causing epidemic juvenile tinea capitis in humans. The clinical features are similar to those of infections caused by *M. audouinii*. Invaded hairs show an ectothrix infection and fluoresce a greenish-yellow under Wood's ultra-violet light. Reported from Asia (including China and Japan), Russia, Eastern Europe and Africa.

RG-2 organism

Morphological description: Colonies are slow growing, forming a waxy, glabrous, convoluted thallus with a cream to buff coloured surface and no reverse pigment (Fig. 59.8). **Note:** Surface pigmentation may vary from cream to yellow to deep red and a flatter white form sometimes occurs. Cultures rapidly become downy and pleomorphic. Microscopic morphology is negative, microconidia or macroconidia are not produced. However, irregular branching hyphae with prominent cross walls ("bamboo hyphae") and chlamydospores are seen. The so-called "bamboo hyphae" is a characteristic of this species (Fig. 59.8).

Key features: Clinical history, culture characteristics and distinctive "bamboo" hyphae.

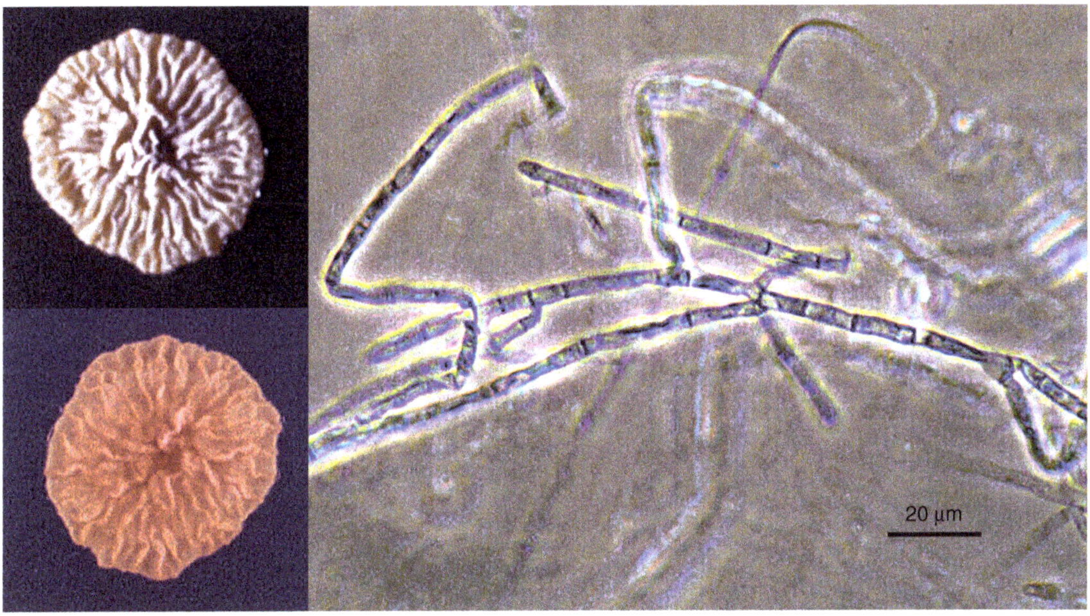

Fig. 59.8. *Microsporum ferrugineum* culture and microscopy showing "bamboo" hyphae.

DOI: 10.1079/9781800622340.0059

60 *Mortierella wolfii* Mehrotra & Baijalm

The genus *Mortierella* is placed in the order Mortierellales (Cavalier-Smith, 1998), separate from the Mucorales and Entomophthorales. The genus contains about 90 recognised species, however *Mortierella wolfii* is the only known pathogenic species being an important causal agent of bovine mycotic abortion, pneumonia and systemic mycosis in New Zealand, Australia, Europe and USA.

RG-2 organism

Morphological description: Cultures are fast growing, white to greyish-white, downy, often with a broadly zonate or lobed (rosette-like) surface appearance and no reverse pigment (Fig. 60.1). Sporangiophores are typically erect, delicate, 80-350 µm in height, 6-20 µm wide at the base, arising from rhizoids or bulbous swellings on the substrate hyphae and terminating with a compact cluster of short terminal branches. Sporangia are usually 15-48 µm in diameter, with transparent walls and a conspicuous collarette is usually present following dehiscence of the sporangiospores. Columellae are generally lacking and sporangiospores are single-celled, short-cylindrical, 6-10 x 3-5 µm, with a double membrane (Fig. 60.1). Chlamydospores with or without blunt appendages (amoeba-like) may be present, zygospores have not been observed. Temperature: grows well at 40-42°C; maximum 48°C.

Key features: Mortierellales, rapid growth at 40°C (thermotolerant), and characteristic delicate terminal branching sporangia without columellae.

References: McGinnis (1980); Rippon (1988); Ellis (2005b); Domsch *et al.* (2007); de Hoog *et al.* (2015); Vandepol *et al.* (2020).

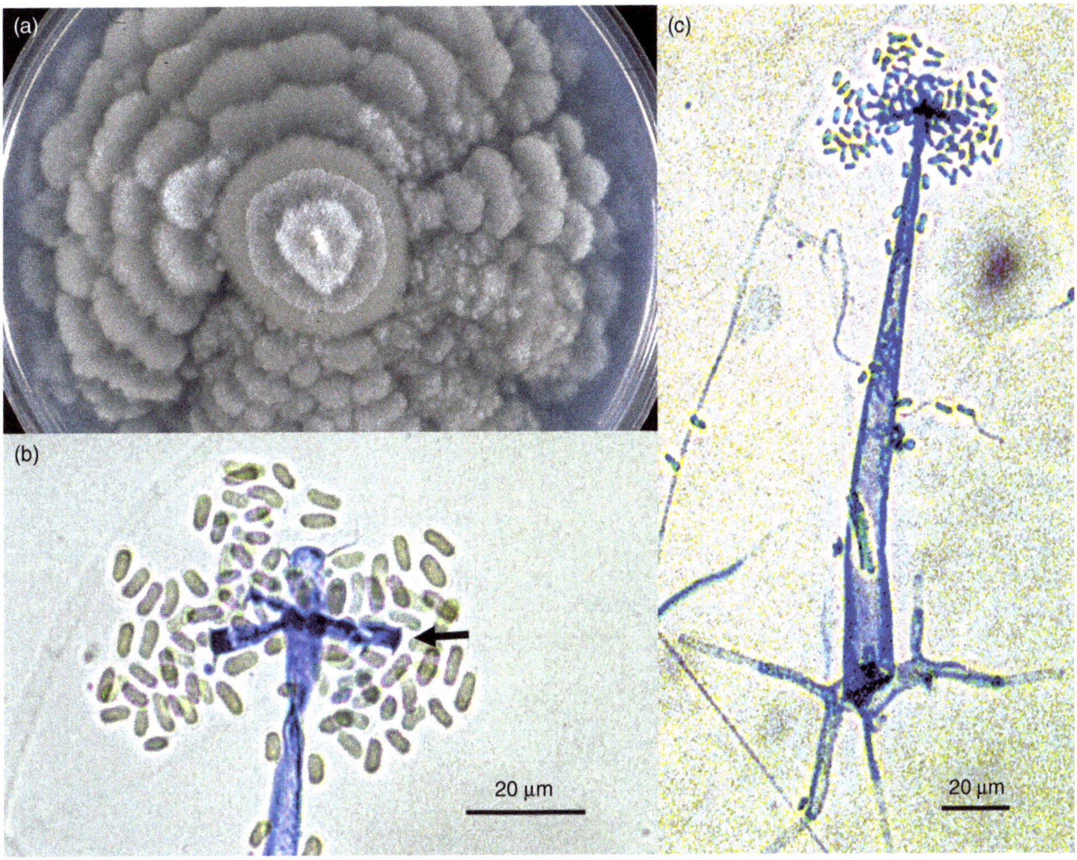

Fig. 60.1. *Mortierella wolfii* showing (a) culture with a typical rosette-like surface appearance, (b) sporangium with terminal branches, with no columella and a conspicuous collarette (arrow), and (c) sporangium with rhizoids arising from a bulbous swelling on the substrate hyphae and terminating with a compact cluster of short terminal branches.

DOI: 10.1079/9781800622340.0060

61 *Mucor* Micheli ex Staint-Amans

The genus *Mucor* contains about 50 recognised taxa, many of which have widespread occurrence and are of considerable economic importance (Zycha *et al.*, 1969; Schipper, 1978; Domsch *et al.*, 2007). However, only a few thermotolerant species are of medical importance and human infections are reported rarely. Most infections reported list *M. circinelloides* and similar species such as *M. indicus*, *M. ramosissimus, M. irregularis* and *M. amphibiorum* as the causative agents. However, *M. hiemalis* and *M. racemosus* have also been reported as infectious agents, although their inability to grow at temperatures above 32°C raises doubt as to their validity as human pathogens and their pathogenic role may be limited to cutaneous infections (Table 61.1) (Scholer *et al.* 1983; Goodman and Rinaldi 1991; Kwon-Chung and Bennett 1992; Walther *et al.*, 2019).

Morphological genus description: Colonies are very fast growing, cottony to fluffy, white to yellow, becoming dark grey, with the development of sporangia. Sporangiophores are erect, simple or branched, forming large (60-300 µm in diameter), terminal, globose to spherical, multispored sporangia, without apophyses and with well-developed subtending columellae (Fig. 61.1). A conspicuous collarette (remnants of the sporangial wall) is usually visible at the base of the columella after sporangiospore dispersal. Sporangiospores are hyaline, grey or brownish, globose to ellipsoidal, and smooth-walled or finely ornamented. Chlamydospores and zygospores may also be present.

Key features: Mucorales, large, spherical, non-apophysate sporangia with pronounced columellae with an inconspicuous collarette at the base of the columella following sporangiospore dispersal.

Molecular identification: ITS sequencing recommended (Walther *et al.* 2012).

References: Schipper (1978); McGinnis (1980); Onions *et al.* (1981); Scholer *et al.* (1983); Rippon (1988); Goodman and Rinaldi (1991); Samson *et al.* (1995); Schipper and Staplers (2003); Ellis (2005b); Domsch *et al.* (2007); Walther *et al.* (2019); de Hoog *et al.* (2020).

Table 61.1. Maximum temperature for growth of the reported pathogenic species of *Mucor*.

Species	Max temp. (°C)	Pathogenicity
M. amphibiorum	36	Animals, principally amphibians
M. circinelloides	37	Animals, occasionally humans
M. hiemalis	30	Questionable cutaneous infections only
M. indicus	42	Humans and animals
M. irregularis	38	Humans
M. racemosus	32	Questionable
M. ramosissimus	36	Humans and animals

61.1. *Mucor amphibiorum* Schipper

RG-2 organism

Morphological description: Colonies are greyish-brown, slightly aromatic and do not grow at 37°C (maximum temperature for growth is 36°C). Sporangiophores are hyaline, erect and mostly unbranched, rarely sympodially branched. Sporangia are dark brown, up to 75 µm in diameter, and are slightly flattened with a diffluent membrane. Columellae are subglobose to ellipsoidal or pyriform, up to 60 x 50 µm, with small collarettes. Sporangiospores are smooth-walled, spherical, and 3.5-5.5 µm in diameter. Zygospores, when formed by compatible mating types, are spherical to slightly compressed, up to 70 x 60 µm in diameter, with stellate projections.

Comment: *Mucor amphibiorum* is distinguished by poor ramification of the sporangiophores and by globose sporangiospores. Ethanol and nitrates are not assimilated (Schipper, 1978; Scholer *et al.*, 1983; de Hoog *et al.*, 2015).

Antifungal susceptibility: (Table 61.2).

Table 61.2. *Mucor amphibiorum* very limited data (Australian national data); MIC µg/mL.

Antifungal	No.	≤0.016	0.03	0.06	0.125	0.25	0.5	1	2	4	8	16	32	≥64
AmB	1						1							
VORI	1										1			
POSA	1						1							
ITRA	1							1						

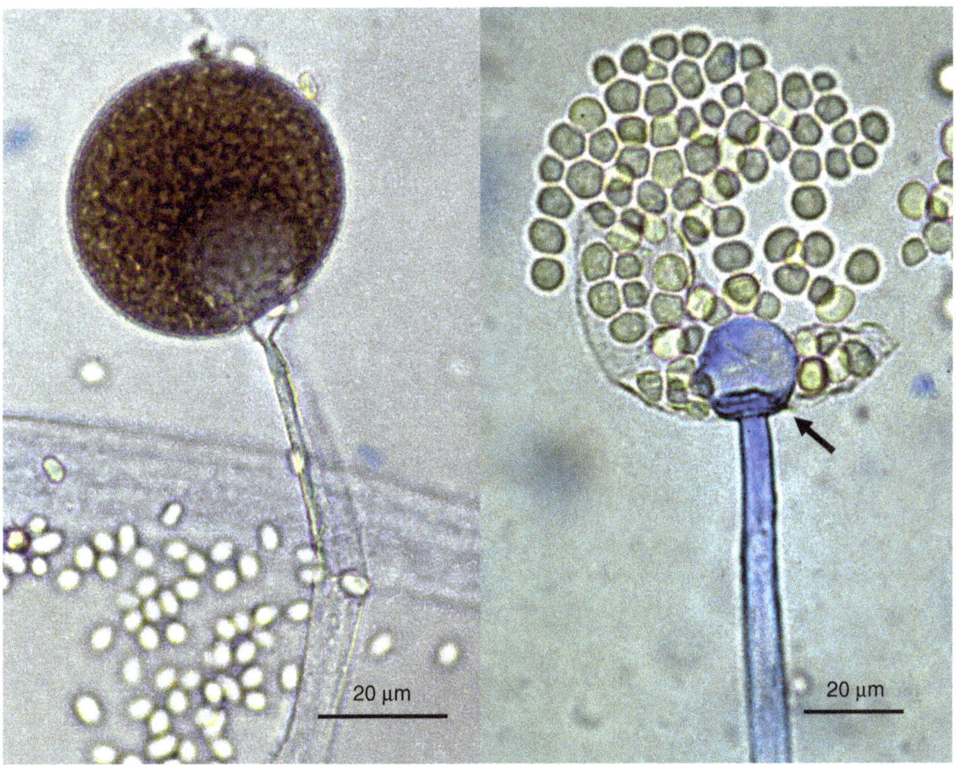

Fig. 61.1. *Mucor* spp. showing sporangia, columellae with inconspicuous collarettes (arrow), and sporangiospores. **Note**: There is no apophysis.

61.2. *Mucor circinelloides* v. Tiegh

RG-1 organism

Mucor circinelloides is a common and variable complex that includes four species: *M. circinelloides, M. lusitanicus, M. griseo-cyanus* and *M. janssenii* (Schipper, 1978; Scholer *et al.*, 1983).

Morphological description: Colonies are floccose, pale greyish-brown (Fig. 61.2) and grow poorly at 37°C (maximum growth temperature 36-40°C). Sporangiophores are hyaline and mostly sympodially branched with long branches erect and shorter branches becoming circinate (coiled). Sporangia are spherical, varying from 20-80 µm in diameter, with small sporangia often having a persistent sporangial wall. Columellae are spherical to ellipsoidal and are up to 50 µm in diameter (Fig. 61.2). Sporangiospores are hyaline, smooth-walled, ellipsoidal, and 4.5-7 x 3.5-5 µm in size. Chlamydospores are generally absent. Zygospores are only produced in crosses of compatible mating types and are reddish-brown to dark brown, spherical with stellate spines, up to 100 µm in diameter and have equal to slightly unequal suspensor cells.

Comment: *Mucor circinelloides* differs from other species of *Mucor* in its formation of short circinated, branched sporangiophores bearing brown sporangia and its ability to assimilate ethanol and nitrates (Schipper, 1976; Scholer *et al.*, 1983; Samson *et al.*, 1995; Schipper and Staplers, 2003; de Hoog *et al.*, 2015).

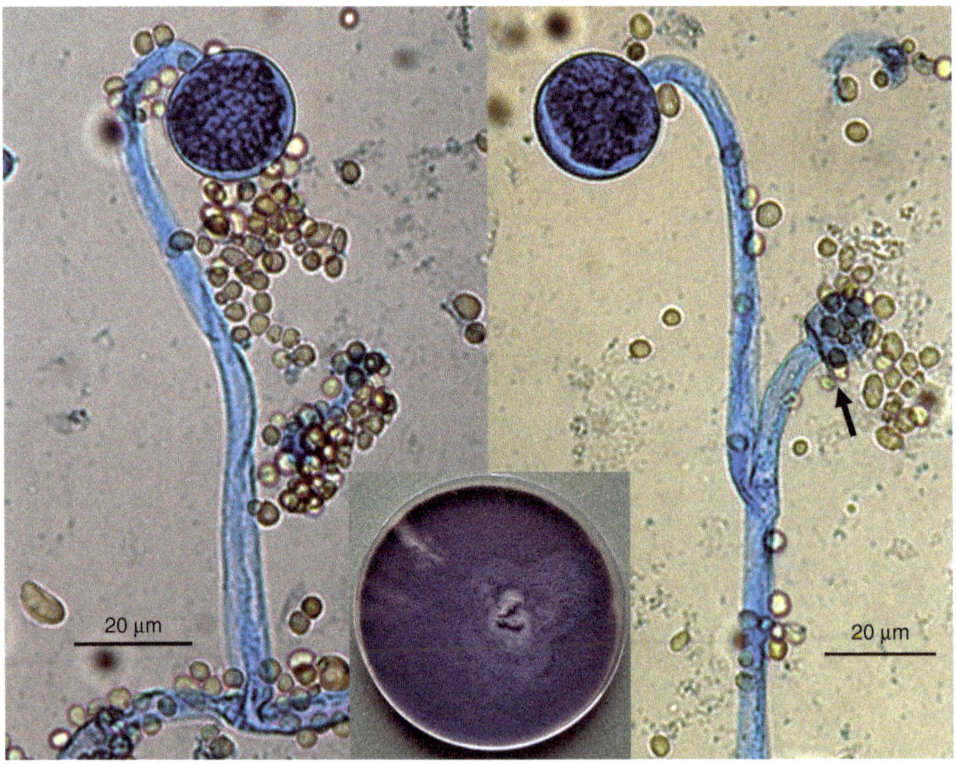

Fig. 61.2. *Mucor circinelloides* culture and sporangia showing circinate sporangiophores, columellae with inconspicuous collarettes (arrow) and sporangiospores.

Antifungal susceptibility: (Table 61.3).

Table 61.3. *Mucor circinelloides*, limited data (Espinel-Ingroff *et al.*, 2015a; Pfaller *et al.*, 2018; and Australian national data); MIC µg/mL.

Antifungal	No.	≤0.016	0.03	0.06	0.125	0.25	0.5	1	2	4	8	16	32	≥64
AmB	143		1	4	16	52	52	18						
ISAV	58							1	9	17	23	9		
VORI	20										20			
POSA	140			2	2	10	28	59	28	5	2	4		
ITRA	89					4	4	23	24	25	5	4		

61.3 *Mucor indicus* Lendner

RG-1 organism

Morphological description: Colonies are characteristically deep yellow, aromatic and have a maximum growth temperature of 42°C. Sporangiophores are hyaline to yellowish, erect or rarely circinate and repeatedly sympodially branched, with long branches. Sporangia are yellow to brown, up to 75 µm in diameter, with diffluent membranes. Columellae are subglobose to pyriform, often with truncate bases, up to 40 µm high. Sporangiospores are smooth-walled, subglobose to ellipsoidal, and 4-5 µm in diameter. Chlamydospores are produced in abundance, especially in light. Zygospores when formed by crosses of compatible mating strains are black, spherical up to 100 µm in diameter, with stellate spines and unequal suspensor cells.

Comment: *Mucor indicus* differs from other species of *Mucor* by its characteristic deep yellow colony colour, growth at over 40°C, assimilating ethanol, but not nitrate, and being thiamine dependent (Schipper, 1978; Schipper and Staplers 2003; de Hoog *et al.*, 2015).

Antifungal susceptibility: (Table 61.4).

Table 61.4. *Mucor indicus* limited data (Espinel-Ingroff *et al.*, 2015a; and Australian national data); MIC µg/mL.

Antifungal	No.	≤0.016	0.03	0.06	0.125	0.25	0.5	1	2	4	8	16	32	≥64
AmB	39		1		3	4	4	16	5		6			
ISAV	5								1	2	2			
VORI	29										26	3		
POSA	29						6	13	9			1		
ITRA	29						3	15	5			6		

61.4. *Mucor irregularis* Stchigel *et al.*

Synonymy: *Rhizomucor variabilis* Zheng & Chen.

RG-1 organism

Morphological description: Colonies are whitish to ochraceous, with buff-ochre reverse. Sporangiophores arising from hyphae or from stolons; rhizoids abundant. Sporangiophores are hyaline, up to 2 mm long, 9-23 µm wide, simple or once branched, with branches terminating at a higher level than the main stems; branches all ending in a sporangium. Sporangia subspherical to spherical, up to 100 µm diameter. Columella spherical, ellipsoidal to cylindrical, about 40 µm wide, sometimes lobed, with or without an apophysis. Sporangiospores are hyaline, smooth-walled, very variable, mostly subspherical to ellipsoidal, 3-11 x 2-7 µm. Chlamydospores are abundant.

Comment: *Mucor irregularis* differs from other species of *Mucor* by having abundant rhizoids of different sizes and sporangiospores of highly variable shape, mostly subspherical to ellipsoidal (Lu *et al.*, 2009, 2013). This species is closely related to *Mucor hiemalis* (Voigt *et al.*, 1999).

61.5. *Mucor ramosissimus* Samutsevich

RG-1 organism

Morphological description: Colonial growth is restricted, greyish and has a maximum growth temperature for 36°C (does not grow at 37°C). Sporangiophores are hyaline, slightly roughened, tapering towards the apex and are erect with repeated sympodial branching. Sporangia are grey to black, globose or somewhat flattened, up to 80 µm in diameter and have very persistent sporangial walls. Columellae are applanate (flattened), up to 40-50 µm in size and are often absent in smaller sporangia. Sporangiospores are faintly brown, smooth-walled, subglobose to broadly ellipsoidal, 5-8 x 4.5-6 µm in size. Oidia (arthroconidia) may be present in the substrate hyphae, chlamydospores and zygospores are absent. Assimilation of ethanol is negative and that of nitrate is positive.

Comment: *Mucor ramosissimus* differs from other species of *Mucor* by its low, restricted growth on any medium, extremely persistent sporangial walls, columellae that are applanate or absent in smaller sporangia (often resembling *Mortierella* species), short sporangiophores that repeatedly branch sympodially as many as 12 times, and the occurrence of racket-shaped enlargements in the sporangiophores (Hesseltine and Ellis 1964b; Schipper, 1976; Scholer *et al.*, 1983; Schipper and Staplers, 2003); de Hoog *et al.*, 2015).

Antifungal susceptibility: (Table 61.5).

Table 61.5. *Mucor ramosissimus* very limited data (Espinel-Ingroff *et al.*, 2015a; and Australian national data); MIC µg/mL.

Antifungal	No.	≤0.016	0.03	0.06	0.125	0.25	0.5	1	2	4	8	16	32	≥64
AmB	21			2	4	5	7	2		1				
VORI	1										1			
POSA	1					1								
ITRA	3					1	2							

DOI: 10.1079/9781800622340.0061

62 *Myrmecridium schulzeri* (Sacc.) Arzanlou *et al.*

Synonymy: *Ramichloridium schulzeri* Stahel ex de Hoog.

Ramichloridium schulzeri was placed in a new genus, *Myrmecridium* by Arzanlou *et al.* (2007). *Myrmecridium schulzeri* is an uncommon soil saprophyte of worldwide distribution. It has also been isolated from plant detritus and as a contaminant of bronchoscopy fluid. It is the causative agent of "Golden Tongue" syndrome reported by Rippon *et al.* (1985).

RG-1 organism

Morphological description: Colonies growing moderately rapidly, consisting of a rather compact, flat, submerged mycelium, pale orange, locally with some powdery, brownish aerial mycelium; reverse pink to orange. Conidiophores are erect, straight, unbranched, thick-walled, reddish-brown, up to 250 µm high, gradually becoming paler towards the apex, of variable length, elongating sympodially during conidiogenesis, with scattered, pimple-shaped denticles which have unpigmented scars and bear conidia (Fig. 62.1). Conidia are subhyaline, smooth-walled or slightly rough-walled, ellipsoidal, obovoidal or fusiform, 6.5-10.0 x 3-4 µm, usually with an acuminate base and unpigmented scars.

Fig. 62.1. *Myrmecridium schulzeri* conidiophores showing sympodial development of conidia.

Molecular identification: ITS and D1/D2 sequencing may be used for accurate species identification (Halliday *et al.*, 2015).

Comment: *Myrmecridium* can be distinguished from other *Ramichloridium*-like fungi by having entirely hyaline vegetative hyphae, and widely scattered, pimple-shaped denticles on the long hyaline rachis. The conidial sheath is visible in lactic acid mounts with bright-field microscopy (Arzanlou *et al.*, 2007).

Antifungal susceptibility: (Table 62.1).

Table 62.1. *Myrmecridium schulzeri* very limited data (Australian national data); MIC µg/mL.

Antifungal	No.	≤0.016	0.03	0.06	0.125	0.25	0.5	1	2	4	8	16	32	≥64
AmB	1						1							
VORI	1							1						
POSA	1							1						
ITRA	1						1							

References: de Hoog (1977); Rippon *et al.* (1985); Arzanlou *et al.* (2007); de Hoog *et al.* (2015).

DOI: 10.1079/9781800622340.0062

63 *Naganishia albida* (Saito) X.Z. Liu *et al.*

Synonymy: *Cryptococcus albidus* (Saito) Skinner; *Cryptococcus diffluens* (Zach) Lodder & Kreger-van Rij.

A phylogenetic review by Liu *et al.* (2015) reclassified *Cryptococcus albidus* as *Naganishia albida*. *N. albida* has variable growth at 37°C and infections in humans are rare. Impaired cellular immunity is the most common risk factor with HIV infection and low CD4 counts a common comorbidity (Morales-Lopez and Garcia-Effron, 2021).

RG-1 organism

Culture: Colonies (SDA) are cream coloured, smooth, mucoid, glabrous, yeast-like.

Microscopy: Globose to ovoid budding yeast-like cells, 3.5-8.8 x 5.5-10.2 µm. Pseudohyphae are absent.

India ink preparation: Positive – distinct thin capsules are present.

Germ tube formation: Negative.

Physiological tests: (+ Positive, - Negative, v Variable, w Weak, s Slow, n not done).

Fermentation: Absent.

Growth reactions:

Glucose	+	L-Sorbose	v	*myo*-Inositol	+
Sucrose	+	L-Rhamnose	v	DL-Lactate	+
Raffinose	+	D-Xylose	+	D-Gluconate	+
Melibiose	–	L-Arabinose	+	2-Keto-D-gluconate	n
Galactose	v	D-Arabinose	v	D-Glucosamine	–
Lactose	v	D-Ribose	+	*N*-Acetyl-D-glucosamine	n
Trehalose	+	Glycerol	v	D-Glucuronate	+
Maltose	+	Erythritol	–	Nitrate	+
Melezitose	+	Ribitol	v	Urease	+
Methyl-α-D-glucoside	v	Galactitol	v	0.1% Cycloheximide	–
Soluble starch	+	D-Mannitol	+	Growth at 35°C	v
Cellobiose	+	D-Glucitol	+	Growth at 37°C	v

Molecular identification: ITS and D1/D2 sequencing may be used for accurate species identification.

MALDI-ToF MS: Able to accurately identify this species (Fraser *et al.*, 2016).

Antifungal susceptibility: (Table 63.1).

Table 63.1. *Naganishia albida* very limited data (Australian national data); MIC µg/mL.

Antifungal	No.	≤0.03	0.06	0.125	0.25	0.5	1	2	4	8	16	32	≥64
AmB	1						1						
FLU	1											1	
VORI	1						1						
POSA	1						1						
ITRA	1												
5FC	1												1

References: Barnett *et al.* (1983); Fonseca *et al.* (2011); de Hoog *et al.* (2020).

DOI: 10.1079/9781800622340.0063

64 *Nannizzia* Stockdale

Based on a recent multilocus phylogenetic study the taxonomy of the dermatophytes has been reviewed (de Hoog *et al.*, 2017). The genus *Nannizzia* now consists of 13 species (Dukik *et al.*, 2019). *Nannizzia fulva, N. gypsea, N. incurvata, N. nana* and *N. persicolor* were previously included in the genus *Microsporum*.

64.1. *Nannizzia fulva* (Uribura) Stockdale

Synonymy: *Microsporum fulvum* Uriburu.

Nannizzia fulva is a geophilic fungus of worldwide distribution which may cause occasional infections in humans and animals. Clinical disease is similar to that caused by *Nannizzia gypsea* but less common. Invaded hairs show a sparse ectothrix infection but do not fluoresce under Wood's ultra-violet light.

RG-1 organism

Morphological description: Colonies are fast growing, flat, suede-like, tawny-buff to pinkish-buff in colour and frequently have a fluffy white advancing edge (Fig. 64.1). A dark red under surface is occasionally seen, otherwise it is colourless to yellow-brown. Abundant thin-walled, elongate, ellipsoidal macroconidia are formed which closely resemble those of *N. gypsea*, except they are longer and more bullet-shaped (clavate) with three to six septa (Fig. 64.1). Numerous spiral hyphae, which are often branched are seen. Numerous pyriform to clavate microconidia are also produced but these are not diagnostic.

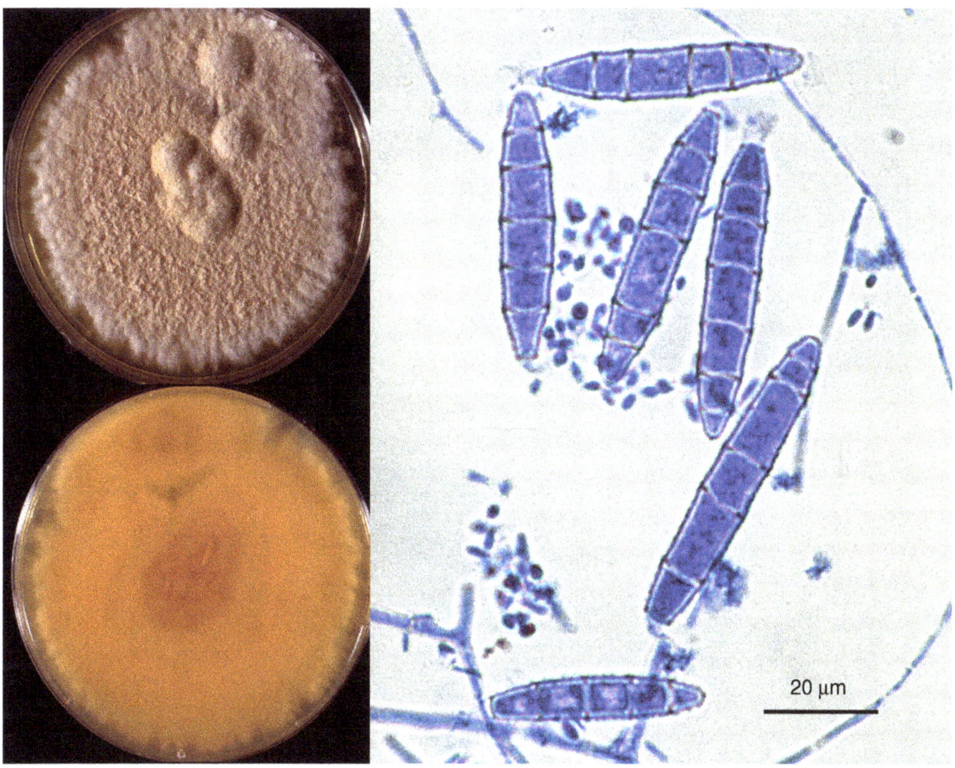

Fig. 64.1. *Nannizzia fulva* culture and macroconidia.

Key features: Macroconidia morphology and culture characteristics.

Molecular identification: ITS sequencing is recommended with comparison to Type strains.

References: Rebell and Taplin (1970); Rippon (1988); Graser *et al.* (2008); Cafarchia *et al.* (2013); de Hoog *et al.* (2015, 2017).

64.2. *Nannizzia gypsea* (Nannizzi) Stockdale

Synonymy: *Microsporum gypseum* (Bodin) Guiart & Grigorakis.

Nannizzia gypsea is a geophilic fungus with a worldwide distribution which may cause infections in animals and humans, particularly children and rural workers during warm humid weather. Usually produces a single inflammatory skin or scalp lesion. Invaded hairs show an ectothrix infection but do not fluoresce under Wood's ultra-violet light.

RG-1 organism

Morphological description: Colonies are usually flat, spreading, suede-like to granular, with a deep cream to tawny-buff to pale cinnamon coloured surface (Fig. 64.2). Many cultures develop a central white downy umbo (dome) or a fluffy white tuft of mycelium and some also have a narrow white peripheral border. A yellow-brown pigment, often with a central darker brown spot, is usually produced on the reverse, however a reddish-brown reverse pigment may be present in some strains. Cultures produce abundant, symmetrical, ellipsoidal, thin-walled, verrucose, four to six-celled macroconidia. The terminal or distal ends of most macroconidia are slightly rounded, while the proximal ends (point of attachment to hyphae) are truncate (Fig. 64.2). Numerous clavate-shaped microconidia are also present, but these are not diagnostic.

Key features: Distinctive macroconidia and culture characteristics.

Molecular identification: ITS sequencing is recommended, especially for the separation of *N. gypsea* and *N. incurvata* which are morphologically similar.

References: Rebell and Taplin (1970); Rippon (1988); Graser *et al.* (2008); Cafarchia *et al.* (2013); de Hoog *et al.* (2015, 2017).

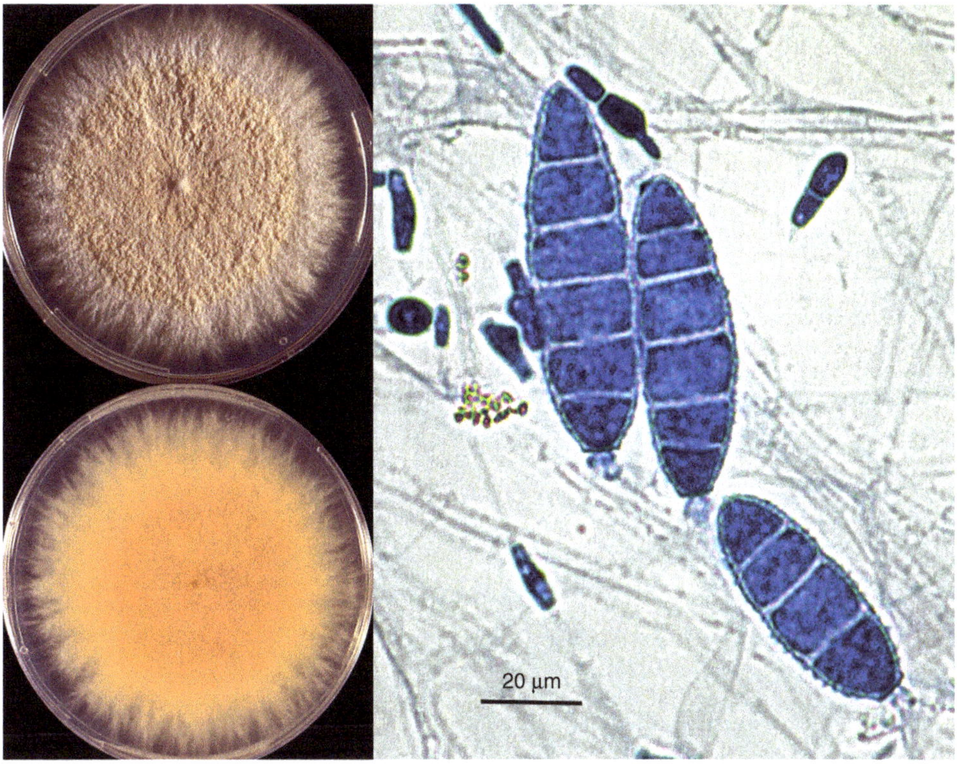

Fig. 64.2. *Nannizzia gypsea* culture and macroconidia. **Note:** Some pyriform to clavate microconidia are also present.

64.3. *Nannizzia nana* (Fuentes) Gräser & de Hoog

Synonymy: *Microsporum nanum* Fuentes.

Nannizzia nana is a geophilic and zoophilic fungus frequently causing chronic non-inflammatory lesions in pigs and a rare cause of tinea in humans. Also present in soil of pig-yards. Infections in man are usually through direct contract with pigs or fomites. Invaded hairs may show a sparse ectothrix or endothrix infection but do not fluoresce under Wood's ultra-violet light. The geographical distribution is worldwide.

RG-2 organism

Morphological description: Colonies are flat, cream to buff in colour with a suede-like to powdery surface texture. Young colonies have a brownish-orange pigment which deepens into a dark reddish-brown with age (Fig. 64.3). Cultures produce numerous small ovoid to pyriform macroconidia with one to three (mostly two) cells, with relatively thin, finely echinulate (rough) walls, and broad truncate bases. Many macroconidia are borne on conidiophores (stalks) which do not stain readily (Fig. 64.3). Occasional clavate microconidia are present, which distinguishes *N. nana* from some species of *Chrysosporium*.

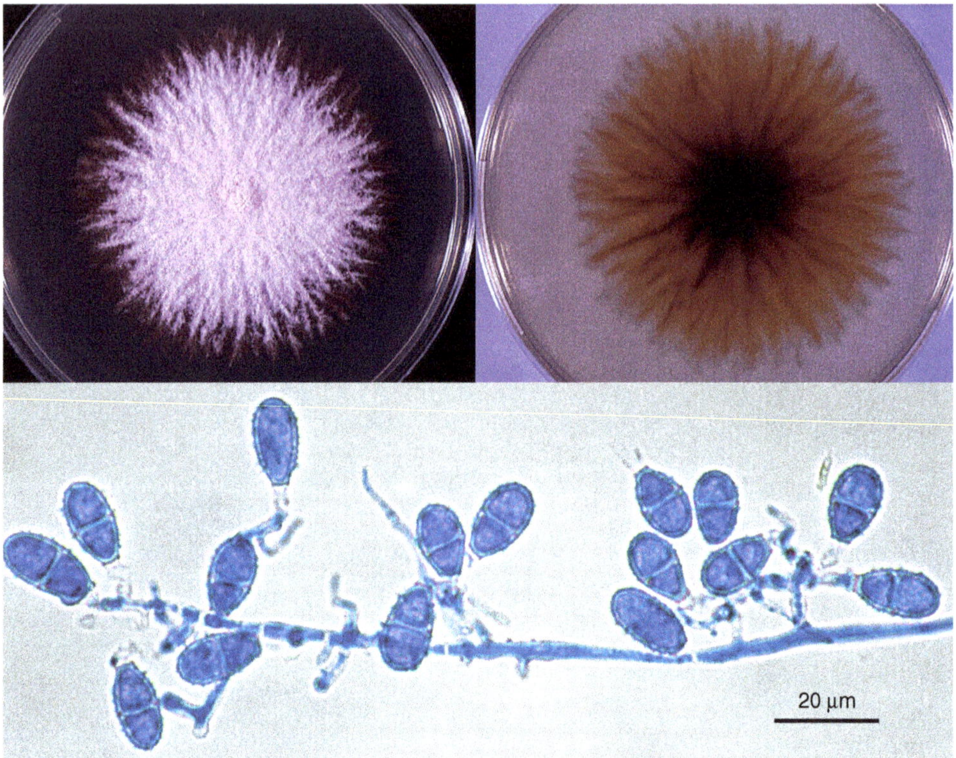

Fig. 64.3. *Nannizzia nana* culture and macroconidia.

Key features: Distinctive macroconidia and culture characteristics.

Molecular identification: ITS sequencing is recommended.

64.4. *Nannizzia persicolor* (Sabouraud) Stockdale

Synonymy: *Microsporum persicolor* (Sabouraud) Guiart & Grigorakis.

Nannizzia persicolor is a zoophilic fungus often occurring as a saprophyte on voles and bats. A rare cause of tinea corporis in humans. Not known to invade hair *in vivo* but produces hair perforations *in vitro*. Distribution: Africa, Australia, Europe and North America.

RG-2 organism

Morphological description: Colonies are generally flat, white to pinkish in colour, with a suede-like to granular texture and peripheral fringe. Reverse pigmentation is orange to red (Fig. 64.4). Macroconidia are thin-walled, cigar-shaped, four to seven-celled, 40-60 x 6-8 µm but are only rarely produced. Microconidia are abundant, spherical to pyriform (Fig. 64.4).

Key features: Microscopic morphology and culture characteristics.

Molecular identification: ITS sequencing is recommended.

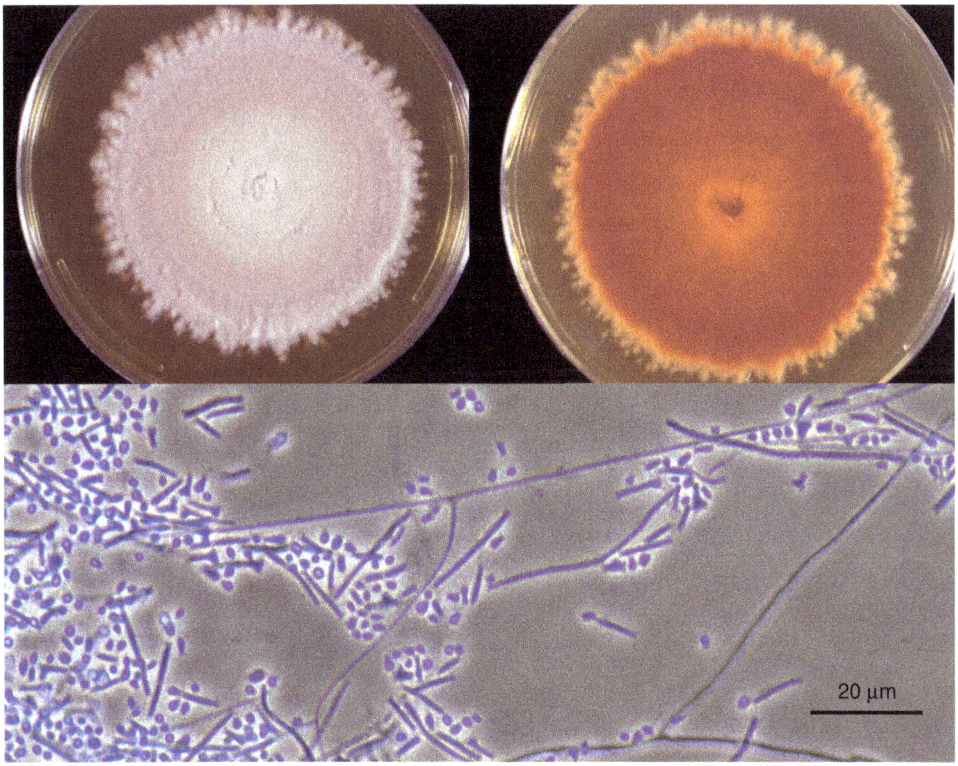

Fig. 64.4. *Nannizzia persicolor* culture and microconidia.

References: Rebell and Taplin (1970); Rippon (1988); Graser *et al.* (2008); Cafarchia *et al.* (2013); de Hoog *et al.* (2015, 2017).

DOI: 10.1079/9781800622340.0064

65 *Neoscytalidium dimidiatum* (Penzig) Crous & Slippers

Synonymy: *Hendersonula toruloidea* Nattras; *Scytalidium dimidiatum* (Penz.) Sutton & Dyko; *Scytalidium hyalinum* Campbell & Mulder.

Neoscytalidium dimidiatum is a recognised agent of onychomycosis and superficial skin infections, especially in tropical regions. The primary isolation of this fungus from clinical specimens may be difficult as isolates are sensitive to cycloheximide (actidione), which is commonly added to primary isolation media used for culturing skin scrapings.

The taxonomy of this species has been very confusing; the conidial state of *Scytalidium dimidiatum* was originally described under the name *Hendersonula toruloidea*. However, it is phylogenetically remote from *Scytalidium* and has now been placed into *Neoscytalidium* (Crous *et al.*, 2006; Machouart *et al.*, 2012). Colourless mutants (previously known as *Scytalidium hyalinum*) often occur and have now been listed as variety *N. dimidiatum* var. *hyalinum* (Madrid *et al.*, 2009).

Note: *Nattrassia mangiferae,* previously considered the teleomorph form of *Scytalidium dimidiatum*, is now considered a distinct species, and has been placed in the genus *Neofusicoccum,* based on the lack of an arthroconidial anamorph (Crous *et al.*, 2006; Machouart *et al.*, 2012).

RG-2 organism

Morphological description: Cultures are effuse, hairy, dark grey to blackish-brown, or white to greyish, with a cream coloured to deep ochraceous-yellow reverse. Colourless mutants often occur. Arthroconidia are typically present in chains of one to two-cells, darkly pigmented, 3.5-5 x 6.5-12 µm, produced by the holothallic fragmentation of undifferentiated hyphae (Fig. 65.1). Pycnidia, only occasionally formed in older cultures are black, ostiolate and contain numerous hyaline flask-shaped phialides. Phialoconidia are at first one-celled and hyaline, later becoming three-celled, brown, with the centre cell darker than the end cells and are ovoid to ellipsoidal in shape.

Molecular identification: ITS and D1/D2 sequencing may be used for accurate species identification (Halliday *et al.*, 2015).

MALDI-ToF MS: Alshawa *et al.* (2012) developed a spectral database for 12 different species of dermatophytes which also included *Neoscytalidium dimidiatum* and *N. dimidiatum* var. *hyalinum*. Correct identification of the species was obtained for 18/21 *Neoscytalidium* isolates (85.7%).

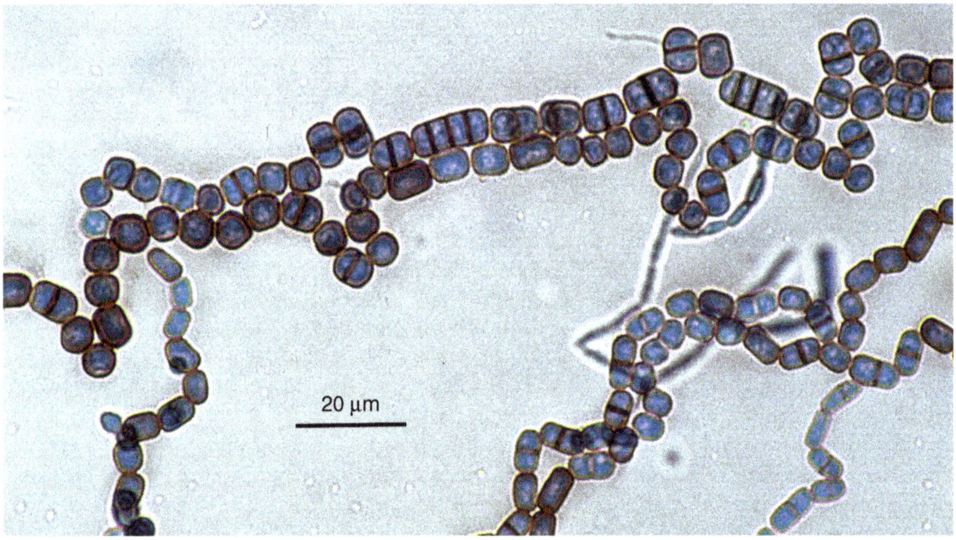

Fig. 65.1. *Neoscytalidium dimidiatum* arthroconidia.

Antifungal susceptibility: (Table 65.1).

Table 65.1. *Neoscytalidium dimidiatum* limited data (Australian national data); MIC µg/mL.

Antifungal	No.	≤0.016	0.03	0.06	0.125	0.25	0.5	1	2	4	8	16	32	≥64
AmB	9			1		1	2	3	2					
VORI	9		2		1	2	1	3						
POSA	9				1	1		1	1		5			
ITRA	9				1			1				7		

References: McGinnis (1980); Moore (1986); Rippon (1988); Frankel and Rippon (1989); Sutton and Dyko (1989); de Hoog *et al.* (2015).

DOI: 10.1079/9781800622340.0065

66 *Ochroconis* de Hoog and Arx

Recently, the genus *Ochroconis* has undergone taxonomic revision, and the most relevant species, *Ochroconis gallopava*, was transferred to the new genus *Verruconis* (Samerpitak *et al.*, 2014). *Ochroconis* species are mesophilic saprobes, with an optimum growth temperature between 15 and 30°C and an inability to grow at 37°C. Nevertheless, some species have been isolated from clinical specimens, such as *Ochroconis tshawytschae, O. mirabilis, O. cordanae, O. olivacea* and *O. ramosa* (Samerpitak *et al.*, 2014; Seyedmousavi *et al.*, 2014; Giraldo *et al.*, 2014b). **Note:** Species described in the literature from clinical cases as *O. constricta* or *O. humicola* are probably *O. musae* (Samerpitak *et al.*, 2014).

Giraldo *et al.* (2014b) reported on the occurrence of *Ochroconis* and *Verruconis* species in clinical specimens from the United States. *Verruconis gallopava* was the most common species (69%), followed by *Ochroconis mirabilis* (22%). Other species isolated were *O. cordanae, O. olivacea* and *O. ramosa*. The most common anatomical site of isolation was the lower respiratory tract (59%), followed by superficial (22%) and deep tissues (20%) (Giraldo *et al.*, 2014b).

RG-1 organisms

Morphological description: Colonies restricted, velvety to funiculose, brown to olivaceous, often with rust-brown reverse. Hyphae smooth- or somewhat rough-walled, pale olivaceous. Conidiophores slightly or conspicuously differentiated, cylindrical, often flexuose, producing conidia on scattered, cylindrical to conical denticles. After detachment an inconspicuous frill often remains both on the denticle and on the conidium base. Conidia one to four-celled, pale olivaceous-brown, smooth- or rough-walled, ellipsoidal, cylindrical, clavate or cuneiform. Maximum growth temperature around 33°C (thermophilic members were transferred to *Verruconis*).

Molecular identification: Giraldo *et al.* (2014b) used sequence analyses of the 18*S*, ITS, D1/2, *ACT* and β-tubulin genes, while Seyedmousavi *et al.* (2014), used ITS sequence analysis to identify species.

Antifungal susceptibility: (Table 66.1).

Table 66.1. *Ochroconis* species very limited data (Australian national data); MIC µg/mL.

Antifungal	No.	≤0.016	0.03	0.06	0.125	0.25	0.5	1	2	4	8	16	32	≥64
AmB	4						1	1			2			
ISAV	1							1			1			
VORI	4									3	1			
POSA	4						3		1					
ITRA	4						1	1	1		1			

References: Samerpitak *et al.* (2014); Seyedmousavi *et al.* (2014); Giraldo *et al.* (2014b); de Hoog *et al.* (2015).

DOI: 10.1079/9781800622340.0066

67 *Onychocola canadensis* Sigler

Onychocola canadensis is an uncommon cause of distal and lateral subungual or white superficial onychomycosis. However, it may sometimes be present in an abnormal-appearing nail as an insignificant finding, not acting as a pathogen.

RG-2 organism

Morphological description: Colonies grow slowly and are velvety to lanose, white to yellowish, with a brownish reverse (Fig. 67.1). Arthroconidia are cylindrical to broadly ellipsoidal, one- or two-celled, hyaline to subhyaline, 4-16 x 2-5 µm in size, forming long chains (Fig. 67.1). Older cultures may show broad, brown, rough-walled hyphae.

Key features: Slow growing, white, arthroconidial mould isolated from nails.

References: Sigler and Congly (1990); Sigler *et al.* (1994); Gupta *et al.* (1998); de Hoog *et al.* 2015).

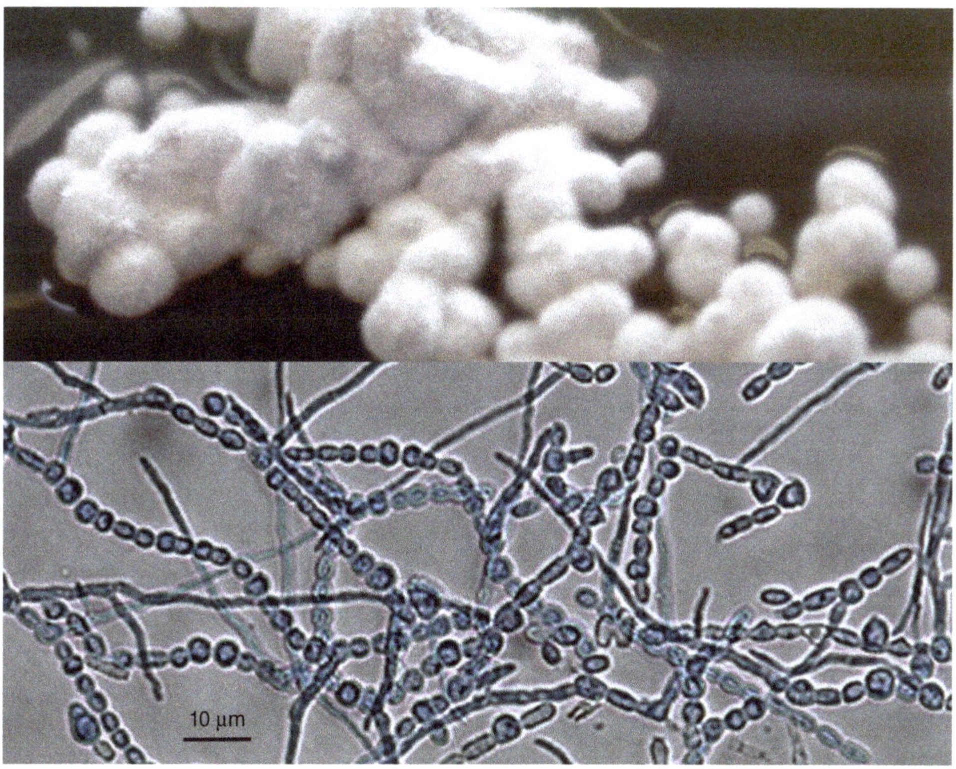

Fig. 67.1. *Onychocola canadensis* culture and arthroconidia.

DOI: 10.1079/9781800622340.0067

68 *Paecilomyces variotii* Bain

The genus *Paecilomyces* may be distinguished from the closely related genus *Penicillium* by having long slender divergent phialides and colonies that are never typically green. *Paecilomyces variotii* is a common environmental mould that is widespread in composts, soils and food products. It is known from substrates including food, indoor air, wood, soil and carpet dust. However, *P. variotii* is an emerging causative agent of mycotic keratitis and of hyalohyphomycosis in the immunocompromised patient. **Note:** *Paecilomyces lilacinus* was transferred to *Purpureocillium lilacinum* (Luangsa-ard *et al.*, 2011); and *Paecilomyces marquandii* has been transferred to *Marquandomyces marquandii* (Mongkolsamrit *et al.*, 2020).

RG-2 organism

Morphological description: Colonies are fast growing, powdery to suede-like, funiculose or tufted, and yellow-brown or sand coloured (Fig. 68.1). Conidiophores bearing dense, verticillately arranged branches bearing phialides. Phialides are cylindrical or ellipsoidal, tapering abruptly into a rather long and cylindrical neck. Conidia are subspherical, ellipsoidal to fusiform, hyaline to yellow, smooth-walled, 3-5 x 2-4 µm and are produced in long divergent chains (Fig. 68.2). Chlamydospores are usually present, singly or in short chains, brown, subspherical to pyriform, 4-8 µm in diameter, thick-walled to slightly verrucose.

Molecular identification: Molecular phylogeny based on 18*S* rDNA sequences was done by Luangsa-ard *et al.* (2011). ITS sequencing is recommended for identification.

Key features: Yellow-brown colony pigmentation, cylindrical phialides, and presence of chlamydospores.

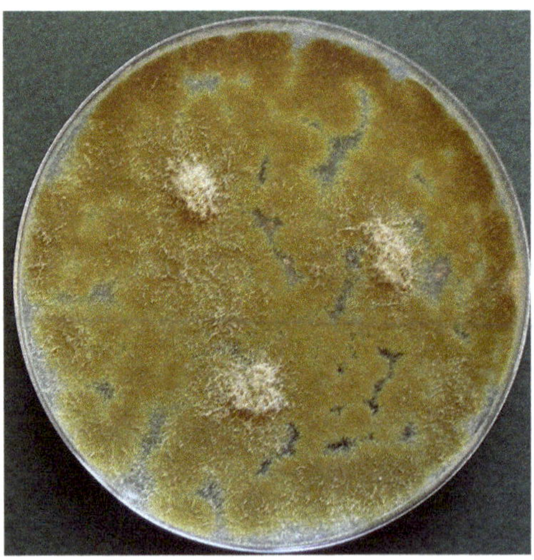

Fig. 68.1. *Paecilomyces variotii* culture.

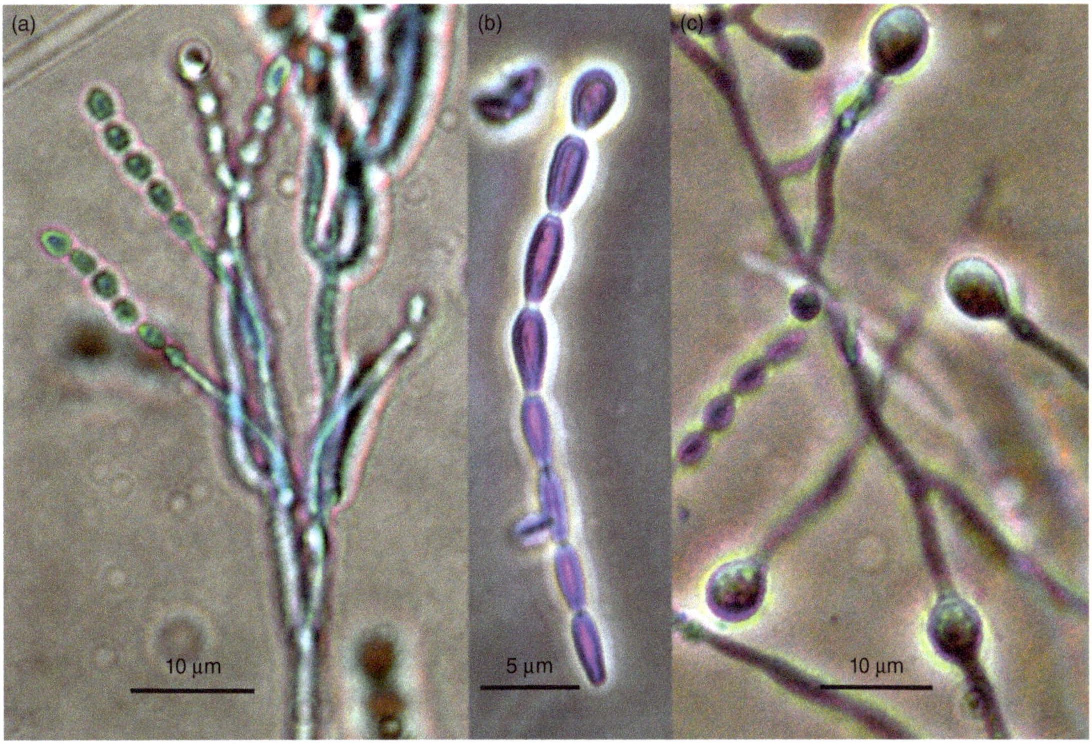

Fig. 68.2. *Paecilomyces variotii* showing (a) conidiophores and phialides, (b) conidia and (c) terminal chlamydospores.

Antifungal susceptibility: (Table 68.1).

Table 68.1. *Paecilomyces variotii*, limited data (Australian national data); MIC µg/mL.

Antifungal	No.	≤0.016	0.03	0.06	0.125	0.25	0.5	1	2	4	8	16	32	≥64
AmB	39			1	8	8	9	7	4		2			
ISAV	8					1		1		1	5			
VORI	37	2			2			1	3	4	24	1		
POSA	36	5	7	7	3	8	4		2					
ITRA	39	4	2	7	9	9	6				1	1		

References: Samson (1974); Domsch *et al.* (2007); Houbraken *et al.* (2010); de Hoog *et al.* (2015).

DOI: 10.1079/9781800622340.0068

69 *Papiliotrema laurentii* (Kufferath) X.Z. Liu *et al.*

Synonymy: *Cryptococcus laurentii* (Kufferath) Skinner.

A phylogenetic review by Liu *et al.* (2015) reclassified *Cryptococcus laurentii* as *Papiliotrema laurentii. P. laurentii* is an emerging opportunistic pathogen, causing disease in patients with impaired cell-mediated immunity (e.g. HIV-infected patients or those with haematologic malignancies), in those using steroids or chemotherapeutic agents, and in those with invasive devices. Human infections due to *P. laurentii* have been reported as fungaemia, central nervous system infection, skin and soft-tissue infections, and pneumonia, among others (Morales-Lopez and Garcia-Effron, 2021).

RG-1 organism

Culture: Colonies (SDA) are cream coloured, often becoming a deeper orange-yellow with age, with a smooth mucoid texture.

Microscopy: Spherical and elongated budding yeast-like cells or blastoconidia, 2.0-5.5 x 3.0-7.0 µm. No pseudohyphae present.

India ink preparation: Positive – narrow but distinct thin capsules are present.

Germ tube formation: Negative.

Physiological tests: (+ Positive, – Negative, v Variable, w Weak, s Slow, n not done).

Fermentation: Absent.

Growth reactions:

Glucose	+	L-Sorbose	v	*myo*-Inositol	+
Sucrose	+	L-Rhamnose	+	DL-Lactate	+
Raffinose	+	D-Xylose	+	D-Gluconate	+
Melibiose	+	L-Arabinose	+	2-Keto-D-gluconate	+
Galactose	+	D-Arabinose	+	D-Glucosamine	+
Lactose	+	D-Ribose	+	*N*-Acetyl-D-glucosamine	+
Trehalose	+	Glycerol	v	D-Glucuronate	+
Maltose	+	Erythritol	+	Nitrate	–
Melezitose	+	Ribitol	+	Urease	+
Methyl-α-D-glucoside	+	Galactitol	v	0.1% Cycloheximide	–
Soluble starch	+	D-Mannitol	+	Growth at 35°C	+
Cellobiose	+	D-Glucitol	+	Growth at 37°C	v

Comment: Some strains of *Papiliotrema laurentii* may develop brown pigment on bird seed agar and turn CGB media somewhat blue, similar to *Cryptococcus gattii*. However *P. laurentii* assimilates both lactose and melibiose while *C. gattii* does not.

Molecular identification: ITS sequencing may be used for accurate species identification.

MALDI-ToF MS: Able to accurately identify this species (Fraser *et al.*, 2016).

Antifungal susceptibility: (Table 69.1).

Table 69.1. *Papiliotrema laurentii*, very limited (Australian national data); MIC µg/mL.

Antifungal	No.	≤0.016	0.03	0.06	0.125	0.25	0.5	1	2	4	8	16	32	≥64
AmB	8				1	2	3	1		1				
FLU	8								1	1		4		2
VORI	7			2		1	2	1			1			
POSA	6		1	1	2	1	1							
ITRA	8		1		2	2	1					2		
5FC	8													8

References: Barnett *et al.* (1983); Fonseca *et al.* (2011); de Hoog *et al.* (2020).

DOI: 10.1079/9781800622340.0069

70 *Paracoccidioides brasiliensis/lutzii* complex

Recently *Paracoccidioides brasiliensis* has been recognised as two species: *P. brasiliensis* and *P. lutzii* (Theodoro *et al.*, 2012; Teixeira *et al.*, 2014). *P. brasiliensis/lutzii* is geographically restricted to areas of South and Central America. The two species are morphologically very similar; conidia of *P. lutzii* are elongated whereas those from *P. brasiliensis* are pyriform. Molecular confirmation is recommended. A novel, uncultivated strain of *P. brasiliensis* reported to cause cutaneous lesions in dolphins has been moved recently to a new species, *Paracoccidioides cetii* (Vilela *et al.*, 2021).

WARNING: RG-3 organism

Cultures of *Paracoccidioides brasiliensis/lutzii* represent a biohazard to laboratory personnel and should be handled with extreme caution in a Class II Biological Safety Cabinet (BSCII).

Molecular identification: ITS sequencing is helpful (Imai *et al.*, 2000) however immunodominant antigen Gp43 (*GP43*) is recommended for barcoding (Vilela *et al.*, 2021).

Morphological description: Colonies at 25°C are slow growing and variable in morphology. Colonies may be flat, wrinkled and folded, glabrous, suede-like or downy in texture, white to brownish with a tan or brown reverse. Microscopically, a variety of conidia may be seen, including pyriform microconidia, arthroconidia and chlamydospores. However, none of these are characteristic of the species, and most strains may grow for long periods of time without the production of conidia.

On BHI blood agar at 37°C, the mycelium converts to the yeast phase and colonies are white to tan, moist and glabrous and become wrinkled, folded and heaped. Microscopically, numerous large, 20-60 µm, round, narrow base budding yeast cells are present. Single and multiple budding occurs, the latter are thick-walled cells that form the classical "steering wheel" or "mickey mouse" structures that are diagnostic for this fungus, especially in methenamine silver-stained tissue sections (Fig. 70.1).

Key features: Clinical history, tissue pathology, culture identification with conversion to yeast phase at 37°C, however molecular identification is now recommended for identification.

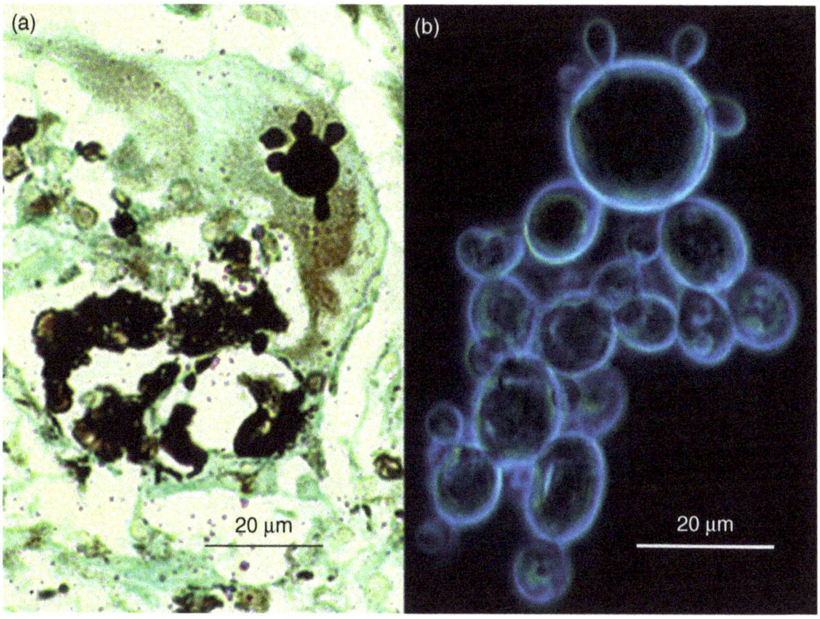

Fig. 70.1. *Paracoccidioides brasiliensis/litzii* showing multiple, narrow base, budding yeast cells (a) in a Grocott-Gomori's methenamine silver-stained tissue section and (b) in a calcofluor white stained wet mount.

Antifungal susceptibility: (Table 70.1).

Table 70.1. *Paracoccidioides brasiliensis*, very limited data (McGinnis *et al.*, 1997); MIC µg/mL. Antifungal susceptibility testing not recommended.

Antifungal	Range	Antifungal	Range
FLU	≤0.125-64	AmB	0.03-4
ITRA	≤0.03-1	VORI	≤0.03-2

References: Chandler *et al.* (1980); McGinnis (1980); Rippon (1988); Bagagli *et al.* (2008); Desjardins *et al.* (2011); de Hoog *et al.* (2015).

DOI: 10.1079/9781800622340.0070

71 *Paraphyton cookei* (Ajello) Y. Graser, Dukik & de Hoog

Synonymy: *Microsporum cookei* Ajello; *Nannizzia cajetana* Borelli.

Based on a recent taxonomic revision of the dermatophytes the genus *Paraphyton* has been created for three species (de Hoog *et al.*, 2017). *Paraphyton cookei* is a geophilic fungus which has also been isolated from the hair of small mammals showing no clinical lesions. Infection has been reported in rodents, dogs and rarely in humans. It is not known to invade hair *in vivo* but produces hair perforations *in vitro*. *P. cookei* has a worldwide distribution.

RG-1 organism

Morphological description: Colonies are flat, spreading, buff to pale brown, powdery to suede-like, with a slightly raised and folded centre and some radial grooves. Reverse pigment dark reddish-brown (Fig. 71.1). Numerous large, 30-65 x 10-15 µm, very thick-walled, echinulate (rough) ellipsoidal macroconidia with predominantly five to six septa but may be from two to eight septa (Fig. 71.1). Occasional spirals may be seen. Moderate numbers of mainly slender clavate with some pyriform microconidia are present.

Confirmatory tests

Vitamin free agar (Trichophyton agar No.1): Good growth indicating no special nutritional requirements, pinkish-buff coloured, suede-like colony with a deep magenta red reverse.

Hair perforation test: Positive.

Key features: The macroconidia of *Paraphyton cookei* are quite characteristic and diagnostic; the thick walls and usually larger size of the macroconidia distinguish it from *Nannizzia gypsea*.

Molecular identification: ITS sequencing is recommended.

References: Rebell and Taplin (1970); Rippon (1988); Graser *et al.* (2008); Cafarchia *et al.* (2013); de Hoog *et al.* (2015, 2017).

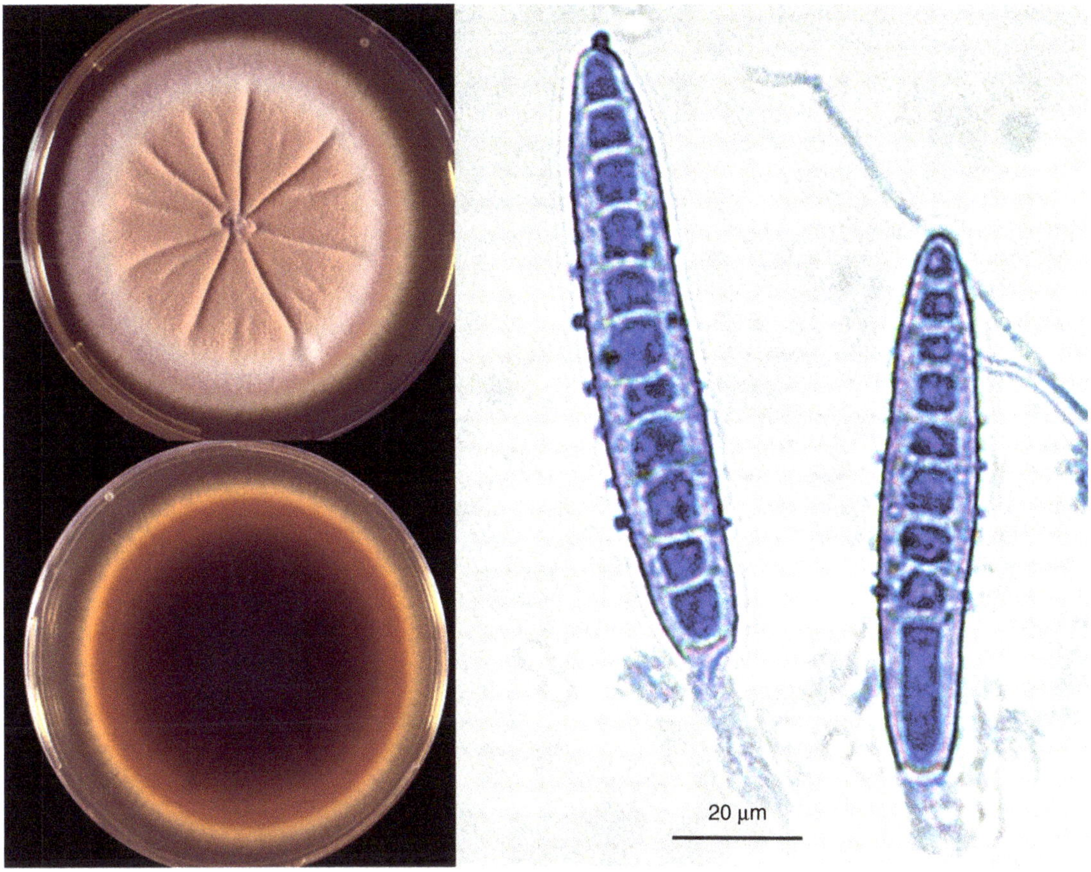

Fig. 71.1. *Paraphyton cookei* culture and macroconidia.

DOI: 10.1079/9781800622340.0071

72 *Penicillium* Link: Fries

Penicillium is a very large and ubiquitous genus which currently contains 354 accepted species (Visagie *et al.*, 2014). Many species are common contaminants on various substrates and are known as potential mycotoxin producers. Correct identification is therefore important when studying possible *Penicillium* contamination of food. Human pathogenic species are rare, however opportunistic infections leading to mycotic keratitis, otomycosis and endocarditis (following insertion of valve prosthesis) have been reported (Lyratzopoulos *et al.*, 2002). **Note:** Species within the subgenus *Biverticillium*, including *Penicillium marneffei* have been transferred to the genus *Talaromyces;* colonies in this genus often have bright colours (Samson *et al.*, 2011b).

RG-1 organisms

Morphological genus description: Colonies are usually fast growing, in shades of green, sometimes white, mostly consisting of a dense felt of conidiophores (Fig. 72.1). Microscopically, chains of single-celled conidia are produced in basipetal succession from a specialised conidiogenous cell called a phialide. The term basocatenate is often used to describe such chains of conidia where the youngest conidium is at the basal or proximal end of the chain. In *Penicillium*, phialides may be produced singly in groups or from branched metulae, giving a brush-like appearance (a penicillus). The penicillus may contain both branches and metulae (penultimate branches which bear a whorl of phialides). All cells between the metulae and the stipes of the conidiophores are referred to as branches. The branching pattern may be either simple (non-branched or monoverticillate), one-stage branched (biverticillate-symmetrical), two-stage branched (biverticillate-asymmetrical) or three- to more-stage branched (Fig. 72.2). Conidiophores are hyaline, smooth or rough-walled. Phialides are usually flask-shaped, consisting of a cylindrical basal part and a distinct neck, or lanceolate (more or less with a narrow basal part tapering to a somewhat pointed apex). Conidia are in long dry chains, divergent or in columns, are globose, ellipsoidal, cylindrical or fusiform, hyaline or greenish, smooth or rough-walled (Fig. 72.3). Sclerotia are produced by some species.

Fig. 72.1. *Penicillium* species culture.

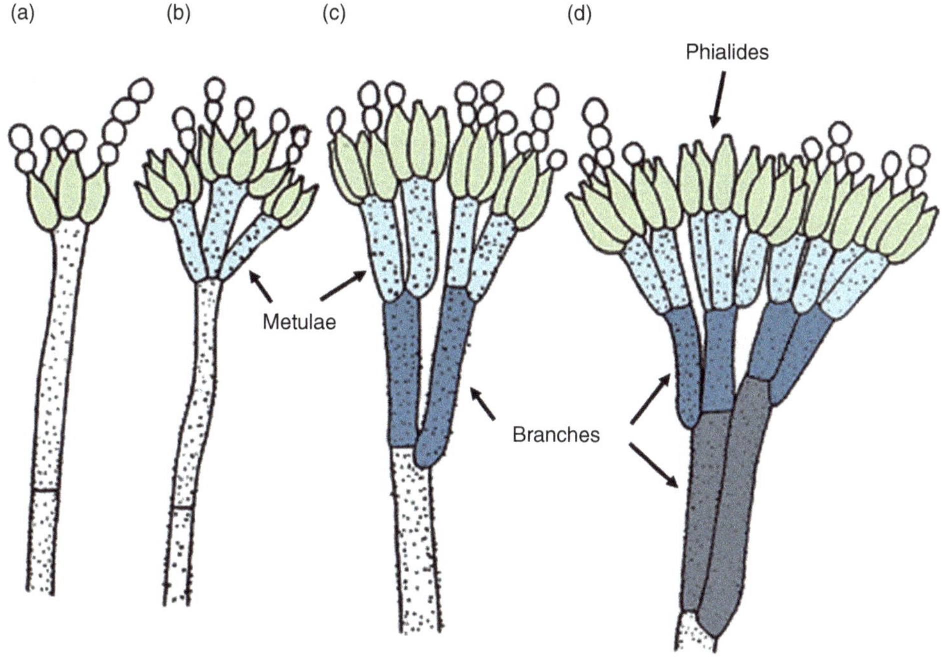

Fig. 72.2. Types of conidiophore branching in *Penicillium* (a) monoverticillate; (b) biverticillate; (c) terverticillate; (d) quaterverticillate (Visagie *et al.*, 2014).

Molecular identification: ITS and/or β-tubulin loci are recommended for identification of *Penicillium* species (Visagie *et al.*, 2014; Yilmaz *et al.*, 2014). ITS sequencing alone is often insufficient to resolve species, and morphological characters may be required for confirmation.

Key features: Flask-shaped phialides arranged in groups from branched metulae forming a penicillus.

Antifungal susceptibility: (Table 72.1).

Table 72.1. *Penicillium* spp. (Australian national data); MIC µg/mL.

Antifungal	No.	<0.016	0.03	0.06	0.125	0.25	0.5	1	2	4	8	16	32	≥64
AmB	70				4	6	11	14	21	9	5			
ISAV	17				1		4	6	2	4				
VORI	68				2	6	12	10	12	5	21			
POSA	67		1	6	16	13	20	6		1	4			
ITRA	70	1		3	13	12	24	11	2		4			

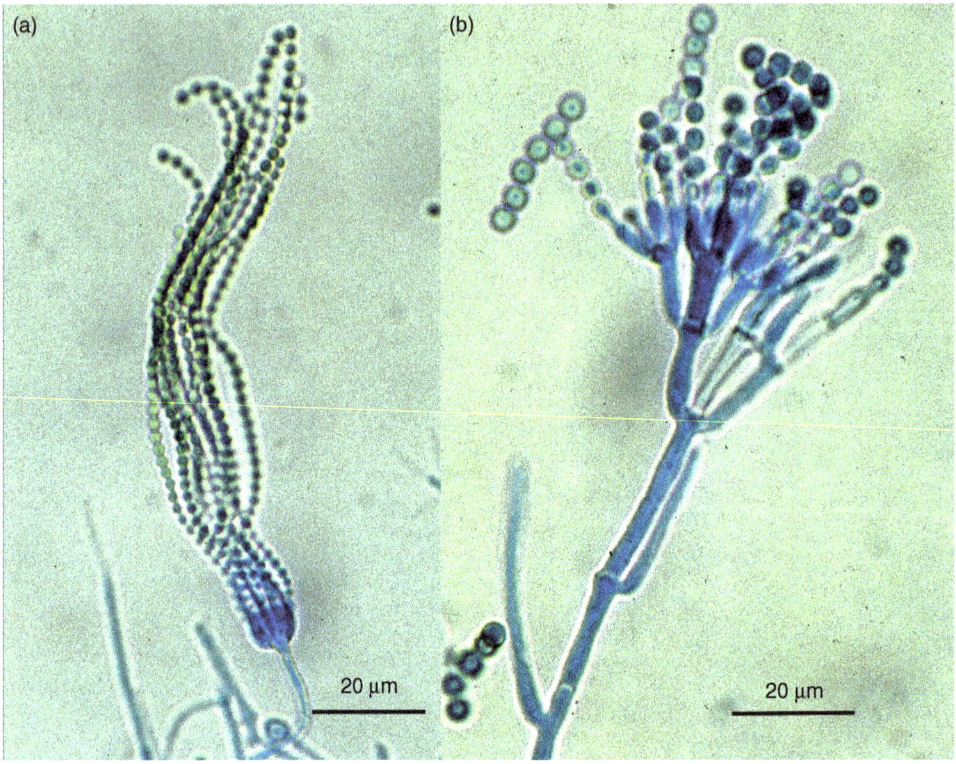

Fig. 72.3. (a) Simple conidiophore of *P. chermesinum* showing long chains conidia and (b) conidiophores of *P. aurantiogriseum* showing two-stage branching.

References: Raper and Thom (1949); Pitt (1979); McGinnis (1980); Onions *et al.* (1981); Ramirez (1982); Samson *et al.* (1995, 2011b); Domsch *et al.* (2007); de Hoog *et al.* (2015).

DOI: 10.1079/9781800622340.0072

73 *Phaeoacremonium parasiticum* (Ajello *et al.*) W. Gams *et al.*

Synonymy: *Phialophora parasiticum* Ajello, Georg & Wang.

The genus *Phaeoacremonium* initially accommodated isolates with features similar to those seen in both *Acremonium* and *Phialophora.* It differs from the former by having pigmented hyphae and conidiophores and from the latter by having indistinct collarettes and warty conidiogenous cells (Revankar and Sutton, 2010).

Phaeoacremonium currently consists of 46 species with *P. parasiticum* and *P. krajdenii* recognised as the predominant species associated with human infections (Mostert *et al.*, 2005). Other species have also been isolated from clinical cases, i.e. *P. alvesii*, *P. amstelodamense*, *P. griseorubrum*, *P. minimum*, *P. rubrigenum*, *P. tardicrescens*, and *P. venezuelense.* Infections caused by *P. parasiticum* include subcutaneous abscesses, thorn-induced arthritis, and disseminated infection (Revankar and Sutton, 2010; Gramaje *et al.*, 2015).

RG-2 organism

Morphological description: Cultures usually slow growing, suede-like with radial furrows, initially whitish-grey becoming olivaceous-grey with age (Fig. 73.1). Hyphae hyaline, later becoming brown and some becoming rough-walled. Phialides are brown, thick-walled, slender, acular to cylindrical slightly tapering towards the tip, 15-50 µm long, often proliferating, with small, funnel-shaped collarettes (Fig. 73.2). Conidia, often in balls, are hyaline, thin-walled, cylindrical to sausage-shaped, 3-6 x 1-2 µm, later inflating. Maximum growth temperature 40°C.

Molecular identification: ITS and/or β-tubulin sequencing (Mostert *et al.*, 2006; Gramaje *et al.*, 2015).

Fig. 73.1. *Phaeoacremonium parasiticum* culture and microscopy of a single phialide.

Key features: The identification of the different *Phaeoacremonium* species can be performed by combining cultural, morphological and ITS, β-tubulin sequence data (Mostert *et al.*, 2006; Gramaje *et al.*, 2015). Relying on morphological characteristics alone can be difficult for species identification.

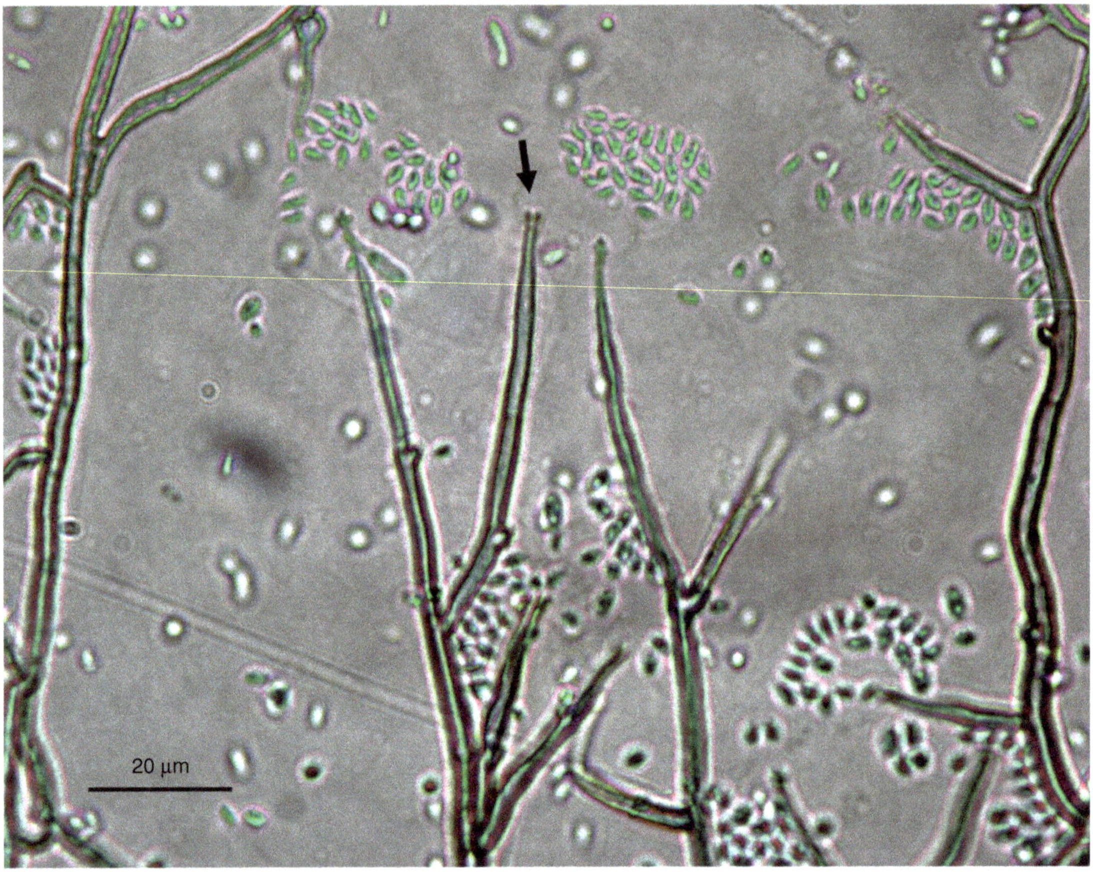

Fig. 73.2. *Phaeoacremonium parasiticum* phialides with small, funnel-shaped collarettes (arrow).

Antifungal susceptibility: (Table 73.1).

Table 73.1. *Phaeoacremonium parasiticum* (Badali *et al.*, 2015; and Australian national data); MIC µg/mL.

Antifungal	No.	≤0.016	0.03	0.06	0.125	0.25	0.5	1	2	4	8	16	32	≥64
AmB	24			1		3	9	5	6					
ISAV	19					1	3	5	6	4				
VORI	24			3	1	6	13	1						
POSA	19					5	8	5	1					
ITRA	24			1			1	3				19		

References: Mostert *et al.* (2005, 2006); Revankar and Sutton (2010); Badali *et al.* (2015); de Hoog *et al.* (2015); Gramaje *et al.* (2015).

DOI: 10.1079/9781800622340.0073

74 *Phialophora verrucosa* Medlar

The genus *Phialophora* contains more than 40 species, most are saprophytes commonly found in soil or on decaying wood. Some human pathogens with phialidic conidiogenesis previously assigned to *Phialophora* have been moved to other genera, namely, *Phaeoacremonium* and *Pleurostoma. Phialophora verrucosa, P. americana*, *P. bubakii*, *P. europaea* and *P. reptans* remain of medical interest (Revankar and Sutton, 2010). Both *P. verrucosa* and *P. americana* produce their conidia from phialides with conspicuous darkened collarettes, however sequencing has demonstrated a close relatedness, suggesting that these species may be synonymous (de Hoog *et al.*, 1999). *P. verrucosa* is primarily an agent of chromoblastomycosis although other reported infections include endocarditis, keratitis, and osteomyelitis (Li *et al.*, 2017).

RG-2 organism

Morphological description: Colonies are slow growing, initially dome-shaped, later becoming flat, suede-like and olivaceous to black in colour. Phialides are flask-shaped or elliptical with distinctive funnel-shaped, darkly pigmented collarettes (Fig. 74.1). Conidia are ellipsoidal, smooth-walled, hyaline, mostly 3.0-5.0 x 1.5-3.0 µm, and aggregate in slimy heads at the apices of the phialide.

Key features: Characteristic flask-shaped phialides with distinctive funnel-shaped, darkly pigmented collarettes.

Molecular identification: ITS sequencing recommended (de Hoog *et al.*, 1999).

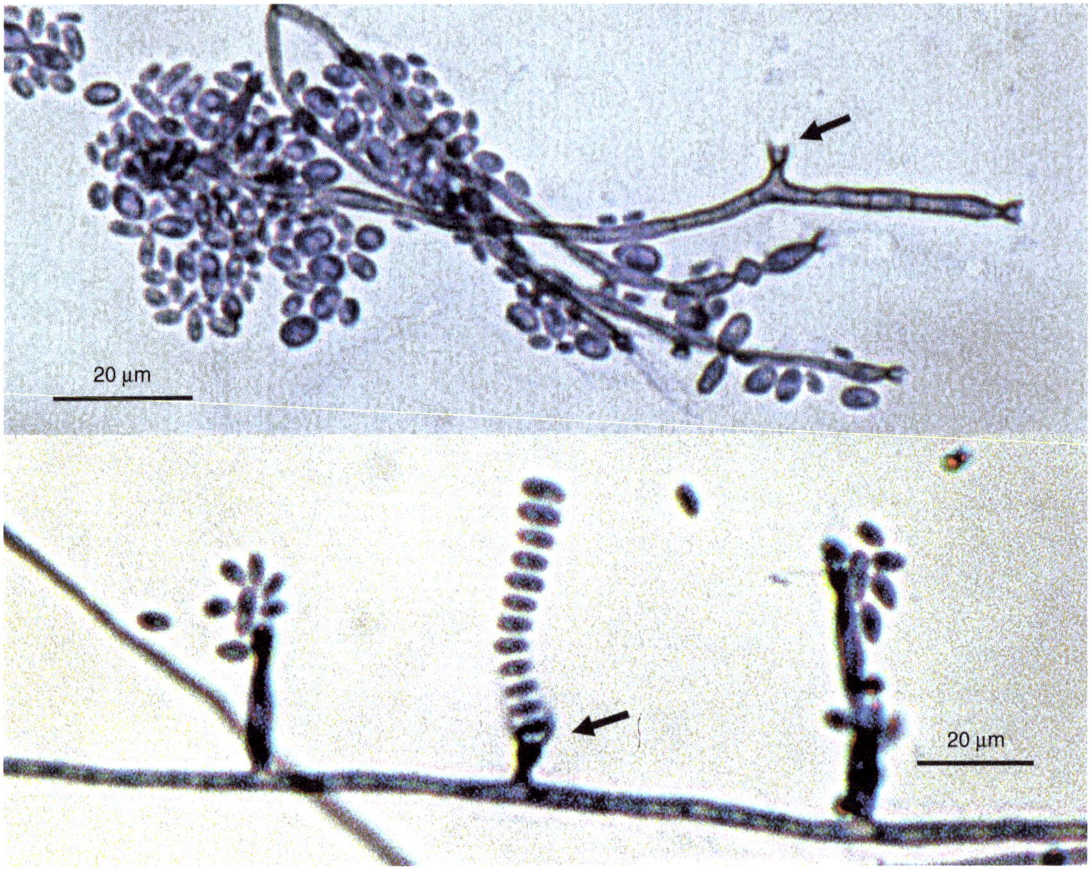

Fig. 74.1. *Phialophora verrucosa* phialides with distinctive funnel-shaped collarettes (arrow) and conidia.

Antifungal susceptibility: (Table 74.1).

Table 74.1. *Phialophora verrucosa* limited data (Australian national data); MIC µg/mL.

Antifungal	No.	≤0.016	0.03	0.06	0.125	0.25	0.5	1	2	4	8	16	≥32
AmB	7						1		4	2			
VORI	7					1	3	3					
POSA	5				2	4	1						
ITRA	7					1	4	2					

References: Ellis (1971); McGinnis (1978b, 1980); de Hoog *et al.* (1999, 2015); Domsch *et al.* (2007); Revankar and Sutton (2010); Li *et al.* (2017).

DOI: 10.1079/9781800622340.0074

75 *Phoma* Saccardo

Members of the genus *Phoma* have a worldwide distribution and are ubiquitous in nature, with over 200 species having been described from soil, as saprophytes on various plants, and as pathogens to plants and humans.

RG-1 organism

Morphological description: Colonies are spreading, greyish-brown, powdery or suede-like and produce large, globose, membranous to leathery, darkly pigmented, ostiolate pycnidia (Fig. 75.1). Conidia are produced in abundance within the pycnidia on narrow thread-like phialides, which are hardly differentiated from the inner pycnidial wall cells. Conidia are globose to cylindrical, one-celled, hyaline, and are usually extruded in slimy masses from the apical ostiole.

Molecular identification: ITS, D1/D2, β-tubulin and 18*S* sequencing has been used to identify *Phoma* species (de Gruyter *et al.*, 2009; Aveskamp *et al.*, 2010). **Note:** Public sequence databases, particularly GenBank, contain many sequences from incorrectly identified species, making identifications of coelomycetous fungi very difficult. Comparison to Type strain sequences, and use of confirmatory morphological studies will be important.

Key features: Ostiolate pycnidia producing masses of slimy, hyaline, single-celled conidia.

References: Punithalingam (1979); McGinnis (1980); Sutton (1980); Rippon (1988); Montel *et al.* (1991); Samson *et al.* (1995); de Hoog *et al.* (2015).

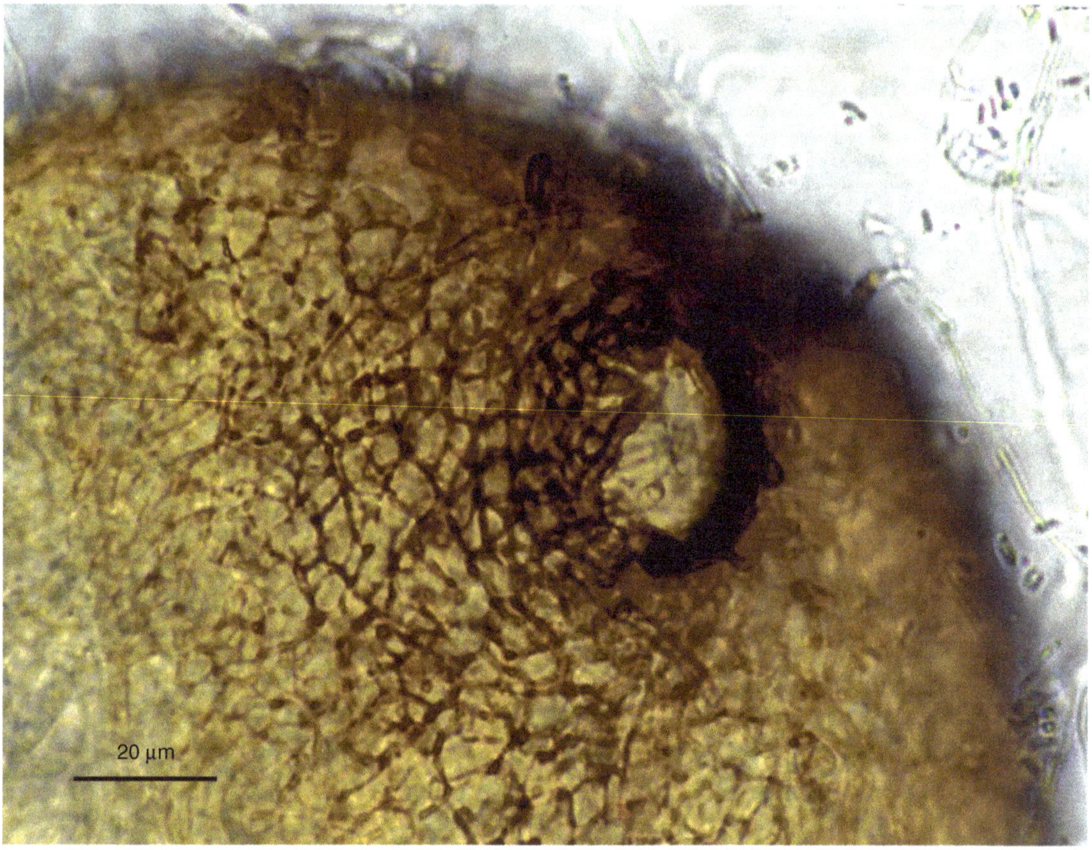

Fig. 75.1. *Phoma* spp. showing a pycnidium with an apical ostiole.

DOI: 10.1079/9781800622340.0075

76 *Pichia* E.C. Hansen

Two species previously classified in the genus *Candida* have now been transferred to the genus *Pichia*. *Pichia kudriavzevii* (formerly *Candida krusei*) and *Pichia norvegensis* (formerly *Candida norvegensis*).

76.1. *Pichia kudriavzevii* Boidin *et al.*

Synonymy: *Candida krusei* (Castellani) Berkhout; *Issatchenkia orientalis* Kudryavtesev.

Pichia kudriavzevii is regularly associated with some forms of infant diarrhoea and occasionally with systemic disease. It has also been reported to colonise the gastrointestinal, respiratory and urinary tracts of patients with granulocytopenia. Environmental isolations have been made from beer, milk products, skin, faeces of animals and birds.

RG-2 organism

Culture: Colonies (SDA) white to cream coloured, smooth, glabrous, yeast-like.

Microscopy: Predominantly small, elongated to ovoid blastoconidia, 2-5 x 4-5 μm.

India ink preparation: Negative - no capsules present.

Dalmau plate culture: Abundant long, wavy, branched pseudohyphae with elongated to ovoid blastoconidia, budding off in verticillate branches.

Molecular identification: ITS sequencing recommended.

MALDI-ToF MS: Able to accurately identify this species.

Germ tube formation: Negative

Physiological tests: (+ Positive, – Negative, v Variable, w Weak, s Slow, n not done).

Fermentation:

Glucose	+	Sucrose	–	Lactose	–
Galactose	–	Maltose	–	Trehalose	–

Growth reactions:

Glucose	+	L-Sorbose	–	*myo*-Inositol	–
Sucrose	–	L-Rhamnose	–	DL-Lactate	+
Raffinose	–	D-Xylose	–	D-Gluconate	–
Melibiose	–	L-Arabinose	–	2-Keto-D-gluconate	–
Galactose	–	D-Arabinose	–	D-Glucosamine	+
Lactose	–	D-Ribose	–	*N*-Acetyl-D-glucosamine	+
Trehalose	–	Glycerol	+	D-Glucuronate	–
Maltose	–	Erythritol	–	Nitrate	–
Melezitose	–	Ribitol	–	Urease	–
Methyl-α-D-glucoside	–	Galactitol	–	0.1% Cycloheximide	v
Soluble starch	–	D-Mannitol	–	Growth at 37°C	+
Cellobiose	–	D-Glucitol	–	Growth at 40°C	+

Key features: Germ tube negative yeast and sugar assimilation pattern. Colonies fuzzy, pale pink on CHROMagar™ Candida.

Antifungal susceptibility: (Table 76.1).

Table 76.1. *Pichia kudriavzevii* (Australian national data); MIC µg/mL. **Note:** *Pichia kudriavzevii* is intrinsically resistant to fluconazole.

Antifungal	No.	≤0.016	0.03	0.06	0.125	0.25	0.5	1	2	4	8	16	32	≥64
AmB	252	1		2	25	47	89	71	16		1			
FLU	286										1	11	123	151
ISAV	80		1	4	12	25	29	9						
VORI	290			5	43	135	83	16	5	2	1			
POSA	260		1	9	36	111	89	11	1		2			
ITRA	252		2	3	31	125	76	13				2		
ANID	231	10	77	100	43	1								
MICA	231	2	1	68	141	19								
5FC	252		2	3	28	114	71	12	1	3	12	5	1	

76.2. *Pichia norvegensis* Leask & Yarrow

Synonymy: *Candida norvegensis* Dietrichson ex van Uden & Buckley.

Pichia norvegensis is a very rare clinical isolate that has been reported as a causative agent of peritonitis and disseminated candidiasis in a patient on CAPD.

RG-1 organism

Culture: Colonies (SDA) white to cream coloured, smooth, glabrous, yeast-like.

Microscopy: Ovoid to ellipsoid, budding blastoconidia, 2.0-4 x 3-10 µm.

India ink preparation: Negative - no capsules present.

Dalmau plate culture: Spherical to ovoid budding yeast cells only. Abundant pseudohyphae produced.

Molecular identification: ITS sequencing recommended.

MALDI-ToF MS: Able to accurately identify this species.

Germ tube formation: Negative.

Physiological tests: (+ Positive, – Negative, v Variable, w Weak, s Slow, n not done).

Fermentation:

Glucose	s	Sucrose	–	Lactose	–
Galactose	–	Maltose	–	Trehalose	–

Growth reactions:

Glucose	+	L-Sorbose	–	*myo*-Inositol	–
Sucrose	–	L-Rhamnose	–	DL-Lactate	w
Raffinose	–	D-Xylose	–	D-Gluconate	–
Melibiose	–	L-Arabinose	–	2-Keto-D-gluconate	–
Galactose	–	D-Arabinose	–	D-Glucosamine	+
Lactose	–	D-Ribose	–	*N*-Acetyl-D-glucosamine	–
Trehalose	–	Glycerol	+	D-Glucuronate	–
Maltose	–	Erythritol	–	Nitrate	–
Melezitose	–	Ribitol	–	Urease	–
Methyl-½-D-glucoside	–	Galactitol	–	0.1% Cycloheximide	–
Soluble starch	–	D-Mannitol	–	Growth at 37°C	+
Cellobiose	+	D-Glucitol	–		

Key features: Germ tube negative yeast, sugar assimilation pattern.

Antifungal susceptibility: (Table 76.2).

Table 76.2. *Pichia norvegensis* very limited data (Australian national data); MIC µg/mL.

Antifungal	No.	≤0.016	0.03	0.06	0.125	0.25	0.5	1	2	4	8	16	32	≥64
AmB	5					1	2	1	1					
FLU	5										1			4
ISAV	1							1						
VORI	4				1		2		1					
POSA	4				1	3								
ITRA	5				1	2	2							
ANID	3		2	1										
MICA	3		1	1	1									
5FC	5							1		1	2		1	

References: Kurtzman (2011c).

DOI: 10.1079/9781800622340.0076

77 *Pleurostoma richardsiae* (Nannf.) Reblova & Jaklitsch

Synonymy: *Phialophora richardsiae* (Nannf.) Conant; *Pleurostomophora richardsiae* (Nannf.) Mostert *et al.*

Pleurostoma richardsiae was separated from *Phialophora* based on molecular data (Vijaykrishna *et al.*, 2004; Reblova *et al.*, 2015). *P. richardsiae* is a soft rot fungus of wood and is an uncommon cause of human infection, usually through traumatic implantation causing subcutaneous phaeohyphomycosis.

RG-2 organism

Morphological description: Colonies grow rapidly, and are powdery to woolly or tufted, greyish-brown with a grey-brown to olivaceous-black reverse. Two conidial types are produced: (1) hyaline conidia which are allantoid or cylindrical, 3-6 x 1.5-2.5 µm in size, formed on inconspicuous, peg-like phialides on thin-walled hyphae; and (2) brown, thick-walled conidia which are spherical to sub-spherical, 2.5-3.5 x 2-3 µm, formed on dark brown, slender, tapering phialides with flaring collarettes (Fig. 77.1).

Key features: *Pleurostoma richardsiae* is characterised microscopically by phialides with prominent flaring collarettes bearing globose, brown conidia while phialides with indistinct collarettes bear pale allantoid to cylindrical conidia.

Molecular identification: ITS sequencing is recommended (Vijaykrishna *et al.*, 2004).

Fig. 77.1. *Pleurostoma richardsiae* showing (a) culture and (b) phialides which produce two types of conidia: (1) hyaline conidia, formed on inconspicuous, peg-like phialides on thin-walled hyphae; and (2) brown, thick-walled conidia formed on dark brown, slender, tapering phialides with flaring collarettes.

Antifungal susceptibility: (Table 77.1).

Table 77.1. *Pleurostoma richardsiae*, limited data (Australian national data); MIC µg/mL.

Antifungal	No.	≤0.016	0.03	0.06	0.125	0.25	0.5	1	2	4	8	16	32	≥64
AmB	15					1	1	5	1	5	2			
ISAV	1						1							
VORI	15				4	8	3							
POSA	13			1	2	3	6	1						
ITRA	15				2	5	6	2						

References: Ellis (1971); McGinnis (1978b, 1980); Vijaykrishna *et al.* (2004); Domsch *et al.* (2007); Revankar and Sutton (2010); de Hoog *et al.* (2015).

DOI: 10.1079/9781800622340.0077

78 *Pneumocystis jirovecii* Frenkel

Pneumocystis was initially considered a protozoan and was subsequently assigned to the kingdom Fungi based on phylogenetic studies of the rRNA (Edman *et al.*, 1988). *Pneumocystis* is included in the phylum Ascomycota, subphylum Taphrinomycotina, and is related to *Saccharomyces cerevisiae* and *Schizosaccharomyces pombe*. Despite these similarities, *Pneumocystis* is considered an "atypical fungus" as it has cholesterol in its cell membrane instead of ergosterol and is difficult to culture (Edman *et al.*, 1988). Initially considered a single species, *Pneumocystis carinii*, it has since been discovered that different species infect different mammals and the species are both phenotypically and genetically distinct (Sinclair *et al.*, 1991). The human form was renamed *Pneumocystis jirovecii* and other species-specific forms have been named *P. murina* in mice, *P. carinii* and *P. wakefieldiae* in rats and *P. oryctolagi* in rabbits (Cushion, 2010).

Pneumocystis jirovecii is ubiquitous and causes *P. jirovecii* pneumonia (PCP) in patients with T cell deficiencies, including those with HIV infection, solid organ transplant recipients and patients with cancer as well as adults and children with other underlying immunological diseases (Alanio *et al.*, 2016). *P. jirovecii* is not cultivable *in vitro* and diagnosis of proven PCP is based on clinical and radiologic criteria and demonstration of *P. jirovecii* by microscopy.

Morphological description: Diagnosis relies on microscopic detection of trophic forms and cysts of *P. jirovecii* in induced sputum, expectorated sputum and bronchoalveolar lavage fluid or washings. The most commonly used stains, i.e. Giemsa, toluidine blue O, calcofluor white and Grocott-Gomori's methenamine silver, stain the cyst form only. Cysts are non-budding, 4-8 µm in diameter, round, oval or collapsed crescent in shape, resembling crushed ping pong balls or parentheses facing one another, often found in honeycomb-like clusters (Fig. 78.1). Immunofluorescent staining exhibits superior sensitivity to conventional microscopy detecting both the cyst and trophic forms, which are 10 x more abundant than the cyst forms in clinical specimens (Alanio *et al.*, 2016). Trophozoites are 1-8 µm long, pleomorphic and contain a single nucleus.

Key features: Intracystic bodies resembling parentheses facing one another or crushed ping pong balls often found in honeycomb-like clusters.

Molecular identification: PCR-based assays are more sensitive than microscopic examination for the detection of *P. jirovecii*, but they cannot distinguish between colonisation and infection. The multi-copy mitochondrial large subunit ribosomal RNA (*mtLSU*) gene is the most common PCR target, but assays targeting the mitochondrial small subunit RNA (*mtSSU*) gene, major surface antigen gene, 18*S* rRNA, ITS, β-tubulin, *DHPS* and *HSP70* genes have also been developed (Lagrou *et al.*, 2021).

References: Edman *et al.* (1988); Sinclair *et al.* (1991); Cushion (2010); Alanio *et al.* (2016); de Hoog *et al.* (2020); Lagrou *et al.* (2021).

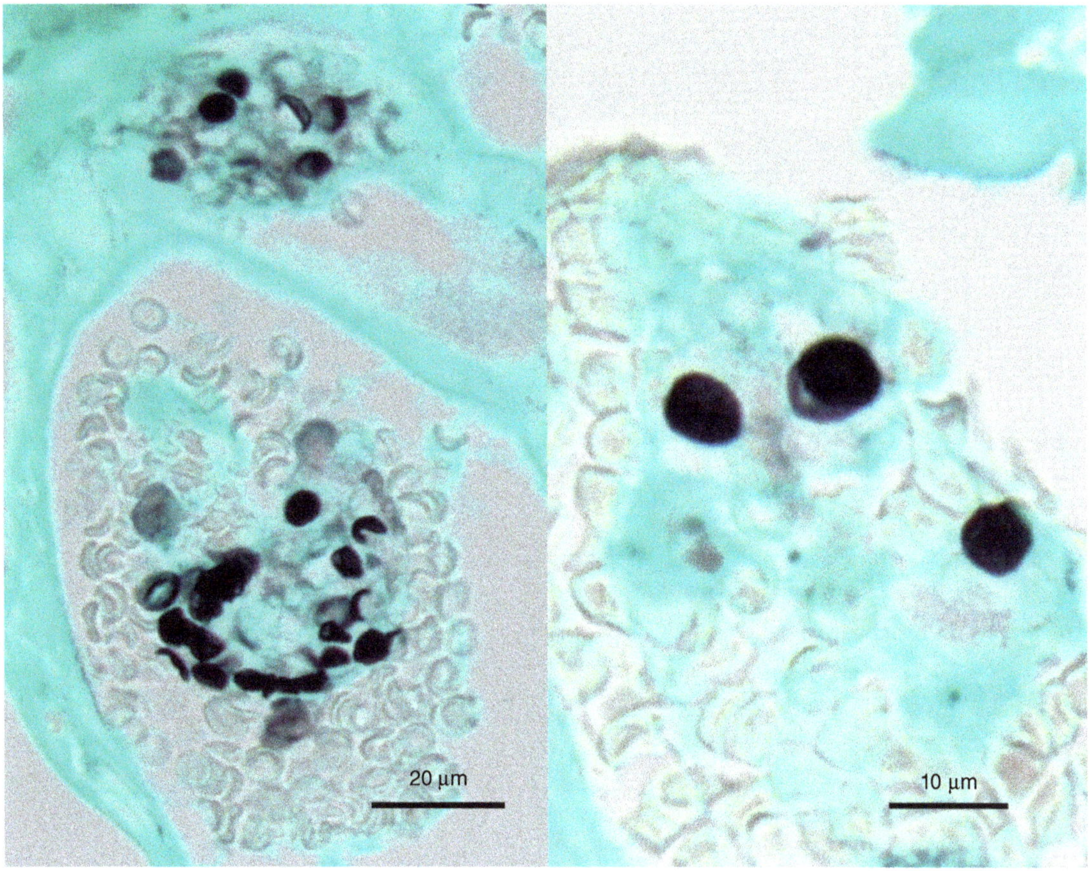

Fig. 78.1. *Pneumocystis jirovecii* cysts in a Grocott-Gomori's methenamine silver-stained tissue section from a lung biopsy (Photograph courtesy Yee Khong, Anatomical Pathology, SA Pathology, Adelaide, Australia).

DOI: 10.1079/9781800622340.0078

79 *Prototheca* Kruger

Prototheca species are not fungi, but achlorophyllous algae with phylogenetic affinities to the genus *Chlorella*. These organisms are commonly handled in mycology laboratories due to their macroscopic similarity to yeasts and susceptibility to antifungal agents. To date only *P. wickerhamii* and *P. ciferrii* (formerly *P. zopfii*) have been involved in human or animal infections (Lass-Florl and Mayr, 2007; Bakuła *et al.*, 2021).

RG-1 organism

Morphological description: Colonies are smooth, moist, white to cream and yeast-like. Cultures are sensitive to cycloheximide (actidione) and optimal growth occurs at 25°C to 30°C. Mycelium and conidia are absent. Vegetative cells are globose to ovoid, hyaline, varying in size from approximately 8-20 µm, and have a relatively thick and highly refractile wall. No budding cells are present; reproduction is by the development of large sporangia (thecae), which contain from 2-20 or more small sporangiospores (endospores or autospores), which are asexually produced by nuclear division, and cleavage of the cytoplasm (Fig. 79.1; Table 79.1).

Molecular identification: ITS, D1/D2 sequencing may be used for accurate species identification (Ahrholdt *et al.*, 2012; Wang *et al.*, 2014).

MALDI-ToF MS: Can provide reliable identification of *Prototheca* species (Ahrholdt *et al.*, 2012; Murugaiyan *et al.*, 2012).

Key features: Achlorophyllous algae reproducing by sporangia (theca) and sporangiospores (autospores). *Prototheca* species can be differentiated by assimilation tests and morphological criteria. The API 20C yeast identification strip may be used for species identification.

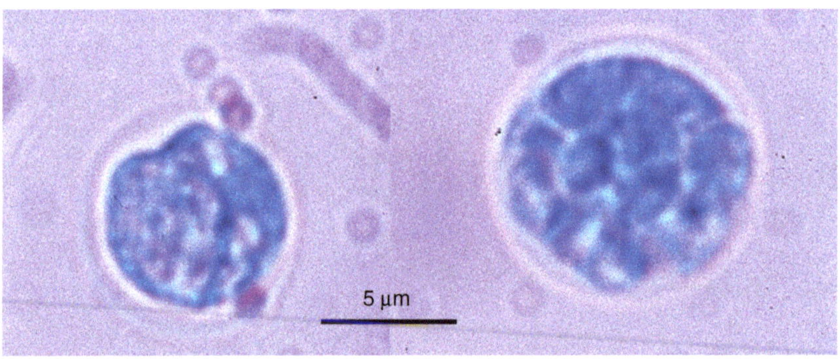

Fig. 79.1. *Prototheca wickerhamii* thecae and autospores.

Table 79.1. Differentiation of *Prototheca* species of medical importance.

	P. wickerhamii	*P. ciferrii*	*P. stagnora*
Colony morphology	Hemispheric, with smooth margin	Flat, rough, central button, corrugated margin	Flat, with smooth margin
Cell diam. µm	3-10	7-30	7-14
Growth at 37°C	+	+	–
Glucose	+	+	+
Trehalose	+	–	–
L-Propanol	–	+	+/–
Acetate (pH5)	–	+	+/–
Galactose	+/–	–	+
Capsule	–	–	+

Antifungal susceptibility: (Table 79.2).

Table 79.2. *Prototheca wickerhamii* (Australian national data); MIC µg/mL.

Antifungal	No.	≤0.016	0.03	0.06	0.125	0.25	0.5	1	2	4	8	16	32	≥64
AmB	12				3	4	3	2						
ISAV	2			1	1									
VORI	10					2	1	5	2					
POSA	10					1	3	5	1					
ITRA	12						2	7			3			

References: Kaplan (1977); McGinnis (1980); Pore (1985); Rippon (1988); Ueno *et al.* (2005); Lass-Flori and Mayr (2007); Wang *et al.* (2014); Jagielski *et al.* (2019); Bakuła *et al.* (2021).

DOI: 10.1079/9781800622340.0079

80 *Pseudopithomyces chartarum* (Berk. & M.A. Curtis) Jun F. Li, Ariyaw. & K.D. Hyde

Synonymy: *Pithomyces chartarum* (Berk. & M.A. Curtis) M.B. Ellis.

The genus *Pseudopithomyces* was introduced by Ariyawansa *et al.* (2015) to accommodate several species which had previously been classified in the genus *Pithomyces*, notably *Pithomyces chartarum*. *Pseudopithomyces* contains 13 species (Ariyawansa *et al.*, 2015) that are commonly found from a very wide range of plant material, also from air, soil, hay, sawn timber and ceiling plaster. *Pseudopithomyces chartarum* (formerly *Pithomyces chartarum*) has long been reported as causing facial eczema of sheep. However, recent molecular evidence has identified at least two additional species *Pseudopithomyces sacchari* (formerly *Pithomyces sacchari*) and *Pseudopithomyces maydicus* (formerly *Pithomyces maydicus*) (da Cunha *et al.*, 2014). Most human isolates are recovered from skin, nail, respiratory and sinus specimens.

RG-1 organism

Morphological description: Colonies are fast growing, suede-like to downy and black. Conidiophores are pale olive, smooth or verrucose, 2.5-10 x 2-3.5 µm. Conidiogenous cells integrated, intercalary or terminal, indeterminate, with one to two loci of similar width in the conidiogenous cells. Conidia are muriform, medium to dark brown, echinulate to verrucose, three (some up to five) euseptate, slightly constricted at the septa, with one or both median cells divided by longitudinal septa, thick-walled, broadly ellipsoidal, apex obtuse, base truncate and characteristically with part of the conidiogenous cell remaining attached as a small pedicel, 18-29 x 10-17 µm (Fig. 80.1).

Key features: Darkly-pigmented, multi-celled conidia produced on small peg-like branches of the vegetative hyphae.

Molecular identification: ITS and D1/D2 sequencing (da Cunha *et al.*, 2014).

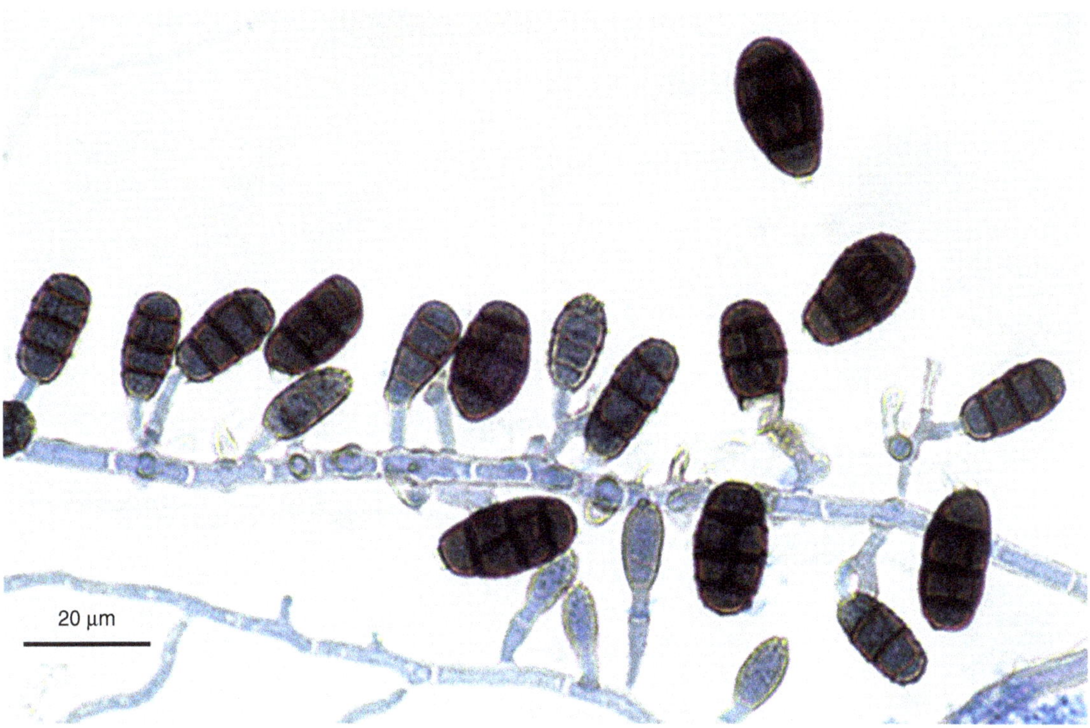

Fig. 80.1. *Pseudopithomyces chartarum* conidiophores and conidia.

Antifungal susceptibility: (Table 80.1).

Table 80.1. *Pseudopithomyces* spp. (Australian national data); MIC µg/mL.

Antifungal	No.	≤0.016	0.03	0.06	0.125	0.25	0.5	1	2	4	8	16	32	≥64
AmB	6				1		2	1	1	1				
FLU	6									1	1	1	1	2
ISAV	2					1	1							
VORI	6		1	1		2	2							
POSA	6		1	1		4								
ITRA	6			1	1	1	3							

References: Ellis (1971, 1976); Domsch *et al*. (2007); Rippon (1988); da Cunha *et al.* (2014); de Hoog *et al.* (2015).

DOI: 10.1079/9781800622340.0080

81 *Purpureocillium lilacinum* (Thom) Luangsa-ard *et al.*

Synonymy: *Paecilomyces lilacinus* (Thom) Samson.

Purpureocillium lilacinum is commonly isolated from soil, decaying vegetation, insects, nematodes and as a laboratory contaminant. It is also a causative agent of infection in humans and other vertebrates (Luangsa-ard *et al.*, 2011). **Note:** *Purpureocillium lilacinum* and *Marquandomyces marquandii* were previously classified within the genus *Paecilomyces*.

RG-1 organism

Morphological description: Colonies are fast growing, suede-like to floccose, vinaceous to violet coloured (Fig. 81.1). Conidiophores are erect 400-600 μm in length, bearing branches with densely clustered phialides. Conidiophore stipes are 3-4 μm wide, yellow to purple and rough-walled. Phialides are swollen at their bases, gradually tapering into a slender neck (Fig. 81.2). Conidia are ellipsoidal to fusiform, smooth-walled to slightly roughened, hyaline to purple in mass, 2.5-3.0 × 2-2.2 μm, and are produced in divergent chains. Chlamydospores are absent. Growth at 38°C.

Molecular identification: ITS sequencing is recommended (Atkins *et al.*, 2005; Luangsa-ard *et al.*, 2011).

Key features: Colony pigmentation, phialides with swollen bases, pigmented and rough-walled conidiophore stipes, absence of chlamydospores and growth at 37°C. **Note:** *Marquandomyces marquandii* differs by having a yellow reverse pigment, smooth conidiophore stipes, presence of chlamydospores, and no growth at 37°C.

Fig. 81.1. *Purpureocillium lilacinum* culture.

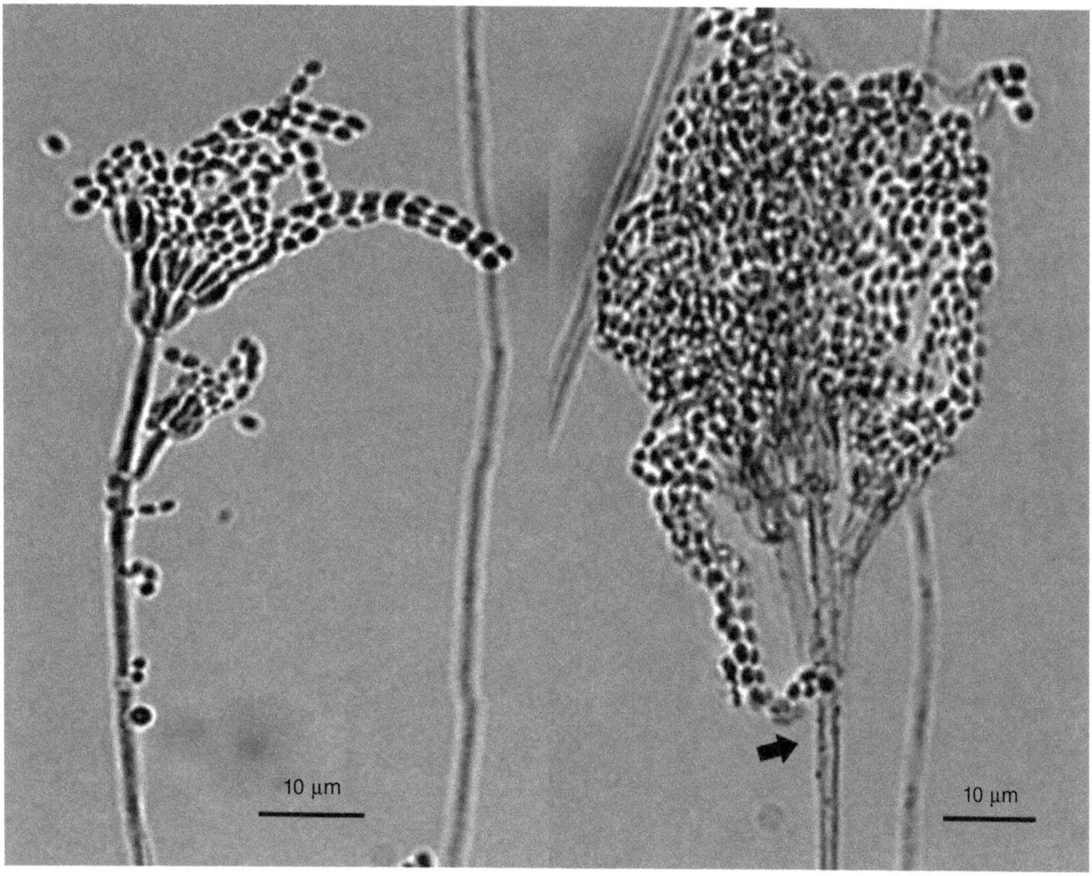

Fig. 81.2. *Purpureocillium lilacinum* microscopy showing conidiophores, phialides and conidia. **Note:** Rough-walled conidiophore (arrow).

Antifungal susceptibility: (Table 81.1).

Table 81.1. *Purpureocillium lilacinum* (Australian national data); MIC µg/mL.

Antifungal	No.	≤0.016	0.03	0.06	0.125	0.25	0.5	1	2	4	8	16	32	≥64
AmB	103				1		1		6	8	87			
ISAV	7					1	6							
VORI	101			8	49	36	5			2	1			
POSA	88			2	8	12	55	10	1					
ITRA	104	1				7	21	47	5	8	2	13		

References: Samson (1974); McGinnis (1980); Onions *et al.* (1981); Rippon (1988); Perdomo *et al.* (2013); Domsch *et al.* (2007); de Hoog *et al.* (2015).

DOI: 10.1079/9781800622340.0081

82 *Quambalaria cyanescens* (de Hoog & de Vries) de Beer *et al.*

Synonymy: *Sporothrix cyanescens* de Hoog & de Vries; *Cerinosterus cyanescens* (de Hoog & de Vries) R.T. Moore.

The genus *Quambalaria* contains five species, including *Q. cyanescens*, *Q. pitereka*, *Q. eucalypti*, *Q. coyrecup* and *Q. simpsonii*. *Quambalaria cyanescens* is a hyaline basidiomycete isolated from a broad range of ecological niches, including air, soil and insect larvae as well as in association with diverse plant sources, including *Corymbia* and *Eucalyptus* species from Australia (Simpson, 2000).

Quambalaria cyanescens appears to be an emerging opportunistic pathogen in immunocompromised or debilitated individuals. It has been isolated from human skin and subcutaneous infections, blood, nosocomial infections in patients with pneumonia, peritonitis and invasive pulmonary infection (Jackson *et al.*, 1996; Tambini *et al.*, 1996; Kuan *et al.*, 2015).

RG-1 organism

Morphological description: Colonies are restricted, farinose or velvety, often compact and somewhat cerebriform, snow-white, and later often exuding a pH-dependent, deep blue/violet pigment into the agar (Fig. 82.1). Conidiogenous cells are undifferentiated, cylindrical, of variable size (1.5-3.0 µm wide), apically with a cluster of small denticles, the cluster often repeatedly proliferating and forming similar clusters. Conidia are hyaline, smooth-walled or finely verrucose, obovoidal, 3-4 µm long, somewhat larger (3.5-6.5 µm long) when bearing secondary conidia (Fig. 82.1).

Molecular identification: ITS and D1/D2 sequencing recommended.

Key features: *Quambalaria cyanescens* is morphologically similar to *Sporothrix*, but the conidial scars are very small, and cultures are thin and fragile. In fresh cultures the diffusible pigment is characteristic. *Sporothrix schenckii* forms tough colonies, which finally become blackish-brown.

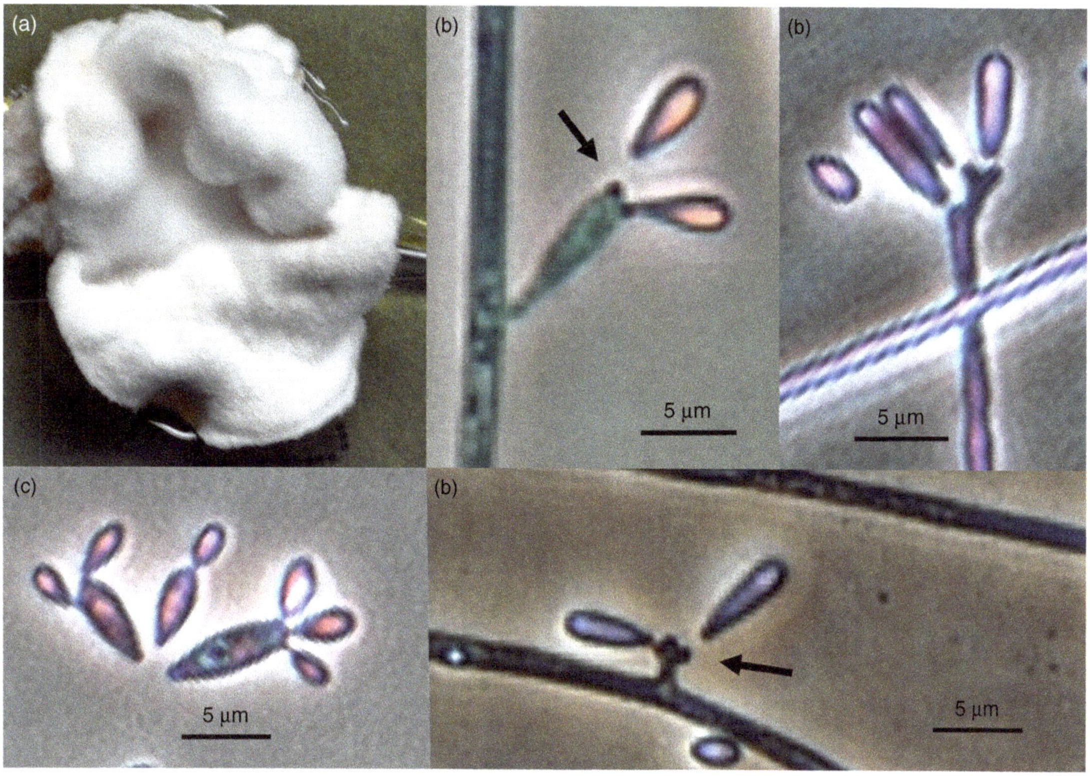

Fig. 82.1. *Quambalaria cyanescens* showing (a) culture, (b) conidiogenous cells with small denticles (arrows) and conidia, and (c) mature conidia bearing secondary conidia.

Antifungal susceptibility: (Table 82.1).

Table 82.1. *Quambalaria cyanescens* (Australian national data); MIC µg/mL.

Antifungal	No.	≤0.016	0.03	0.06	0.125	0.25	0.5	1	2	4	8	16	32	≥64
AmB	25				1		7	4	12	1				
ISAV	2	1		1										
VORI	25	10	9	5				1						
POSA	24	3	8	12				1						
ITRA	25	1	4	10	9		1							

References: de Hoog and de Vries (1973); de Beer *et al.* (2006); Simpson (2000); de Hoog *et al.* (2015).

DOI: 10.1079/9781800622340.0082

83 *Rhinocladiella* Nannfeldt

Rhinocladiella contains six to eight species, with five species of medical interest: *R. aquaspersa, R. atrovirens, R. basitona, R. mackenziei* (formerly *Ramichloridium mackenziei*) and *R. similis*. *Rhinocladiella mackenziei* is a frequently fatal neurotropic organism with a geographic concentration to individuals residing in or immigrating from Middle Eastern countries (Revankar and Sutton, 2010).

83.1. *Rhinocladiella atrovirens* Nannfeldt

Morphologically *Rhinocladiella atrovirens* is identical to *R. similis,* but the species are phylogenetically distinct. In the past, some cases reported under the name *R. atrovirens* may have concerned *R. similis* (Revankar and Sutton, 2010; Taj-Aldeen *et al.*, 2010).

RG-1 organism

Morphological description: Colonies are restricted, velvety or lanose, olivaceous, often slightly mucoid at the centre; reverse dark olivaceous-green to blackish (Fig. 83.1). Conidiophores are short, brown, thick-walled. Conidiogenous cells are cylindrical, intercalary or free, 9-19 × 1.6-2.2 μm; denticulate rachis up to 15 μm

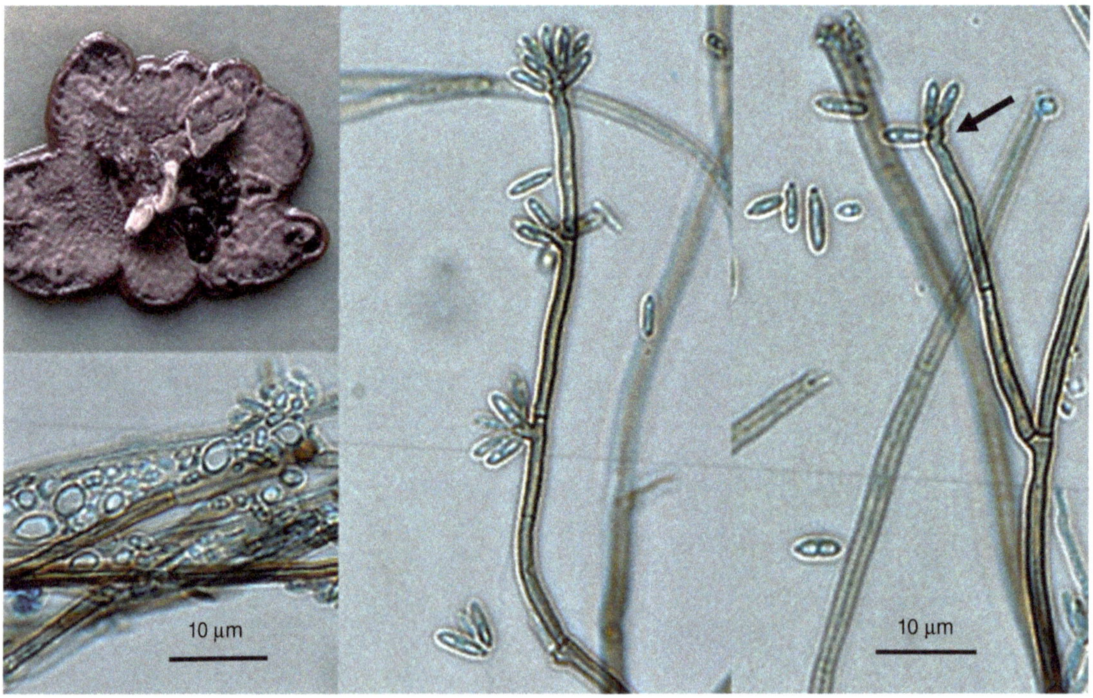

Fig. 83.1. *Rhinocladiella atrovirens* culture and conidiophores showing a terminal denticulate rachis (arrow), conidia and budding yeast cells.

long, with crowded, flat or butt-shaped, unpigmented conidial denticles (Fig. 83.1). Conidia are hyaline, thin- and smooth-walled, short-cylindrical, with truncate basal scars, 3.7-5.5 x 1.2-1.8 μm. Budding cells, if present, are hyaline, thin-walled, broadly ellipsoidal, 3.0-4.3 x 1.7-2.5 μm. Germinating cells are inflated, spherical to subspherical, 4.5-6.0 μm. An annellidic *Exophiala*-like synanamorph may be present.

Molecular identification: ITS sequencing recommended for accurate species identification (Taj-Aldeen *et al.*, 2010).

Antifungal susceptibility: (Table 83.1).

Table 83.1. *Rhinocladiella atrovirens* very limited data (Australian national data); MIC μg/mL.

Antifungal	No.	≤0.016	0.03	0.06	0.125	0.25	0.5	1	2	4	8	16	32	≥64
AmB	2								1		1			
ISAV	1										1			
VORI	2							1	1					
POSA	2					1	1							
ITRA	2						2							

83.2. *Rhinocladiella mackenziei* (Campbell & Al-Hedaithy) Arzanlou & Crous

Rhinocladiella mackenziei is an extremely rare, neurotropic organism that causes fatal brain lesions, mostly in patients who are immunocompromised or who have underlying metabolic diseases. The species is typically restricted to the Middle East, in an arid zone between Israel and Pakistan (Kanj *et al.*, 2001; Khan *et al.*, 2002; Taj-Aldeen *et al.*, 2010), with a single autochthonous case in India (Badali *et al.*, 2010c). Cases in the USA and Europe invariably were found in patients originating from the Middle East (Sutton *et al.*, 1998; Revankar and Sutton, 2010; de Hoog *et al.*, 2015).

WARNING: RG-3 organism

Cultures of *Rhinocladiella mackenziei* represent a potential biohazard to laboratory personnel and must be handled with extreme caution in a Class II Biological safety cabinet (BSCII).

Morphological description: Colonies growing moderately rapidly, velvety, olivaceous-brown. Conidiophores arising at right angles from creeping hyphae, stout, thick-walled, brown, 3.0-4.5 μm wide, 10-25 μm long, apically with short-cylindrical denticles. Conidia brown, ellipsoidal, 8.5-12.0 x 4-5 μm, with prominent, 1 μm wide basal scar.

Molecular identification: ITS and D1/D2 sequencing is recommended for accurate species identification (Taj-Aldeen *et al.*, 2010).

Antifungal susceptibility: (Table 83.2).

Table 83.2. *Rhinocladiella mackenziei*, very limited data (Taj-Aldeen *et al.*, 2010); MIC μg/mL.

Antifungal	No.	≤0.016	0.03	0.06	0.125	0.25	0.5	1	2	4	8	16	32	≥64
AmB	7		1					3			3			
VORI	3	1	1				1							
POSA	5	2	2			1								
ITRA	7	5	1		1									

References: de Hoog (1977, 1983); Schell *et al.* (1983); Revankar and Sutton (2010); Taj-Aldeen *et al.* (2010).

DOI: 10.1079/9781800622340.0083

84 *Rhizomucor* Lucet & Costantin

The genus *Rhizomucor* is distinguished from *Mucor* by the presence of stolons and poorly developed rhizoids at the base of the sporangiophores and by the thermophilic nature of its two species: *R. miehei* and *R. pusillus*. Both of these species are potential human and animal pathogens and were originally classified in the genus *Mucor. Rhizomucor pusillus* is cosmopolitan and both *R. miehei* and *R. pusillus* have been reported as pathogens to humans and animals, the latter to a greater extent.

References: Cooney and Emerson (1964); Schipper (1978); McGinnis (1980); Ellis and Keane (1981); Scholer *et al.* (1983); Schipper and Staplers (2003); Ellis (2005b); Domsch *et al.* (2007); de Hoog *et al.* (2015).

Identification of most mucormycetous fungi is based primarily on the morphology of the sporangia, i.e. arrangement and number of sporangiospores, shape, colour, presence or absence of columellae and apophyses, as well as the arrangement of the sporangiophores and the presence or absence of rhizoids. Growth temperature tests can also be especially helpful in identifying and differentiating members of the genera *Rhizomucor, Rhizopus* and *Lichtheimia.*

84.1. *Rhizomucor miehei* (Cooney and Emerson) Schipper

Synonymy: *Mucor miehei* Lindt.

This species has been reported as a rare cause of bovine mastitis (Scholer *et al.*, 1983) and is similar in most respects to *Rhizomucor pusillus.*

RG-1 organism

Morphological description: All strains are homothallic forming numerous zygospores, which are reddish-brown to blackish-brown, globose to slightly compressed, up to 50 µm in diameter, with stellate warts and equal suspensor cells. Colony colour is a dirty grey rather than brown, and sporangia have spiny walls, are up to 50-60 µm in diameter, with columellae rarely larger than 30 µm in diameter. Growth is stimulated by thiamine, with no assimilation of sucrose and maximum temperature of growth is 54-58°C.

Key features: Growth at 45°C, the formation of numerous zygospores, a dirty grey culture colour and a partial growth requirement for thiamine.

Antifungal susceptibility: (Table 84.1).

Table 84.1. *Rhizomucor miehei*, very limited (Australian national data); MIC µg/mL.

Antifungal	No.	≤0.016	0.03	0.06	0.125	0.25	0.5	1	2	4	8	16	32	≥64
AmB	11				3	2	3	2	1					
VORI	11								3	5	3			
POSA	11				4	4	1	2						
ITRA	11				4	5	1	1						

84.2. *Rhizomucor pusillus* (Lindt) Schipper

Synonymy: *Mucor pusillus* Lindt.

This species is a rare human pathogen. It has been reported from cases of pulmonary, disseminated and cutaneous types of infection. It is more often associated with animal disease, especially bovine abortion. *Rhizomucor pusillus* has a worldwide distribution and is commonly associated with compost heaps.

RG-2 organism

Morphological description: Cultures are characterised by compact, low growing (2-3 mm high), grey to greyish-brown coloured mycelium and by the development of typical sympodially branched, hyaline to yellow-brown sporangiophores (8-15 µm in diameter), always with a septum below the sporangium. Sporangia are globose (40-60 µm in diameter), each possessing an oval or pear-shaped columella (20-30 µm), often with a collarette (Fig. 84.1). Sporangiospores are hyaline, smooth-walled, globose to subglobose, occasionally oval (3-5 µm), and are often mixed with crystalline remnants of the sporangial wall. Chlamydospores are absent. Zygospores are rough-walled, reddish-brown to black, 45-65 µm in diameter and may be produced throughout the aerial hyphae in matings between compatible isolates. Temperature growth range: minimum 20-27°C; optimum 35-55°C; maximum 55-60°C. There is positive assimilation of sucrose and no thiamine dependence.

Molecular identification: ITS sequencing recommended for accurate species identification (Dannaoui, 2009).

Key features: Growth at 45°C (thermophilic), poorly developed stolons and rhizoids, branching sporangiophores with a septum below the sporangium, dark coloured sporangia without apophyses and smooth-walled globose to subglobose sporangiospores.

Antifungal susceptibility: (Table 84.2).

Table 84.2. *Rhizomucor pusillus* (Espinel-Ingroff *et al.*, 2015a; and Australian national data); MIC µg/mL.

Antifungal	No.	≤0.016	0.03	0.06	0.125	0.25	0.5	1	2	4	8	16	32	≥64
AmB	37			3	8	9	12	1	2		1			
VORI	4									4				
POSA	37			1	5	13	10	7	1					
ITRA	19				4	5	5	3	1				1	

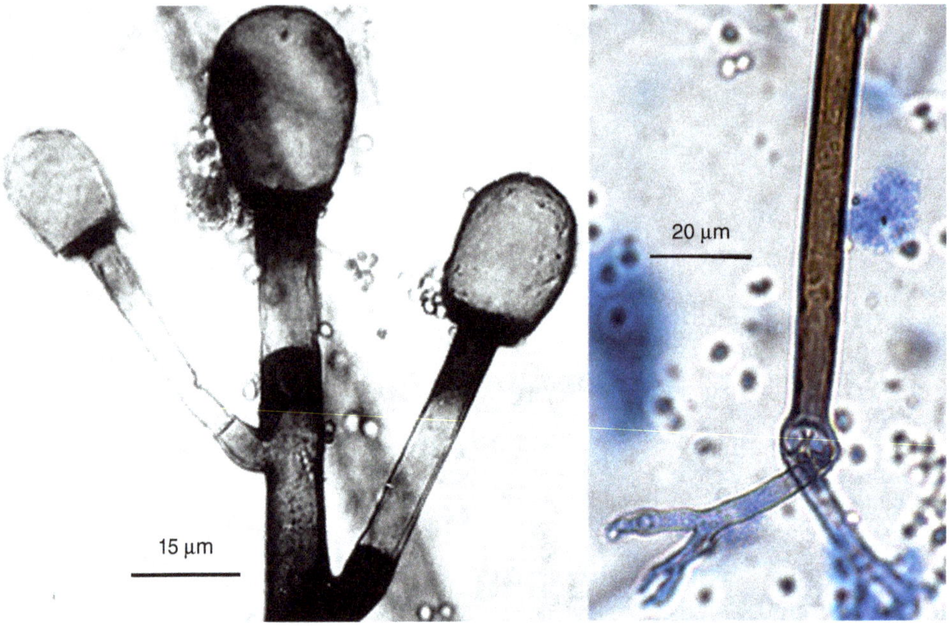

Fig. 84.1. *Rhizomucor pusillus* sporangiophores, columellae and primitive rhizoids. **Note:** Septa below the sporangia.

DOI: 10.1079/9781800622340.0084

85 *Rhizopus* Ehrenberg ex Corda.

A molecular phylogeny-based taxonomy of the genus *Rhizopus* by Abe *et al.* (2010) recognised eight species: *R. caespitosus, R. delemar, R. homothallicus, R. microsporus, R. arrhizus* (formerly *R. oryzae*), *R. reflexus, R. schipperae* and *R. stolonifer*. Based on these results and confirmed by Dolatabadi *et al.* (2014) the varieties of *Rhizopus microsporus* (*R. microsporus* var. *microsporus, R. microsporus* var. *oligosporus* and *R. microsporus* var. *rhizopodiformis*) have been reduced to synonyms. In addition, *R. azygosporus* has also been reduced to a synonym of *R. microsporus*. Finally, the controversy surrounding, the validity of the names *R. oryzae* vs *R. arrhizus* has been resolved in favour of the latter (Ellis, 1985; de Hoog *et al.*, 2015). Thus, the important medical pathogens have now been reduced to just *R. arrhizus* and *R. microsporus*. These two species are most common causative agents of mucormycosis, accounting for some 60% of the reported cases.

Morphological genus description: The genus *Rhizopus* is characterised by the presence of stolons and pigmented rhizoids, the formation of sporangiophores, singly or in groups from nodes directly above the rhizoids, and apophysate, columellate, multispored, generally globose sporangia. After spore release the apophyses and columella often collapse to form an umbrella-like structure. Sporangiospores are globose to ovoid, one-celled, hyaline to brown and striate in many species. Colonies are fast growing and cover an agar surface with a dense cottony growth that is at first white becoming grey or yellowish-brown with sporulation.

Molecular identification: ITS sequencing is recommended but sequences must be compared to those of quality-controlled reference strains using current species names (Alvarez *et al.*, 2009; Abe *et al.*, 2010).

References: McGinnis (1980); Onions *et al.* (1981); Scholer *et al.* (1983); Schipper (1984); Schipper and Stalpers (1984, 2003); Yuan and Jong (1984); Ellis (1985, 1986); Rippon (1988); Kwon-Chung and Bennett (1992); Samson *et al.* (1995); Schipper *et al.* (1996); Ellis (2005b); Domsch *et al.* (2007); Alvarez *et al.* (2009); Abe *et al.* (2010); Dolatabadi *et al.* (2014); de Hoog *et al.* (2015); Walther *et al.* (2019).

85.1. *Rhizopus arrhizus* Fischer

Synonymy: *Rhizopus oryzae* Went & Prinsen Geerligs.

RG-2 organism

Morphological description: Colonies are very fast growing, about 5-8 mm high, with some tendency to collapse, white cottony at first becoming brownish-grey to blackish-grey depending on the amount of sporulation (Fig. 85.1). Sporangiophores up to 1500 µm in length and 18 µm in width, smooth-walled, non-septate, simple or branched, arising from stolons opposite rhizoids usually in groups of three or more. Sporangia are globose, often with a flattened base, greyish-black, powdery in appearance, up to 175 µm in diameter and many-spored. Columellae and apophysis together are globose, subglobose or oval, up to 130 µm in height and soon collapsing to an umbrella-like form after spore release (Fig. 85.2). Sporangiospores are angular, subglobose to ellipsoidal, with ridges on the surface, and up to 8 µm in length. No growth at 45°C; good growth at 40°C.

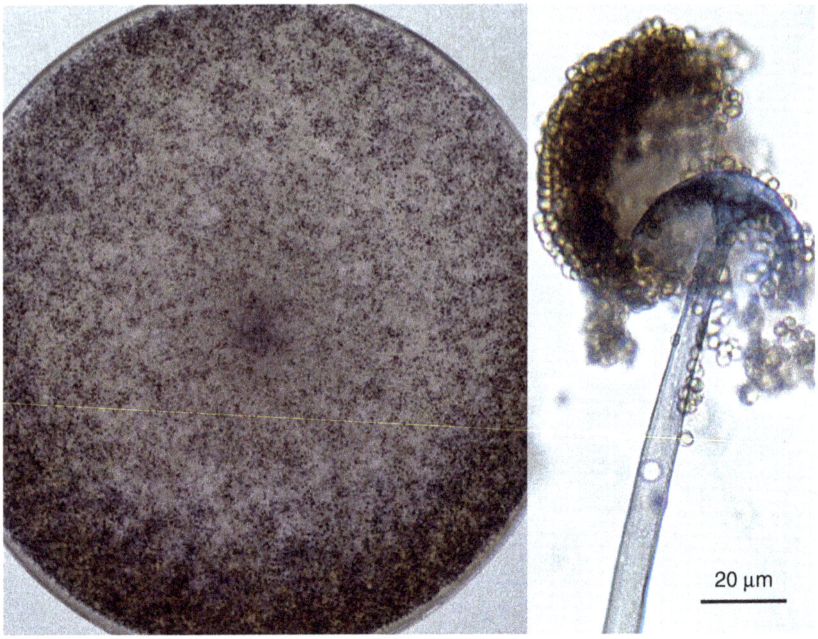

Fig. 85.1. *Rhizopus arrhizus* culture and a sporangium with a collapsed columella and head of sporangiospores.

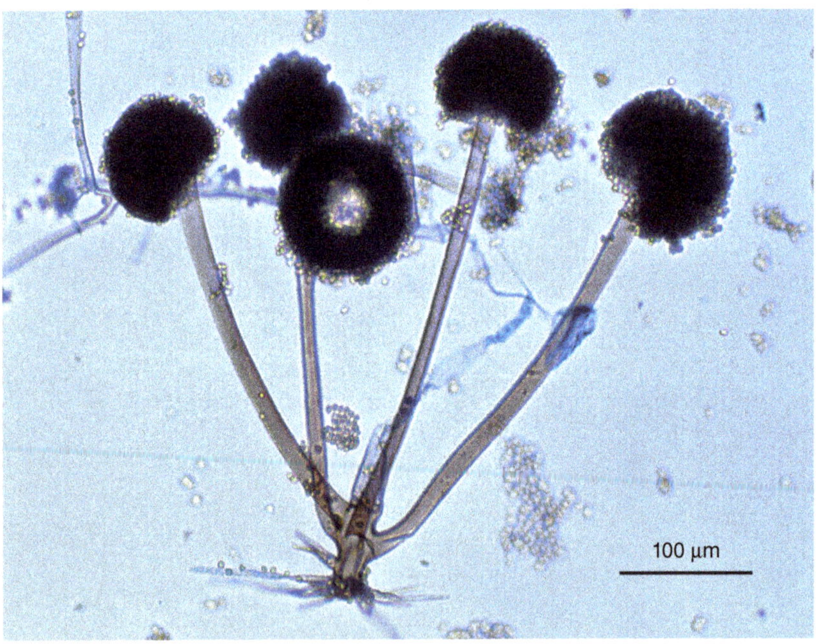

Fig. 85.2. *Rhizopus arrhizus* sporangiophores, rhizoids and sporangia.

Antifungal susceptibility: (Table 85.1).

Table 85.1. *Rhizopus arrhizus* (Australian national data); MIC µg/mL.

Antifungal	No.	≤0.016	0.03	0.06	0.125	0.25	0.5	1	2	4	8	16	32	≥64
AmB	42			1		5	12	10	6		8			
ISAV	11					1	2	2	4	1	1			
VORI	41								1	2	12	21	5	
POSA	40						6	13	14	7				
ITRA	42			1		4	10	12	6	1	1	7		

85.2. *Rhizopus microsporus* v. Tiegh

Synonymy: *Rhizopus azygosporus* Yuan & Jong; *Rhizopus. microsporus* var. *microsporus* Tiegh; *Rhizopus microsporus* var. *oligosporus* (Saito) Schipper & Stalpers; *Rhizopus microsporus* var. *rhizopodiformis* (Cohn) Schipper & Staples; *Rhizopus microsporus* var. *chinensis* (Saito) Schipper & Stalpers.

RG-2 organism

Morphological description: Colonies are dark greyish-brown, up to 10 mm high producing simple rhizoids. Sporangiophores are brownish, up to 400 µm high and 10 µm wide, and may be produced in groups of 1-4, usually in pairs. Sporangia are greyish-black, spherical, up to 100 µm in diameter. Columellae are subglobose to globose to conical comprising 80% of the sporangium (Fig. 85.3). Sporangiospores are angular to broadly ellipsoidal or subglobose, up to 5-9 µm in length and are distinctly striate. Chlamydospores may be present. Zygospores, formed by crosses of compatible mating strains, are dark red–brown, spherical, up to 100 µm in diameter, with stellate projections and unequal suspensor cells. Some strains may be homothallic and produce azygospores. There is good growth at 45°C, with a maximum of 50-52°C.

Antifungal susceptibility: (Table 85.2).

Table 85.2. *Rhizopus microsporus* (Australian national data); MIC µg/mL.

Antifungal	No.	≤0.016	0.03	0.06	0.125	0.25	0.5	1	2	4	8	16	32	≥64
AmB	109				12	31	45	12	8	1				
ISAV	16						2	7	5	1	1			
VORI	111								3	27	77	4		
POSA	109				3	10	53	31	11		1			
ITRA	109				1	3	24	54	9	5		12		1

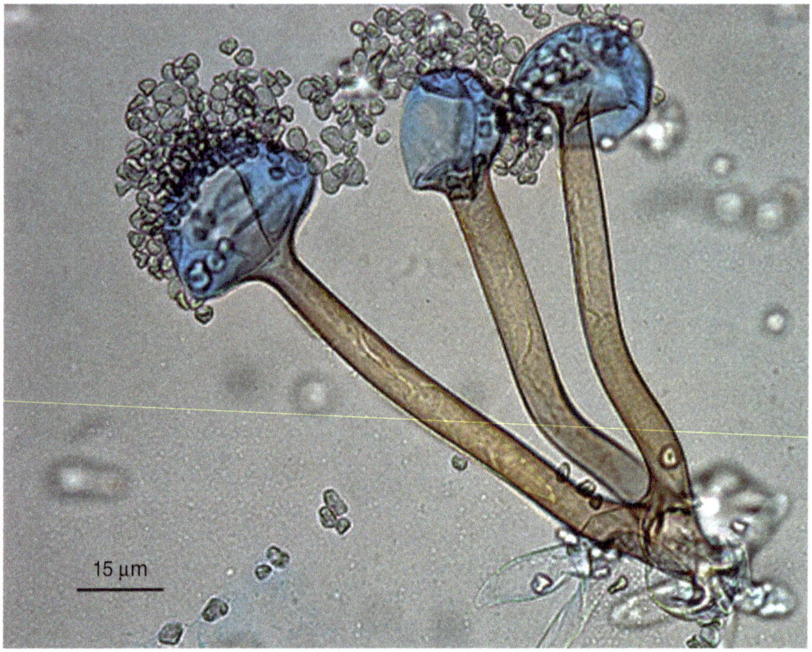

Fig. 85.3. *Rhizopus microsporus* sporangia showing sporangiospores, columellae, sporangiophores and rhizoids.

DOI: 10.1079/9781800622340.0085

86 *Rhodotorula* Harrison

Rhodotorula species are common environmental basidiomycetous yeasts, which can be found in soil, ocean and lake water, fruit juice and milk, and on shower curtains and toothbrushes. Today, the genus contains 46 species of which three have been described as rare human pathogens: *R. mucilaginosa* (formerly *R. rubra*), *R. glutinis* and *R. minuta* (Arendrup *et al.*, 2014).

Rhodotorula mucilaginosa is a common airborne contaminant of skin, lungs, urine and faeces and is a known cause of fungal peritonitis in patients on continuous ambulatory peritoneal dialysis (CAPD). This is usually due to saprophytic colonisation of catheters or dialysis machinery and removal of the source of contamination usually leads to clearing of the symptoms. This species accounts for the majority of infections (74-79%) followed by *R. glutinis* (7.7%) (Tuon and Costa 2008; Arendrup *et al.*, 2014).

Molecular identification: ITS or D1/D2 sequencing is recommended (Duboc de Almeida *et al.*, 2008; Tuon and Costa, 2008; Arendrup *et al.*, 2014).

MALDI-ToF MS: Reliably identifies clinically relevant *Rhodotorula* species.

References: McGinnis (1980); Barnett *et al.* (1983); Rippon (1988);Duboc de Almeida *et al.* (2008); Tuon and Costa (2008); Sampaio (2011); Spiliopoulou *et al.* (2012); Arendrup *et al.* (2014).

86.1. *Rhodotorula glutinis* (Fresenius) Harrison

RG-1 organism

Morphological description: Colonies are coral-red to salmon coloured or slightly orange, smooth to wrinkled, highly glossy to semi-glossy. Mucoid to pasty to slightly tough, yeast-like colonies. Ovoidal to globose or more elongate budding yeast-like cells or blastoconidia, 2.3-5.0 × 4.0-10.0 µm.

India ink preparation: Small capsules present.

Germ tube formation: Negative.

Physiological tests: (+ Positive, – Negative, v Variable, w Weak, s Slow, n not done).

Fermentation: Absent.

Growth reactions:

Glucose	+	L-Sorbose	+	*myo*-Inositol	–
Sucrose	+	L-Rhamnose	–	DL-Lactate	–
Raffinose	+,s	D-Xylose	s	D-Gluconate	+
Melibiose	–	L-Arabinose	+	2-Keto-D-gluconate	n
Galactose	+	D-Arabinose	+	D-Glucosamine	–
Lactose	–	D-Ribose	+	*N*-Acetyl-D-glucosamine	n
Trehalose	+	Glycerol	+	D-Glucuronate	–
Maltose	+	Erythritol	–	Nitrate	–
Melezitose	+	Ribitol	+	Urease	+
Methyl-α-D-glucoside	+	Galactitol	–	0.1% Cycloheximide	+
Soluble starch	–	D-Mannitol	+	Growth at 35°C	–
Cellobiose	+	D-Glucitol	+	Growth at 37°C	–

Molecular identification: Requires sequencing of the ITS and/or D1/D2 regions.

Key features: Germ tube negative yeast, sugar assimilation pattern. Growth on nitrate is positive (*R. mucilaginosa* growth on nitrate is negative). Common saprophyte, however cases of fungaemia have been reported.

Antifungal susceptibility: (Table 86.1).

Table 86.1. *Rhodotorula glutinis* (Diekema *et al.*, 2005; and Australian national data); MIC µg/mL. **Note:** *Rhodotorula* species are intrinsically resistant to echinocandins and fluconazole (Arendrup *et al.*, 2014).

Antifungal	No.	≤0.016	0.03	0.06	0.125	0.25	0.5	1	2	4	8	16	32	≥64
AmB	38			1	2	4	3	28						
FLU	38													38
VORI	36					1	1	12	12	9	1			
POSA	34						2	9	18	4	1			
ITRA	38			1		1	3	9	11	2	3	8		
5FC	38			17	12	8		1						

86.2. *Rhodotorula mucilaginosa* (Jorgensen) Harrison

Synonymy: *Rhodotorula rubra* (Demme) Lodder.

RG-1 organism

Morphological descriptions: Colonies are coral-pink, usually smooth, sometimes reticulate, rugose or corrugated, moist to mucoid, yeast-like colonies (Fig. 86.1). Spherical to elongate budding yeast-like cells or blastoconidia, 2.5-6.5 x 6.5-14.0 µm.

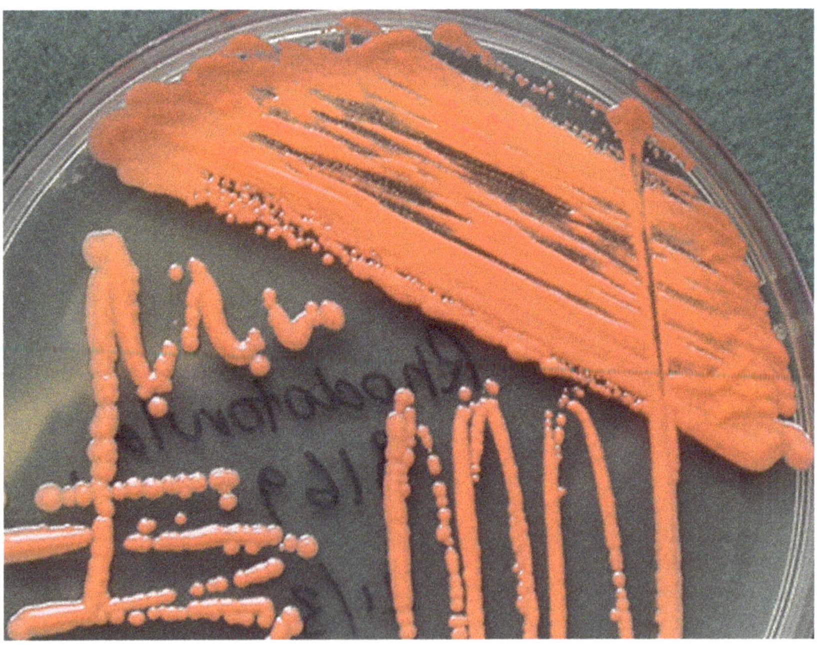

Fig. 86.1. *Rhodotorula mucilaginosa* culture.

India ink preparation: Small capsules present.

Germ tube formation: Negative.

Physiological tests: (+ Positive, – Negative, v Variable, w Weak, s Slow, n not done).

Fermentation: Absent.

Growth reactions:

Glucose	+	L-Sorbose	v	*myo*-Inositol	–
Sucrose	+	L-Rhamnose	–	DL-Lactate	v
Raffinose	+	D-Xylose	+	D-Gluconate	+
Melibiose	–	L-Arabinose	+	2-Keto-D-gluconate	n
Galactose	+,s	D-Arabinose	+	D-Glucosamine	–
Lactose	–	D-Ribose	+	*N*-Acetyl-D-glucosamine	n
Trehalose	+	Glycerol	+	D-Glucuronate	–
Maltose	+	Erythritol	–	Nitrate	–
Melezitose	+	Ribitol	+	Urease	+
Methyl-α-D-glucoside	v	Galactitol	–	0.1% Cycloheximide	–
Soluble starch	–	D-Mannitol	v	Growth at 35°C	v
Cellobiose	+,s,w	D-Glucitol	v	Growth at 37°C	–

Molecular identification: Requires sequencing of the ITS and/or D1/D2 regions.

Key features: Germ tube negative yeast and sugar assimilation pattern. Growth on nitrate is negative (*R. glutinis* growth on nitrate is positive). Common saprophyte however cases of peritonitis and fungaemia have been reported.

Antifungal susceptibility: (Table 86.2).

Table 86.2. *Rhodotorula mucilaginosa* (Diekema *et al.*, 2005; and Australian national data; MIC µg/mL. **Note:** *Rhodotorula* species are intrinsically resistant to echinocandins and fluconazole (Arendrup *et al.*, 2014).

Antifungal	No.	≤0.016	0.03	0.06	0.125	0.25	0.5	1	2	4	8	16	32	≥64
AmB	77				2	3	29	39	3		1			
FLU	77											1	8	68
ISAV	18	1	1	1	5	1	4	3	2					
VORI	76				9	2	6	11	15	23	9	1		
POSA	76			1		5	7	15	34	2	8	4		
ITRA	77			1		9	11	28	10	2	2	14		
5FC	77			24	31	17	4	1						

DOI: 10.1079/9781800622340.0086

87 *Saccharomyces cerevisiae* Meyen ex Hansen

Saccharomyces cerevisiae, commonly known as baker's yeast, may be found as a harmless and transient digestive commensal and coloniser of mucosal surfaces of normal individuals. The anamorphic state of *S. cerevisiae* is sometimes referred to as *Candida robusta*. This species is phylogenetically closely related to *Candida glabrata* and shares many clinical and microbiological characteristics to this species (Arendrup *et al.*, 2014). *S. cerevisiae* may be involved in mucosal infections like vaginitis, and in bloodstream infections, particularly in fluconazole-exposed patients.

Note: *Saccharomyces boulardii*, a genetically similar subtype that is used as a probiotic for prevention and treatment of various sorts of diarrhoea and recurrent *Clostridium difficile*-associated diarrhoea should be avoided in immunocompromised hosts (Enache-Angoulvant and Hennequin, 2005; Arendrup *et al.*, 2014).

RG-1 organism

Morphological description: Colonies are white to cream, smooth, glabrous and yeast-like. Large globose to ellipsoidal budding yeast-like cells or blastoconidia, 3.0-10.0 × 4.5-21.0 µm. No capsules present on India Ink preparation.

Germ tube formation: Negative.

Physiological tests: (+ Positive, – Negative, v Variable, w Weak, s Slow, n not done).

Fermentation:

Glucose	+	Sucrose	+	Lactose	–
Galactose	v	Maltose	+	Trehalose	–

Growth reactions:

Glucose	+	L-Sorbose	–	*myo*-Inositol	–
Sucrose	+	L-Rhamnose	–	DL-Lactate	v
Raffinose	+	D-Xylose	–	D-Gluconate	v
Melibiose	–	L-Arabinose	–	2-Keto-D-gluconate	–
Galactose	v	D-Arabinose	–	D-Glucosamine	–
Lactose	–	D-Ribose	–	*N*-Acetyl-D-glucosamine	–
Trehalose	+	Glycerol	v	D-Glucuronate	–
Maltose	+	Erythritol	–	Nitrate	–
Melezitose	v	Ribitol	–	Urease	–
Methyl-α-D-glucoside	–	Galactitol	–	0.1% Cycloheximide	–
Soluble starch	–	D-Mannitol	–	Growth at 30°C	+
Cellobiose	–	D-Glucitol	–	Growth at 37°C	v

Molecular identification: ITS and D1/D2 sequencing is recommended (McCullough *et al.*, 1988).

Antifungal susceptibility: (Table 87.1).

Table 87.1. *Saccharomyces cerevisiae* (Australian national data); MIC µg/mL.

Antifungal	No.	≤0.016	0.03	0.06	0.125	0.25	0.5	1	2	4	8	16	32	≥64
AmB	85		1	1	9	20	28	33	3					
FLU	96						4	7	17	16	11	24	13	4
ISAV	12		3	1	2	3	1	2						
VORI	92		4	12	23	12	27	10	3	1				
POSA	89			1	6	21	19	29	11		2			
ITRA	96			2	15	25	26	22	2	1		3		
ANID	86	2	4	7	47	20	6							
MICA	86	1	4	55	24	2								
5FC	96		8	78	7					1	1	1		

References: McGinnis (1980); Barnett *et al.* (1983); Rippon (1988); Vaughan-Martini and Martini (2011); de Hoog *et al.* (2015).

DOI: 10.1079/9781800622340.0087

88 *Saksenaea vasiformis* complex

The genus *Saksenaea* is characterised by the formation of flask-shaped sporangia with columellae and simple, darkly pigmented rhizoids. It is an emerging human pathogen (Holland, 1997) that is most often associated with cutaneous or subcutaneous lesions after trauma. Until recently, *Saksenaea vasiformis* was the only known species and appears to have a worldwide distribution in association with soil.

However, recent phylogenetic studies have identified additional species, namely, *S. erythrospora, S. dorisiae, S. loutrophoriformis, S. oblongispora* and *S. trapezispora* (Alvarez *et al.*, 2010b; Crous *et al.*, 2016, 2017; Labuda *et al.*, 2019). So far, all species (except *S. dorisiae*) have been isolated from clinical samples.

88.1. *Saksenaea vasiformis* Saksena

RG-2 organism

Morphological description: Colonies are fast growing, downy, white with no reverse pigment made up of broad, non-septate hyphae typical of a mucormycetous fungus. Sporangia are typically flask-shaped with a distinct spherical venter and long-neck, arising singly or in pairs from dichotomously branched, darkly pigmented rhizoids. Columellae are prominent and dome-shaped (Fig. 88.1). Sporangiospores are small, oblong, 1-2 x 3-4 µm, and are discharged through the neck following the dissolution of an apical mucilaginous plug. Good growth at 42°C; no growth at 50°C.

Key features: Unique flask-shaped sporangia, failure to sporulate on primary isolation media.

Molecular identification: ITS sequencing typically required for identification and differentiation between species due to lack of sporulation on routine media (Alvarez *et al.*, 2010b; Walther *et al.*, 2012; Halliday *et al.*, 2015).

Comment: Laboratory identification of this fungus may be difficult or delayed because of the mould's failure to sporulate on primary isolation media or on subsequent subculture onto potato dextrose agar. Sporulation may be stimulated by using the agar block method described by Ellis and Ajello (1982), Ellis and Kaminski (1985) and Padhye and Ajello (1988), although this may still take a period of days to weeks. Failure to sporulate also prohibits antifungal susceptibility testing (Fig. 88.2).

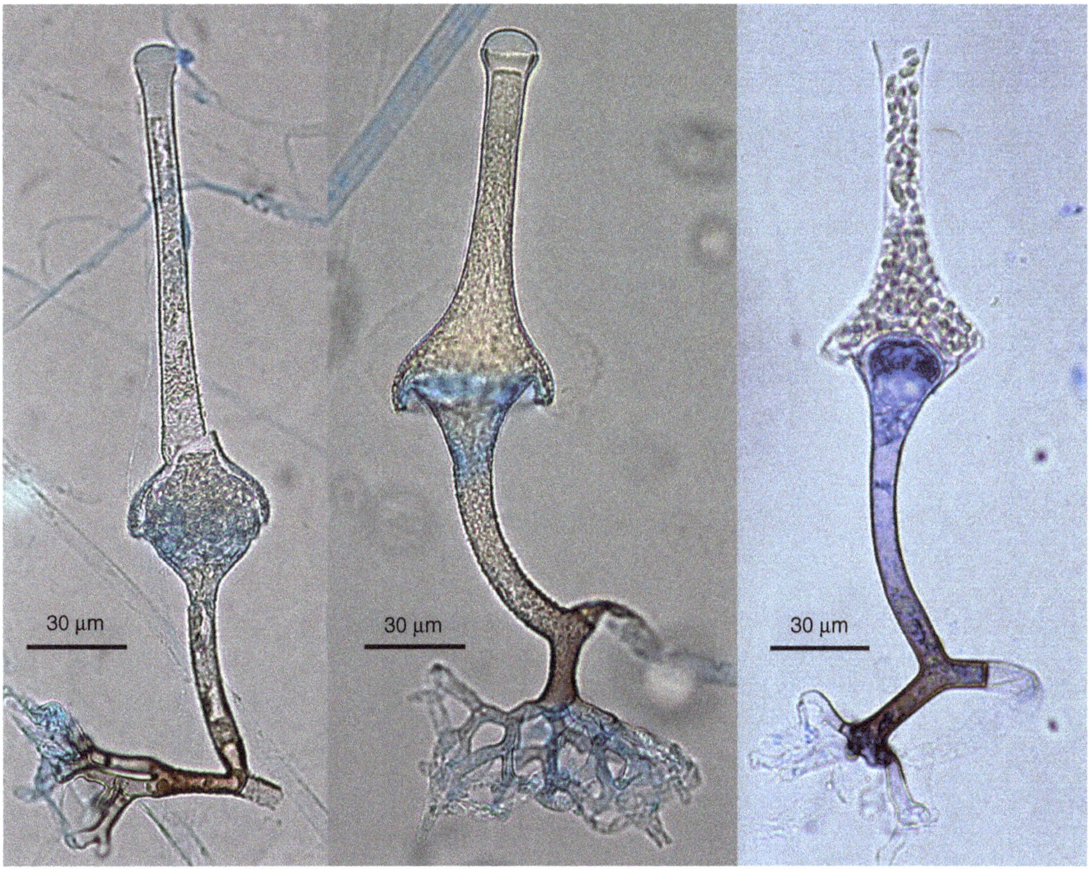

Fig. 88.1. *Saksenaea vasiformis* sporangia.

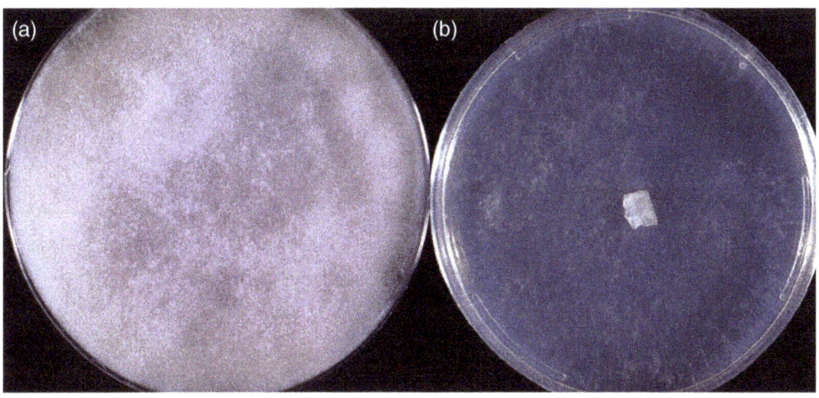

Fig. 88.2. *Saksenaea vasiformis* cultures showing (a) rapid non-sporulating growth on potato dextrose agar, and (b) the agar block method to induce sporulation. A small block of agar is cut from the culture grown on potato dextrose agar and is placed in the centre of petri dish containing 1% agar in distilled water and incubated at 26°C. Sporangia may be found by examining growth at the periphery of the petri dish at intervals of up to 4 weeks.

Antifungal susceptibility: Antifungal susceptibility testing is rarely successful due to poor sporulation, even when sporulation is stimulated (Table 88.1).

Table 88.1. *Saksenaea vasiformis*, very limited data (Sun *et al.,* 2002; and Australian national data); MIC µg/mL.

Antifungal	Range	MIC_{90}	Antifungal	Range	MIC_{90}
AmB	0.125-2	2	VORI	0.5-4	4
ITRA	0.016-0.03	0.03	POSA	0.016-2	0.25

References: Saksena (1953); Ellis and Hesseltine (1966b); Ajello *et al.* (1976); Ellis and Ajello (1982); Ellis and Kaminski (1985); Pritchard *et al.* (1986); Padhye *et al.* (1988); Padhye and Ajello (1988); Goldschmied-Reouven *et al.* (1989); Ellis (2005b); Walther *et al.* (2019); Davidson *et al.* (2020); de Hoog *et al.* (2020).

DOI: 10.1079/9781800622340.0088

89 *Sarocladium* W. Gams & D. Hawksw.

Based on a recent molecular phylogenetic study the taxonomy of *Acremonium* was reviewed and some medically important species have been transferred to *Sarocladium*, i.e. *S. kiliense* (formerly *A. kiliense*) and *S. strictum* (formerly *A. strictum*). Although both genera are morphologically similar, they are phylogenetically distant. *Sarocladium* can be morphologically differentiated from *Acremonium* by its elongated phialides rising solitary on vegetative hyphae or on conidiophores that are sparsely or repeatedly branched, the production of abundant adelophialides and elongated conidia (Glenn *et al.*, 1996; Summerbell *et al.*, 2011; Giraldo *et al.*, 2015).

89.1. *Sarocladium strictum* (W. Gams) Summerbell

Sarocladium strictum is commonly found in soil and plant debris. Cutaneous, CAPD-related peritonitis and invasive infections in immunosuppressed patients have been reported.

RG-1 organism

Morphological description: Colonies growing rapidly, moist to slimy, pink or orange; reverse remaining colourless or turning pink to orange (Fig. 89.1). Conidiophores simple, occasionally branched. Phialides slender, arising from submerged or slightly fasciculate aerial hyphae, 20-65 x 1.4-2.5 µm (Fig. 89.1). Submerged sporulation frequently from reduced phialides. Conidia grouped in slimy heads, cylindrical or ellipsoidal, 3.3-5.5 x 0.9-1.8 µm, hyaline.

Molecular identification: Summerbell *et al.* (2011) revised the genus and ITS and/or D1/D2 sequencing is recommended for phylogenetic analysis and sequence-based identification (Giraldo *et al.*, 2015).

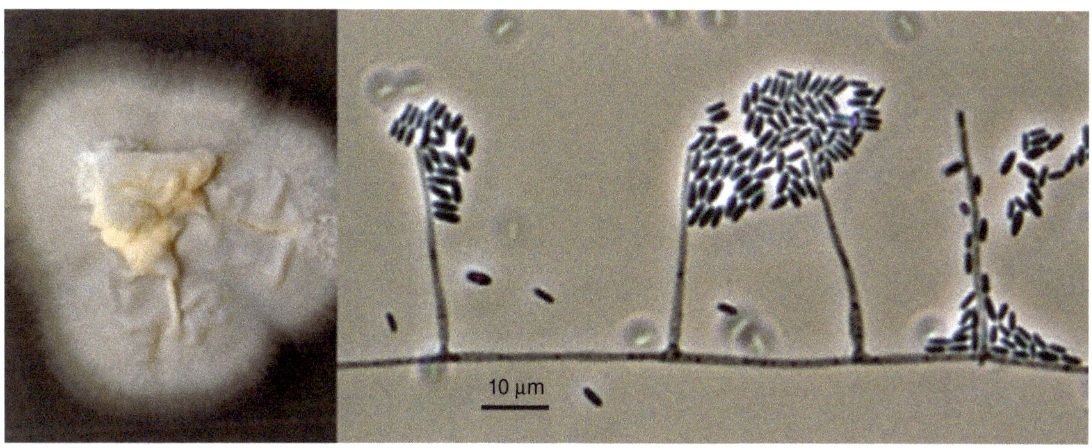

Fig. 89.1. *Sarocladium strictum* culture and slender phialides with conidia in slimy heads.

Antifungal susceptibility: (Table 89.1).

Table 89.1. *Sarocladium strictum* limited data (Australian national data); MIC μg/mL.

Antifungal	No.	≤0.016	0.03	0.06	0.125	0.25	0.5	1	2	4	8	16	32	≥64
AmB	6						1		3		2			
VORI	6				1	1	1	1	2					
POSA	6					1	1	2	1		1			
ITRA	6						1		1	1				

References: Glenn *et al.* (1996); Summerbell *et al.* (2011); Giraldo *et al.* (2015); de Hoog *et al.* (2015).

DOI: 10.1079/9781800622340.0089

90 *Scedosporium* Sacc. ex Castell. & Chalm.

The taxonomy of this genus has undergone significant change based on phylogenetic data; *Scedosporium apiospermum* and *Scedosporium boydii* (formerly *Pseudallescheria boydii*) are now recognised as separate species and along with *S. aurantiacum* are the principal human pathogens (Lackner *et al.*, 2014a). Most infections are mycetomas, the remainder include infections of the eye, ear, central nervous system, internal organs and more commonly the lungs. *Scedosporium dehoogii* and *S. minutisporum* are mainly isolated from environmental samples and have been rarely reported from clinical cases (Gilgado *et al.*, 2005; Rainer and de Hoog, 2006; Kaltseis *et al.*, 2009).

Scedosporium prolificans was found to be phylogenetically and morphologically distinct from other *Scedosporium* species and has been transferred to the genus *Lomentospora* as *L. prolificans* (Lennon *et al.*, 1994; Lackner *et al.*, 2014a).

Morphology-based identification of *Scedosporium* species has become increasingly unreliable and molecular identification methods are now recommended. The conidial states of *S. apiospermum* and *S. boydii* are morphologically indistinguishable; although the latter is homothallic and produces ascocarps. *S. aurantiacum* also exhibits similar conidial morphology but most strains produce a pale to bright yellow diffusible pigment on potato dextrose agar. Identifications made on the basis of morphology alone should be reported as *Scedosporium apiospermum* complex.

Molecular identification: Recommended genetic markers are ITS and β-tubulin (Lackner *et al.*, 2012; Chen *et al.*, 2021).

MALDI-ToF MS: A comprehensive "in-house" database of reference spectra allows accurate identification of *Scedosporium* and *Lomentospora* species (Lau *et al.*, 2013; Sitterle *et al.*, 2014).

References: McGinnis (1980); Campbell and Smith (1982); Rippon (1988); Gilgado *et al.* (2005); Rainer and de Hoog (2006); Guarro *et al.* (2006); Domsch *et al.* (2007); Heath *et al.* (2009); de Hoog *et al.* (2015).

90.1. *Scedosporium apiospermum* (Saccardo) Castellani and Chalmers

Synonymy: *Pseudallescheria apiosperma* Gilgado, Gene, Cano & Guarro.

RG-2 organism

Morphological description: Colonies are fast growing, greyish-white, suede-like to downy with a greyish-black reverse (Fig. 90.1). Numerous single-celled, pale brown, broadly clavate to ovoid conidia, 4-9 x 6-10 µm, rounded above with truncate bases are observed. Conidia are borne singly or in small groups on elongate, simple or branched conidiophores or laterally on hyphae (Fig. 90.2). Conidial development can be described as annellidic, although the annellations (ring-like scars left at the apex of an annellide after conidial secession) are extremely difficult to see. Erect synnemata may be present in some isolates (Fig. 90.1). Optimum temperature for growth is 30-37°C.

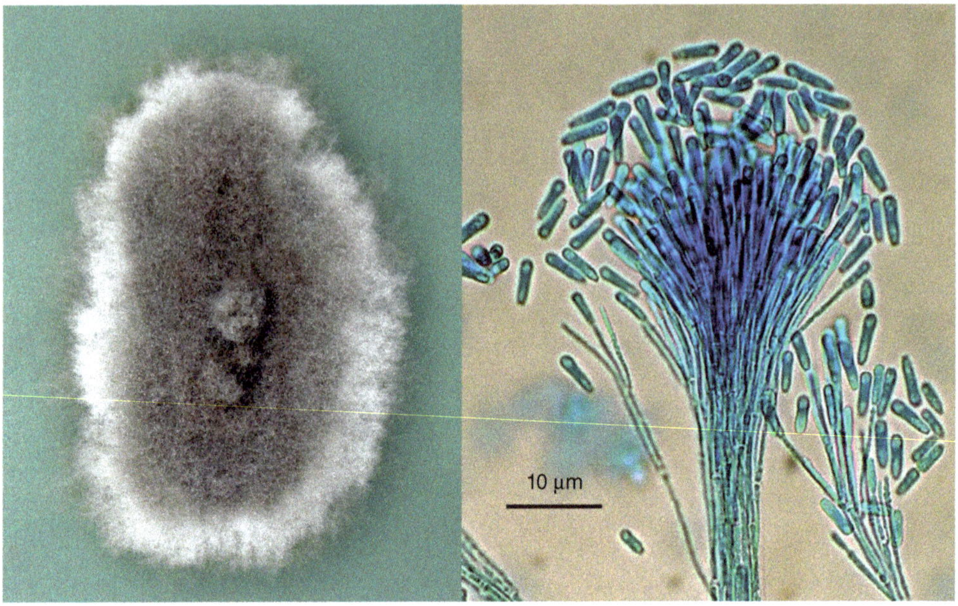

Fig. 90.1. *Scedosporium apiospermum* culture and synnemata.

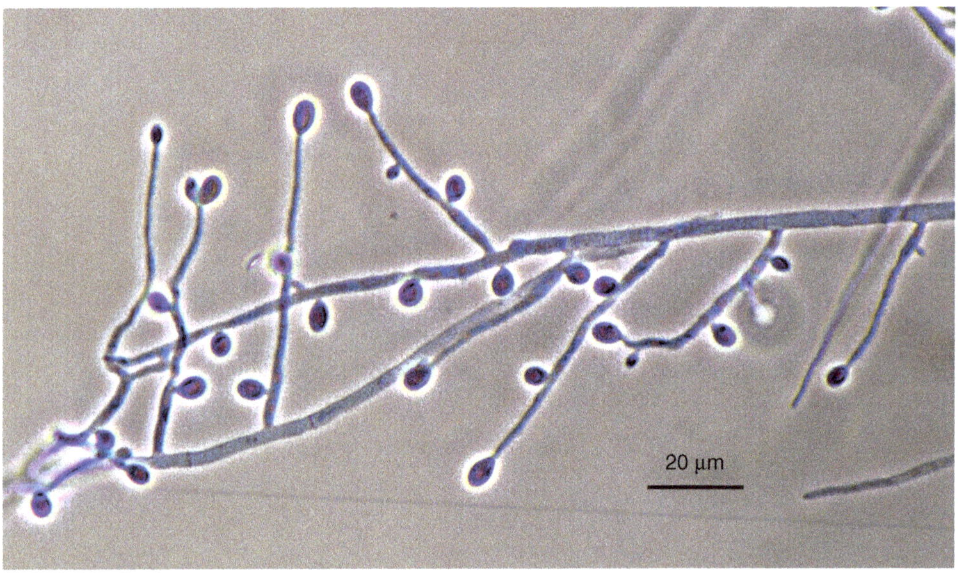

Fig. 90.2. *Scedosporium apiospermum* conidiophores and conidia.

Antifungal susceptibility: (Table 90.1).

Table 90.1. *Scedosporium apiospermum* (Australian national data); MIC µg/mL.

Antifungal	No.	≤0.016	0.03	0.06	0.125	0.25	0.5	1	2	4	8	16	32	≥64
AmB	406					1	4	13	64	102	222			
ISAV	80					1	4	9	21	30	15			
VORI	402		6	31	83	123	135	20	1	1	2			
POSA	351			2	6	34	93	173	40	1	2			
ITRA	406			2	7	34	137	168	30	9	2	17		

90.2. *Scedosporium aurantiacum* Gilgado et al.

RG-2 organism

Morphological description: Most isolates produce a light to bright yellow diffusible pigment on potato dextrose agar (PDA) after a few days incubation (Fig. 90.3). Conidiogenous cells and conidia are similar in shape and size to *S. apiospermum*, and the two can best be distinguished by genetic analysis. Conidiogenous cells arising from undifferentiated hyphae are cylindrical to slightly flask-shaped, producing slimy heads of one-celled, smooth-walled, subhyaline, obovoid or sub-cylindrical conidia, 5-14 × 2-5 µm (Fig. 90.4). Erect synnemata may be present in some isolates, but the teleomorph is unknown. Optimum temperature for growth 37-40°C, max 45°C.

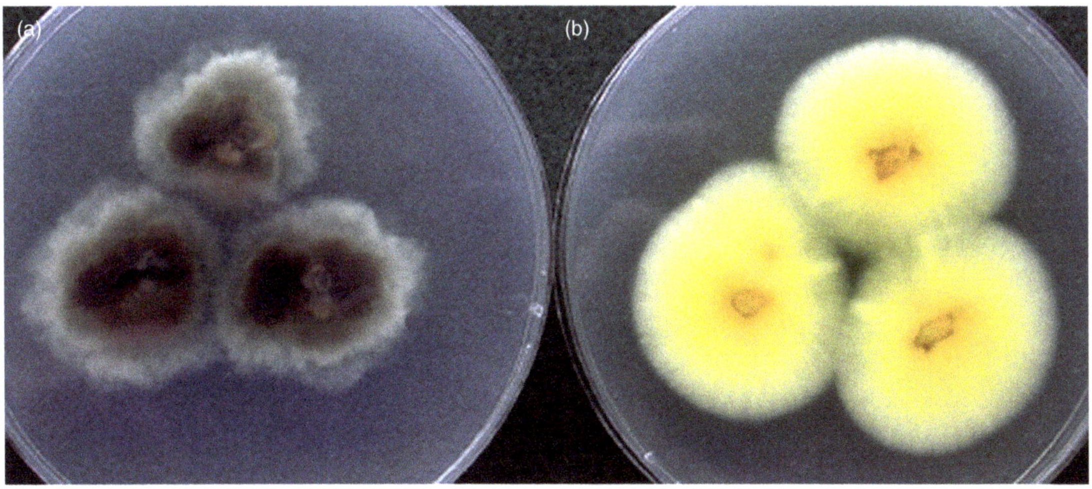

Fig. 90.3. Culture reverse (PDA) of (a) *Scedosporium apiospermum* and (b) *S. aurantiacum* showing production of a light yellow diffusible pigment that is typical of *S. aurantiacum*.

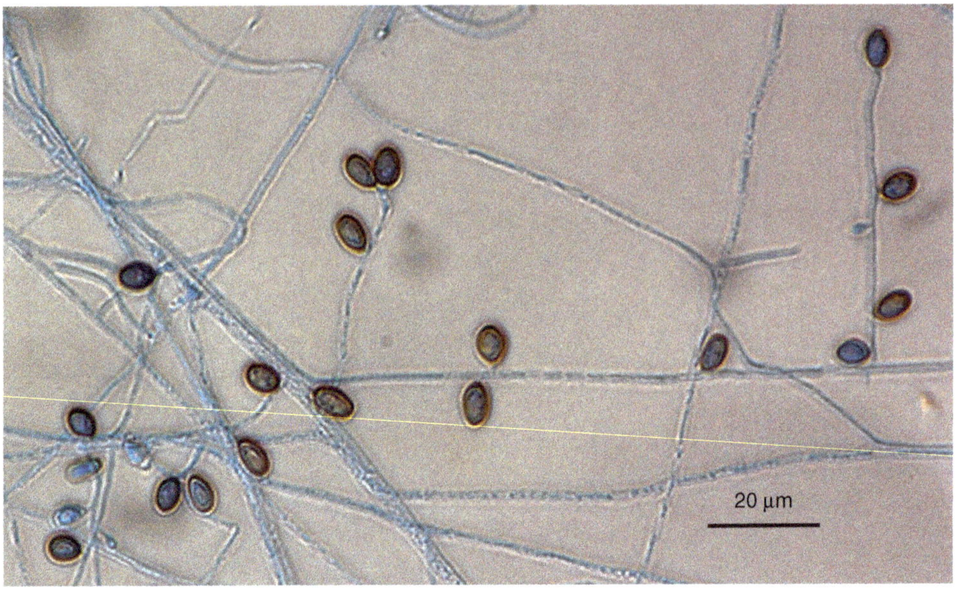

Fig. 90.4. *Scedosporium aurantiacum* conidiophores and conidia.

Antifungal susceptibility: (Table 90.2).

Table 90.2. *Scedosporium aurantiacum* (Australian national data); MIC µg/mL.

Antifungal	No.	≤0.016	0.03	0.06	0.125	0.25	0.5	1	2	4	8	16	32	≥64
AmB	27					1			1	8	17			
ISAV	13						1		1	7	4			
VORI	27				6	6	12	1		1	1			
POSA	27						5	16	5		1			
ITRA	27						5	14	3			5		

90.3. *Scedosporium boydii* (Shear) Gilgado *et al.*

Synonymy: *Pseudallescheria boydii* (Shear) McGinnis, A.A. Padhye & Ajello.

RG-2 organism

Morphological description: Colonies are fast growing, greyish-white, suede-like to downy with a greyish-black reverse (Fig. 90.5). Numerous single-celled, pale brown, broadly clavate to ovoid conidia, 4-9 × 6-10 µm, rounded above with truncate bases are observed. Conidia are borne singly or in small groups on elongate, simple or branched conidiophores or laterally on hyphae. Cleistothecia (non-ostiolate ascocarps) are yellow-brown to black, spherical, 50-200 µm in diameter, and are mostly submerged in the agar and are composed of irregularly interwoven brown hyphae (Fig. 90.5). When crushed cleistothecia release numerous, faintly brown, ellipsoidal ascospores, 4-5 × 7-9 µm in size. Erect synnemata may be present in some isolates. Optimum temperature for growth is 30-37°C.

Note: *Scedosporium boydii* is homothallic and is recognised by smaller cleistothecia (50-200 µm) whereas *S. apiospermum* is heterothallic (requires mating of 2 strains) and has larger cleistothecia, 140-480 µm in size (Gilgado *et al.*, 2010).

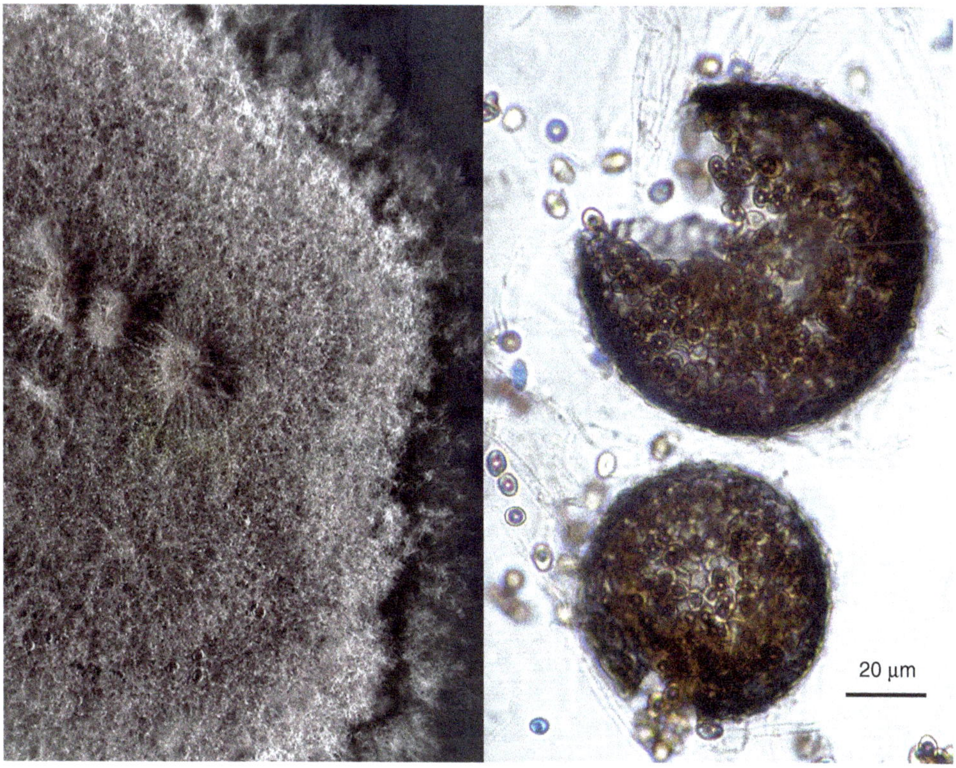

Fig. 90.5. *Scedosporium boydii* culture and cleistothecia with ascospores.

Antifungal susceptibility: (Table 90.3).

Table 90.3. *Scedosporium boydii* vs. *S. apiospermum* (Lackner *et al.*, 2014b); MIC µg/mL.

	Scedosporium boydii		*Scedosporium apiospermum*	
Antifungal	Range	MIC_{90}	Range	MIC_{90}
AmB	0.5–>16	≥16	0.5–>16	≥16
VORI	0.125–2	2	0.25–>8	2
POSA	0.125–>16	≥16	0.25–>16	≥16
ITRA	0.125–>16	≥16	0.25–>16	≥16

DOI: 10.1079/9781800622340.0090

91 *Schizophyllum commune* Fries

Schizophyllum commune is a common basidiomycetous bracket fungus growing on rotten wood, and is an occasional human pathogen, principally associated with sinusitis, allergic bronchopulmonary mycosis and as a contaminant from respiratory specimens. However, the introduction of DNA-based fungal or MALDI-ToF MS identification in the clinical laboratory has seen many more cases of invasive fungal rhinosinusitis from *S. commune* being identified (Chowdhary *et al.*, 2013a, 2014a,b; Michel *et al.*, 2015).

RG-1 organism

Morphological description: Colonies on 2% malt extract agar are spreading, woolly, whitish to pale greyish-brown, soon forming macroscopically visible fruiting bodies. Some isolates may take up to 12 weeks to form fruiting bodies. Fruit bodies (basidiocarps) are sessile, kidney-shaped, lobed with split gills on the lower side (Fig. 91.1). Hyphae are hyaline, wide and have clamp connections. Basidia bear four basidiospores on erect sterigmata. Basidiospores hyaline, smooth-walled, elongate with lateral scar at lower end, 6-7 x 2-3 µm.

Note: Many clinical isolates of *Schizophyllum commune* are monokaryotic and do not show clamp connections, therefore any white, rapidly growing, sterile isolate showing good growth at 37°C with tolerance to benomyl, susceptibility to cycloheximide, and a pronounced odour should be suspected of being *S. commune* (Sigler *et al.*, 1995).

Molecular identification: Sequencing of the ITS and D1/D2 regions is recommended (Buzina *et al.*, 2001; Won *et al.*, 2012; Chowdhary *et al.*, 2013b; Michel *et al.*, 2015), however the number of well identified nucleotide sequences from this species in the GenBank database remains limited.

MALDI-ToF MS: Chowdhary *et al.* (2014b), Huguenin *et al.* (2015), Michel *et al.* (2015), provide identification procedures, however the number of mass spectral profiles to be found in MALDI-ToF libraries remains limited.

Antifungal susceptibility: (Table 91.1).

Table 91.1. *Schizophyllum commune* (Chowdhary *et al.*, 2013b); MIC µg/mL.

Antifungal	Range	MIC_{90}	Antifungal	Range	MIC_{90}
AmB	0.03-2	1	FLU	2-64	64
ITRA	0.03-8	1	VORI	0.06-2	0.5

References: Morton and Smith (1963); McGinnis (1980); Rippon (1988); Sigler *et al.* (1995); de Hoog *et al.* (2015); Michel *et al.* (2015).

Fig. 91.1. *Schizophyllum commune* basidiocarps growing on malt extract agar.

DOI: 10.1079/9781800622340.0091

92 *Scopulariopsis* Bain

Most members of the genus *Scopulariopsis* are soil fungi, which are frequently isolated from food, paper and other materials. They also occur as laboratory contaminants. Several species have been reported as causative agents of onychomycosis and hyalohyphomycosis (Sandoval-Denis *et al.*, 2013). Phylogenetic studies have separated some *Scopulariopsis* species into the genus *Microascus* (Sandoval-Denis *et al.*, 2016). The most common species seen in the clinical laboratory is *S. brevicaulis,* followed by *Microascus gracilis, S. brumptii, S. candida* complex, *Microascus cinereus* and *M. cirrosus* (Sandoval-Denis *et al.*, 2013).

RG-2 organism for species isolated from humans

Morphological description: Colonies are fast growing, varying in colour from white, cream, grey, buff to brown, black, but are predominantly light brown. Microscopic morphology shows chains of single-celled conidia produced in basipetal succession from by a specialised conidiogenous cell called an annellide. Once again, the term basocatenate can be used to describe such chains of conidia where the youngest conidium is at the basal end of the chain. In *Scopulariopsis*, annellides may be solitary, in groups, or organised into a distinct penicillus. Conidia are globose to pyriform, usually truncate, with a rounded distal portion, smooth to rough, and hyaline to brown in colour (Fig. 92.1).

Key features: Conidia often shaped like light globes, basocatenate arising from annellides.

Molecular identification: Sequencing of the ITS, D1/D2 and *TEF-1* can easily identify the most common clinically relevant species (Sandoval-Denis *et al.*, 2013, 2016).

Antifungal susceptibility: (Table 92.1).

Table 92.1. *Scopulariopsis* spp. limited data (Australian national data); MIC µg/mL.

Antifungal	No.	≤0.016	0.03	0.06	0.125	0.25	0.5	1	2	4	8	16	32	≥64
AmB	18						1	1	7	2	7			
ISAV	2								1		1			
VORI	18		1			1		2	5	1	8			
POSA	18		1		1	1	1	3	4		7			
ITRA	18		1		1			5	1			10		

References: Morton and Smith (1963); McGinnis (1980); Rippon (1988); Samson *et al.* (1995); Domsch *et al.* (2007); Sandoval-Denis *et al.* (2013, 2016); de Hoog *et al.* (2020).

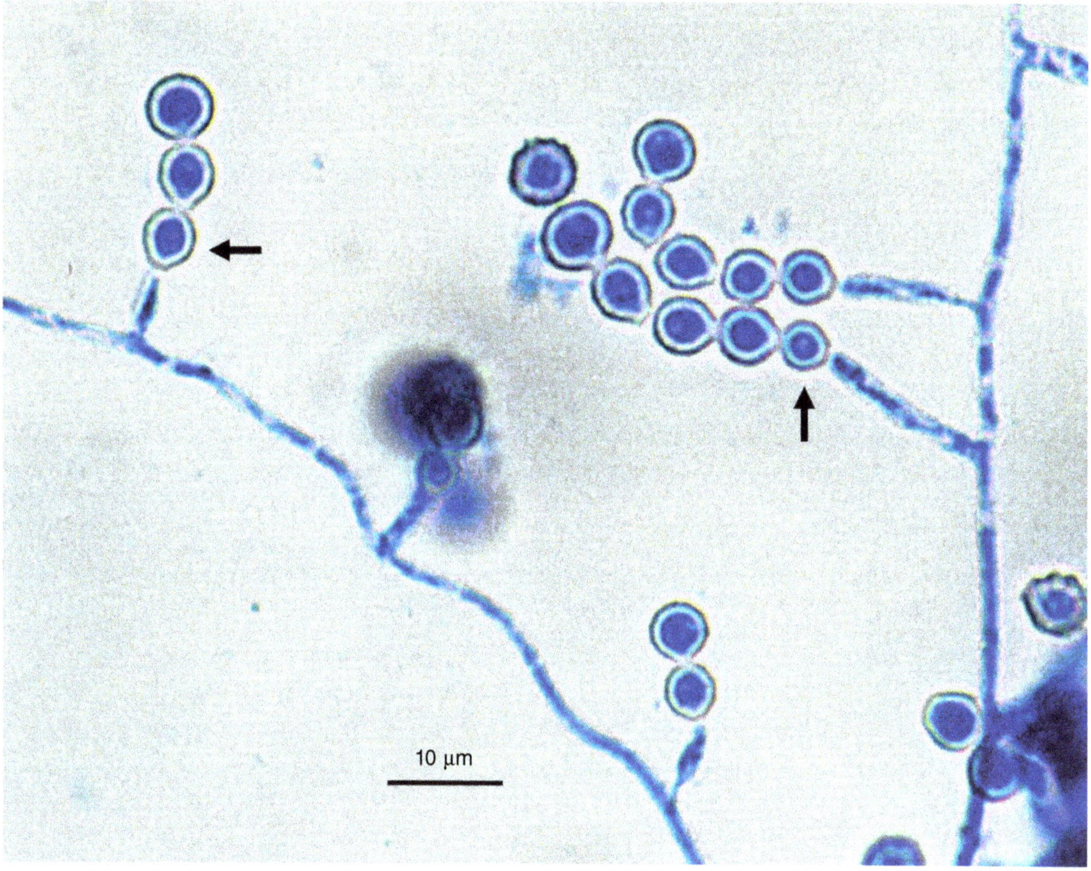

Fig. 92.1. *Scopulariopsis brevicaulis* conidiophores (annellides) and conidia. **Note:** Basipetal chain of conidia from an annellide with the youngest at the base (arrows).

DOI: 10.1079/9781800622340.0092

93 *Sepedonium* Link ex Greville

Sepedonium species are common soil fungi and parasites of fleshy fungi, however Yogo *et al.* (2014) reported an intra-abdominal infection in an immunosuppressed patient. *Sepedonium* species closely resemble *Histoplasma capsulatum*. **Note:** For laboratory safely, if there is any possibility or suspicion of *Histoplasma capsulatum* identification should be performed by DNA sequencing.

RG-1 organism

Morphological description: Colonies are moderately fast growing, usually white to golden-yellow, suede-like to downy, becoming fluffy with age. Conidiophores are hyaline and non-specialised, resembling short branches of the vegetative hyphae. Conidia are terminal, solitary, or in clusters, one-celled, globose to ovoid, 7 to 17 μm, hyaline to amber, smooth to verrucose and usually with a thick wall (Fig. 93.1).

Key features: Large, thick-walled, one-celled, verrucose, globose, terminal macroconidia from non-specialised conidiophores, resembling *Histoplasma capsulatum*.

Molecular identification: ITS and D1/D2 sequencing may be used for accurate species identification (Halliday *et al.*, 2015).

References: McGinnis (1980); Rippon (1988); Yogo *et al.* (2014).

DOI: 10.1079/9781800622340.0093

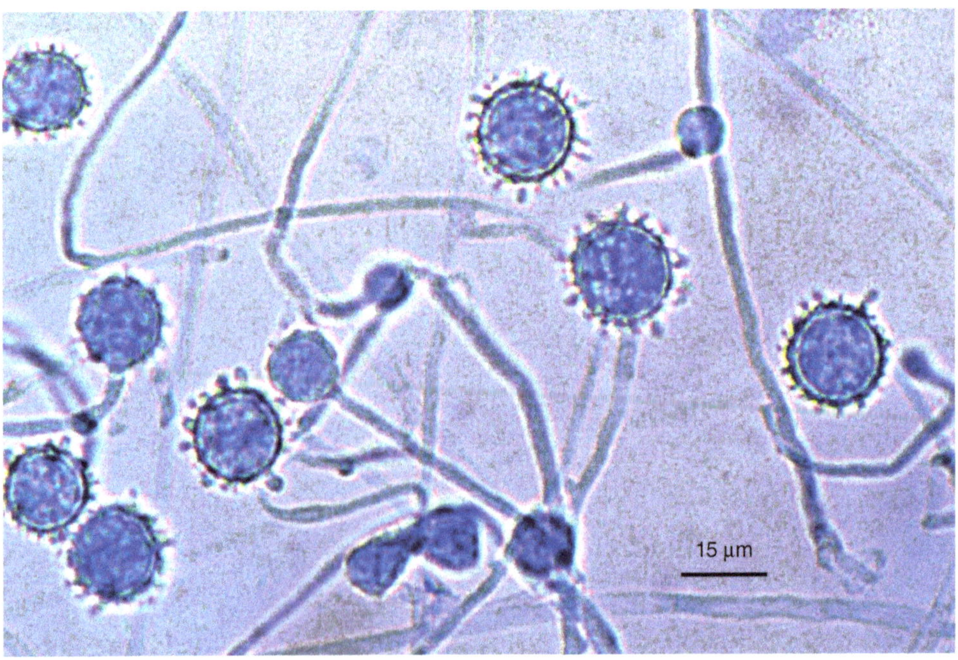

Fig. 93.1. *Sepedonium* spp. showing large, globose, thick-walled, one-celled, verrucose, terminal conidia.

94 *Sporothrix schenckii* complex

It is now recognised that *Sporothrix schenckii* is a complex of five distinct species: *S. schenckii sensu stricto, S. brasiliensis, S. globosa, S. mexicana* and *S. luriei* (Marimon *et al.*, 2007; Barros *et al.*, 2011; Romeo *et al.*, 2011; Oliveira *et al.*, 2013; Zhang *et al.*, 2015b).

Sporothrix schenckii complex is a dimorphic fungus and has a worldwide distribution, particularly in tropical and temperate regions. It is commonly found in soil and on decaying vegetation and is a well-known pathogen of humans and animals. Sporotrichosis is primarily a chronic mycotic infection of the cutaneous or subcutaneous tissues and adjacent lymphatics characterised by nodular lesions which may suppurate and ulcerate. Infections are caused by the traumatic implantation of the fungus into the skin, or very rarely, by inhalation into the lungs. Secondary spread to articular surfaces, bone and muscle is not infrequent, and the infection may also occasionally involve the central nervous system, lungs or genitourinary tract (Rodrigues *et al.*, 2020b).

94.1. *Sporothrix schenckii* Hektoen & Perkins

RG-2 organism

Morphological description: Colonies at 25°C, are slow growing, moist and glabrous, with a wrinkled and folded surface. Some strains may produce short aerial hyphae and pigmentation may vary from white to cream to black (Fig. 94.1). Conidiophores arise at right angles from thin septate hyphae and are usually solitary, erect and tapered toward the apex. Conidia are formed in clusters on tiny denticles by sympodial proliferation at the apex of the conidiophore, their arrangement often suggestive of a flower (Fig. 94.2). As the culture ages, conidia are subsequently formed singly along the sides of both conidiophores and undifferentiated hyphae. Conidia are ovoid or elongated, 3-6 × 2-3 µm, hyaline, one-celled and smooth-walled. In some

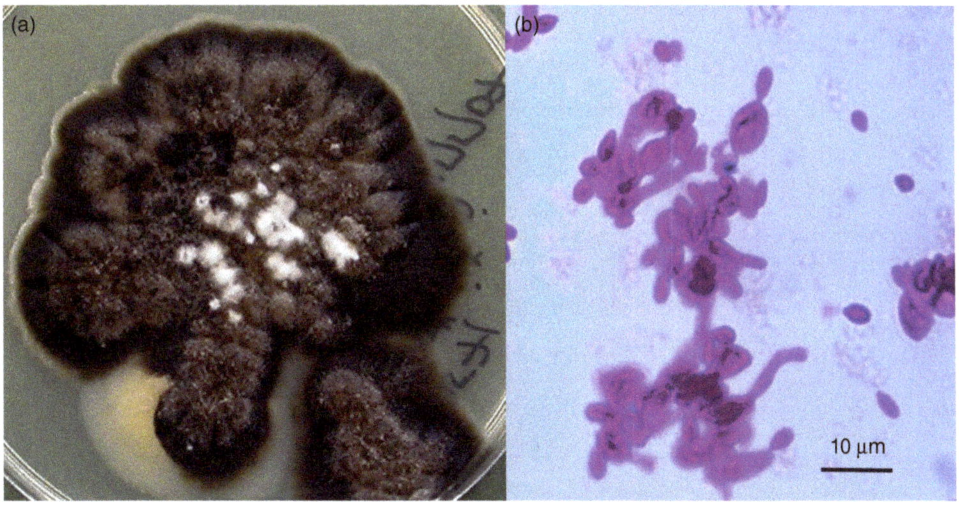

Fig. 94.1. *Sporothrix schenckii* showing (a) mould-like culture at 25°C and (b) budding yeast cells in BHI broth at 37°C.

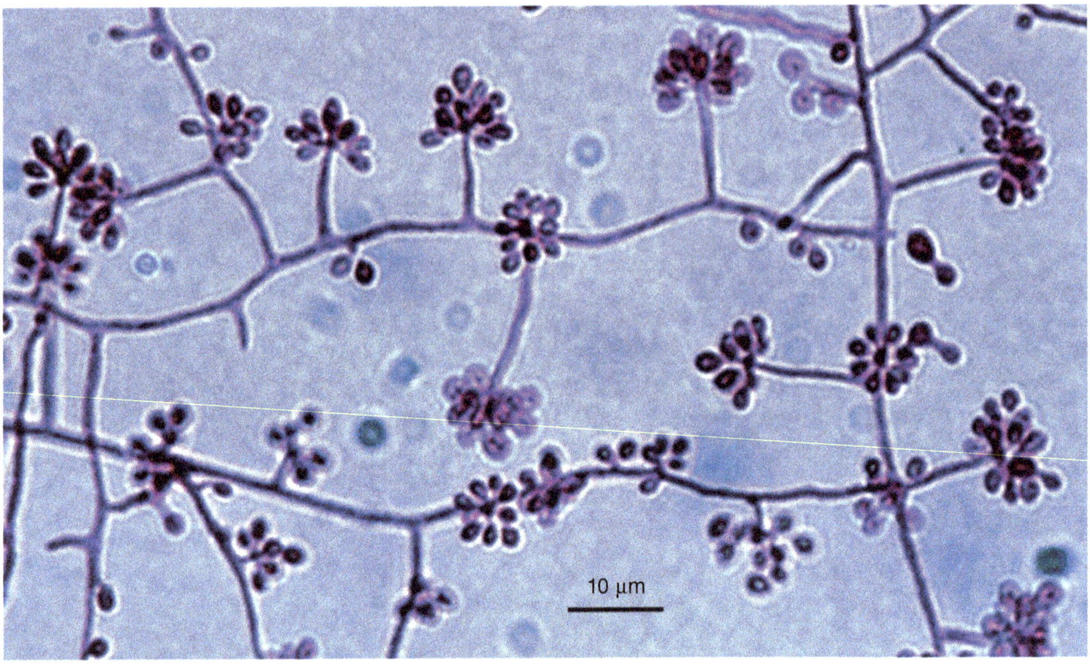

Fig. 94.2. *Sporothrix schenckii* showing conidiophores producing conidia in clusters on tiny denticles by sympodial proliferation, their arrangement often suggestive of a flower.

isolates, solitary, darkly-pigmented, thick-walled, one-celled, obovate to angular conidia may also be observed along the hyphae. On brain heart infusion (BHI) blood agar at 37°C, colonies are glabrous, white to greyish-yellow and yeast-like consisting of spherical or oval budding yeast cells (Fig. 94.1).

Molecular identification: DNA sequencing using ITS, D1/D2, β-tubulin, *Cal* and chitin synthase genes is recommended for species identification (Marimon *et al.*, 2007; Barros *et al.*, 2011; Romeo *et al.*, 2011; Oliveira *et al.*, 2013; Zhang *et al.*, 2015b).

MALDI-ToF MS identification: Oliveira *et al.* (2015) established a protocol and reference database for the identification of the *Sporothrix* species complex.

Key features: Thermal dimorphism and clusters of ovoid, denticulate conidia produced sympodially on short conidiophores.

Antifungal susceptibility: (Table 94.1).

Table 94.1. *Sporothrix schenckii* complex (Australian national data); MIC µg/mL.

Antifungal	No.	≤0.016	0.03	0.06	0.125	0.25	0.5	1	2	4	8	16	32	≥64
AmB	27					1		2	4	11	9			
ISAV	2								2					
VORI	27					1	1	7	9	6	3			
POSA	27					1	11	13	2					
ITRA	27					2	16	9						

References: McGinnis (1980); de Hoog *et al.* (1985, 2015); Rippon (1988); Domsch *et al.* (2007); Espinel-Ingroff *et al.* (2017); Rodrigues *et al.* (2020b).

DOI: 10.1079/9781800622340.0094

95 *Stemphylium* Wallroth

Most species of *Stemphylium* are plant pathogens with occasional isolates from soil, they are rarely seen in the clinical laboratory.

RG-1 organism

Morphological description: Colonies are rapid growing, brown to olivaceous-black or greyish and suede-like to floccose. Microscopically, solitary, darkly pigmented, terminal, multicellular conidia (dictyoconidia) are formed on a distinctive conidiophore with a darker terminal swelling. The conidiophore proliferates percurrently through the scar where the terminal conidium (poroconidium) was formed. Conidia are pale to mid-brown, oblong, rounded at the ends, ellipsoidal, obclavate or subspherical and are smooth or in part verrucose (Fig. 95.1). *Stemphylium* should not be confused with *Ulocladium* which produces similar dictyoconidia from a sympodial conidiophore, rather than from a percurrent conidiogenous cell as in *Stemphylium.*

Molecular identification: ITS sequencing is recommended (Woudenberg *et al.,* 2013).

Key features: Darkly pigmented, dictyoconidia from the swollen end of a percurrent conidiophore.

References: Ellis (1971, 1976); Rippon (1988); de Hoog *et al.* (2015).

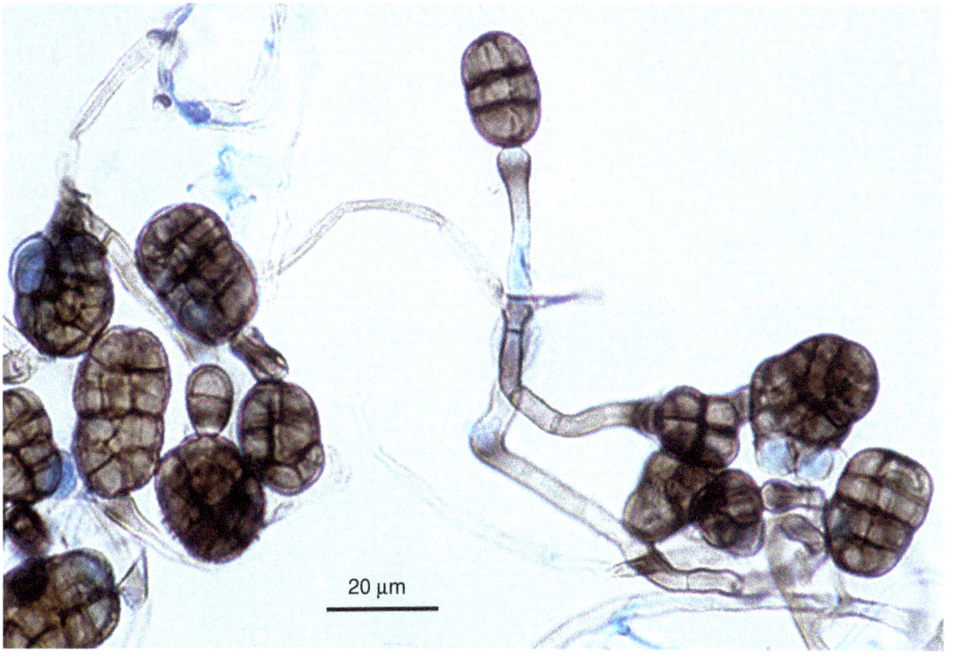

Fig. 95.1. *Stemphylium* spp. conidiophore and conidia.

DOI: 10.1079/9781800622340.0095

96 *Syncephalastrum racemosum* Cohn

The genus *Syncephalastrum* is characterised by the formation of cylindrical merosporangia on a terminal swelling of the sporangiophore. Sporangiospores are arranged in a single row within the merosporangia. *Syncephalastrum racemosum* is the type species of the genus and a potential human pathogen; however, well-documented cases are lacking. It is found mainly from soil and dung in tropical and subtropical regions. It can also be a difficult laboratory aerial contaminant. The sporangiophore and merosporangia of *Syncephalastrum* species may also be mistaken for *Aspergillus* species if the microscopic morphology is not examined carefully.

RG-2 organism

Morphological description: Colonies are very fast growing, cottony to fluffy, white to light grey, becoming dark grey with the development of sporangia. Sporangiophores are erect, stolon-like, often producing adventitious rhizoids, and show sympodial branching (racemose branching) producing curved lateral branches. The main stalk and branches form terminal, globose to ovoid vesicles, which bear finger-like merosporangia directly over their entire surface. At maturity merosporangia are thin-walled, evanescent and contain five to ten (up to 18) globose to ovoid, smooth-walled sporangiospores (merospores) (Fig. 96.1). Optimum growth temperature 20-40°C.

Key features: Mucorales, producing sympodially branching sporangiophores with terminal vesicles bearing merosporangia.

Molecular identification: ITS sequencing is recommended for barcoding (Voigt *et al.*, 1999).

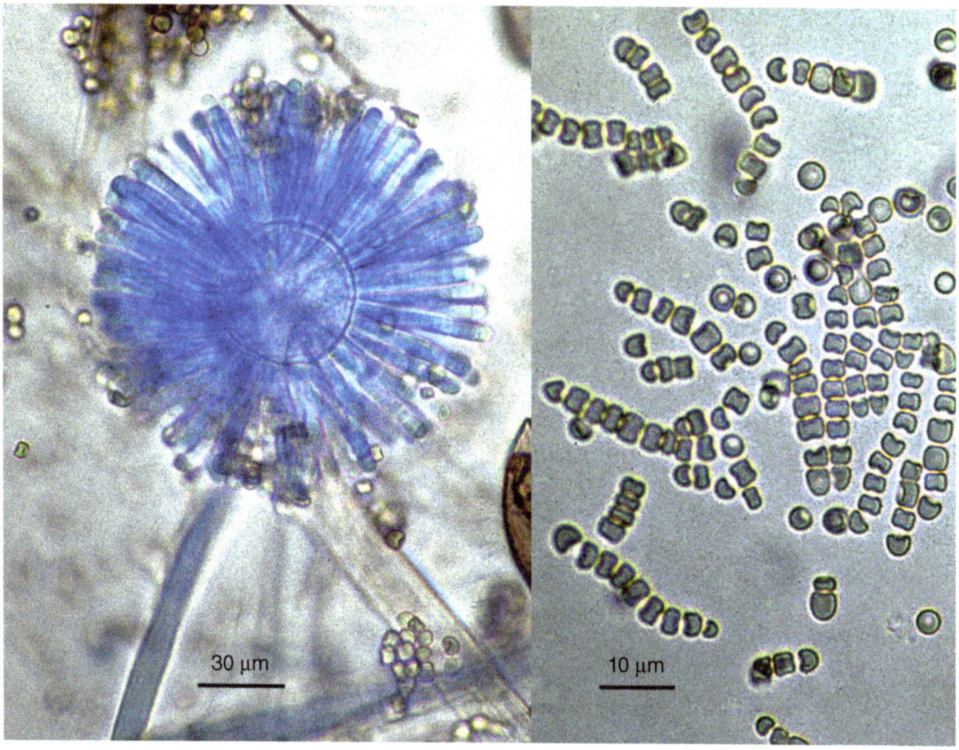

Fig. 96.1. *Syncephalastrum racemosum* microscopy showing a terminal vesicle bearing finger-like merosporangia and sporangiospores.

Antifungal susceptibility: (Table 96.1).

Table 96.1. *Syncephalastrum racemosum* (Espinel-Ingroff *et al.*, 2015a; and Australian national data); MIC µg/mL.

Antifungal	No.	≤0.016	0.03	0.06	0.125	0.25	0.5	1	2	4	8	16	32	>64
AmB	41		8	16	3	11	3							
ISAV	1							1						
VORI	6							1		4	1			
POSA	42		1	2	6	12	13	5	1	2				
ITRA	32		4	3	7	9	6		1			2		

References: McGinnis (1980); Onions *et al.* (1981); Rippon (1988); Samson *et al.* (1995); Ellis (2005b); Domsch *et al.* (2007); de Hoog *et al.* (2015).

DOI: 10.1079/9781800622340.0096

97 *Talaromyces marneffei* (Segretain *et al.*) Samson *et al.*

Synonymy: *Penicillium marneffei* Segretain *et al.*

The genus *Talaromyces* has been redefined based upon phylogenetic analysis of the ITS and RPB1 loci, with *Penicillium* subgenus *Biverticillata* being moved into the genus *Talaromyces* (Samson *et al.*, 2011b). The genus contains 88 species that were placed into seven sections based on a multigene phylogeny of the ITS, β-tubulin and *RPB2* genes (Yilmaz *et al.*, 2014). *T. marneffei* is the only known dimorphic species in the genus, producing filamentous growth at 25°C and a yeast phase at 37°C (Andrianopoulos, 2002).

WARNING: RG-3 organism

Cultures of *Talaromyces marneffei* may represent a biohazard to laboratory personnel and should be handled with caution in a class II Biological Safety Cabinet (BSCII). *T. marneffei* exhibits thermal dimorphism and is endemic in Southeast Asia and the southern region of China.

Molecular identification: ITS sequencing is recommended for identification, as well as β-tubulin as a secondary molecular marker for identification (Yilmaz *et al.*, 2014).

Morphological description: Colonies at 25°C are fast growing, suede-like to downy, white with yellowish-green conidial heads. Colonies become greyish-pink to brown with age and produce a diffusible brownish-red to wine-red pigment (Fig. 97.1). Conidiophores generally biverticillate and sometimes monoverticillate; hyaline, smooth-walled and bear terminal verticils of 3-5 metulae, each bearing 3-7 phialides. Phialides are acerose to flask-shaped (Fig. 97.1). Conidia are globose to subglobose, 2 to 3 μm in diameter, smooth-walled and are produced in basipetal succession from the phialides.

On brain heart infusion (BHI) blood agar incubated at 37°C, colonies are rough, glabrous, tan coloured and yeast-like. Microscopically, yeast cells are spherical to ellipsoidal, 2-6 μm in diameter, and divide by fission rather than budding (Fig. 97.1). Numerous short hyphal elements are also present.

Histopathology: Tissue sections show small, oval to elliptical yeast-like cells, 3 μm in diameter, either packed within histiocytes or scattered through the tissue. Occasional, large, elongated sausage-shaped cells, up to 8 μm long, with distinctive septa may be present.

Key features: *Talaromyces marneffei* is the only dimorphic species of *Talaromyces,* which grows as a yeast phase at 37°C. It produces a red soluble pigment on general media and conidiophores have flask-shaped to acerose phialides.

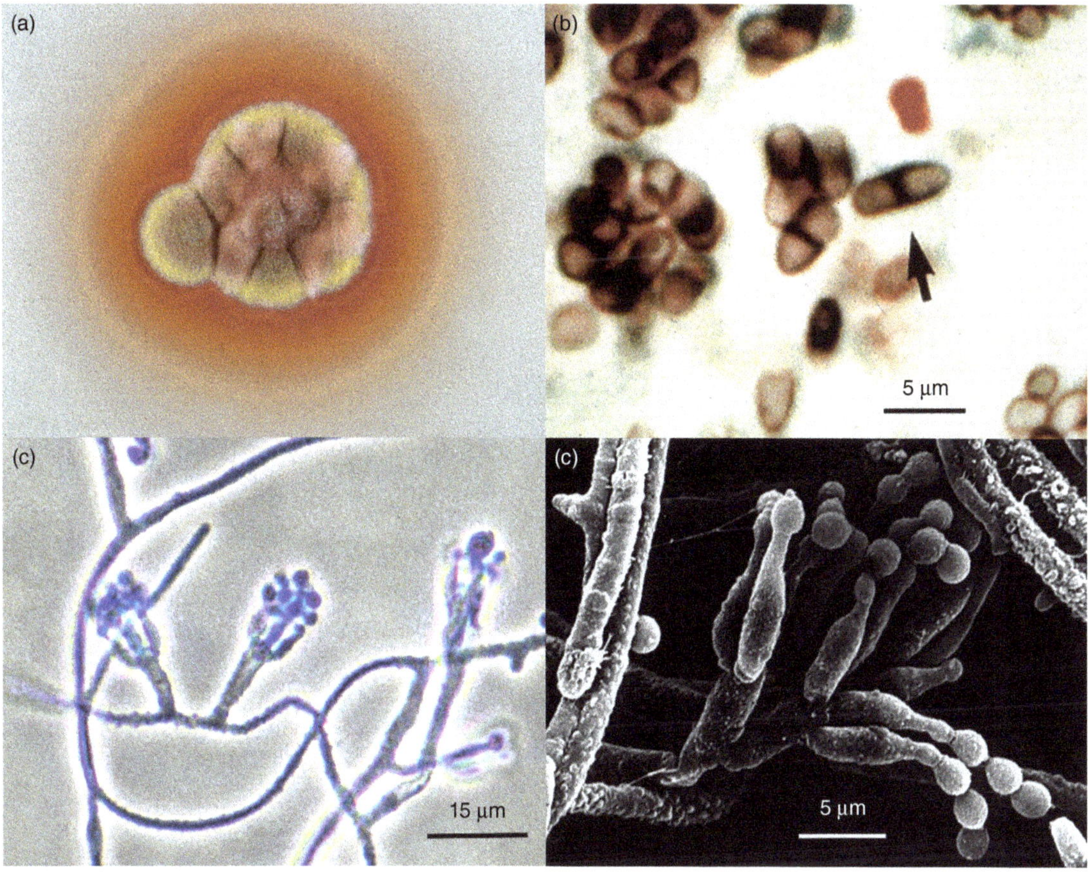

Fig. 97.1. *Talaromyces marneffei* showing (a) culture with red diffusible pigment, (b) a Giemsa-stained touch smear of a skin lesion showing typical septate yeast-like cells (arrow) and (c) phialides and conidia.

Antifungal susceptibility: (Table 97.1).

Table 97.1. *Talaromyces marneffei*, very limited data (Australian national data); MIC µg/mL.

Antifungal	No.	≤0.016	0.03	0.06	0.125	0.25	0.5	1	2	4	8	16	32	≥64
AmB	5				2	2	1							
VORI	5	5												
POSA	4	4												
ITRA	5	5												

References: Pitt (1979); Ramirez (1982); Andrianopoulos (2002); Samson *et al.* (2011b); Visagie *et al.* (2014); Yilmaz *et al.* (2014); de Hoog *et al.* (2015).

DOI: 10.1079/9781800622340.0097

98 *Torulaspora delbrueckii* (Linder) Linder.

Synonymy: *Candida colliculosa* (Hartmann) S.A. Meyer & Yarrow; *Debaryomyces delbrueckii* (Lindner) Kudryavtsev.

Torulaspora delbrueckii is widely distributed in nature and is a common food and beverage spoilage organism. It is a rare cause of candidaemia.

RG-1 organism

Culture: Colonies (SDA) white to cream coloured, smooth, glabrous, yeast-like.

Microscopy: Spherical to ellipsoidal budding blastoconidia, 2-6 x 3-7 µm in size. Ascospores may be produced on 5% malt extract or cornmeal agar after 5-30 days at 25°C.

India ink preparation: Negative - no capsules present.

Dalmau plate culture: Budding yeast cells only. No pseudohyphae or true hyphae produced.

Molecular identification: ITS sequencing recommended.

MALDI-ToF MS: Able to accurately identify this species.

Germ tube formation: Negative

Physiological tests: (+ Positive, - Negative, v Variable, w Weak, s Slow, n not done).

Fermentation:

Glucose	+	Sucrose	v	Lactose	–
Galactose	v	Maltose	v	Trehalose	v

Growth reactions:

Glucose	+	L-Sorbose	v	*myo*-Inositol	–
Sucrose	v	L-Rhamnose	–	DL-Lactate	v
Raffinose	v	D-Xylose	v	D-Gluconate	v
Melibiose	–	L-Arabinose	–	2-Keto-D-gluconate	+
Galactose	v	D-Arabinose	–	D-Glucosamine	–
Lactose	–	D-Ribose	–	*N*-Acetyl-D-glucosamine	–
Trehalose	s	Glycerol	v	D-Glucuronate	–
Maltose	v	Erythritol	–	Nitrate	–
Melezitose	v	Ribitol	v	Urease	–
Methyl-α-D-glucoside	v	Galactitol	–	0.1% Cycloheximide	–
Soluble starch	–	D-Mannitol	+	Growth at 30°C	+
Cellobiose	–	D-Glucitol	v	Growth at 37°C	v

Key features: Asci containing one to four spheroidal ascospores, growth at 37°C and a variable sugar assimilation profile.

Antifungal susceptibility: (Table 98.1).

Table 98.1. *Torulaspora delbrueckii*, very limited data (Australian national data); MIC µg/mL.

Antifungal	No.	≤0.016	0.03	0.06	0.125	0.25	0.5	1	2	4	8	16	32	≥64
AmB	1						1							
FLU	1									1				
VORI	1			1										
POSA	1				1									
ITRA	1				1									
ANID	1	1												
MICA	1		1											
5FC	1				1									

References: Kurtzman (2011d).

DOI: 10.1079/9781800622340.0098

99 *Trematosphaeria grisea (MacKinnon et al.)* S.A. Ahmed *et al.*

Synonymy: *Madurella grisea* Mackinnon, Ferrada and Montemayer.

Trematosphaeria grisea (formerly *Madurella grisea*) is a soil fungus and is a known causative agent of mycetoma (de Hoog *et al.*, 2004a, 2012; Desnos-Ollivier *et al.*, 2006).

RG-2 organism

Morphological description: Colonies are slow growing, dark, leathery, folded with radial grooves and with a light brown to greyish surface mycelium. With age, colonies become dark brown to reddish-brown and have a brownish-black reverse. Microscopically, cultures are sterile, although hyphae of two widths have been described, thin at 1-3 µm in width or broad at 3-5 µm in width. The optimum temperature for growth of *T. grisea* is 30°C; this fungus does not grow at 37°C.

Histopathology: Grains of *Trematosphaeria grisea* (tissue microcolonies) are black, round to lobed, soft to firm, up to 1.0 mm, with two distinctive zones, a hyaline to weakly pigmented central zone and a deeply pigmented periphery (Fig. 99.1).

Trematosphaeria grisea can be distinguished from *Madurella mycetomatis* by its inability to grow at 37°C, and its ability to assimilate sucrose but not lactose.

Key features: Black grain mycetoma, no growth at 37°C, no diffusible brown pigment produced on culture and absence of conidia.

References: Chandler *et al.* (1980); Kwon-Chung and Bennett (1992); Ahmed *et al.* (2014b); Tanaka *et al.* (2015); Suetrong *et al.* (2011).

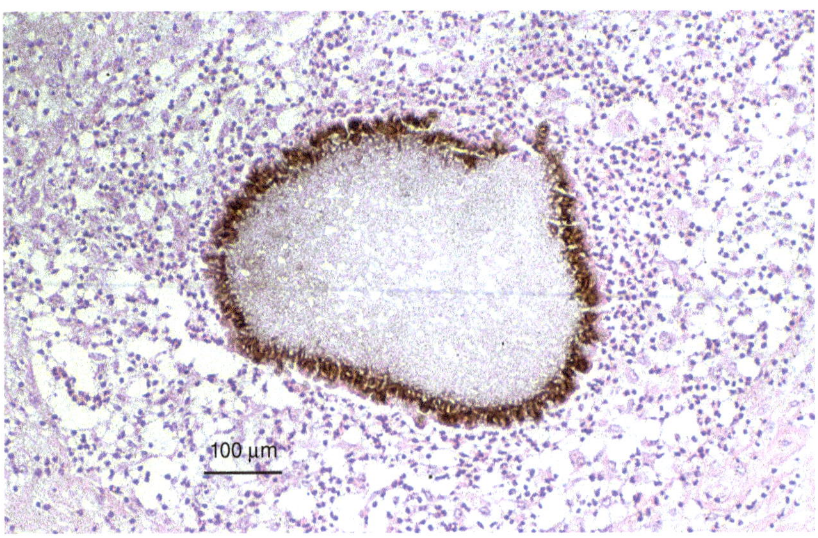

Fig. 99.1. *Trematosphaeria grisea* showing a typical "black grain" in a haematoxylin and eosin-stained tissue section.

DOI: 10.1079/9781800622340.0099

100 *Trichoderma Persoon* ex Grey

Trichoderma is a very common genus especially in soil and decaying wood. *Gliocladium* (with strongly convergent phialides) and *Verticillium* (with straight and moderately divergent phialides) are closely related genera. *Trichoderma* infections in humans have been associated with several risk factors, particularly peritoneal dialysis, organ transplantation, and haematologic disorders (Sandoval-Denis *et al.*, 2014b).

RG-1 organism

Morphological descriptions: Colonies are fast growing, at first white and downy, later developing yellowish-green to deep green compact tufts, often only in small areas or in concentric ring-like zones on the agar surface. Conidiophores are repeatedly branched, irregularly verticillate, bearing clusters of divergent, often irregularly bent, flask-shaped phialides (Fig. 100.1). Conidia are mostly green, sometimes hyaline, with smooth or rough walls and are formed in slimy conidial heads (gloiospora) clustered at the tips of the phialides.

Key features: Repeatedly branched conidiophores bearing clusters of divergent, flask-shaped phialides.

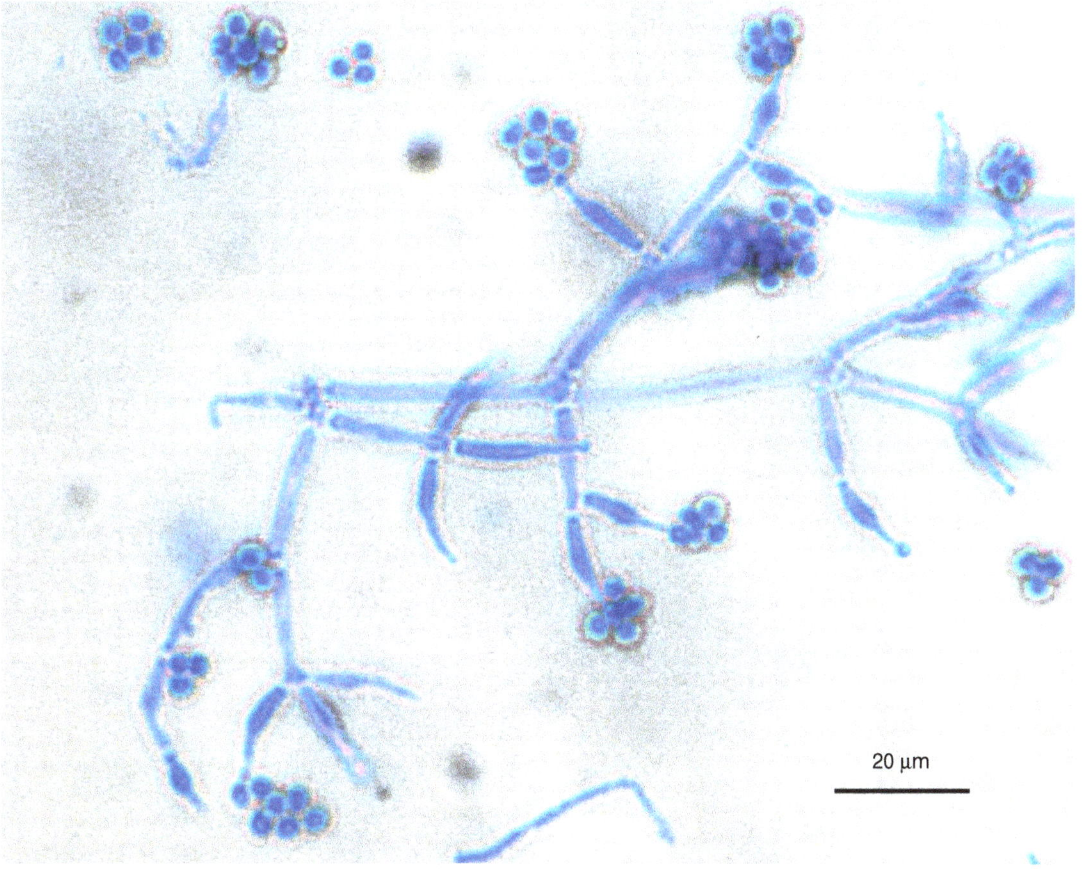

Fig. 100.1. *Trichoderma harzianum* microscopy showing flask-shaped phialides and conidia.

Molecular identification: Species identification is based on multilocus sequence data using ITS, *TEF-1*, Chi18-5 and *ACT* genes (Sandoval-Denis *et al.* 2014b).

Antifungal susceptibility: (Table 100.1).

Table 100.1. *Trichoderma* spp., very limited data (Australian national data); MIC μg/mL.

Antifungal	No.	≤0.016	0.03	0.06	0.125	0.25	0.5	1	2	4	8	16	32	≥64
AmB	7						2	4			1			
VORI	6				1	2	1	2						
POSA	5							4	1					
ITRA	7							5	1		1			

References: McGinnis (1980); Rippon (1988); Samson *et al.* (1995); Domsch *et al.* (2007); Sandoval-Denis *et al.* (2014b); de Hoog *et al.* (2015).

DOI: 10.1079/9781800622340.0100

101 *Trichophyton* Malmsten

Rippon (1988) accepted 22 species and four varieties in the genus *Trichophyton* based on morphology. DNA sequences now play a prominent role in delineating phylogenetic relationships, and as such species concepts in *Trichophyton* have changed. Sixteen species are now recognised in the genus. The descriptions and species concepts provided in this publication are based upon a combination of traditional morphological criteria and the current recognised phylogenetic species (de Hoog *et al.*, 2017).

Laboratory identification: The genus *Trichophyton* is characterised morphologically by the development of both smooth-walled macro- and microconidia. Macroconidia are mostly borne laterally directly on the hyphae or on short pedicels, and are thin- or thick-walled, clavate to fusiform, and range from 4-8 x 8-50 µm in size. Macroconidia are few or absent in many species. Microconidia are spherical, pyriform to clavate or of irregular shape and range from 2-3 x 2-4 µm in size. The presence of microconidia differentiates this genus from *Epidermophyton*, and the smooth-walled, mostly sessile macroconidia differentiates it from *Lophophyton, Microsporum, Nannizzia* and *Paraphyton.*

In practice, two groups may be recognised on direct microscopy:

1. Those species that usually produce microconidia. Macroconidia may or may not be present i.e. *T. rubrum, T. interdigitale, T. mentagrophytes, T. equinum, T. erinacei, T. tonsurans* and to a lesser extent *T. verrucosum*, which may produce conidia on some media. In these species the shape, size and arrangement of the microconidia is the most important character. Culture characteristics are also useful.
2. Those species that usually do not produce conidia. Chlamydospores or other hyphal structures may be present, but microscopy is generally non-diagnostic, i.e. *T. verrucosum, T. violaceum, T. concentricum, T. schoenleinii* and *T. soudanense.* Culture characteristics and clinical information such as the site, appearance of the lesion, geographic location, travel history, animal contacts and even occupation are most important.

Many laboratories have used growth on additional media and/or confirmatory tests to help differentiate between species of *Trichophyton*, especially isolates of *T. rubrum, T. interdigitale, T. mentagrophytes* and *T. tonsurans.* These include growth characteristics on media such as Littman oxgall agar, lactritmel agar, potato dextrose agar, Sabouraud's agar with 5% Salt, 1% peptone agar, bromocresol purple-milk solids glucose agar (BCP), Trichophyton agars No. 1-5, hydrolysis of urea and hair perforation tests.

Molecular identification: ITS and *TEF-1* sequencing is recommended for accurate species identification (Graser *et al.,* 1998, 1999a,b, 2000, 2008; Katiraee *et al.,* 2021; Irinyi *et al.*, 2015; Mirhendi *et al.*, 2015; Diongue *et al.*, 2019).

MALDI-ToF MS: Methods reported by Erhard *et al.* (2008); Cassagne *et al.* (2011); L'Ollivier *et al.* (2013); Nenoff *et al.* (2013); Packeu *et al.* (2013, 2014); Calderaro *et al.* (2014). A comprehensive "in-house" database of reference spectra allows accurate identification.

References: Rebell and Taplin (1970); Ajello (1977); Vanbreusegham *et al.* (1978); McGinnis (1980); Rippon (1988); Kane *et al.* (1997); Domsch *et al.* (2007); Chen *et al.* (2011); de Hoog *et al.* (2017, 2020).

101.1. *Trichophyton concentricum* Blanchard

Trichophyton concentricum is an anthropophilic fungus, which causes chronic widespread non-inflammatory tinea corporis known as tinea imbricata because of the concentric rings of scaling it produces. It is not known to invade hair. Distribution is restricted to the Pacific Islands of Oceania, South East Asia and Central and South America.

RG-2 organism

Morphological description: Colonies are slow growing, raised and folded, glabrous becoming suede-like, mostly white to cream coloured, but sometimes orange-brown coloured, often deeply folded into the agar which may produce splitting of the medium in some cultures. Reverse is buff to yellow-brown to brown in colour. Cultures consist of broad, much-branched, irregular, often segmented, septate hyphae which may have "antler" tips resembling *Trichophyton schoenleinii.* Chlamydospores are often present in older cultures (Fig. 101.1). Microconidia and macroconidia are not usually produced, although some isolates will produce occasional clavate to pyriform microconidia. **Note:** Hyphal segments may artificially resemble macroconidia.

Confirmatory tests

Hydrolysis of urea: Negative after 7 days.

Vitamin free agar (Trichophyton agar No.1): Growth occurs on vitamin free agar (T1) but is usually slightly better on media containing thiamine i.e. T3 = T1 + thiamine and inositol, and T4 = T1 + thiamine. The slight enhancement of growth in the presence of thiamine helps to distinguish *T. concentricum* from *T. schoenleinii*, although this does not occur in all strains.

Hair perforation test: Negative at 28 days.

Molecular identification: Species identification supported by ITS sequencing (Graser *et al.*, 1999a; Chen *et al.*, 2011).

Key features: Clinical disease, geographical distribution and culture characteristics.

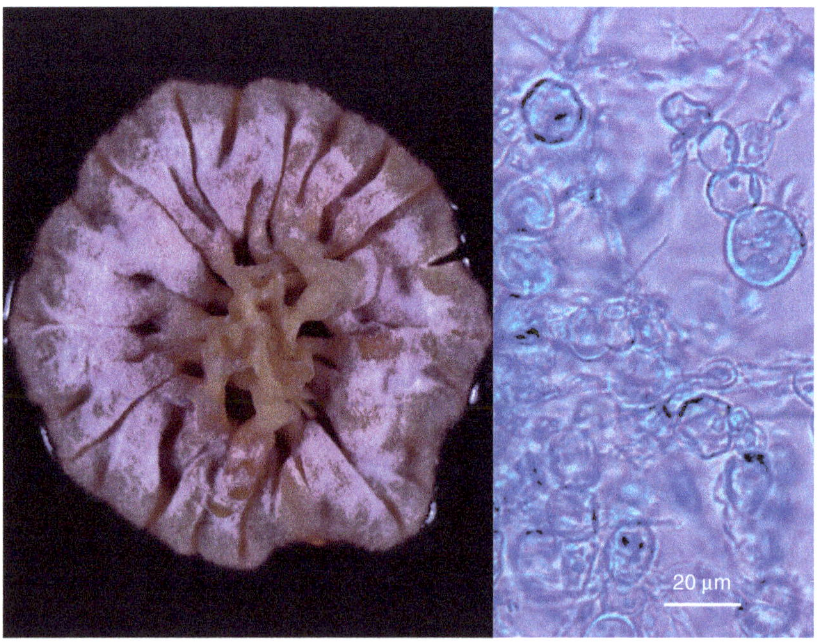

Fig. 101.1. *Trichophyton concentricum* culture showing a typical heaped and folded colony and microscopy showing the formation of typical balloon-shaped chlamydospores. **Note:** Microconidia and macroconidia are usually not produced.

101.2. *Trichophyton equinum* (Matruchot & Dassonville) Gedoelst

Synonomy: *Trichophyton equinum* var. *autotrophicum* J.M.B. Smith, Jolly, Georg & Connole

Trichophyton equinum is a zoophilic fungus causing ringworm in horses and rare infections in humans. It has a worldwide distribution except for the *autotrophicum* strain which is restricted to Australia and New Zealand. Invaded hairs show an ectothrix infection but do not fluoresce under Wood's ultra-violet light.

RG-2 organism

Morphological description: Colonies are usually flat, but some may develop gentle folds or radial grooves, white to buff in colour, suede-like to downy in texture, and are similar to *T. mentagrophytes*. Cultures usually have a deep-yellow submerged fringe and reverse, which later becomes dark red in the centre (Fig. 101.2). Microscopically, abundant microconidia which may be clavate to pyriform and sessile or spherical and stalked are formed laterally along the hyphae (Fig. 101.3). Macroconidia are only rarely produced, but when present are clavate, smooth, thin-walled and of variable size. Occasional nodular bodies may be present, and the microconidia often undergo a transformation to produce abundant chlamydospores in old cultures.

Confirmatory tests

Hydrolysis of urea: Positive in 4-5 days.

Nutritional tests on Trichophyton agars: Most strains require nicotinic acid for growth except those from Australia and New Zealand, which are autotrophic. T1 = vitamin free agar, T5 = vitamin free + nicotinic acid agar (Fig. 101.4).

Hair perforation test: Negative; but positive for the *autotrophicum* strains.

Molecular identification: Species identification supported by ITS sequencing (Graser *et al.*, 1999a; Chen *et al.*, 2011; Kandemir *et al.*, 2020).

Key features: Microscopic morphology, culture characteristics, nicotinic acid requirement and clinical lesions in horses. Most strains require nicotinic acid for growth except the autotrophic strains.

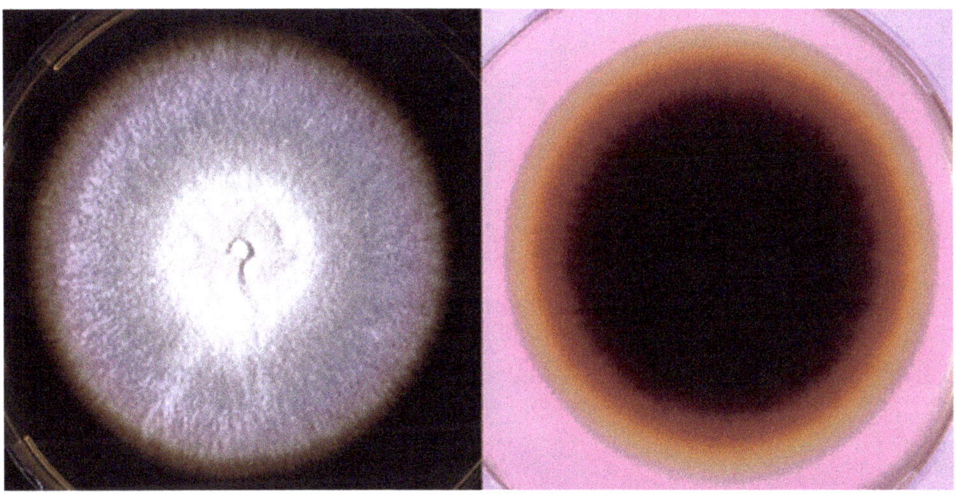

Fig. 101.2. *Trichophyton equinum* culture.

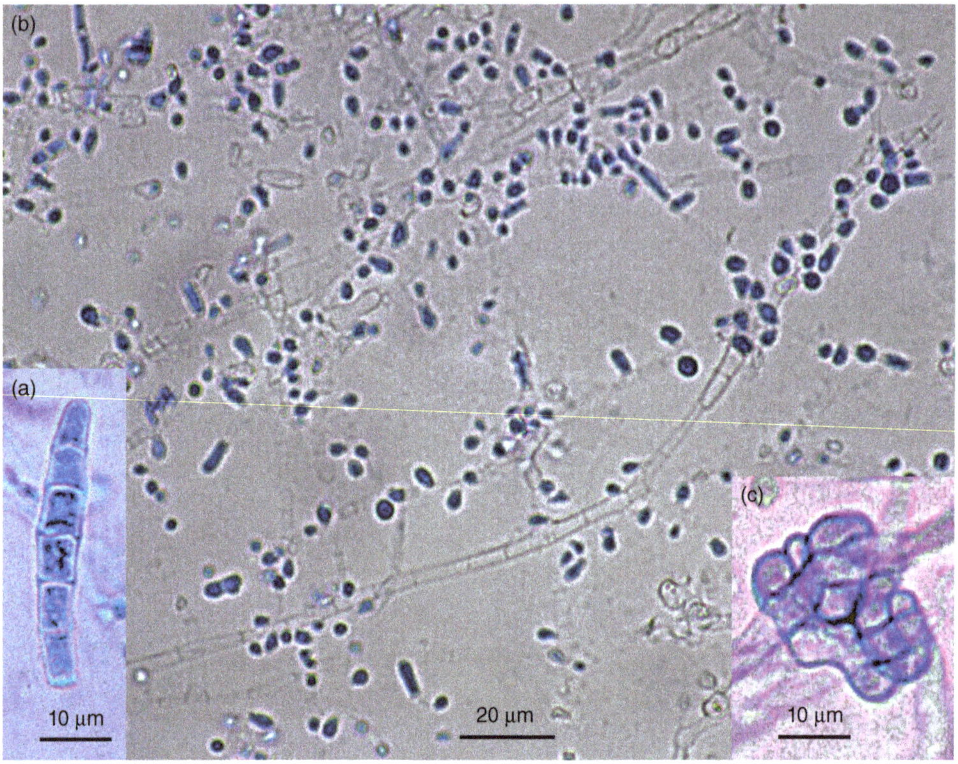

Fig. 101.3. *Trichophyton equinum* showing (a) macroconidium, (b) microconidia and (c) nodular bodies.

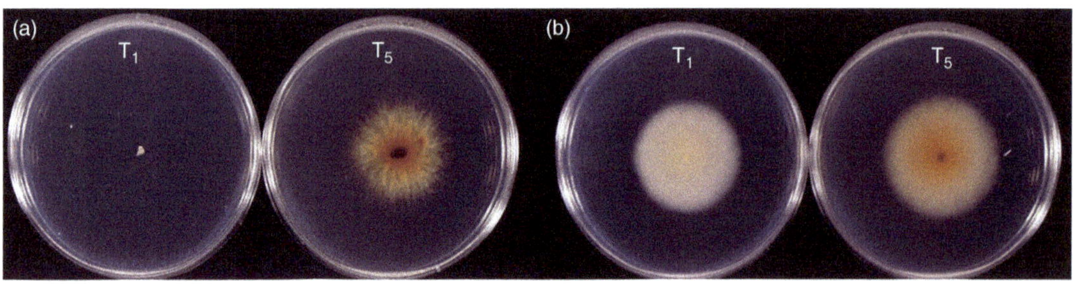

Fig. 101.4. *Trichophyton equinum* showing (a) no growth on T1 but good growth on T5; and *T. equinum* var. *autotrophicum* strain showing (b) good growth on both T1 and T5.

101.3. *Trichophyton erinacei* (Smith & Marples) Quaife

Synonymy: *Trichophyton mentagrophytes* var. *erinacei* J.M.B. Smith & Marples; *Trichophyton proliferans* English & Stockdale.

Trichophyton erinacei is a zoophilic fungus associated with hedgehogs and the epidermal mites, which they often harbour. Human infections occur most frequently on the exposed parts of the body; but tinea of the scalp and nails can also occur. Invaded hairs show an ectothrix infection but do not fluoresce under Wood's ultra-violet light. The distribution of this fungus is New Zealand and Europe.

RG-2 organism

Morphological description: Colonies are white, flat, powdery, sometimes downy to fluffy with a brilliant lemon-yellow reverse (Fig. 101.5). Numerous large clavate microconidia are borne on the sides of hyphae. Macroconidia are smooth-walled, two- to six-celled, clavate, variable in size, and may have terminal appendages. Macroconidia are much shorter than those seen in *Trichophyton interdigitale* (Fig. 101.6).

Fig. 101.5. *Trichophyton erinacei* culture showing its typical brilliant lemon-yellow reverse pigment.

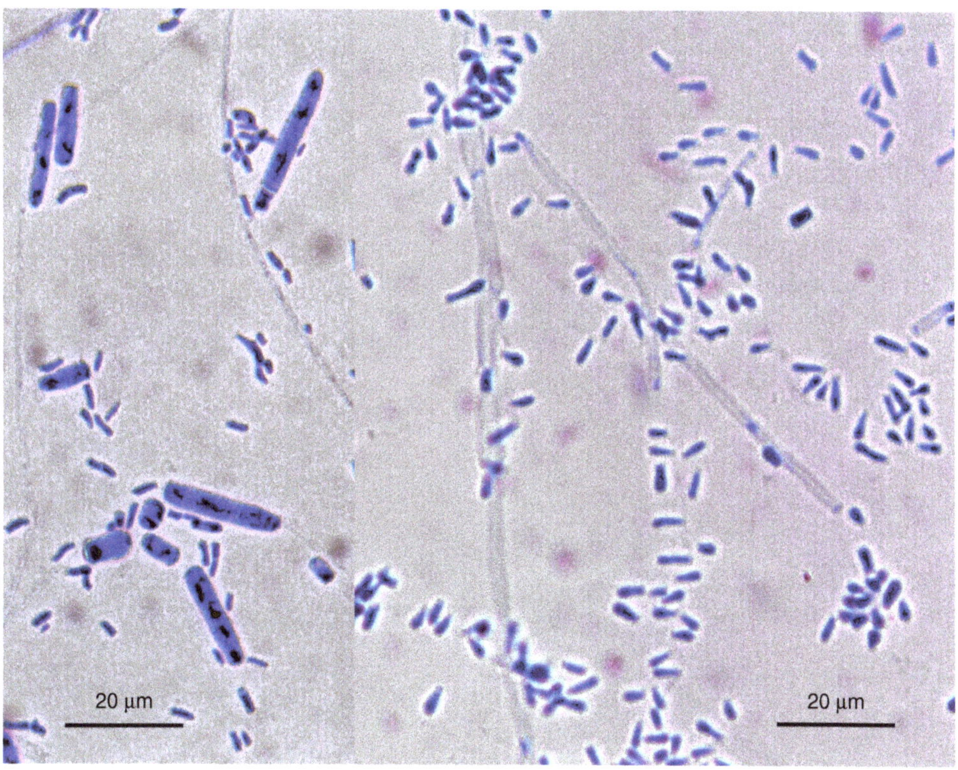

Fig. 101.6. *Trichophyton erinacei* macroconidia and microconidia.

Confirmatory tests

Hydrolysis of urea: Negative at 7 days.

Vitamin free agar (Trichophyton agar No.1): Good growth indicating no special nutritional requirements. Colonies are white suede to powdery with no reverse pigment.

Hair perforation test: Positive.

Molecular identification: Species identification based on ITS sequencing (Graser *et al.*, 1999a; Chen *et al.*, 2011).

Key features: Include microscopic morphology and culture characteristics; slender clavate microconidia and brilliant lemon-yellow reverse pigment on Sabouraud's dextrose agar; and negative hydrolysis of urea.

Trichophyton erinacei is generally distinguished from zoophilic isolates of *Trichophyton interdigitale* by (a) its microscopic morphology showing numerous large slender clavate microconidia borne on the sides of hyphae and its smooth, thin-walled clavate macroconidia; (b) its brilliant lemon-yellow reverse pigment on Sabouraud's dextrose agar; (c) its lack of reverse pigment on Sabouraud's salt agar; and (d) its negative hydrolysis of urea.

101.4. *Trichophyton interdigitale* Priestley

Synonymy: *Trichophyton mentagrophytes* var. *interdigitale* (Priestley) Moraes, Anais Bras.

Trichophyton interdigitale is an anthropophilic fungus which is a common cause of tinea pedis, particularly the vesicular type, tinea corporis, and sometimes superficial nail plate invasion in humans. It is not known to invade hair *in vivo* but produces hair perforations *in vitro*. Distribution is worldwide. This species may be regarded as a clonal offshoot of the zoophilic *T. mentagrophytes* (de Hoog *et al.* 2017).

Morphological description: Colonies are usually flat, white to cream in colour with a powdery to suede-like surface and yellowish to pinkish-brown reverse pigment (Fig. 101.7), often becoming a darker red-brown with age. Numerous subspherical to pyriform microconidia (Fig. 101.8), occasional spiral hyphae and spherical chlamydospores are present, the latter being more abundant in older cultures (Fig. 101.9). Occasional slender, clavate, smooth-walled, multiseptate macroconidia are also present in some cultures.

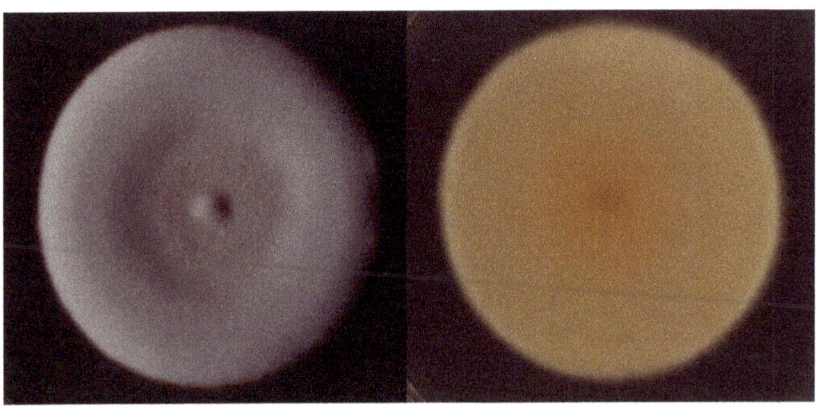

Fig. 101.7. *Trichophyton interdigitale* culture.

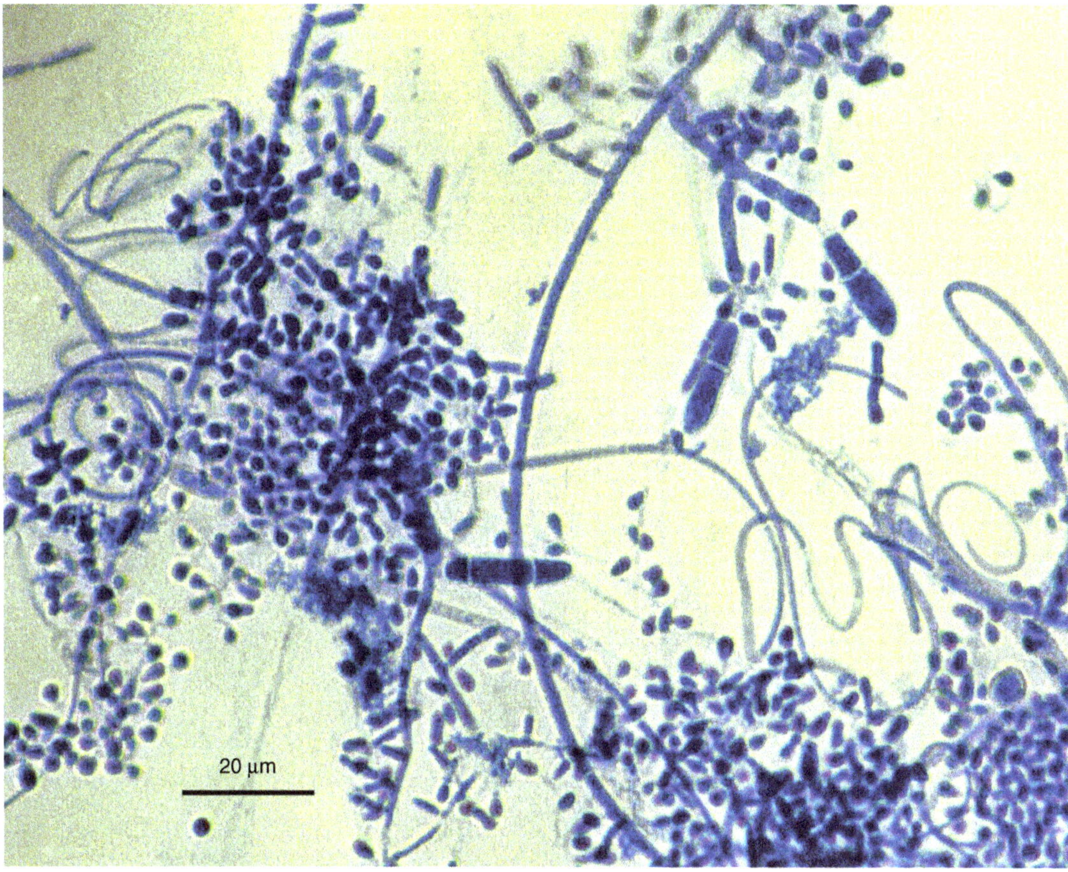

Fig. 101.8. *Trichophyton interdigitale* microconidia, macroconidia and spiral hyphae.

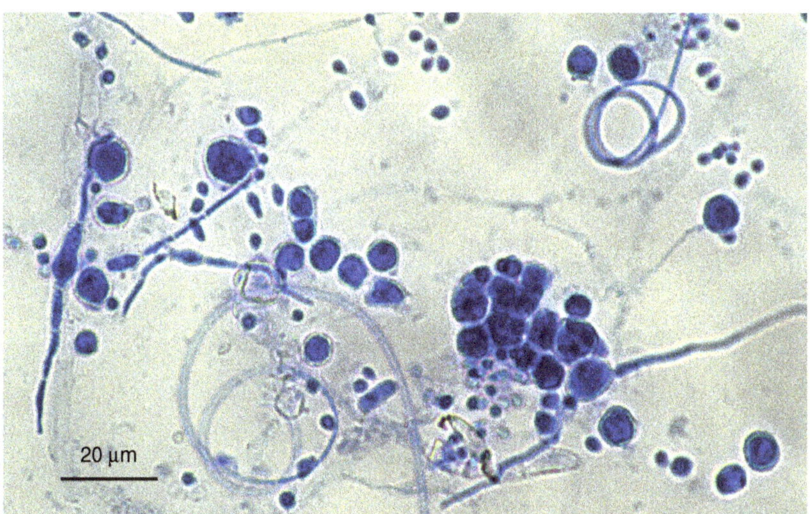

Fig. 101.9. *Trichophyton interdigitale* chlamydospores.

Confirmatory tests

Littman oxgall agar: Raised white downy colony with no reverse pigment.

Lactritmel agar: Macroscopic and microscopic features as described above.

Sabouraud's dextrose agar with 5% salt: Heaped and folded, buff coloured, suede-like surface with a dark reddish-brown submerged fringe and brown reverse.

1% Peptone agar: Flat, white to cream, suede-like surface with raised white downy centre. No reverse pigment.

Vitamin free agar (Trichophyton agar No.1): Good growth indicating no special nutritional requirements, flat cream powdery surface with central downy tuft. Reverse pale pinkish-brown.

Hydrolysis of urea: Positive within 7 days (usually 3 to 5 days).

Hair perforation test: Positive (Fig. 101.10).

Key features: Culture characteristics, microscopic morphology and *in vitro* perforation of human hair.

Trichophyton interdigitale can be distinguished from *T. rubrum* and *T. mentagrophytes* by: (a) its culture characteristics and microscopic morphology on Sabouraud's dextrose agar and/or lactritmel agar; (b) its growth and colony morphology on Sabouraud's salt agar (colonies of *T. interdigitale* and *T. mentagrophytes* unlike *T. rubrum*, grow very well on this medium and usually produce a distinctive dark reddish-brown reverse pigment); (c) a positive urease test (within 7 days), a positive hair perforation test and the production of a yellow-brown to pinkish-brown reverse pigment on pigment stimulation media like lactritmel and Trichophyton No.1 agars; (d) *T. interdigitale* demonstrates profuse growth and alkalinity on BCP milk solids agar; and (e) on 1% peptone agar *T. interdigitale* has a suede-like to downy surface whereas *T. mentagrophytes* has a characteristic granular appearance.

Note: The recently proposed species, *Trichophyton indotineae* is closely related to *T. interdigitale,* and likely represents another clonal offshoot of *T. mentagrophytes*. It is associated with terbinafine resistance, and is increasingly being reported in India (Kano *et al*., 2020; Tang *et al*., 2021). While separation of *T. indotineae* and *T. interdigitale* is not reliable from morphological characteristics, *T. indotineae* typically has a negative urea hydrolysis result after 7 days, while *T. interdigitale* and *T. mentagrophytes* are typically positive (Tang *et al*., 2021).

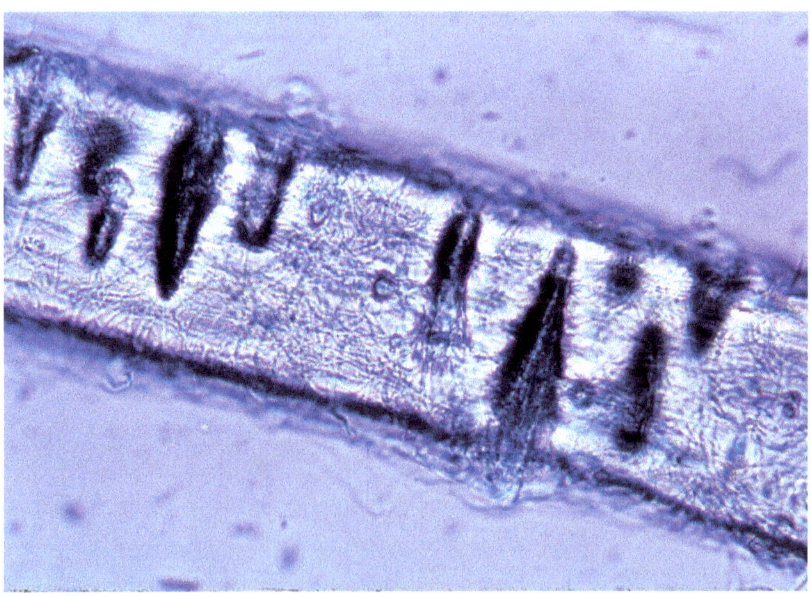

Fig. 101.10. *Trichophyton interdigitale* positive hair perforation test.

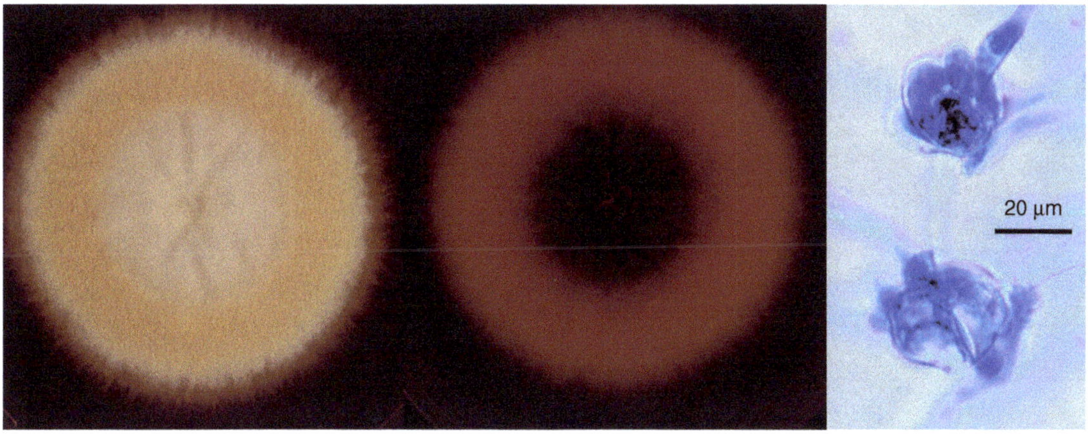

Fig. 101.11. *Trichophyton interdigitale* var. *nodulare* (a dysgonic nodular variant) culture and "nodular" bodies.

Supplementary description for *Trichophyton interdigitale* var. *nodulare*: A dysgonic variant of *T. interdigitale*. Cultures show distinctive bright yellow to apricot coloured colonies with a suede-like to powdery surface and a bright yellow-brown to orange reverse; and microscopically characteristic "nodular bodies" are observed in the vegetative hyphae. Usually, no conidia are seen but some isolates, especially with subculture may produce subspherical to pyriform microconidia similar to those of *T. interdigitale* (Fig. 101.11).

101.5 *Trichophyton mentagrophytes* (Robin) Blanchard

Synonymy: *Trichophyton mentagrophytes* var. mentagrophytes (Robin) Sabour.

Trichophyton mentagrophytes is a zoophilic fungus with a worldwide distribution and a wide range of animal hosts including mice, guinea pigs, kangaroos, cats, horses, sheep and rabbits. Produces inflammatory skin or scalp lesions in humans, particularly in rural workers. Kerion of the scalp and beard may occur. Invaded hairs show an ectothrix infection but do not fluoresce under Wood's ultra-violet light. Distribution is worldwide.

RG-2 organism

Morphological description: Colonies are generally flat, white to cream in colour, with a powdery to granular surface. Some cultures show central folding or develop raised central tufts or pleomorphic suede-like to downy areas. Reverse pigmentation is usually a yellow-brown to reddish-brown colour (Fig. 101.12). Numerous single-celled microconidia are formed, often in dense clusters. Microconidia are hyaline, smooth-walled, and are predominantly spherical to subspherical in shape, however occasional clavate to pyriform forms may occur (Fig. 101.13). Varying numbers of spherical chlamydospores, spiral hyphae and smooth, thin-walled, clavate-shaped, multi-celled macroconidia may also be present (Fig. 101.14).

Confirmatory tests

Littman oxgall agar: Raised greyish-white, suede-like to downy colony. Some cultures may show a diffusible yellow to brown pigment.

Lactritmel agar: Cultures are flat, white to cream in colour, with a powdery to granular surface. Some cultures develop raised central tufts or pleomorphic downy areas. Reverse pigmentation is yellow-brown to pinkish-brown to red-brown. Microscopic morphology similar to that described above, with predominantly spherical microconidia, often formed in dense clusters, and varying numbers of spherical chlamydospores, spiral hyphae and smooth, thin-walled, clavate, multiseptate macroconidia.

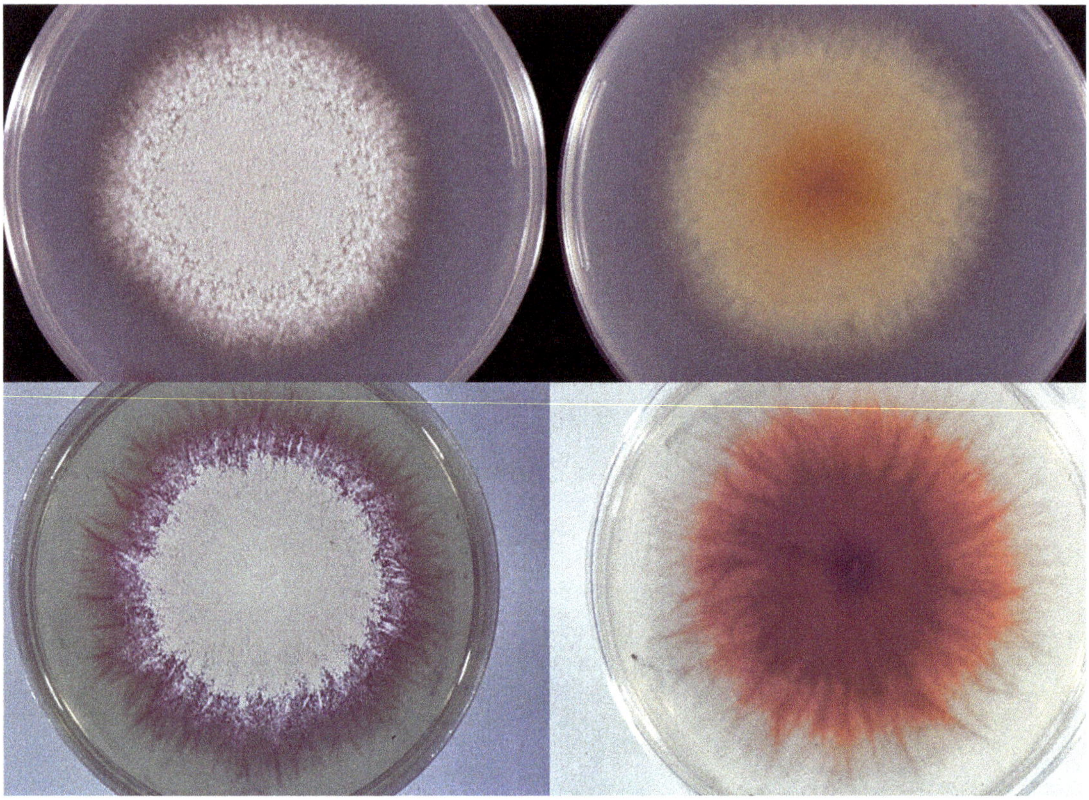

Fig. 101.12. *Trichophyton mentagrophytes* culture characteristics.

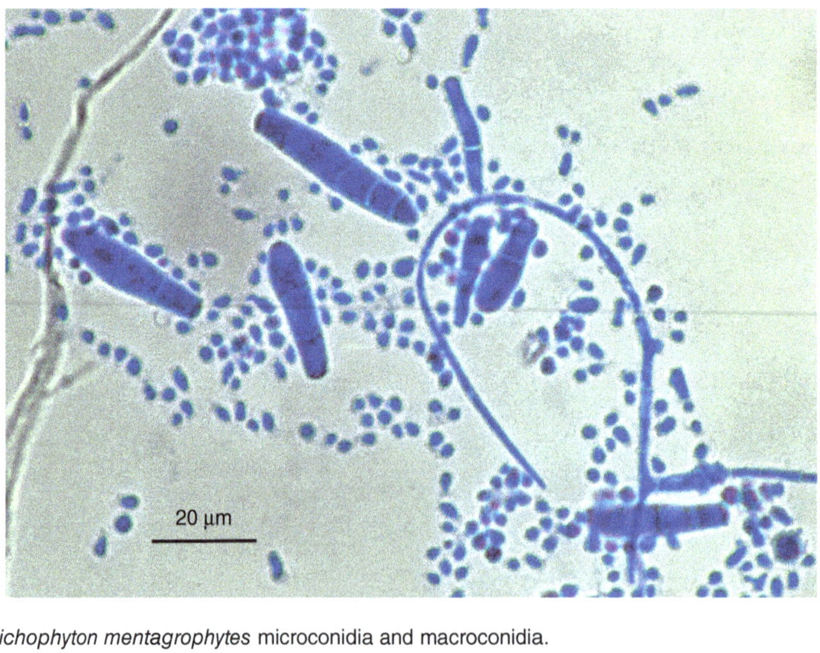

Fig. 101.13. *Trichophyton mentagrophytes* microconidia and macroconidia.

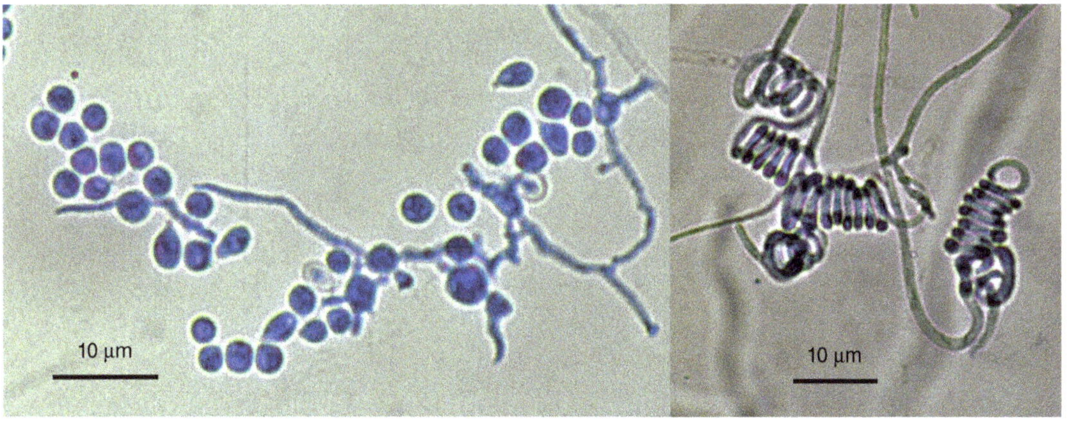

Fig. 101.14. *Trichophyton mentagrophytes* chlamydospores and hyphal spirals.

Sabouraud's dextrose agar with 5% salt: Cultures are heaped and folded, buff to brown in colour, with a suede-like surface texture and characteristically have a very dark reddish-brown submerged peripheral fringe and reverse pigmentation.

1% Peptone agar: Flat, cream coloured, powdery to granular colony with no reverse pigment.

Hydrolysis of urea: Positive within 7 days (usually 3 to 5 days).

Vitamin free agar (Trichophyton agar No.1): Good growth indicating no special nutritional requirements. Cultures are flat, cream coloured, with a powdery to suede-like surface, and have a reddish-brown reverse pigmentation.

Hair perforation test: Positive within 14 days.

Key features: Culture characteristics, microscopic morphology and clinical disease with known animal contacts. *Trichophyton mentagrophytes* can be distinguished from *T. interdigitale* by: (a) its granular appearance on the 1% Peptone agar; (b) its microscopic morphology of more spherical microconidia and generally greater numbers of macroconidia; and (c) a yellow to brown diffusible pigment is often seen on the Littman oxgall agar. Both *T. interdigitale* and *T. mentagrophytes* demonstrate profuse growth and alkalinity on BCP agar.

101.6. *Trichophyton quinckeanum* (Zopf) MacLeod & Munde

Synonymy: *T. mentagrophytes* var. *quinckeanum*.

Trichophyton quinckeanum causes "mouse favus" on mice, and this is seen as thick, yellow, saucer-shaped crusted lesions up to 1 cm in diameter called scutula. Invaded hairs are rarely seen but they may show either ectothrix or endothrix infection. Infected human hairs do not fluoresce under Wood's ultra-violet light, but very occasional hairs from experimental lesions in guinea pigs may show a pale yellow fluorescence. The geographical distribution of this dermatophyte is difficult to establish, but it's probably worldwide. It is often associated with mice plagues in the Australian Wheat Belt.

RG-2 organism

Morphological description: Colonies are generally flat, white to cream in colour, with a powdery to granular surface. Some cultures show central folding or develop raised central tufts or pleomorphic suede-like to downy areas. Reverse pigmentation is usually a yellow-brown to reddish-brown colour (Fig. 101.15). Numerous microconidia, predominantly slender clavate when young, are borne laterally along the sides of

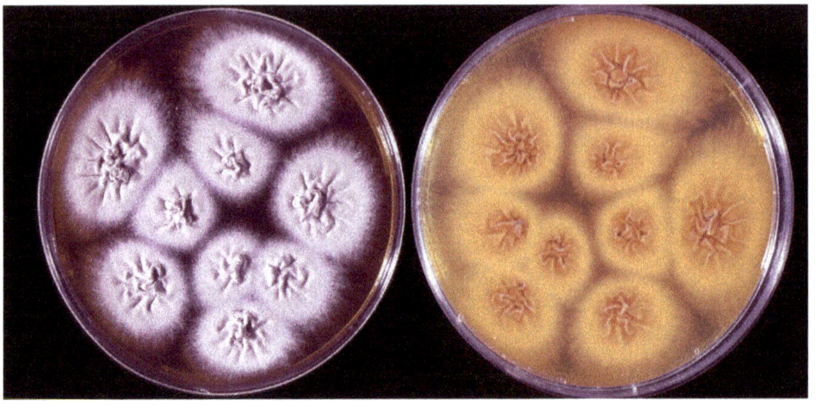

Fig. 101.15. *Trichophyton quinckeanum* culture.

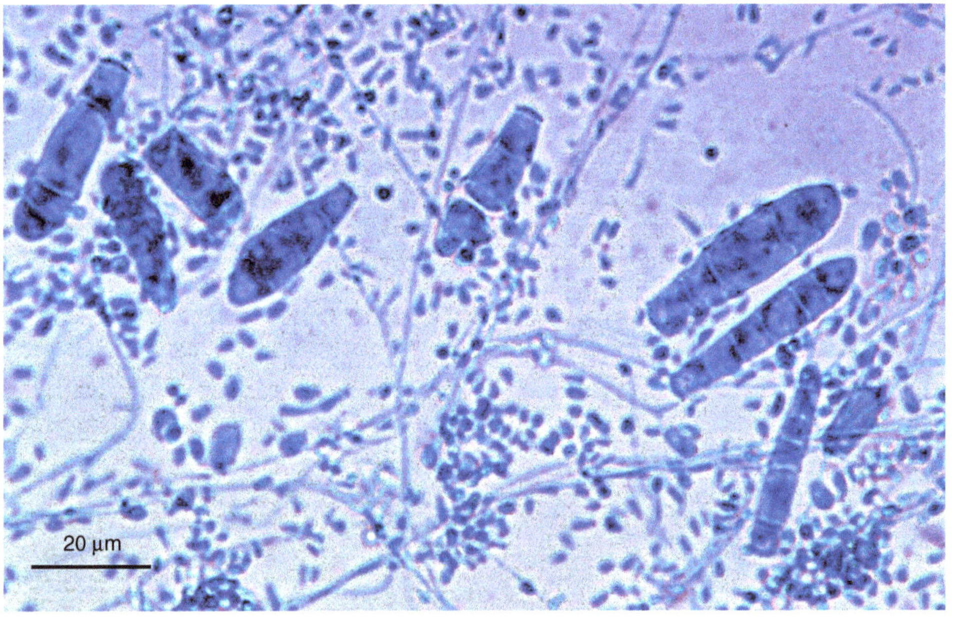

Fig. 101.16. *Trichophyton quinckeanum* microconidia and macroconidia.

hyphae. With age the microconidia become broader and pyriform to spherical in shape. Occasional to moderate numbers of smooth-walled, thin-walled, multiseptate, clavate to cigar shaped macroconidia may be present (Fig. 101.16). Varying numbers of spherical chlamydospores and spiral hyphae may also be present.

Confirmatory tests

Littman oxgall agar: Raised, dome-like bluish-grey, suede-like colony with a narrow flat, greyish-white, suede-like border. No diffusible or reverse pigment should be present.

Sabouraud's dextrose agar with 5% NaCl: Heaped and much folded white suede-like colony with very pale yellow-brown reverse. No submerged fringe.

Hydrolysis of urea: Positive within 7 days (usually very rapid 2-3 days).

Vitamin free agar (Trichophyton agar No.1): Flat, white to cream, suede-like colony with either no reverse pigment or a pale yellow-brown reverse. i.e. no special nutritional requirements.

Hair perforation test: Positive in 7 to 10 days.

Key features: Culture characteristics, microscopic morphology and rapid urease test.

Trichophyton quinckeanum may be distinguished from *T. mentagrophytes* by (a) its characteristic culture appearance on Littman Oxgall agar (i.e. raised, dome-like, bluish-grey suede-like colony with a narrow flat, greyish-white, suede-like border and no diffusible or reverse pigment); and on Sabouraud's salt agar (typically heaped and folded white suede-like colony, but with no distinctive dark reddish-brown submerged fringe and reverse pigment as seen on *T. mentagrophytes*); (b) microscopic morphology showing numerous slender clavate with some pyriform microconidia and moderate numbers of smooth thin-walled, clavate macroconidia; (c) a rapid urease test, usually 2 to 3 days.

101.7. *Trichophyton rubrum* (Castellani) Semon

Synonymy: *Trichophyton fischeri* Kane; *Trichophyton raubitschekii* Kane, Salkin, Weitzman & Smitka; *Trichophyton kanei* Summerbell; all varieties of *T. rubrum*.

Trichophyton rubrum is an anthropophilic fungus that has become the most widely distributed dermatophyte of humans. It frequently causes chronic infections of skin, nails and rarely scalp. Granulomatous lesions may sometimes occur. Infected hairs do not fluoresce under Wood's ultraviolet light, and microscopically may show endothrix or ectothrix type of invasion.

Morphologically *T. rubrum* exhibits a spectrum of overlapping characters; for example culture surface texture may vary from downy to suede-like; culture surface pigmentation may vary from white to cream to deep red; culture reverse pigmentation may vary from colourless to yellowish to yellow-brown to wine red; numbers of microconidia range from none to scanty to many; shape of microconidia vary from slender clavate to pyriform; numbers of macroconidia range from none to scanty to many and may or may not have terminal projections. This is why so many varieties or synonyms have been described in the past. However, molecular evidence by Graser *et al.* (1999b) reveals little variation between strains of *T. rubrum* and determined that the species is largely clonal.

The majority of isolates, especially those causing tinea pedis and onychomycosis, are characterised by the production of scanty to moderate numbers of slender clavate microconidia and no macroconidia (formerly the "downy strain"). Some isolates, usually from cases of tinea corporis, are characterised by the production of moderate to abundant numbers of clavate to pyriform microconidia and moderate to abundant numbers of thin-walled, cigar-shaped macroconidia (formerly the "granular strain").

It should be stressed that intermediate strains do occur as many culture and morphological characteristics overlap.

Note: *Trichophyton rubrum* produces both red and yellow pigments. Culture colours may range from none to dark red to dark yellow, with all combinations in between (Fig. 101.17).

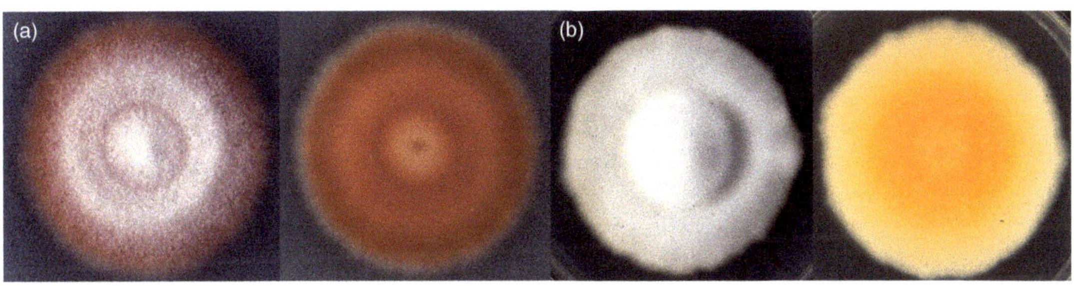

Fig. 101.17. *Trichophyton rubrum* culture grown on (a) lactritmel agar that promotes red pigmentation and (b) on mycobiotic agar that shows an underlying yellow pigmentation.

RG-2 organism

Morphological description: Colonies are mostly flat to slightly raised, white to cream, suede-like to downy, with either no reverse pigment or a yellow-brown to wine-red reverse (Fig. 101.18). Most cultures show scanty to moderate numbers of slender clavate to pyriform microconidia (Fig. 101.19). Macroconidia are usually absent, but when present are smooth, thin-walled multiseptate, slender and cylindrical to cigar-shaped (Fig. 101.20). Older cultures may show numerous chlamydospores with a few clavate to pyriform microconidia.

Note: On primary isolation some cultures may lack reverse pigmentation and fail to produce microconidia. These need to be subcultured onto media like lactritmel agar or potato dextrose agar, which stimulate pigmentation and sporulation. If sporulation still fails subculture the fungus onto Trichophyton agar No.1.

Fig. 101.18. *Trichophyton rubrum* culture morphology; (a) "downy strain" with typical wine-red reverse; (b) "Y variant" with both yellow and red pigmentation; (c) "*flava* variant" with yellow pigmentation; (d) "granular strain" with red surface and reverse pigmentation.

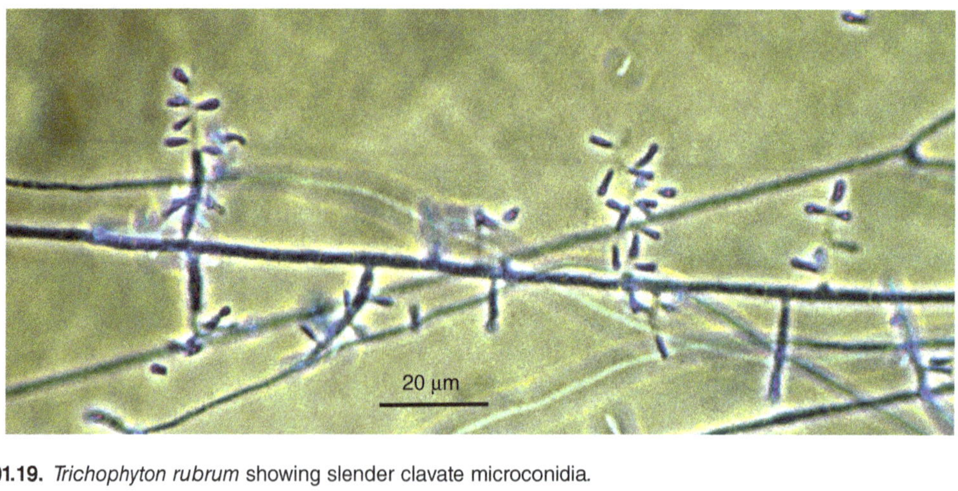

Fig. 101.19. *Trichophyton rubrum* showing slender clavate microconidia.

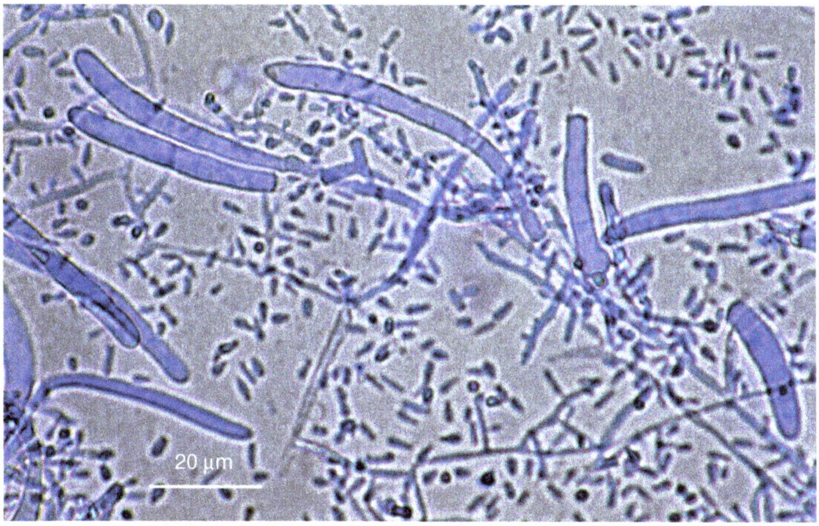

Fig. 101.20. *Trichophyton rubrum* macroconidia.

Confirmatory tests

Littman oxgall agar: Raised, greyish-white, suede-like to downy colony with no reverse pigment. Some cultures may show a greenish-yellow diffusible pigment.

Lactritmel agar: Flat, white to rose-pink, downy to granular colonies with a deep wine-red reverse pigment.

Sabouraud's dextrose agar with 5% salt: Very stunted, white to cream, downy to glabrous colony with a pale yellow-brown reverse pigment.

1% Peptone agar: Flat, white to cream, downy to glabrous colony with no reverse pigment.

BCP agar: Restricted colony growth and neutral (unchanged) pH. Colonies typically demonstrate red pigment on reverse

Hydrolysis of urea: Typically, negative at 7 days (some may be positive).

Vitamin free agar (Trichophyton agar No.1): Good growth indicating no special vitamin requirements. Colonies are flat, white to cream, suede-like to downy with a deep wine-red reverse pigment.

Hair perforation test: Negative at 28 days.

Molecular identification: Sequencing of the ITS region (Graser *et al.*, 1999a; Irinyi *et al.*, 2015), and *TEF-1* gene (Mirhendi *et al.*, 2015) are recommended for identification.

Key features: Include clinical history, culture characteristics, microscopic morphology and failure to perforate hair *in vitro*.

101.8. *Trichophyton schoenleinii* (Lebert) Langeron & Milochevitch

Trichophyton schoenleinii is an anthropophilic fungus causing favus in humans. Favus is a chronic, scarring form of tinea capitis characterised by saucer-shaped crusted lesions or scutula and permanent hair loss. Invaded hairs remain intact and fluoresce a pale greenish-yellow under Wood's ultraviolet light. Favus was once common in Eurasia and North Africa, however its incidence is now in decline.

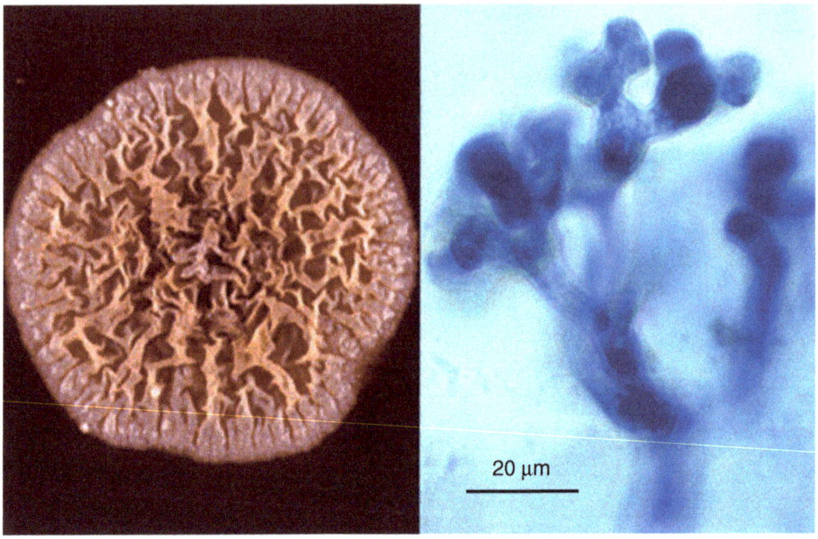

Fig. 101.21. *Trichophyton schoenleinii* culture and microscopy showing "favic chandeliers".

RG-2 organism

Morphological description: Colonies are slow growing, waxy or suede-like with a deeply folded honeycomb-like thallus and some sub-surface growth. The thallus is cream coloured to yellow to orange-brown (Fig. 101.21). Cultures are difficult to maintain in their typical convoluted form, and rapidly become flat and downy. No reverse pigmentation is present. No macroconidia and microconidia are seen in routine cultures, however numerous chlamydospores may be present in older cultures. However, characteristic antler-like "nail head" hyphae also known as "favic chandeliers" may be observed. A few distorted clavate microconidia may be formed by some isolates when grown on polished rice grains.

Key features: Clinical history, culture characteristics and microscopic morphology showing "favic chandeliers".

101.9. *Trichophyton soudanense* Joyeux

Trichophyton soudanense is an anthropophilic fungus that is a frequent cause of tinea capitis in Africa. Invaded hairs show an endothrix infection but do not fluoresce under Wood's ultra-violet light. Distribution is mainly in Africa with imported cases now reported from Europe, Brazil, Australia and USA.

RG-2 organism

Morphological description: Colonies are slow-growing with a flat to folded, suede-like surface. Often there is a broad fringe of submerged growth. Surface mycelium and reverse pigment are characteristically a deep apricot-orange in colour. Microscopically, the hyphae often show reflexive or right-angle branching (Fig. 101.22). Pyriform microconidia may occasionally be present and numerous chlamydospores are often found in older cultures.

Note: *Trichophyton soudanense* demonstrates profuse growth and alkalinity on BCP agar. A thin halo of clearing usually appears in the milk solids around the colony edge at 7-10 days.

Key features: Clinical history, culture characteristics and microscopic morphology showing reflexive hyphal branching and endothrix invasion of hair.

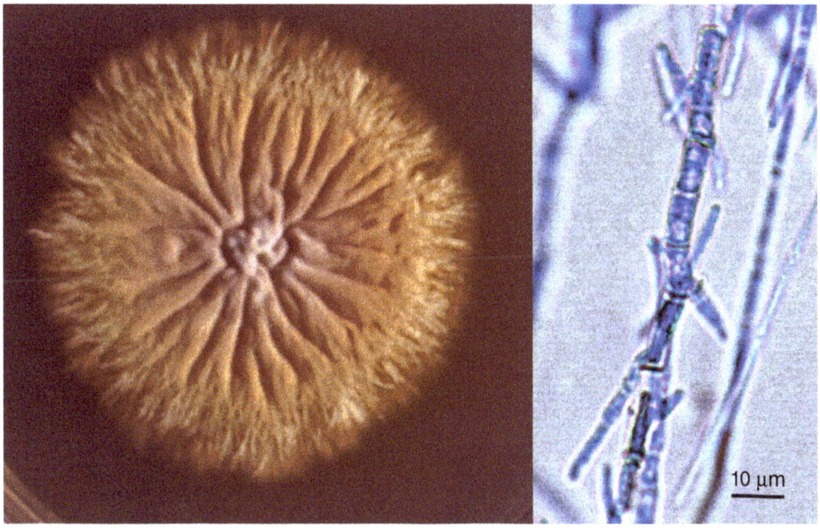

Fig. 101.22. *Trichophyton soudanense* culture and microscopy showing "reflexive" hyphal branching.

101.10. *Trichophyton tonsurans* Malmsten

Trichophyton tonsurans is an anthropophilic fungus with a worldwide distribution, which causes inflammatory or chronic non-inflammatory finely scaling lesions of skin, nails and scalp. It is a common cause of tinea capitis in the Australian Aborigine, African, and African American populations. Invaded hairs show an endothrix infection and do not fluoresce under Wood's ultra-violet light.

RG-2 organism

Morphological description: Colonies show considerable variation in texture and colour. They may be suede-like to powdery, flat with a raised centre or folded, often with radial grooves. The colour may vary from pale buff to yellow, (the *sulfureum* form which resembles *Epidermophyton floccosum*), to dark brown. The reverse colour varies from yellow-brown to reddish-brown to deep mahogany (Fig. 101.23). Hyphae are relatively broad, irregular, much branched with numerous septa. Numerous characteristic microconidia varying in size and shape from long clavate to broad pyriform, are borne at right angles to the hyphae, which often remain unstained by lactophenol cotton blue (Fig. 101.24). Very occasional smooth, thin-walled, irregular, clavate macroconidia may be present on some cultures. Numerous swollen giant forms of microconidia and chlamydospores are produced in older cultures.

Confirmatory Tests

Mycosel agar: Chlamydospore-like structures produced by 5 days is characteristic of *T. tonsurans* (Mochizuki *et al.* 2013).

Hydrolysis of urea: Positive at 5 days.

Nutritional Tests on Trichophyton agars: Results demonstrate a partial requirement for thiamine. T1 = vitamin free agar, T4 = vitamin free + thiamine agar (Fig. 101.25).

Hair Perforation Test: Positive within 14 days.

Molecular identification: ITS sequencing is recommended for identification (Graser *et al.* 1999a; Yoshida *et al.*, 2006; Kandemir *et al.*, 2020).

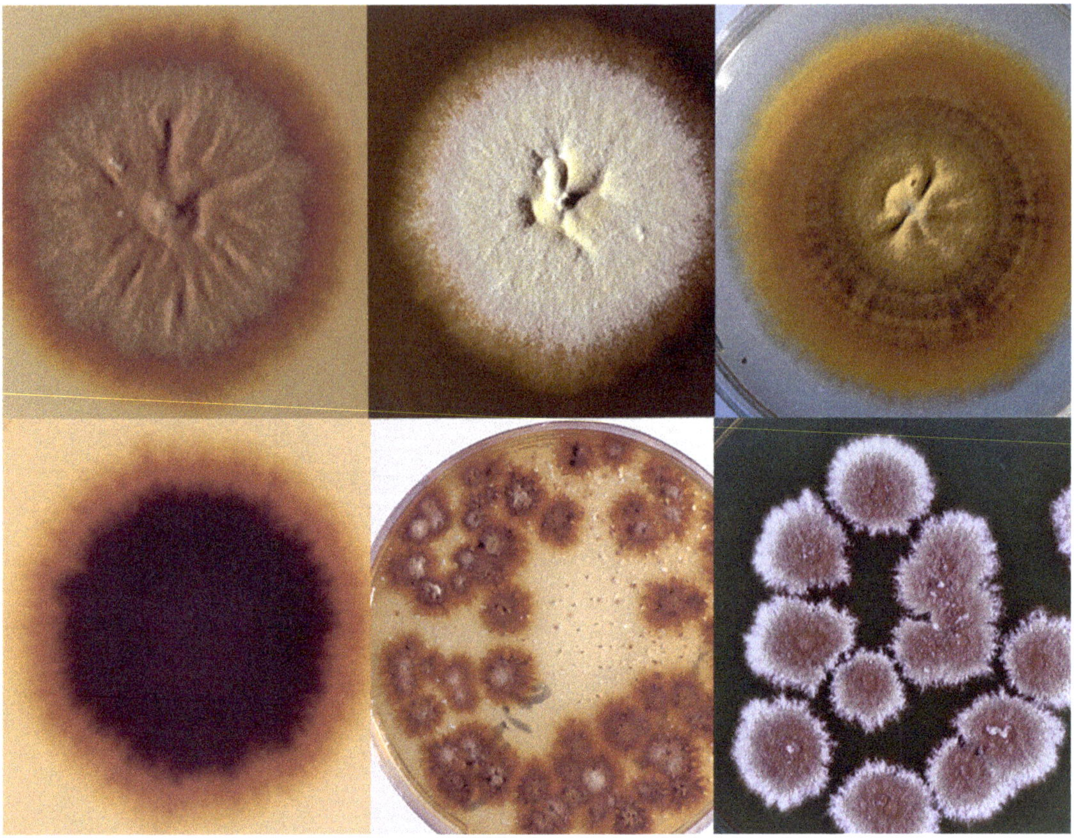

Fig. 101.23. *Trichophyton tonsurans* cultures show considerable variation in texture and colour. The colour may vary from pale buff to yellow to dark brown. The reverse colour varies from yellow-brown to reddish-brown to deep mahogany.

Key features: Microscopic morphology, culture characteristics, endothrix invasion of hairs and partial thiamine requirement.

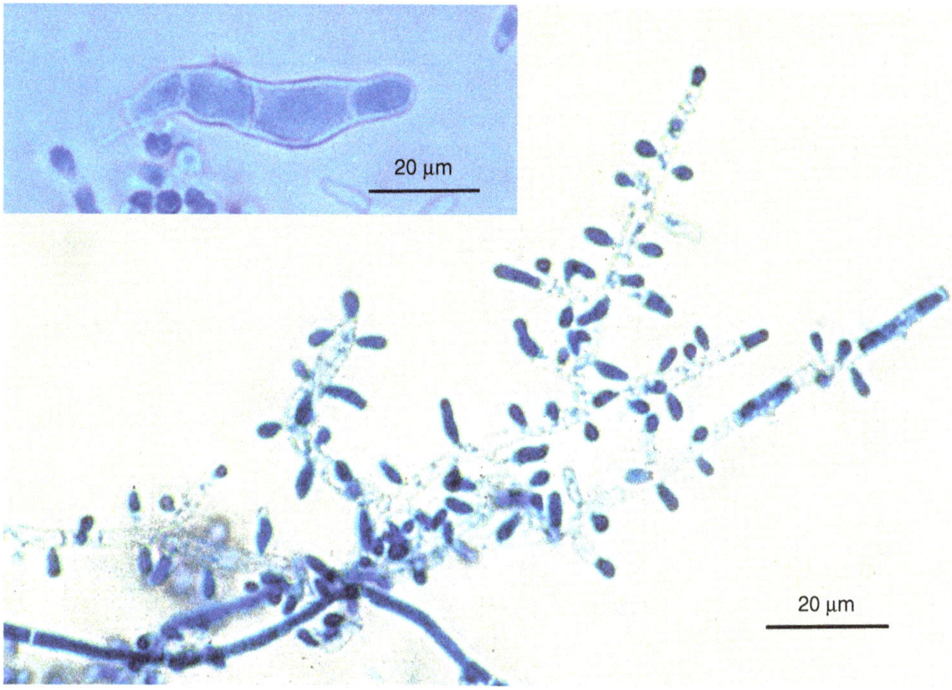

Fig. 101.24. *Trichophyton tonsurans* microconidia and macroconidia. **Note:** The broad unstained hyphae.

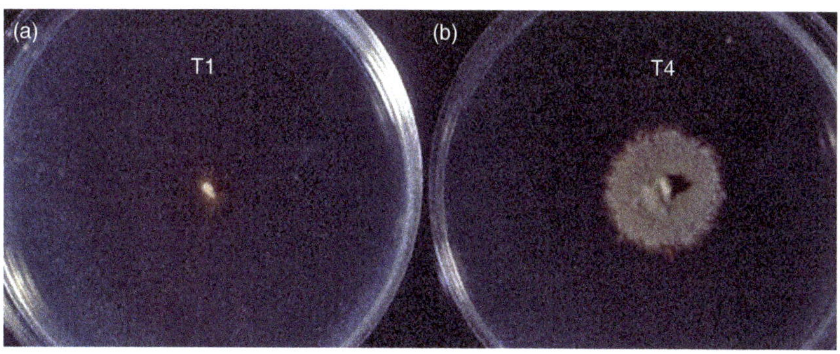

Fig. 101.25. *Trichophyton tonsurans* showing good growth on T4 agar indicating a requirement for thiamine.

References: Rebell and Taplin (1970); Ajello (1977); Vanbreusegham *et al.* (1978); Rippon (1988); McGinnis (1980); Kane *et al.* (1997); Domsch *et al.* (2007); de Hoog *et al.* (2015).

101.11. *Trichophyton verrucosum* Bodin

Trichophyton verrucosum is a zoophilic fungus causing ringworm in cattle. Infections in humans result from direct contact with cattle or infected fomites and are usually highly inflammatory involving the scalp, beard or exposed areas of the body (i.e. nails, skin). Invaded hairs show an ectothrix infection and fluorescence under Wood's ultra-violet light has been noted in cattle but not in humans. Geographic distribution is worldwide.

RG-2 organism

Morphological description: Colonies are slow growing, small, button or disk-shaped, white to cream coloured, with a suede-like to velvety surface, a raised centre, and flat periphery with some submerged growth (Fig. 101.26). Reverse pigment may vary from non-pigmented to yellow. Broad, irregular hyphae with many terminal and intercalary chlamydospores. Chlamydospores are often in chains (Fig. 101.28). The tips of some hyphae are broad and club-shaped, and occasionally divided, giving the so-called "antler" effect. When grown on thiamine-enriched media, occasional strains produce clavate to pyriform microconidia borne singly along the hyphae (Fig. 101.27). Macroconidia are only rarely produced, but when present have a characteristic tail or string bean shape (Fig. 101.28).

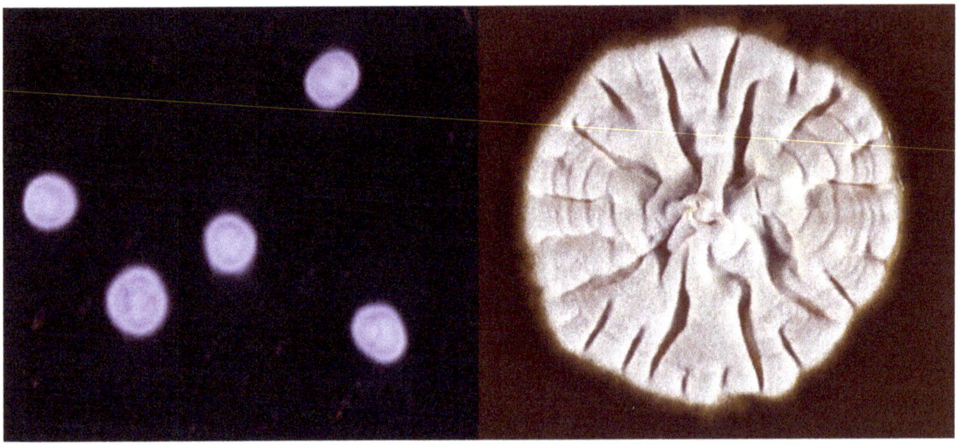

Fig. 101.26. *Trichophyton verrucosum* showing young button-shaped colonies and a mature culture.

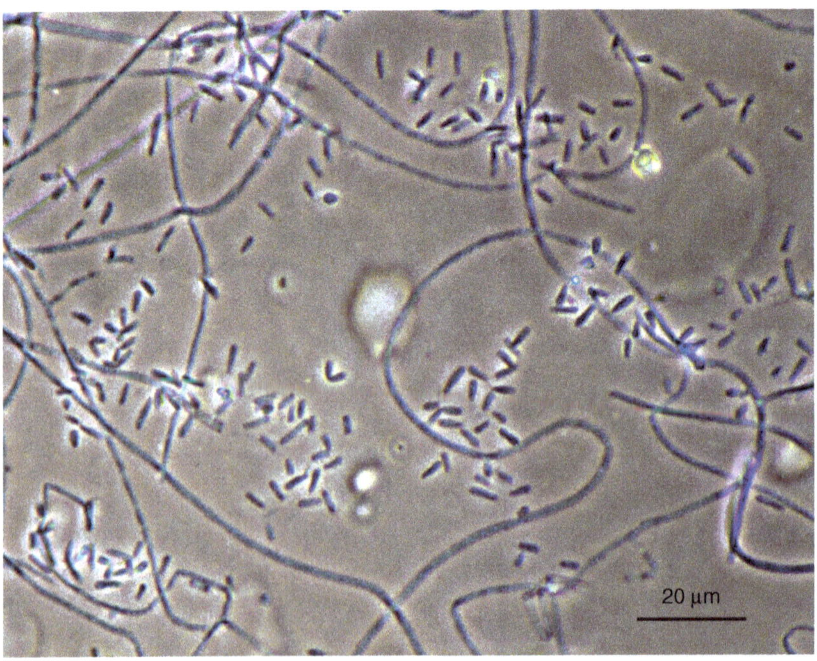

Fig. 101.27. *Trichophyton verrucosum* microscopy showing clavate to pyriform microconidia.

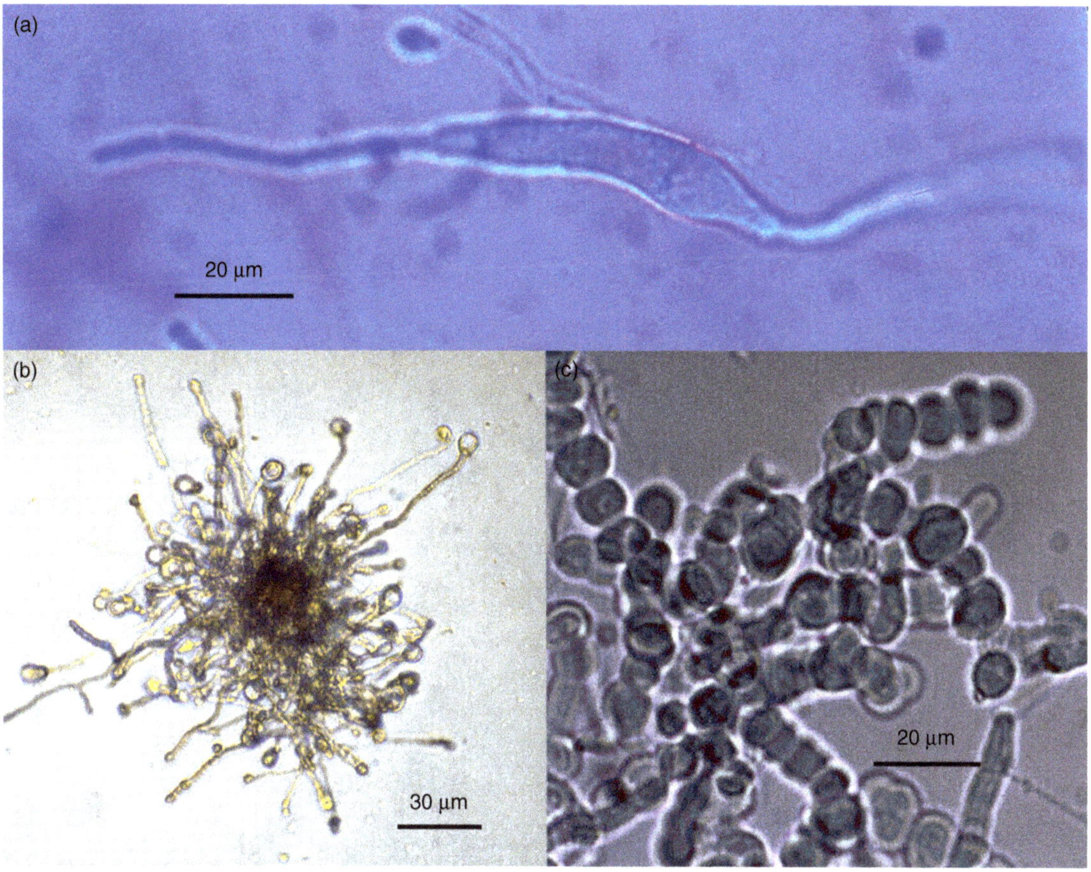

Fig. 101.28. *Trichophyton verrucosum* showing (a) characteristic rat tail or string bean shaped macroconidia, (b) characteristic terminal vesicles at the tips of hyphae and (c) chains of chlamydospores.

Confirmatory Tests

Growth at 37°C: Unlike other dermatophytes growth is enhanced at 37°C.

Nutritional requirements: All strains require thiamine and approximately 80% require thiamine and inositol. There is no growth on casein vitamin free agar (T1), minimal submerged growth on T1 + inositol (T2), good growth on T1 + inositol and thiamine (T3) and good growth on T1 + thiamine only (T4).

All strains produce typical chains of chlamydospores, often referred to as "chains of pearls", especially when grown on BCP agar at 37°C (Fig. 101.28). Also when grown at 25°C on milk solids glucose agar a halo of peripheral clearing of milk solids occurs within 7 days.

Microscopic examination of young four to five day old colonies, grown from a very small inoculum, on Sabouraud's dextrose agar containing 0.5% yeast extract and incubated at 30°C, show characteristic terminal vesicles (not chlamydospores) at the tips of hyphae (Fig. 101.28). The number of vesicles produced is greater from primary inoculations of skin scrapings or hairs.

Key features: Culture characteristics and requirements for thiamine and inositol, large ectothrix invasion of hair, clinical lesions and history.

101.12. *Trichophyton violaceum* Sabouraud apud Bodin

Synonyms: *Trichophyton yaoundei* Cochet & Doby-Dubois; *Trichophyton glabrum* Sabour; all varieties of *T. violaceum*.

Trichophyton violaceum is an anthropophilic fungus causing inflammatory or chronic non-inflammatory finely scaling lesions of skin, nails, beard and scalp, producing the so-called "black dot" tinea capitis. Distribution is worldwide, particularly in Southern Europe, Central America, Eastern Asia and North Africa. Invaded hairs show an endothrix infection and do not fluoresce under Wood's ultra-violet light.

RG-2 organism

Morphological description: Colonies are very slow growing, glabrous or waxy, heaped and folded and a deep violet in colour (Fig 101.29). Cultures often become pleomorphic, forming white sectors and occasional non-pigmented strains may occur. Hyphae are relatively broad, tortuous, much branched and distorted. Young hyphae usually stain well in lactophenol cotton blue, whereas older hyphae stain poorly and show small central fat globules and granules. Usually no conidia are seen, although occasional pyriform microconidia have been observed on enriched media. Numerous chlamydospores are usually present, especially in older cultures (Fig 101.29).

Nutritional requirements: *Trichophyton violaceum* has a partial nutrient requirement for thiamine. There is minimal growth on casein vitamin-free agar (T1), and slightly better growth on vitamin-free agar plus thiamine (T4). The partial requirement for thiamine separates this organism from *T. rubrum* and other species that may produce purple pigmented colonies.

Molecular identification: Differentiated from *T. rubrum* by microsatellite data (Ohst *et al.*, 2004) and ITS sequencing. White mutants, sometimes referred to as *Trichophyton glabrum*, were recognised by ITS sequencing (Woldeamanuel *et al.*, 2005).

Key features: Culture characteristics, partial thiamine requirement and endothrix hair invasion.

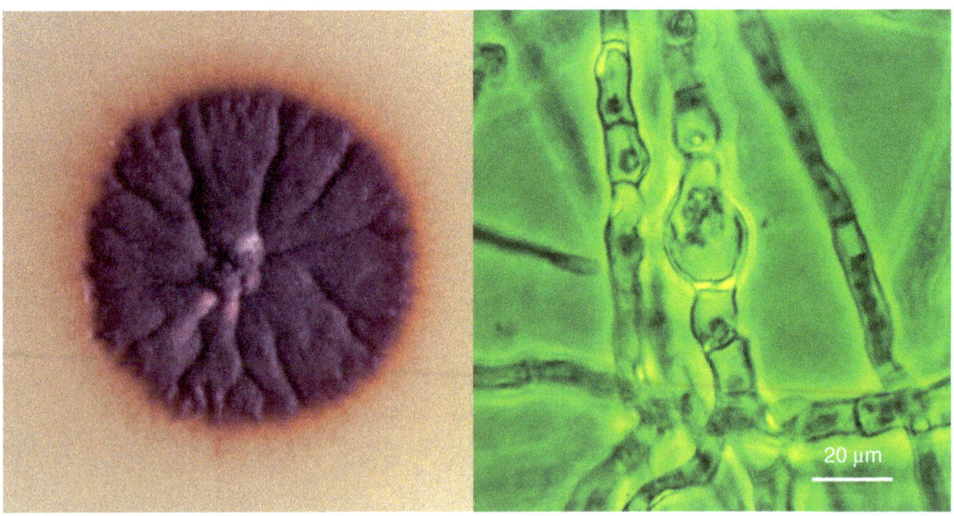

Fig. 101.29. *Trichophyton violaceum* culture and hyphal chlamydospores.

DOI: 10.1079/9781800622340.0101

102 *Trichosporon* Behrend

The taxonomy of *Trichosporon* has been redefined: *Trichosporon cutaneum*, *T. dermatis*, *T. jirovecii* and *T. mucoides* have now been transferred to the new genus *Cutaneotrichosporon*, while *Trichosporon domesticum*, *T. loubieri* and *T. mycotoxinovorans* have been included in the re-defined genus *Apiotrichum* (Liu *et al.*, 2015).

Trichosporon species are urease-positive, non-encapsulated basidiomycetous yeasts characterised by the development of hyaline, septate hyphae that fragment into oval or rectangular arthroconidia. Some blastoconidia are also seen. The colonies are usually raised and have a waxy appearance, which develop radial furrows and irregular folds. They are widely distributed in the environment. Many have different habitats and usually occupy narrow ecologic niches. Some are soil borne and others are associated with humans and animals (Colombo *et al.*, 2011; Sugita, 2011; Arendrup *et al.*, 2014).

Table 102.1. Key phenotypic characteristics of medically important species.

1	Growth with melibiose	2
	No growth with melibiose	3
2	Tolerant to cycloheximide	*Cutaneotrichosporon mucoides*
	Not tolerant to cycloheximide	*Cutaneotrichosporon cutaneum*
3	Growth with myo-inositol, no growth with L-arabinose	*Trichosporon inkin*
	No growth with myo-inositol, growth with L-arabinose	4
4	Colony very slow growth; thallus consisting of clumps of meristematic cells	*Trichosporon asteroides*
	Colonies and microscopy otherwise	5
5	Appressoria present in slide cultures	*Trichosporon ovoides*
	Appressoria absent in slide cultures	6
6	Arthroconidia barrel-shaped; thallus not meristematic	*Trichosporon asahii*
	Arthroconidia elongate, or thallus meristematic	*Trichosporon asteroides*

Trichosporon asahii, *T. asteroides*, *T. inkin* and *T. ovoides* are the four most common clinical species (Table 102.1), however, *T. coremiiforme*, *T. dohaense*, *T. faecale*, *T. japonicum* and *T. lactis* have also been reported from human and animal infections (Rodriguez-Tudela *et al.*, 2005; Chagas-Neto *et al.*, 2008; Colombo *et al.*, 2011). Importantly, all species are resistant to echinocandins (Arendrup *et al.*, 2014).

Trichosporon species are a minor component of normal skin flora and are widely distributed in nature. They are regularly associated with the soft nodules of white piedra and have been involved in a variety of opportunistic infections in the immunosuppressed patient. Disseminated infections are most frequently (75%) caused by *T. asahii* (Arendrup *et al.*, 2014) and have been associated with leukaemia, organ transplantation, multiple myeloma, aplastic anaemia, lymphoma, solid tumours and AIDS. Disseminated infections are often fulminate and widespread, with lesions occurring in the liver, spleen, lungs and gastrointestinal tract. Infections in non-immunosuppressed patients include endophthalmitis after surgical extraction of cataracts, endocarditis usually following insertion of prosthetic cardiac valves, peritonitis in patients on continuous ambulatory peritoneal dialysis (CAPD), and intravenous drug abuse.

Note: Genus identification is mandatory for clinical management and should be performed and provided in a timely manner. Species identification remains difficult and requires molecular analysis or MALDI-ToF MS with an extensive database (Arendrup *et al.*, 2014).

Molecular identification: ITS and D1/D2 sequencing is required for accurate species identification (Arendrup *et al.*, 2014).

MALDI-ToF MS: A promising identification tool to accurately identify species (with an extensive database) (Kolecka *et al.*, 2013).

Comment: The API 20C yeast identification system is recommended for sugar assimilation tests.

References: Gueho *et al.* (1992); Rodriguez-Tudela *et al.* (2005); Chagas-Neto *et al.* (2008); Sugita (2011); Guo *et al.* (2011); Xiao *et al.* (2013); de Hoog *et al.* (2015).

102.1. *Trichosporon asahii* Akagi ex Sugita *et al.*

RG-2 organism

Morphological description: Colonies are white to cream coloured, powdery, suede-like to farinose with radial furrows and irregular folds. Budding cells and lateral conidia are absent. Arthroconidia are barrel-shaped (Fig. 102.1). Appressoria absent. This species assimilates L-arabinose but not melibiose. Growth at 37°C. Most common species, especially from invasive infections.

Physiological tests: (+ Positive, – Negative, v Variable, w Weak, s Slow. n not done).

Fermentation: Absent.

Growth reactions:

Glucose	+	Cellobiose	+	D-Mannitol	v
Sucrose	v	L-Sorbose	v	D-Glucitol	v
Raffinose	–	L-Rhamnose	+	*myo*-Inositol	v
Melibiose	–	D-Xylose	+	DL-Lactate	v
Galactose	+	L-Arabinose	+	D-Gluconate	+
Lactose	+	D-Arabinose	+	2-Keto-D-gluconate	+
Trehalose	+	D-Ribose	+	D-Glucosamine	+
Maltose	+	Glycerol	v	*N*-Acetyl-D-glucosamine	n
Melezitose	v	Erythritol	v	D-Glucuronate	+
Methyl-α-D-glucoside	+	Ribitol	v	0.1% Cycloheximide	+
Soluble starch	v	Galactitol	–	Growth at 37°C	+

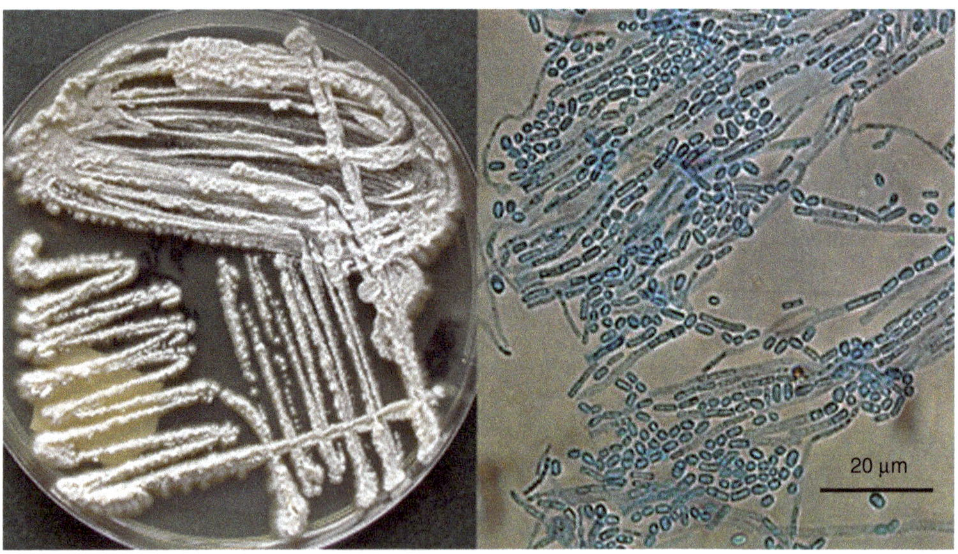

Fig. 102.1. *Trichosporon asahii* culture and arthroconidia.

Antifungal susceptibility: (Table 102.2).

Table 102.2. *Trichosporon asahii* (Australian national data); MIC µg/mL.

Antifungal	No.	≤0.016	0.03	0.06	0.125	0.25	0.5	1	2	4	8	16	32	≥64
AmB	46				4	4	14	7	15	2				
FLU	46							1	4	22	17	2		
ISAV	9			1	4	3		1						
VORI	44	2		12	20	9	1							
POSA	43			1	1	28	10	3						
ITRA	46				5	35	5	1						

102.2. *Trichosporon asteroides* (Rischin) Ota

RG-2 organism

Morphological description: Colonies are restricted, dry, cream coloured, cerebriform, with radial furrows and irregular folds. The meristematic form is punctiform, brownish and consists of hyphae which swell and become multiseptate which may fall apart into smaller packets. Budding cells and lateral conidia are absent. Arthroconidia are elongate and hyphae are often present. Appressoria absent. This species assimilates L-arabinose but not myo-inositol. Growth at 37°C is variable. Uncommon species usually associated with superficial infections.

Physiological tests: (+ Positive, – Negative, v Variable, w Weak, s Slow. n not done).

Fermentation: Absent.

Growth reactions:

Glucose	+	Cellobiose	+	D-Mannitol	-
Sucrose	+	L-Sorbose	+	D-Glucitol	v
Raffinose	-	L-Rhamnose	v	*myo*-Inositol	v
Melibiose	-	D-Xylose	+	DL-Lactate	+
Galactose	+	L-Arabinose	+	D-Gluconate	+
Lactose	+	D-Arabinose	v	2-Keto-D-gluconate	+
Trehalose	+	D-Ribose	+	D-Glucosamine	v
Maltose	+	Glycerol	+	*N*-Acetyl-D-glucosamine	n
Melezitose	+	Erythritol	+	D-Glucuronate	+
Methyl-α-D-glucoside	+	Ribitol	v	0.1% Cycloheximide	+
Soluble starch	+	Galactitol	-	Growth at 37°C	+

102.3. *Trichosporon inkin* (Oho ex Ota) do Carmo-Sousa & van Uden

RG-2 organism

Morphological description: Colonies are restricted, white, finely cerebriform with a granular covering, without marginal zone, often cracking the media (Fig. 102.2). Budding cells and lateral conidia absent. Arthroconidia are long cylindrical. Appressoria present in slide cultures. Sarcinae present on media with high sugar-content. This species assimilates myo-inositol but not melibiose and is tolerant to 0.01% (variable tolerance to 0.1%) cycloheximide. Growth at 37°C. Usually associated with white piedra on pubic hairs.

Fig. 102.2. *Trichosporon inkin* culture.

Physiological tests: (+ Positive, - Negative, v Variable, w Weak, s Slow. n not done).

Fermentation: Absent.

Growth reactions:

Glucose	+	Cellobiose	+	D-Mannitol	+
Sucrose	+	L-Sorbose	v	D-Glucitol	–
Raffinose	–	L-Rhamnose	–	*myo*-Inositol	+
Melibiose	–	D-Xylose	+	DL-Lactate	+
Galactose	+	L-Arabinose	–	D-Gluconate	+
Lactose	+	D-Arabinose	v	2-Keto-D-gluconate	+
Trehalose	+	D-Ribose	+	D-Glucosamine	–
Maltose	+	Glycerol	v	*N*-Acetyl-D-glucosamine	n
Melezitose	+	Erythritol	+	D-Glucuronate	+
Methyl-α-D-glucoside	+	Ribitol	–	0.1% Cycloheximide	–
Soluble starch	+	Galactitol	–	Growth at 42°C	+

Antifungal susceptibility: (Table 102.3).

Table 102.3. *Trichosporon inkin* (Australian national data); MIC µg/mL.

Antifungal	No.	≤0.016	0.03	0.06	0.125	0.25	0.5	1	2	4	8	16	32	≥64
AmB	5					2	1	1		1				
FLU	5							1	2	1	1			
VORI	5		1		2	2								
POSA	5				1	1	2	1						
ITRA	5				1	2	2							

102.4. *Trichosporon ovoides* Behrend

RG-2 organism

Morphological description: Colonies are restricted, white, granular, folded at the centre, with a flat marginal zone. Budding cells and lateral conidia absent. Arthroconidia are cylindrical. Appressoria present in slide cultures. This species does not assimilate melibiose but tolerates 0.1% cycloheximide. Growth at 37°C is variable. Uncommon species usually associated with superficial infections, like white piedra.

Physiological tests: (+ Positive, – Negative, v Variable, w Weak, s Slow. n not done).

Fermentation: Absent.

Growth reactions:

Glucose	+	Cellobiose	+	D-Mannitol	+
Sucrose	+	L-Sorbose	v	D-Glucitol	v
Raffinose	v	L-Rhamnose	+	*myo*-Inositol	+
Melibiose	–	D-Xylose	+	DL-Lactate	+
Galactose	+	L-Arabinose	v	D-Gluconate	+
Lactose	+	D-Arabinose	v	2-Keto-D-gluconate	+
Trehalose	+	D-Ribose	+	D-Glucosamine	v
Maltose	+	Glycerol	v	*N*-Acetyl-D-glucosamine	n
Melezitose	v	Erythritol	v	D-Glucuronate	+
Methyl-α-D-glucoside	+	Ribitol	–	0.1% Cycloheximide	+
Soluble starch	+	Galactitol	–	Growth at 37°C	v

DOI: 10.1079/9781800622340.0102

103 *Trichothecium roseum* (Persoon) Link

Trichothecium roseum has a worldwide distribution and is often isolated from decaying plant substrates, soil, seeds of corn, and food-stuffs (especially flour products). It is occasionally isolated as a saprophyte in the clinical laboratory.

RG-1 organism

Morphological description: Colonies are moderately fast growing, flat, suede-like to powdery, initially white but becoming rosy-pink or orange with age. The conidiophores are indistinguishable from the vegetative hyphae until the first conidium is produced. They are erect, unbranched, often septate near the base, more or less rough-walled, bearing basipetal zig-zag (alternating) chains of conidia at the apex. **Note:** The conidiophore is progressively shortened with the formation of each conidium i.e. retrogressive conidial development. Conidia are two-celled ellipsoidal to pyriform, with an obliquely truncate basal scar, hyaline, smooth to delicately roughened and thick-walled (Fig. 103.1).

Comment: *Trichothecium roseum* should not be confused with *Nannizzia nana*. Colonies of the latter may be pinkish-buff in colour and also produce ovoid to pear-shaped, mostly two-celled macroconidia with thin, verrucose walls. However, *N. nana* usually produces a red-brown reverse pigment, and the two-celled macroconidia are sessile and formed singly, they are not produced in basipetal chains as in *T. roseum*.

Molecular identification: Summerbell *et al.* (2011) revised the genus using D1/D2 sequences for phylogenetic analysis and sequence-based identification. ITS sequencing may not differentiate *Trichothecium* from *Fusarium* and other species of the Hypocreales (Inacio *et al.*, 2011).

Key features: Basipetal, zig-zag chains of two-celled conidia showing with retrogressive development where the conidiophore becomes progressively shorter.

References: McGinnis (1980); Rippon (1988); Samson *et al.* (1995); Domsch *et al.* (2007); Summerbell *et al.* (2011).

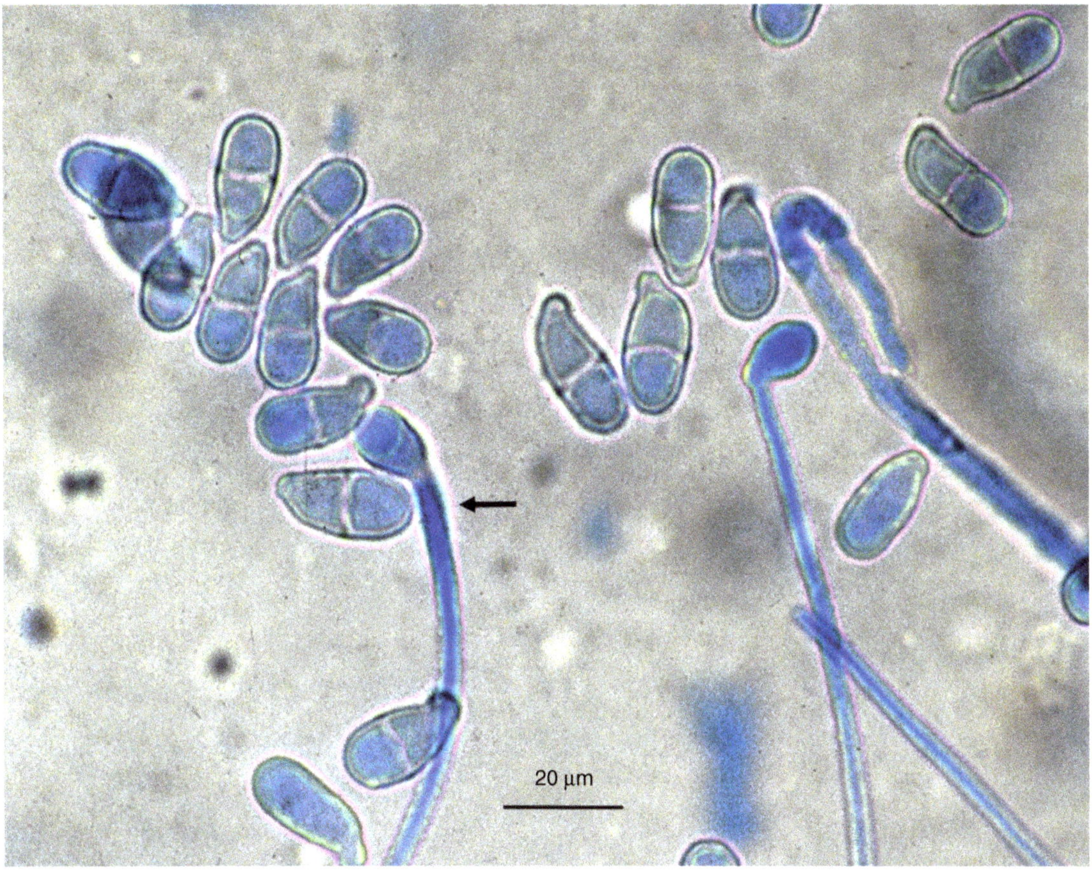

Fig. 103.1. *Trichothecium roseum* conidiophores showing retrogressive conidial development. **Note:** The long conidiophores that progressively shortened with the formation of each conidium (arrow).

DOI: 10.1079/9781800622340.0103

104 *Ulocladium* Preuss

Ulocladium is a ubiquitous genus containing common saprophytes from soil, wood and decaying plants. Some species are plant pathogens and food spoilers. Human infections are rare, although a human case of keratitis has been reported (Badenoch *et al.*, 2006).

RG-1 organism

Morphological description: Colonies are rapid growing, brown to olivaceous-black or greyish and suede-like to floccose. Microscopically, numerous, usually solitary, multi-celled conidia (dictyoconidia) are formed through a pore (poroconidia) by a sympodially elongating geniculate conidiophore (Fig. 104.1). Conidia are typically obovoid (narrowest at the base), dark brown and often rough-walled. Seven species have been described, all being saprophytes.

Molecular identification: ITS sequencing is sufficient for genus identification to genus level (Badenoch *et al.*, 2006; Woudenberg *et al.*, 2013).

Antifungal susceptibility: (Table 104.1).

Table 104.1. *Ulocladium* spp. limited data (Australian national data); MIC µg/mL.

Antifungal	No.	≤0.016	0.03	0.06	0.125	0.25	0.5	1	2	4	8	16	32	≥64
AmB	2								1	1				
VORI	2					1					1			
POSA	1						1							
ITRA	2				1			1						

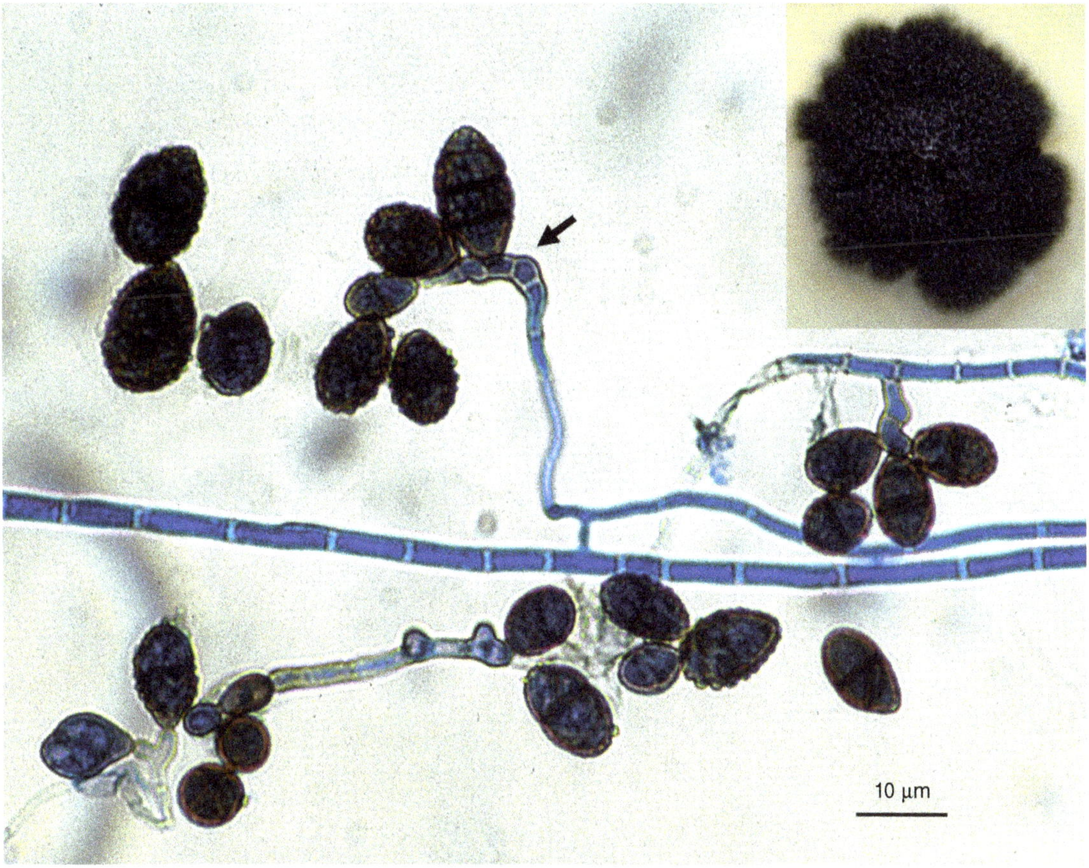

Fig. 104.1. *Ulocladium* spp. culture, conidiophore and conidia. **Note:** Sympodially elongating geniculate conidiophore (arrow).

References: Ellis (1971, 1976); Rippon (1988); Samson *et al.* (1995); Domsch *et al.* (2007); de Hoog *et al.* (2015).

DOI: 10.1079/9781800622340.0104

105 *Veronaea botryosa* Ciferri & Montemartini

This genus is very similar to *Rhinocladiella*; however, the conidia are typically two-celled. Occasional human infections have been reported.

RG-2 organism

Morphological description: Colonies grow rapidly and are suede-like to downy, greyish-brown to blackish-brown. Conidiophores are erect, straight or flexuose, occasionally branched and are usually geniculate, due to the sympodial development of the conidia (Fig. 105.1). They are smooth-walled, pale to medium olivaceous-brown, up to 250 µm long and 2-4 µm wide. Conidia are pale brown, two-celled, cylindrical with a truncated base, smooth-walled or slightly verrucose, 5-12 x 3-4 µm.

Molecular identification: Arzanlou *et al.* (2007) used D1/D2 and ITS sequence data in a phylogenetic revision.

Antifungal susceptibility: (Table 105.1).

Table 105.1. *Veronaea botryosa* limited data (Australian national data); MIC µg/mL.

Antifungal	No.	≤0.016	0.03	0.06	0.125	0.25	0.5	1	2	4	8	16	32	≥64
AmB	2										2			
VORI	2									2				
POSA	2					1	1							
ITRA	2						1		1					

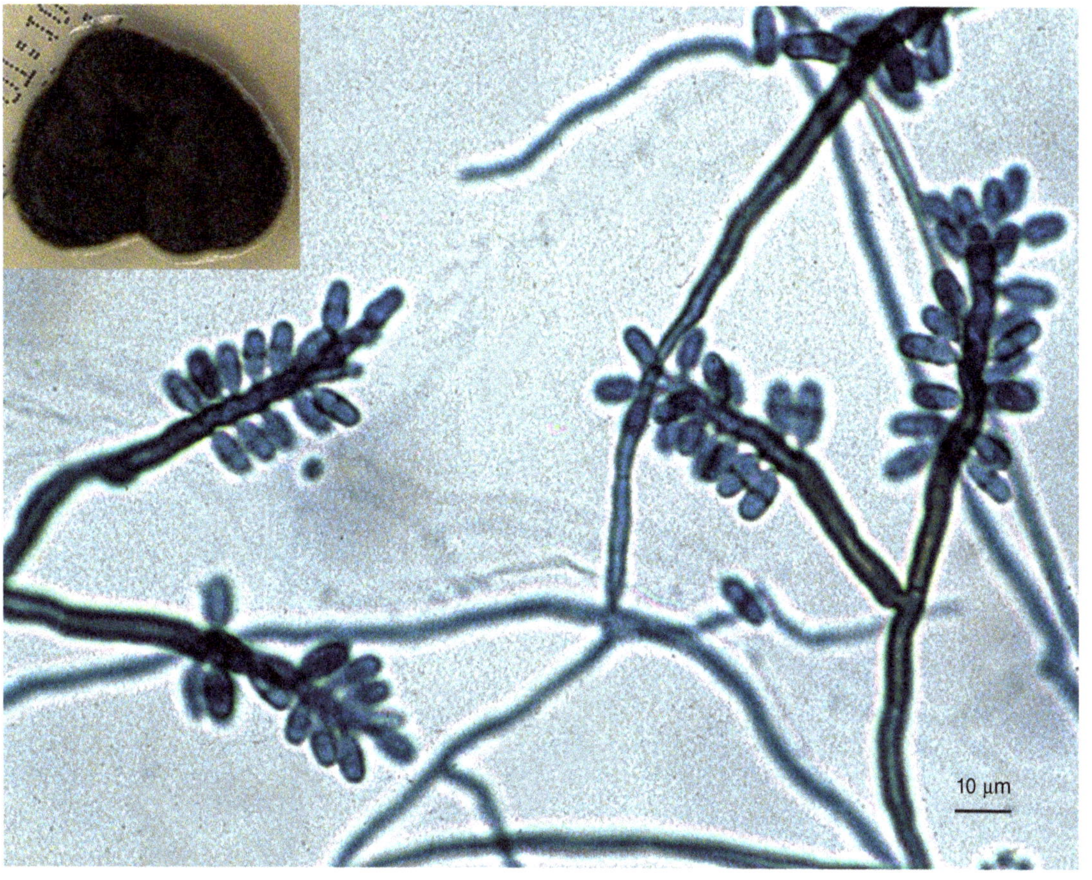

Fig. 105.1. *Veronaea botryosa* culture, conidiophores and sympodial development of conidia.

References: Ellis (1971); Revankar and Sutton (2010); Badali *et al.* (2013); de Hoog *et al.* (2020).

DOI: 10.1079/9781800622340.0105

106 *Verruconis gallopava* (W.B. Cooke) Samerpitak & de Hoog

Synonymy: *Ochroconis gallopava* (W.B. Cooke) de Hoog.

Verruconis species are thermophilic, with *Verruconis gallopava* occurring in hot environments, such as thermal soils, broiler house litter, hot springs and self-heated waste (Samerpitak *et al.*, 2014). *Verruconis gallopava* is neurotropic and is a recognised agent of human brain infections, as well as in poultry, wild birds, dogs and cats (Seyedmousavi *et al.*, 2014). Occasional human pulmonary infections in immunocompetent hosts have also been reported (Giraldo *et al.*, 2014b; Samerpitak *et al.*, 2014; Seyedmousavi *et al.*, 2014).

RG-2 organism

Morphological description: Colonies are smooth to suede-like, dry, flat, tobacco-brown to brownish-black with a dark brown diffusible pigment (Fig. 106.1). Hyphae are brown with relatively thick walls. Conidiophores are mostly cylindrical to acicular, sometimes poorly differentiated, bearing a few conidia at the tip. Conidia are two-celled, subhyaline to pale brown, smooth-walled to verrucose, cylindrical to clavate, constricted at the septum, 11-18 x 2.5-4.5 µm in size, with the apical cell wider than the basal cell (Fig. 106.1). A remnant of a denticle may also be seen at the conidial base. Optimum growth at 35°C, tolerant to 40°C.

Molecular identification: ITS sequencing can identify species. Additional loci include β-tubulin, *ACT* and the D1/D2 region (Giraldo *et al.*, 2014b; Seyedmousavi *et al.*, 2014).

Antifungal susceptibility: (Table 106.1).

Table 106.1. *Verruconis gallopava* (Australian national data); MIC µg/mL.

Antifungal	No.	≤0.016	0.03	0.06	0.125	0.25	0.5	1	2	4	8	16	32	≥64
AmB	20				1	3	4	8	4					
ISAV	3						1				2			
VORI	20				2	4	4	6	2		2			
POSA	19		4	2	6	6		1						
ITRA	20		3		3	11	3							

References: McGinnis (1980); de Hoog (1983); Domsch *et al.* (2007); Giraldo *et al.* (2014b); Samerpitak *et al.* (2014); Seyedmousavi *et al.* (2014).

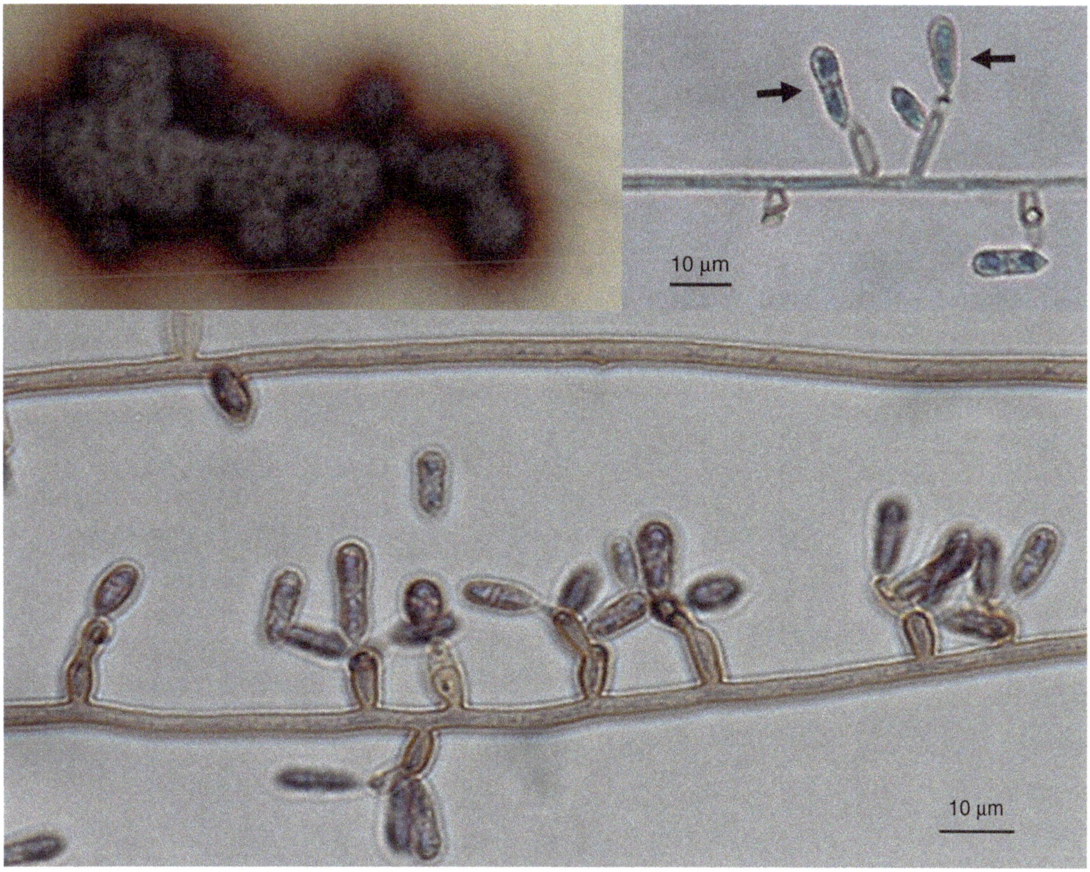

Fig. 106.1. *Verruconis gallopava* culture, conidiophores and conidia. **Note:** Conidia are two-celled, constricted at the septum (arrows), with the apical cell wider than the basal cell and a remnant of a denticle may also be seen at the conidial base.

DOI: 10.1079/9781800622340.0106

107 *Verticillium* Nees ex Link

Members of this genus are often isolated from the environment. *Verticillium* wilt is a fungal disease of many economically important plants, especially tomatoes. *Verticillium dahliae*, *V. albo-atrum*, *V. longisporum*, *V. nubilum*, *V. tricorpus* and *Musicilliium theobromae* (formerly *V. theobromae*) and are the principal plant pathogens. *Verticillium* has been reported as a rare agent of mycotic keratitis in humans.

RG-1 organism

Morphological description: Colonies are fast growing, suede-like to downy, white to pale yellow in colour, becoming pinkish-brown, red, green, or yellow with a colourless, yellow, or reddish-brown reverse. Conidiophores are usually well differentiated and erect, verticillately branched over most of their length, bearing whorls of slender awl-shaped divergent phialides (Fig. 107.1). Conidia are hyaline or brightly coloured, mostly one-celled, and are usually borne in slimy heads (glioconidia).

Molecular identification: ITS, *ACT*, *TEF-1 GAPDH* and tryptophan synthase genes have been used to identify all recognised *Verticillium* species (Inderbitzin *et al.*, 2013).

Key features: Verticillate branched conidiophores bearing whorls of awl-shaped, divergent phialides.

References: McGinnis (1980); Rippon (1988); Samson *et al.* (1995); Domsch *et al.* (2007); de Hoog *et al.* (2015).

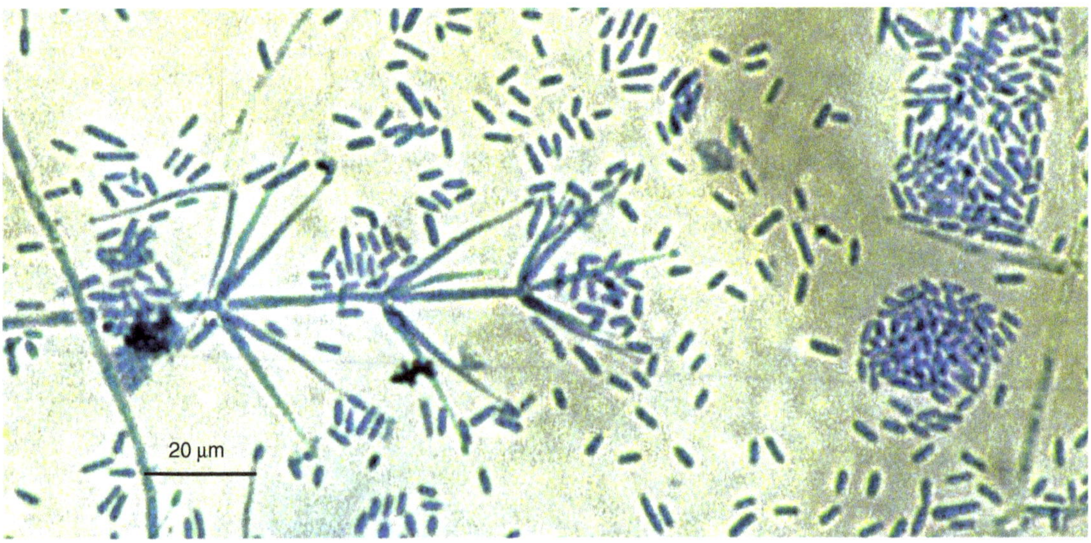

Fig. 107.1. *Verticillium* spp. microscopy showing verticillately branched conidiophores bearing whorls of slender awl-shaped divergent phialides and conidia.

DOI: 10.1079/9781800622340.0107

108 *Wickerhamomyces anomalus* (E.C. Hansen) Kurtzman *et al.*

Synonymy: *Candida pelliculosa* Redaelli.

Wickerhamomyces anomalus is widespread in nature. It has been reported from cases of candidaemia and catheter-related infections in humans and has been isolated from soil, grains, fruit and warm-blooded animals.

RG-1 organism

Culture: Colonies (SDA) white to cream coloured, smooth, glabrous, yeast-like.

Microscopy: Spherical to ellipsoidal budding blastoconidia, 2-4 x 2-6 µm. Pseudohyphae may be present. Asci when present, contain one to four hat-shaped ascospores.

India ink preparation: Negative - no capsules present.

Dalmau plate culture: Spherical to ellipsoidal budding yeast cells and abundant pseudohyphae in most strains.

Molecular identification: ITS sequencing is recommended for species identification.

MALDI-ToF MS: Able to accurately identify this species.

Germ tube formation: Negative.

Physiological tests: (+ Positive, - Negative, v Variable, w Weak, s Slow, n not done).

Fermentation:

Glucose	+	Sucrose	+	Lactose	-
Galactose	v	Maltose	v	Trehalose	-

Growth reactions:

Glucose	+	L-Sorbose	–	*myo*-Inositol	–
Sucrose	+	L-Rhamnose	–	DL-Lactate	+
Raffinose	+	D-Xylose	v	D-Gluconate	v
Melibiose	-	L-Arabinose	v	2-Keto-D-gluconate	–
Galactose	v	D-Arabinose	–	D-Glucosamine	–
Lactose	–	D-Ribose	v	*N*-Acetyl-D-glucosamine	–
Trehalose	+	Glycerol	+	D-Glucuronate	n
Maltose	+	Erythritol	+	Nitrate	+
Melezitose	+	Ribitol	v	Urease	–
Methyl-α-D-glucoside	+	Galactitol	–	0.1% Cycloheximide	–
Soluble starch	+	D-Mannitol	+	Growth at 30°C	+
Cellobiose	+	D-Glucitol	+	Growth at 37°C	v

Key features: Germ tube negative yeast and sugar assimilation pattern.

Antifungal susceptibility: (Table 108.1).

Table 108.1. *Wickerhamomyces anomalus* (Diekema *et al.*, 2009; and Australian national data); MIC µg/mL.

Antifungal	No.	≤0.016	0.03	0.06	0.125	0.25	0.5	1	2	4	8	16	32	≥64
AmB	44				3	14	21	5	1					
FLU	44							3	10	24	7			
ISAV	2			2										
VORI	44	1	1	1	25	13	3							
POSA	44			1	2	8	5	14	12	2				
ITRA	4				1	3								
ANID	18	15	2	1										
MICA	18	6	10	2										
5FC	4			2	2									

References: Kurtzman (2011e).

DOI: 10.1079/9781800622340.0108

109 *Yarrowia lipolytica* (Wickerham *et al.*) van der Walt & von Arx.

Synonymy: *Candida lipolytica* (F.C. Harrison) Diddens & Lodder.

Yarrowia lipolytica is widespread in nature and is commonly found on substrates high in lipids. It is a rare cause of candidaemia.

RG-1 organism

Culture: Colonies (SDA) white to cream coloured, smooth, glabrous, yeast-like.

Microscopy: Spherical, ellipsoidal to elongate budding blastoconidia, 3-5 x 3-15 µm.

India ink preparation: Negative - no capsules present.

Dalmau plate culture: Pseudohyphae and true hyphae are produced.

Germ tube formation: Negative.

Physiological tests: (+ Positive, – Negative, v Variable, w Weak, s Slow, n not done).

Fermentation: Absent.

Growth reactions;

Glucose	+	L-Sorbose	v	*myo*-Inositol	–
Sucrose	–	L-Rhamnose	–	DL-Lactate	+
Raffinose	–	D-Xylose	–	D-Gluconate	v
Melibiose	–	L-Arabinose	–	2-Keto-D-gluconate	–
Galactose	v	D-Arabinose	–	D-Glucosamine	–
Lactose	–	D-Ribose	v	*N*-Acetyl-D-glucosamine	+
Trehalose	–	Glycerol	+	D-Glucuronate	n
Maltose	–	Erythritol	+	Nitrate	–
Melezitose	–	Ribitol	v	Urease	–
Methyl-α-D-glucoside	–	Galactitol	–	0.1% Cycloheximide	–
Soluble starch	–	D-Mannitol	+	Growth at 35°C	v
Cellobiose	w,-	D-Glucitol	+	Growth at 37°C	–

Molecular identification: ITS sequencing recommended.

MALDI-ToF MS: Able to accurately identify this species.

Key features: Germ tube negative yeast and sugar assimilation pattern.

Antifungal susceptibility: (Table 109.1).

Table 109.1. *Yarrowia lipolytica* limited data (Diekema *et al.*, 2009; and Australian national data); MIC µg/mL.

Antifungal	No.	≤0.016	0.03	0.06	0.125	0.25	0.5	1	2	4	8	16	32	≥64
AmB	20				1		1	8	5	4	1			
FLU	20						1	1	10	6	2			
VORI	20	1	8	8	2			1						
POSA	20				2	3	10	4		1				
ITRA	4				2	2								
ANID	14				2	4	5	2	1					
MICA	14					7	5	2						
5FC	4								1		1	2		

References: Kurtzman (2011f).

DOI: 10.1079/9781800622340.0109

110 Microscopy stains and techniques

110.1. KOH with Calcofluor white or Blankophor BA

For the direct microscopic examination of skin scrapings, hairs, nails and other clinical specimens for fungal elements. This is a very rapid and sensitive method, however a fluorescence microscope fitted with filters to give an excitation with ultraviolet light below 400 nm wavelength is required. Calcofluor white (M2R powder, Polysciences Inc.) or Blankophor BA (Indulor Leverkusen GmbH & Co. KG) are used as whitening agents by the paper industry and selectively bind to cellulose and chitin. The dye fluoresces as it is exposed to ultraviolet light (Fig. 110.1).

Prepare two solutions: (A) Dissolve 10 g of KOH in 90 mL distilled water and then add 10 mL of glycerol; (B) Dissolve 0.5 g of Calcofluor white powder in 50 mL of distilled water by gentle heating. Store this solution in a light proof glass bottle; some precipitation will occur over time. Evans blue (0.02 g) may be added to solution B to enhance contrast.

Method for making microscopic mounts: Mix one drop of each solution on the centre of a clean microscope slide. Place the specimen in the solution and cover with a coverslip, squash the preparation with the butt of the inoculation needle and then blot off the excess fluid. Gently heat the slide and examine microscopically for the presence of fungal elements that fluoresce a chalk-white or brilliant apple-green colour, depending on the filters used.

References: Hageage and Harrington (1984); Harrington and Hageage (2003).

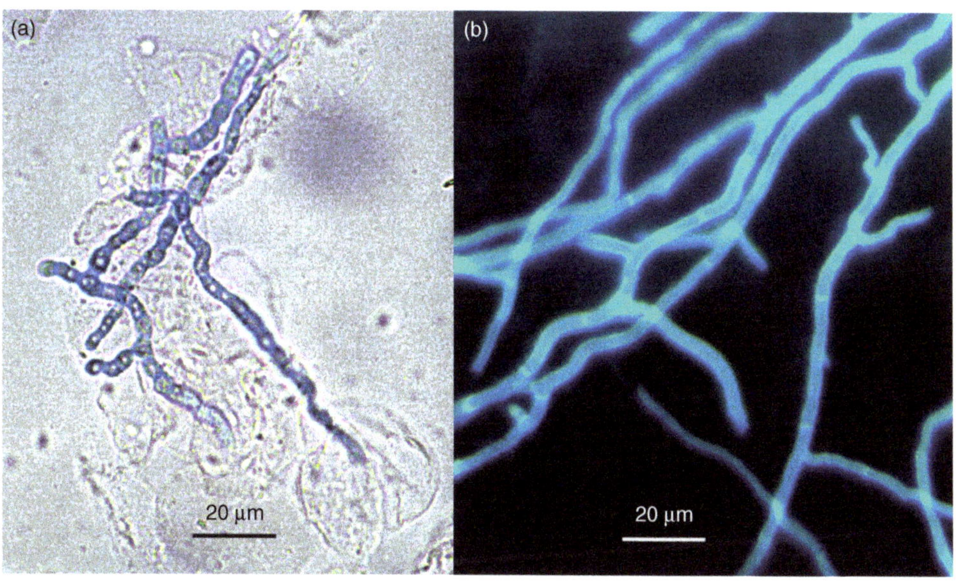

Fig. 110.1. Direct microscopy of fungal hyphae (a) KOH Parker Quink ink stain and (b) KOH with calcofluor white showing fungal hyphae that fluoresce a chalk-white colour.

110.2. KOH with Chlorazol Black

For the direct microscopic examination of skin scrapings, hairs, nails and other clinical specimens for fungal elements. **Note:** Parker Quink permanent ink with solvex is no longer available, fluorescence microscopy using Calcofluor white (or Blankophor BA) with 10% KOH is now the method of choice for direct microscopy of skin, hair and nail samples.

Method: Dissolve 10 g of potassium hydroxide (KOH) in 80 mL of distilled water and add 10 mL of glycerol and 10 mL of Chlorazol Black E (Polysciences, Inc). The glycerol prevents crystallisation of the reagent and prevents the specimen from drying out.

Method for making direct microscopy mounts: Mount a drop or small portion of the specimen in a drop of KOH-based stain, cover with a coverslip, squash the preparation with the butt of the inoculation needle and then blot off the excess fluid. **Note:** Specimens may take from 20 minutes to several hours for nail clippings for the specimen to clear so that fungal elements can be detected. Larger pieces of specimen such as nail clippings or fibrous tissue, should be digested in 10% KOH overnight prior to mounting on the slide. Negative specimens should be kept and re-examined the next day to avoid reporting false negative results due to delayed clearance and staining of the specimen.

110.3. India ink mounts

For the direct microscopic examination of CSF or bronchoalveolar lavage for *Cryptococcus neoformans* and *C. gattii*. India ink is a negative stain that provides a black, semi opaque background to enable the capsule of the yeast to be clearly visible.
Method: Place a drop of the specimen on a glass slide, followed by a drop of India ink, mix well with a loop, and cover with a coverslip. **Note:** The best quality brands are "Pelikan" or "Talens" India ink, which is what is used by tattooists. The slide should be viewed immediately by bright field (not phase contrast) microscopy. Beware of artefacts such as white blood cells that can look similar to cryptococcal cells.

110.4. Lactophenol cotton blue (LPCB)

For the staining and microscopic identification of fungi. This stain contains phenol which will kill the fungus and lactic acid that acts as a clearing agent and helps preserve the fungal structures. Cotton blue is an aniline dye that stains the chitin in the fungal cell walls, which adds colour to the fungal preparation thereby enhancing and contrasting the structures. The stain can be purchased commercially, or it can be prepared in the laboratory, although this requires two days.

Method: This stain is prepared over two days.

On the first day, dissolve 0.05 g of Cotton blue (Aniline blue) in 20 mL distilled water. Leave overnight to eliminate insoluble dye.

On the second day, wearing gloves add 20 g of phenol crystals ($C_6H_5O_4$) to 20 mL of lactic acid ($CH_3CHOH\ COOH$) in a glass beaker and place on a magnetic stirrer until the phenol is dissolved. Then add 40 mL of glycerol.

Filter the Cotton blue and distilled water solution into the phenol/glycerol/lactic acid solution. Mix and store at room temperature.

110.5. Microscopic mounts and squash preparations

Using sterile technique, remove a small portion of the colony with an inoculation needle and mount in a drop of lactophenol cotton blue on a clean microscope slide. Cover with a coverslip, squash the preparation with the butt of the inoculation needle and then blot off the excess fluid. **Note:** Cellotape flag preparations or slide culture preparations are superior methods for the examination of fungal structures.

110.6. Cellotape flag preparations

An excellent technique for the rapid mounting of sporulating fungi because it keeps more of the reproductive structures intact (Fig. 110.2).

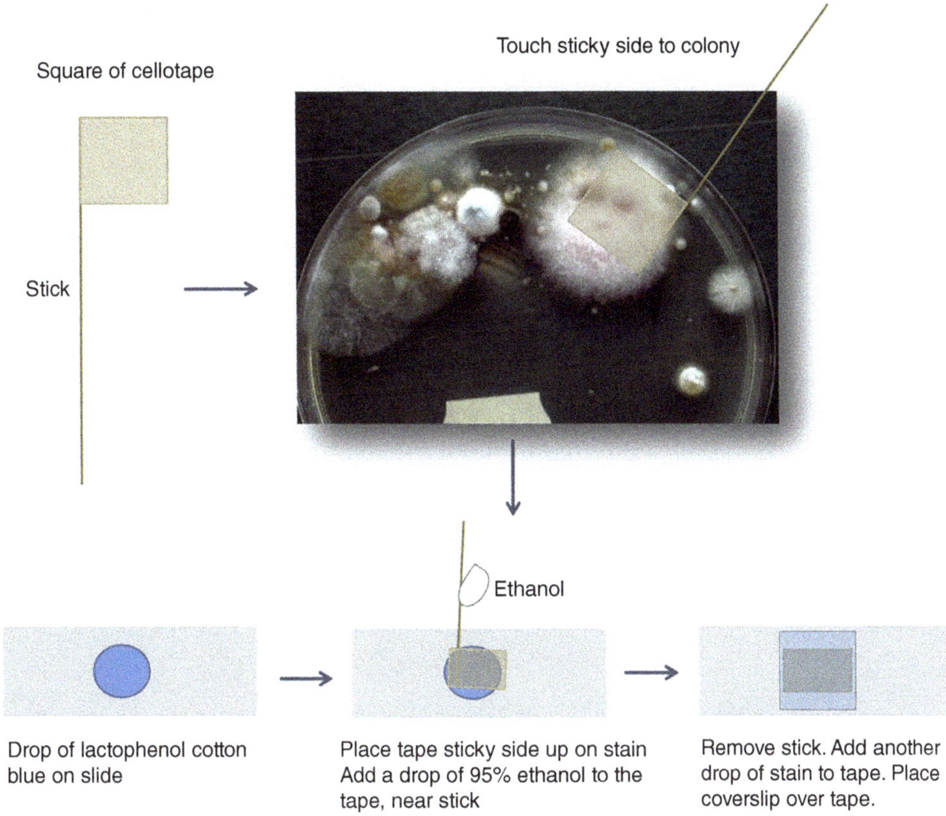

Fig. 110.2. Diagrammatic instruction for making a cellotape mount.

Method: Using clear 2 cm wide cellotape and a wooden applicator stick (orange stick) make a small cellotape flag (2 x 2 cm). Gently press the sticky side of the flag onto the surface of the culture. Remove and apply a drop of 95% alcohol to the sticky side of the flag, this acts as a wetting agent and dissolves the adhesive glue holding the flag to the applicator stick. Place the flag onto a small drop of lactophenol cotton blue on a clean glass slide (sticky side up or down, doesn't matter), remove the applicator stick and add another drop of stain. Then cover with a coverslip and gently press, mopping up any excess stain. **Note:** Best viewed within 2 hours; flag mounts only last for a few days before the phenol/lactic acid destroys the cellotape.

110.7. Slide culture preparations

To accurately identify many fungi, it is essential to observe the precise arrangement of the conidiophores and the way in which the conidia are produced (conidial ontogeny).

Riddell's (1950) simple method of slide culturing permits fungi to be studied virtually *in situ* with as little disturbance as possible (Fig. 110.3). A simple modification of this method using a single agar plate is described below (Fig. 110.4).

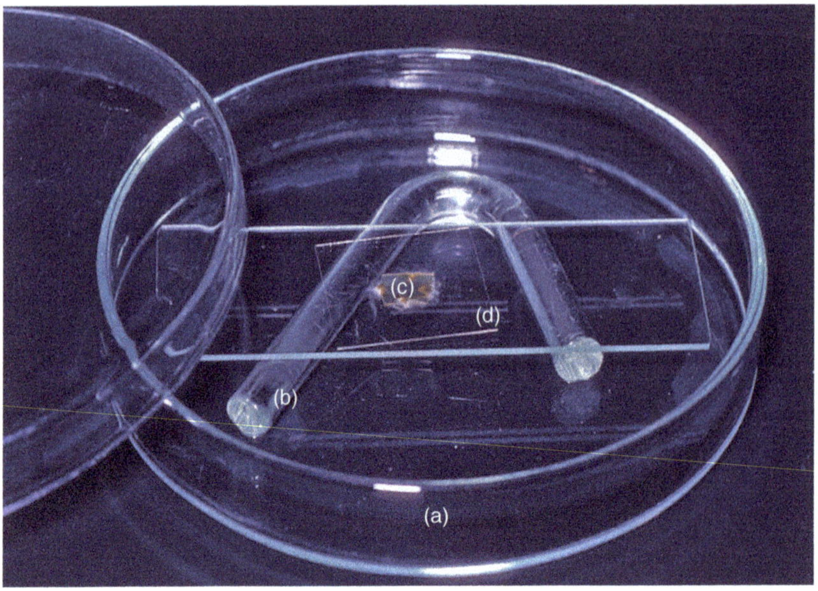

Fig. 110.3. Riddell's (1950) slide culturing method required using an agar block of media transferred to a glass slide and placed in a moist chamber (a) petri dish, (b) bent glass rod to support the slide above some water, (c) agar square cut out and inoculated on 4 sides, (d) cover slip.

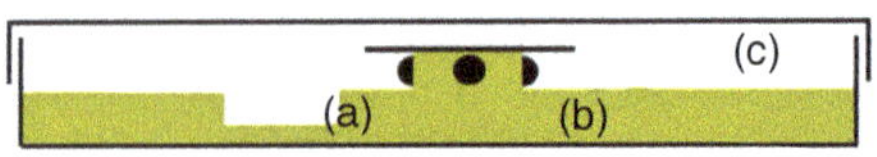

Fig. 110.4. A simple slide culture method using a single agar plate (a) square agar block is cut out and flipped up onto the agar surface, (b) all 4 sides are inoculated with the mould and (c) a coverslip is placed on top.

Method: One plate potato dextrose is recommended; however, some fastidious fungi may require less nutrient-rich media to induce sporulation like cornmeal agar or Czapek dox agar.

Using a sterile blade cut out an agar block (7 x 7 mm) small enough to fit under a coverslip. Flip the block up onto the surface of the agar plate. Inoculate the four sides of the agar block with spores or mycelial fragments of the fungus. Place a flamed coverslip centrally upon the agar block. Incubate the plate at 26°C until growth and sporulation have occurred.

Remove the coverslip from the agar block. Apply a drop of 95% alcohol as a wetting agent. Gently lower the coverslip onto a small drop of lactophenol cotton blue on a clean glass slide. The slide can be left overnight to dry and later sealed with nail polish. When sealing with nail polish use a coat of clear polish followed by one coat of coloured polish.

DOI: 10.1079/9781800622340.0110

111 Specialised culture media

111.1. Bird seed agar

For the selective isolation of *Cryptococcus neoformans* and *C. gattii*.

Ingredients: *Guizotia abyssinica* (niger seed) 50 g; Glucose 1 g; KH_2PO_4 (potassium dihydrogen orthophosphate) 1 g; Creatinine 1 g; Bacto-agar (BD) 15 g; Distilled water 1000 mL.

Additives: To each 500 mL bottle add Penicillin G (20 units/mL) 1 mL; Gentamicin (40 mg/mL) 1 mL.

Method: Grind seeds of *Guizotia abyssinica* as finely as possible with an electric grinder and add to 1000 mL distilled water in a stainless-steel jug. Boil for 30 minutes, pass through filter paper and adjust volume to 1000 mL. Add remaining ingredients except Bacto-agar to filtrate and dissolve. Cool to room temperature and pH to 5.5. Dispense into 500 mL bottles. Add 7.5 g Bacto-agar to each 500 mL reagent bottle. Autoclave at 121°C for 20 minutes. Cool to 48°C and add 0.5 mL Penicillin G and 0.5 mL Gentamicin to each 500 mL bottles of agar. Mix gently and pour into 90 mm plastic petri dishes.

111.2. Bromocresol purple milk solids glucose agar (BCP)

For the differentiation of *Trichophyton* species (Fischer and Kane 1971; Summerbell *et al.*,1988).

Solution A: Distilled water 1000 mL; skim milk powder (Carnation brand is best) 80 g; bromcresol (or bromocresol) purple (1.6% solution in ethanol) 2 mL. Dissolve in 2 litre flask and autoclave 121°C for 10 minutes.

Solution B: Glucose 40 g; distilled water 200 mL. Dissolve and autoclave at 121°C for 10 minutes.

Solution C: Bacto-agar 30 g; distilled water 800 mL. Soak for 15 minutes in 3 litre flask and autoclave at 121°C for 10 minutes.

Final mixture: To make media add solution A and B to solution C. Adjust final pH to 6.6. Aseptically dispense for slopes.

111.3. Creatinine dextrose bromothymol blue thymine (CDBT) agar

For differentiation of *C. neoformans* serotypes A and D (Irokanulo *et al.*, 1994).

Solution A: Creatinine 1 g; dextrose 0.5 g; KH_2PO_4 1 g; $MgSO_4$ $7H_2O$ (epsomite) 0.5 g; thymine 0.1 g; distilled water 980 mL. Dissolve ingredients in small beaker and adjust pH to 5.6. Store at 4°C.

Solution B (aqueous bromothymol blue): Bromothymol blue 0.4 g; 0.01N NaOH 64 mL; distilled water 36 mL. Dissolve the bromothymol blue in the NaOH and add to the water.

Final mixture: To prepare a litre of medium, mix 980 mL of solution A with 20 mL of solution B. Add 20 g of Bacto-agar and autoclave to 121°C for 15 minutes, cool to 48°C and dispense as plates.

111.4. Canavanine glycine bromothymol blue (CGB) agar

For differentiation of *C. neoformans* and *C. gattii* (Kwon-Chung *et al.*, 1982).

Solution A: Glycine (Univar) 10 g; KH_2PO_4 1 g; $MgSO_4$ 1 g; thiamine HCl 1 mg; L-canavanine sulfate 30 mg; distilled water 100 mL. Dissolve ingredients in small beaker and adjust pH to 5.6. Filter sterilise solution using 0.22 µm filter. Store at 4°C.

Solution B (aqueous bromothymol blue): Bromothymol blue 0.4 g; 0.01N NaOH 64 mL; distilled water 36 mL. Dissolve the bromothymol blue in the NaOH and add to the water.

Final mixture: To prepare a litre of the medium, mix 880 mL of distilled water, 20 mL of solution B, and 20 g of Bacto-agar (BD 214010) in a large flask. Autoclave to 121°C for 15 minutes and cool to 48°C. Add 100 mL of solution A and stir thoroughly before pouring into petri dishes or dispensing aseptically into tubes to make slopes. A positive test with CGB medium indicates a change from pH from 5.8 (greenish-yellow) to at least 7.0 (cobalt blue).

111.5. Cornmeal agar

For the routine cultivation and identification of fungi.

Ingredients: Cornmeal agar 8.5 g; distilled water 500 mL.

Method: Mix dry ingredients into 100 mL H_2O, boil remaining water. Add boiling water to mixture and bring to boil. Dispense for slopes as required. Autoclave at 121°C for 10 minutes, remove and slope or pour for plates as required.

111.6. Cornmeal glucose sucrose yeast extract agar

To stimulate sporulation in mucormycetous fungi, especially *Saksenaea* and *Apophysomyces*.

Ingredients: Cornmeal agar 17 g; dextrose (glucose) 2 g; sucrose 3 g; yeast extract 1 g; distilled water 1000 mL.

Method: Soak dry ingredients in 100 mL H_2O, then boil remaining water. Add boiling water to mixture and bring to the boil. Dispense for slopes as required. Autoclave at 121°C for 10 minutes, remove and slope or pour for plates as required.

111.7. Czapek dox agar

For the routine cultivation of fungi, especially *Aspergillus, Penicillium* and non-sporulating moulds.

Ingredients: Czapek dox agar (Oxoid CM97) 45.4 g; distilled water 1000 mL.

Method: Soak the ingredients in a small amount of water. Bring remaining water to boil, add to soaking ingredients and bring to the boil again, stirring continuously. Dispense for slopes as required. Autoclave at 121°C for 10 minutes, remove and slope or pour for plates as required.

111.8. Modified Dixon's agar

For primary isolation and cultivation of *Malassezia* species.

Ingredients: Malt extract 9 g; Bacto Tryptone 1.5 g; ox-bile desiccated 5 g; Tween 40 2.5 mL; oleic acid 0.5 g; glycerol 0.5 mL; Bacto-agar 3 g; distilled water 250 mL.

Method: Soak ingredients for 15 minutes in some water. Bring remaining water to boil, add to other ingredients. Bring to boil once more. Dispense for slopes. Autoclave at 121°C for 10 minutes and then slope.

111.9. Hair perforation test for dermatophytes

To distinguish between dermatophytes species, particularly *Trichophyton mentagrophytes* and its variants.

Ingredients: Autoclaved blonde prepubertal hair that has been cut into 1 cm pieces. Sterile distilled water 5 mL in a suitable vial. **Note:** Perforations are easiest to see on blonde hair, and prepubertal hair has no fatty acids that might inhibit fungal growth.

Method: Place hair in water in vial. Inoculate with small fragments of the test fungus. Incubate at room temperature.

Individual hairs are removed at intervals up to 4 weeks and examined microscopically in lactophenol cotton blue. Isolates of *T. mentagrophytes* produce marked localised areas of pitting and marked erosion, whereas those of *T. rubrum* do not.

111.10. Lactritmel agar

For enhanced sporulation and the production of pigment by dermatophytes, especially *Trichophyton rubrum*.

Ingredients: Skimmed milk powder 7 g; honey 10 g; Cornmeal agar 17 g; Chloramphenicol 0.05 g; distilled water 1000 mL.

Method: Weigh skimmed milk powder into stainless-steel jug. Slowly add some water, mixing milk into smooth paste. Continue adding small quantities of water until powder is dissolved (about 150mL). Weigh other ingredients into skimmed milk and allow to soak. Boil remaining water, and with it wash out honey from beaker. Add to other ingredients and boil. Dispense for slopes if required. Autoclave for 10 minutes at 121°C. Dispense for plates or slopes as required. **Note:** Do not filter or adjust pH.

111.11. Littman oxgall agar

A confirmatory test for the identification of *Trichophyton* species.

Ingredients: Littman oxgall agar 27.5 g; distilled water 500 mL.

Method: Soak agar in 100 mL of water in stainless steel jug. Boil remaining 400 mL in a separate jug. When water has boiled add to soaking agar and bring to the boil, stirring constantly. Dispense for slopes. Autoclave for 10 minutes at 121°C, remove and slope.

111.12. Malt extract agar

For the routine cultivation and identification of fungi.

Ingredients: Malt extract 20 g; Bacto-agar 20 g; distilled water 1000 mL.

Method: Dissolve malt extract in a plastic beaker and adjust to pH 6.5 with NaOH. Soak agar in small quantity of solution. Bring remaining solution to the boil, stirring constantly. Add to the soaking agar. Bring to boil, stirring constantly. Dispense for slopes as required. Autoclave at 121°C for 10 minutes, remove and dispense for slopes or pour for plates as required.

111.13. 1% Peptone agar

For the cultivation and differentiation of fungi.

Ingredients: Tryptone peptone 5 g; Bacto-agar 10 g; distilled water 500 mL.

Method: Soak agar and peptone in a small amount of water. Boil remaining water, add this to soaking ingredients and bring to boil again. Dispense for slopes. Autoclave for 10 minutes at 121°C, then dispense for slopes or plates as required.

111.14. Potato dextrose agar

For the routine cultivation and identification of fungi.

Ingredients: Potato dextrose agar 39 g; distilled water 1000 mL.

Method: Soak potato dextrose agar in small amount of the water in a stainless-steel jug. Boil remaining water, add to soaking ingredients, bring to the boil, stirring constantly. Dispense for slopes as required. Autoclave at 121°C for 15 minutes. Remove and dispense for slopes or plates as required.

111.15. Rice grain slopes

To induce sporulation and for differentiation of *M. audouinii* and *M. canis*.

Method: Place approximately 1/2 teaspoon rice grains into wide neck 20 mL vials. Add 8 mL distilled water to each vial. Cap the vials, then slope on racks ensuring rice grains are evenly distributed. Autoclave racks at 121°C for 15 minutes.

111.16. Sabouraud's dextrose agar (SDA) with cycloheximide (0.05%), chloramphenicol, gentamicin and yeast extract

For the primary isolation and cultivation of dermatophytes.

Ingredients: Sabouraud's dextrose agar 65 g; cycloheximide (actidione) 0.5 g; chloramphenicol 0.05 g; yeast extract 5 g; gentamicin (40 mg/mL) 0.56 mL; distilled water 1000 mL.

Method: Soak all ingredients, except gentamicin, in 100 mL water. Boil remaining water, add to soaking ingredients, and bring to boil to dissolve, stirring well to prevent from charring. Add the gentamicin. Mix well. Dispense for slopes if required. Autoclave at 121°C for 10 minutes. Remove and dispense for slopes or plates as required.

111.17. Sabouraud's dextrose agar with chloramphenicol and gentamicin

For the primary isolation and cultivation of yeasts and moulds.

Ingredients: Sabouraud's dextrose agar 65 g; chloramphenicol 0.05 g; gentamicin (40 mg/mL) 0.56 mL; distilled water 1000 mL.

Method: Soak all ingredients, except gentamicin, in 100 mL water. Boil remaining water, add to soaking ingredients, and bring to boil to dissolve, stirring well to prevent from charring. Add the gentamicin. Mix well. Dispense for slopes if required. Autoclave at 121°C for 10 minutes. Remove and dispense for slopes or plates as required.

111.18. Sabouraud's dextrose agar with 5% NaCl

For the cultivation and differentiation of dermatophytes especially *T. rubrum* from *T. mentagrophytes*.

Ingredients: Sabouraud's dextrose agar 32.5 g; sodium chloride NaCl 25 g; Distilled distilled water 500 mL.

Method: Soak ingredients in approximately 100mL water. Bring remaining water to boil, add to soaking ingredients, then bring back to the boil. Dispense for slopes. Autoclave at 121°C for 10 minutes, then dispense for slopes.

111.19. Tap water agar

For the stimulation of sporulation in *Apophysomyces* and *Saksenaea* isolates.

Ingredients: Bacto-agar 15 g; distilled water 1000 mL.

Method: Add agar to water in stainless-steel jug, allow to soak. Autoclave at 121°C for 10 minutes, remove and dispense for slopes or plates as required.

111.20. Urea agar slopes with 0.5% glucose

For the differentiation of urease producing organisms.

Urea glucose broth base: Urea, broth base 0.9 g; glucose 5 g; distilled water 450 mL. Add the urea broth base and glucose to the distilled water in a 500mL beaker. Dispense in 5 x 90 mL amounts. Autoclave at 121°C for 20 mins. When cool store at 4°C.

Method to make slopes: 40% Urea solution 10 mL; Bacto-agar 3 g; distilled water 100 mL. Add agar to 100 mL of distilled water in a 250 mL pyrex bottle. Autoclave at 121°C for 15 minutes and cool to 50°C in a water bath. Add 90 mL of the urea glucose broth base and 10 mL of 40% urea solution to the autoclaved agar solution and dispense in 3 mL aliquots. Angle the vials on racks to form slopes.

111.21. Vitamin Free Agar (Trichophyton Agar No. 1)

For the differentiation of *Trichophyton* species.

Ingredients: Trichophyton agar No. 1 11.8 g; distilled water 200 mL.

Method: Add agar to water in stainless-steel jug, allow to soak. Bring to boil to dissolve, stirring constantly. Once boiled remove immediately to avoid discolouration. Dispense for slopes. Autoclave at 121°C for 10 minutes, remove and dispense for slopes.

DOI: 10.1079/9781800622340.0111

Glossary

For additional definitions of mycological terminology, the reader is referred to Ainsworth and Bisby's *Dictionary of the Fungi* (Kirk *et al.*, 2008).

Acerose: Needle-like and stiff, like a pine needle.

Acervulus (pl. acervuli): A small, saucer or disk-shaped conidial fruiting body (embedded in host plant tissue), conidiophores are formed on a cushion-like structure at the bottom of the cavity.

Acicular: Slender and pointed, needle-shaped.

Acrogenous: Development at the apex. Conidia born at the tip of the conidiophore.

Acropetal: Describes a chain of conidia in which the youngest conidium is at the tip and the oldest is at the base.

Acropleurogenous: Formed at the end and on the sides. Conidia developing at the tip and along the sides of the conidiophore.

Aculaete: Bearing narrow spines.

Acuminate: Gradually narrowing to a point.

Adelophialide: Reduced form of phialide lacking a basal septum. Seen in the general *Phialemonium* and *Lecythophora.*

Adiaspore: A large spherical chlamydospore produced in the lungs of humans and animals by the enlargement of an inhaled conidium of *Blastomyces* (*Emmonsia*) *parvus.*

Aerial mycelium: Hyphal elements growing above the agar surface.

Allantoid: Slightly curved with rounded ends; sausage-like in form.

Ameroconidium (pl. ameroconidia): A one-celled conidium.

Ampulliform: Flask-like in form.

Anamorph: An asexual state of a fungus.

Annellide: A specialised conidiogenous cell producing conidia in basipetal succession by a series of short percurrent proliferations (annellations). The tip of an annellide increases in length and becomes narrower as each subsequent conidium is formed.

Annelloconidium (pl. annelloconidia): A conidium produced by an annellide.

Anthropophilic: Of dermatophytes, that are primarily parasitic on man. They are unable to colonise other animals and they have no other environmental sources.

Apiculus: A short, sharp projection from a conidium.

Apophysis: A swelling. The term is primarily applied to the funnel-shaped swelling of a sporangiophore, immediately below the columella, seen in Mucorales.

Apothecium (pl. apothecia): A cup-like or saucer-like ascoma in which the hymenium is exposed at maturity.

Applanate: Flattened.

Appressorium (pl. appressoria): A swelling on a germ tube or hypha, especially for attachment in an early stage of infection (by many fungal plant pathogens), whether or not morphologically differentiated from vegetative hyphae, as long as the structure adheres to and penetrates the host.

Arthric: Thallic conidial ontogeny involving the conversion and subsequent disarticulation of a determinant conidiogenous hypha.

Arthro- (prefix): Jointed.

Arthroconidium (pl. arthroconidia): A thallic conidium released by either the splitting of a double septum or by the fragmentation or lysis of a disjunctor cell.

Ascocarp: A fruiting body containing asci and ascospores.

Ascomycota: A group of fungi that reproduce sexually by the endogenous formation of ascospores in an ascus.

Ascomycetous: Referring to the Ascomycota.

Asco- (prefix): Pertaining to an ascus.

Ascoma pl. ascomata: An ascus-containing structure, ascocarp.

Ascospore: A spore produced in an ascus by "free cell formation".

Ascus (pl. asci): A sac-like cell containing ascospores. Asci are characteristic of the Ascomycota.

Aseptate: Lacking septa, often pertaining to the hyphae seen in zygomycetous fungi (also see coenocytic).

Avellaneous: Hazel, wood-brown, pale brown in colour.

Awl-shaped: Tapering to slender stiff point.

Ballistoconidium (pl. ballistoconidia): A conidium that is forcibly discharged; as in *Basidiobolus* and *Conidiobolus*.

Bamboo-hyphae: Straight hypha with darkened septa at regular intervals.

Basidioma: A basidium-producing organ; basidiocarp.

Basidiomycota: A group of fungi that reproduce sexually by the exogenous formation of basidiospores from a basidium.

Basidiospore: A propagative cell (typically a ballistospore) containing one or two haploid nuclei produced, after meiosis, on a basidium. The colour, form and ornamentation of the basidiospore are fundamental to basidiomycete classification and an essential part of any specific description.

Basidium (pl. basidia): The cell or organ, diagnostic for basidiomycetes, from which, after karyogamy and meiosis, basidiospores (generally 4) are produced externally each on an extension (sterigma) of its wall.

Basipetal: Describes a chain of conidia having the youngest conidia at the base of the chain, the oldest conidium is at the apex.

Basocatenulate: A chain of conidia having the youngest conidia at the base of the chain, the oldest conidium is at the apex.

Bipolar budding: Blastoconidia developing at the opposite poles of a parent cell.

Biseriate: In two series, i.e. phialides arising from metulae as in the genus *Aspergillus*.

Biverticillate: (of a penicillus), having branching at two levels, i.e. having metulae bearing phialides as in the genus *Penicillium*.

Blastic: A form of conidial development where there is a recognisable enlargement or "blowing out" of a conidial initial before being delimited by a septum.

Blastocatenate: A chain of conidia having the youngest cell at the tip.

Blastoconidium (pl. blastoconidia): An asexual conidium that forms by a blowing out or budding process.

Bud: A young conidium. Usually used to denote the young blastoconidia of yeasts.

Budding: Asexual multiplication by the production of a small outgrowth or bud from a parent cell.

Butyrous: Buttery.

Capsule: A hyaline mucopolysaccharide sheath around the cell wall of certain yeasts e.g. *Cryptococcus* and *Rhodotorula.*

Catenulate: Conidia arranged in chains.

Cerebriform: Resembling the brain in texture or structure.

Chlamydospore: A thick-walled, thallic conidium formed within the vegetative hyphae. Chlamydospores function as organs of perennation (survival from one year to the next by vegetative means) rather than dissemination.

Cicatrised: (of conidiogenous cells and conidia), having thickened scars.

Cigar shaped: Cylindrical but gradually becoming narrower towards both ends.

Circinate: Twisted round, coiled.

Circumscissile: Opening or cracking along a circle.

Clamp connection: A specialised hyphal bridge over a septum in the basidiomycetous fungi.

Clavate: Club-like; narrowing in the direction of the base.

Cleistothecium (pl. cleistothecia): An enclosed ascocarp containing randomly dispersed asci.

Coenocytic: Infrequently septate, multi-nucleate hyphae as in the zygomycetous fungi.

Collarette: A small collar. Usually, a remnant of a cell wall present at the tip of a phialide, or around a sporangiophore.

Columella (pl. columellae): A sterile dome-like structure at the tip of a sporangiophore or within a sporangium.

Columnar: Forming a column.

Conidiogenous cell: A cell that forms conidia.

Conidiophore: A specialised hypha upon which conidia develop.

Conidium (pl. conidia): An asexual reproductive propagule formed in any manner that does not involve cytoplasmic cleavage. Conidia function as organs of dissemination.

Cottony: Having a loose and coarse texture.

Cylindrical: Cylindric, having parallel walls and circular cross-section.

Dematiaceous: A dark brown, greenish-grey or black colour.

Dermatophyte: A fungus with the ability to utilise keratin to infect hair, nail and skin of animals and humans.

Denticle: A small projection or peg on which conidia are produced.

Determinate conidiophore: Growth ceasing with the production of terminal conidia. The conidiophore does not alter in length after the formation of conidia.

Deuteromycetes: An artificial subdivision to accommodate those fungi where only the asexual state is known.

Dichotomous: A type of hyphal branching, frequently successively, into two more or less equal arms.

Dictyoconidium (pl. dictyoconidia): A conidium with both longitudinal and transverse septa; a muriform conidium.

Didmoconidium (pl. didymoconidia): A two-celled conidium.

Dimorphic: Having two different morphological forms.

Disjunctor cell: An empty cell that fragments and/or undergoes lysis to release a conidium.

Distoseptate: (of conidial septation), having the individual cells each surrounded by a sac-like wall distinct from the outer wall, as in *Drechslera*, cf. euseptate.

Divergent: Tending to develop in different directions.

Dolipore septum: A characteristic septum found in the Basidiomycetes that flares out near the pore to form an elongate channel.

Double septum: A two layered septum that may undergo centripetal separation (schizolysis) to release a conidium.

Downy: Having a short and dense mycelial texture.

Dysgonic: A slow growing variant.

Echinulate: Having sharply pointed spines; spinose.

Ectothrix: Natural hair invasion by a dermatophyte characterised by arthroconidia on the outside of the hair shaft.

Effuse: Spread out, radiate.

Elliptical: Oval, with a symmetric curve.

Elongate: Lengthened.

Endospore: A spore produced within a spherule.

Endothrix: Natural hair invasion by a dermatophyte characterised by the development of arthroconidia within the hair shaft only.

Enteroblastic: A mode of blastic conidium ontogeny in which inner cell wall layers of the conidiogenous cell are involved in conidium development.

Erect: Upright.

Euseptate: Of conidial septation, having cells separated by multilayered walls of similar structure to lateral walls, cf. distoseptate.

Evanescent: Having a short existence; fugacious.

Exudate: Droplets of fluid formed on the surface of a colony.

Falcate: Curved like a sickle.

Farinose: Covered with a sort of white, mealy powder.

Flexuous: Wavy.

Floccose: Fluffy or cottony.

Foot cell: A basal cell of a conidiophore as seen in *Aspergillus* and *Fusarium.*

Funiculose: (of hyphae), aggregated into rope-like strands.

Fusiform: Spindle-shaped, tapering toward the end.

Geniculate: Bent like a knee.

Geophillic: Of dermatophytes, that normally inhabit the soil where they are believed to decompose keratinaceous debris. Some species may cause infections in animals and man following contact with soil.

Germ tube: The initial hypha that develops from a conidium or spore.

Glabrous: Smooth.

Globose: Spherical or almost so.

Glioconidia: Conidia aggregated in slimy heads at the tip of an annellide or phialide.

Guttulate: Containing one or more oil droplets.

Gymnothecium (pl. gymnothecia): A non-ostiolate ascocarp composed of loosely interwoven hyphae and containing randomly dispersed asci; this arrangement allows many open spaces for the release of ascospores formed within the fruiting structure.

Heterothallic: A fungus that requires mating between two compatible strains for sexual reproduction to occur.

Hilum: A mark or scar, especially that on a spore at the point of attachment to a conidiogenous cell or sterigma. A scar at the base of a conidium.

Holoblastic: A mode of blastic conidium ontogeny in which all the cell wall layers of the conidiogenous cell are involved in conidium development.

Holothallic: A mode of thallic conidium ontogeny in which all the cell wall layers of the conidiogenous cell are involved in conidium development.

Homothallic: A fungus capable of sexual reproduction on a single thallus.

Hülle cells: Thick-walled cells with characteristic thin-walled pores, usually associated with cleistothecia of *Aspergillus nidulans.*

Hyaline: Transparent or nearly so; translucent; frequently used in the sense of colourless.

Hyalo-: A prefix (of conidia) meaning hyaline or brightly coloured, especially for groups of anamorphic fungi.

Hyalohyphomycosis: A mycotic infection of humans or animals caused by a non-dematiaceous fungus. cf. phaeohyphomycosis.

Hypha (pl. hyphae): A single filament of a fungus.

Hyphomycetes: A class of mycelial moulds which reproduce asexually by conidia on hyphae or aggregations of hyphae.

Hymenium (pl. hymenia): A layer of cells producing basidiospores.

Indeterminate: (1) Having the edge not well-defined, especially of fruit-bodies and leaf-spots; (2) (of conidiophores), continuing growth indefinitely.

Intercalary: (1) (of growth), between the apex and the base; (2) (of cells, spores, etc.), between two cells.

Keratinophilic: (1) Capable of decomposing keratin; (2) of many fungi causing superficial mycoses in humans; see dermatophyte.

Lageniform: Swollen at the base, narrowed at the top; like a Florence flask.

Lanceolate: Lance-shaped.

Lanose: Like wool; covered with short hair-like processes.

Lateral: At the side.

Lenticular: Shaped like a double convex lens.

Limoniform: Lemon-like in form.

Macroconidium (pl. macroconidia): The larger of two different types of conidia produced by a fungus in the same manner.

Macronematous: Having a conidiophore that is morphologically different from the vegetative hyphae.

Meristematic: Pertaining to an area (meristem) of the hyphae or of other specialised structures from which new growth occurs.

Merosporangium (pl. merosporangia): A cylindrical outgrowth from the swollen end of a sporangiophore in which a chain-like series of sporangiospores is generally produced.

Mesophilic: Of fungi that grow at temperatures in the range of 5-35°C with optimum temperatures for growth between 25-30°C.

Metula (pl. metulae): A sterile cell below the phialides of some *Aspergillus* and *Penicillium* species.

Microconidium (pl. microconidia): The smaller of two different types of conidia produced by a fungus in the same manner.

Micronematous: Having a conidiophore that is not morphologically different from the vegetative hyphae.

Mucoid: Sticky or slimy.

Multiseptate: Having several septa.

Multipolar budding: Blastoconidia developing at different sites on the surface of a parent cell.

Mitosporic fungi: (Deuteromycotina, Deuteromycetes, Fungi Imperfecti, asexual fungi, conidial fungi, mitosporic fungi) These are fungi that are unable to produce sexual spores, reproduction is by mitosis (asexual cell division) which results in the daughter cells having the same genetic makeup as the mother cell.

Muriform: A conidium with both longitudinal and transverse septa, cf. dictyoconidia.

Mycelium (pl. mycelia): The mass of hyphae making up the thallus of a fungus.

Neurotropic: Tending to attack or affect the nervous system preferentially.

Niger: Black.

Nodular bodies: (of dermatophytes), rounded bodies made up of massed hyphae.

Nodulose: (of conidia), having broad-based, blunt, wart-like excrescences.

Nonseptate: Without septa.

Obclavate: Club-shaped in reverse; widest at the base.

Obpyriform: Pear-shaped in reverse; the distal region is larger.

Oidium (pl. oidia): A flat-ended asexual spore formed by the breaking up (usually centripetally) of a hypha into cells, as in *Dipodascus geotrichum*; arthroconidia.

Olivaceous: Olive-grey colour.

Ostiole: An opening or pore in an ascocarp or a pycnidium.

Oval: Widely elliptical.

Ovoid: Egg-shaped.

Papilla (pl. papillae): A small rounded nipple-shaped projection.

Pectinate: Like the teeth of a comb.

Pedicel: A slender stalk.

Pedicellate: Having a pedicel.

Pellicle: A film-like or skin-like surface growth.

Penicillus (pl. penicilli): The brush-like conidiophore seen in *Penicillium* and related genera.

Percurrent: Conidiogenous cell growth where a new conidium grows out through the open end of the conidiogenous cell after the dehiscence of the previous conidium.

Perennation: The ability of organisms, particularly fungi and plants, to survive from one germinating season to another, especially under unfavourable conditions such as drought or winter.

Peridium: The outer wall of an ascocarp surrounding the asci; also the wall of a sporangium.

Perithecium (pl. perithecia): An enclosed ascocarp characterised by an apical ostiole and by asci arranged in a basal tuft or hymenium layer.

Phaeo-: A prefix meaning darkly pigmented.

Phaeohyphomycosis: A mycotic infection in humans or other animals caused by a dematiaceous fungus.

Phialide: A specialised conidiogenous cell that produces conidia in basipetal succession without increasing in length.

Phialidic: Conidiogenesis in which each conidium originates by the laying down of new wall material not from existing walls or layers of the wall of the conidiogenous cell (a phialide). A basipetal succession of conidia is formed.

Phialoconidium (pl. phialoconidia): A conidium produced from a phialide.

Phragmoconidium (pl. phragmoconidia): A conidium having two or more transverse septa.

Pleomorphic: Having more than one form.

Pleurogenous: Borne on the sides of a conidiophore or hyphae.

Polyphyletic: Having arisen in separate evolutionary lines. A group of organisms derived from more than one common evolutionary ancestor or ancestral group and therefore not suitable for placing in the same taxon.

Poroconidium (pl. poroconidia): A conidium produced through a small pore in a conidiogenous cell.

Pseudohyphae: A string of elongated blastoconidia formed in some yeasts that resemble a hypha-like filament.

Pseudoparenchyma: A mass of hyphae arranged together to form a tissue-like structure.

Pseudoseptum (obsolete term): A protoplasmic or vacuolar membrane looking like a septum (= distoseptum, distoseptate).

Punctiform: Very small, but seen without a lens.

Pycnidium (pl. pycnidia): An asexual fruiting body containing conidia.

Pyriform: Pear-shaped.

Rachis: A geniculate or zig-zag extension of a sympodial proliferating conidiogenous cell bearing conidia.

Rachiform: Narrow based geniculate sympodial development.

Raduliform: Broad based geniculate sympodial development.

Racquet hyphae: A hypha composed of a number of cells swollen at one end resembling a tennis racquet.

Retrogressive conidial development: The conidiogenesis cell becomes shorter during the successive development of conidia.

Rhexolytic: Conidial liberation in which an intermediate, supporting cell is scarified; (circumscissile splitting of the cell below the basal conidial septum) cf. schizolytic.

Rhizoids: A short branching root-like hyphae seen in some Zygomycetes.

Sarcina (pl. sarcinae): A swollen, meristematic clump of cells.

Sclerotium (pl. sclerotia): A hard mass composed of intertwining mycelium, usually with thick pigmented walls that are resistant to adverse environmental conditions. It will germinate to produce new hyphae under favourable conditions or in response to a chemical stimulant from a prospective host.

Schizolytic: Secession of conidia involving a splitting of the delimiting septum so that one half of the cross wall becomes the base of the seceding conidium and the other half remains at the apes of the conidiogenous cell. cf. rhexolytic.

Scutulum (pl. scutula): the compact mass of hyphae, keratinous material, and hair found in favus.

Semimacronematous: Having a conidiophore that is only slightly morphologically different from the vegetative hyphae.

Septum (pl. septa): A cross wall in a hypha.

Serrate: Edged with teeth, like a saw.

Seta (pl. setae): A stiff hair, generally thick-walled and dark in colour.

Solitary: Alone.

Spherule: A sporangium-like structure in *Coccidioides.*

Spinulose: Covered in small spines.

Spiral hyphae: A hypha ending in a spiral or helical coil, as in *Trichophyton.*

Sporangiolum (pl. sporangiola): A small sporangium producing a small number of sporangiospores.

Sporangiophore: A specialised hypha that bears a sporangium.

Sporangiospore: An asexual spore produced within a sporangium.

Sporangium (pl. sporangia): A sac-like structure producing asexual spores endogenously by cytoplasmic cleavage.

Spore: a reproductive propagule formed by either meiosis or mitosis. However, if by asexual means, cleavage of cytoplasm is usually involved.

Sporodochium (pl. sporodochia): A cushion-shaped mass of hyphae bearing conidiophores.

Stellate: Star-shaped

Sterigma (pl. sterigmata): A small, pointed structure upon which a basidiospore forms.

Stolon: A running hypha from which rhizoids and sporangiospores arise.

Striate: Having lines or minute furrows.

Subglobose: Not quite round or spherical.

Sympodial: A mode of conidiogenous cell growth which results in the development of conidia on a geniculate or zig-zag rachis.

Synanamorph: One of the asexual forms of propagation of a fungus which forms several kinds of conidia. When a single fungus produces multiple morphologically distinct anamorphs, these are called synanamorphs.

Synnema (pl. synnemata): A group of erect conidiophores that are cemented together producing conidia at the apex and/or along the sides.

Teleomorph: The sexual state of a fungus.

Thallic: A mode of conidial ontogeny where a conidium is formed from a pre-existing hyphal segment or cell.

Thermophilic: Making active growth at high temperature. A *thermophilic fungus* is defined as one that has minimum growth at 20°C or above and a maximum growth at 50°C or above.

Toruloid: Having swellings at intervals.

Truncate: Cut off sharply.

Tuberculate: Having small wart-like structures.

Umbo: A central swelling like the boss at the centre of a shield; especially one on top of a pileus above the stipe.

Uniserate: Phialides arising directly from a vesicle as in *Aspergillus*.

Verrucose: Having many small rounded warts.

Verruculose: Delicately verrucose.

Verticillate: Having branches arranged in verticils or whorls.

Vesicle: A swollen cell.

Villi (pl. villae): Long soft hairs.

Woronin bodies: Rounded or elongated-oval highly refractive bodies in the cells of certain ascomycetous fungi, particularly in association with septa. Woronin bodies seal septal pores in response to damage of adjacent cells.

Zoophilic: Of dermatophytes, that are primarily parasitic on animals and infections may be transmitted to humans following contact with the animal host.

Zygospores: A thick-walled sexual spore formed by the fusion of two similar gametangia; characteristic of the Zygomycetes.

References

Abe, A., Asano, K. and Sone, T. (2010) A molecular phylogeny-based taxonomy of the genus *Rhizopus*. *Bioscience, Biotechnology and Biochemistry*, 74, 1325-1331.

Abliz, P., Fukushima, K., Takizawa, K., *et al.* (2003) Rapid identification of the genus *Fonsecaea* by PCR with specific oligonucleotide primers. *Journal of Clinical Microbiology,* 41, 873-876.

Abliz, P., Fukushima, K., Takizawa, K., *et al.* (2004) Specific oligonucleotide primers for identification of *Cladophialophora carrionii,* a causative agent of chromoblastomycosis. *Journal of Clinical Microbiology*, 42, 404-407.

Adam, R.D., Paquin, M.L., Petersen, E.A., *et al.* (1986) Phaeohyphomycosis caused by the fungal genera *Bipolaris* and *Exserohilum*. A report of 9 cases and review of the literature. *Medicine*, 65, 203-217.

Adler, A., Yaniv, I., Samra, Z., *et al.* (2006) *Exserohilum*: an emerging human pathogen. *European Journal of Clinical Microbiology and Infectious Diseases*, 25, 247-253.

Ahmed, S.A., van den Ende, G., Fahal, B., *et al.* (2014a) Rapid identification of black grain eumycetoma causative agents using Rolling Circle Amplification. *PLoS Neglected Tropical Diseases*, 8, e3368.

Ahmed, S.A., van de Sande, W.W.J., Stevens, D.A., *et al.* (2014b) Revision of agents of black-grain eumycetoma in the order Pleosporales. *Persoonia*, 33, 141-154.

Ahmed, E.A., Nour, B.Y.M. and Abakar, A.D. (2020) The genus *Madurella*: Molecular identification and epidemiology in Sudan. *PLoS Neglected Tropical Diseases*, 14(7), e0008420.

Ahrholdt, J., Murugaiyan, J., Straubinger, R.K., *et al.* (2012) Epidemiological analysis of worldwide bovine, canine and human clinical *Prototheca* isolates by PCR genotyping and MALDI-TOF mass spectrometry proteomic phenotyping. *Medical Mycology*, 50, 234-243.

Ajello, L., Dean, D.F. and Irwin, R.S. (1976) The zygomycete *Saksenaea vasiformis* as a pathogen of humans with a critical review of the aetiology of zygomycosis. *Mycologia*, 68, 52-62.

Ajello, L. (1977) Taxonomy of the dermatophytes: a review of their imperfect and perfect states. In Iwata, K. (ed.) *Recent Advances in Medical and Veterinary Mycology*, pp. 289-297. University Park Press, Baltimore, Maryland.

Alanio, A., Hauser, P., Lagrou, K., *et al.* (2016) ECIL guidelines for the diagnosis of *Pneumocystis jirovecii* pneumonia in patients with haematological malignancies and stem cell transplant recipients. *Journal of Antimicrobial Chemotherapy*, 71, 2386-2396.

Alatruey-Izquierdo, A., Hoffman, K., de Hoog, G.S., *et al.* (2010) Species recognition and clinical relevance of the Zygomycetous genus *Lichtheimia* (syn, *Absidia* pro parte, *Mycocladus*). *Journal of Clinical Microbiology*, 48, 2154-2170.

Alcoba-Florez, J., Mendez-Alvarez, S., Cano, J., *et al.* (2005) Phenotypic and molecular characterization of *Candida nivariensis* sp. nov., a possible new opportunistic fungus. *Journal of Clinical Microbiology*, 43, 4107-4111.

Alcorn, J.L. (1983) Genetic concepts in *Drechslera, Bipolaris* and *Exserohilum*. *Mycotaxon*, 17, 1-86.

Al-Hatmi, A.M.S., van Diepeningen, A.D., Curfs-Breuker, I., *et al.* (2015) Specific antifungal susceptibility profiles of opportunists in the *Fusarium fujikuroi* complex. *Journal of Antimicrobial Chemotherapy*, 70, 1068-1071.

Al-Hatmi, A.M.S, Balkhair, A., Al-Busaidi, I., *et al.* (2021) *Basidiobolus omanensis* sp. nov. causing angioinvasive abdominal Basidiobolomycosis. *Journal of Fungi*, 7(8), 653.

Al-Mohsen, I.Z., Sutton, D.A., Sigler, L., *et al.* (2000) *Acrophialophora fusispora* brain abscess in a child with acute lymphoblastic leukemia: review of cases and taxonomy. *Journal of Clinical Microbiology*, 38, 4569-4576.

Alshawa, K., Beretti, J.L., Lacroix, C., *et al.* (2012) Successful identification of clinical dermatophyte and *Neoscytalidium* species by matrix-assisted laser desorption ionization-time of flight mass spectrometry. *Journal Clinical of Microbiology*, 50, 2277-2281.

Alvarez, E., Sutton, D.A., Cano, J., *et al.* (2009) Spectrum of zygomycete species identified in clinically significant specimens in the United States. *Journal of Clinical Microbiology*, 47, 1650-1656.

Alvarez, E., Stchigel, A.M., Cano, J., *et al.* (2010a) Molecular phylogenetic diversity of the emerging mucoralean fungus *Apophysomyces*: proposal of three new species. *Revista Iberoamericana de Micologia*, 27, 80-89.

Alvarez, E., Garcia-Hermoso, D., Sutton, D.A., *et al.* (2010b) Molecular phylogeny and proposal of two new species of the emerging pathogenic fungus *Saksenaea*. *Journal of Clinical Microbiology*, 48, 4410-4416.

Ames, L.M. (1963) *A monograph of the Chaetomiaceae*. *U.S. Army Research and Development Serial*, 2, 1-125.

Andrianopoulos, A. (2002) Control of morphogenesis in the human fungal pathogen *Penicillium marneffei*. *International Journal of Medical Microbiology*, 292, 331-347.

Arastehfar, A., Fang, W., Al-Hatmi, A.M.S., *et al.* (2019) Unequivocal identification of an underestimated opportunistic yeast species, *Cyberlindnera fabianii,* and its close relatives using a dual-function PCR and literature review of published cases. *Medical Mycology*, 57, 833-840.

Arendrup, M.C., Boekhout, T., Akova, M., *et al.* (2014) ESCMID and ECMM joint clinical guidelines for the diagnosis and management of rare invasive yeast infections. *Clinical Microbiology and Infection*, 20, Suppl 3, 76-98.

Ariyawansa, H.A., Hyde, K.D., Buyck, *et al.* (2015) Fungal diversity notes 111-252 taxonomic and phylogenetic contributions to fungal taxa. *Fungal Diversity*, 75, 27-274.

Arrieta-Aguirre, I., Menendez-Manjon, P., Cuetara, M.S., *et al.* (2017) Molecular identification of *Saprochaete capitata* in human blood and paraffinized tissue samples. *Journal of Clinical Microbiology*, 55, 2556-2559.

Arzanlou, M., Groenewald, J.Z., Gams. W., *et al.* (2007) Phylogenetic and morphotaxonomic revision of *Ramichloridium* and allied Genera. *Studies in Mycology*, 58, 57-93.

Asadzadeh, M., Ahmad, S., Al-Sweih, N., *et al.* (2009) Rapid molecular differentiation and genotypic heterogeneity among *Candida parapsilosis* and *Candida orthopsilosis* strains isolated from clinical specimens in Kuwait. *Journal of Medical Microbiology*, 58, 745-752.

Atkins, S.D., Clark, I.M., Pande, S., *et al.* (2005) The use of real-time PCR and species-specific primers for the identification and monitoring of *Paecilomyces lilacinus*. *FEMS Microbiology Ecology*, 51, 257-264.

Aveskamp, M.M., de Gruyter, J., Woudenberg, J.H.C., *et al.* (2010) Highlights of the *Didymellaceae*: A polyphasic approach to characterise *Phoma* and related pleosporalean genera. *Studies in Mycology*, 65, 1-60.

Badali, H., Gueidan, C., Najafzadeh, M.J., *et al.* (2008) Biodiversity of the genus *Cladophialophora*. *Studies in Mycology*, 61, 175-191.

Badali, H., de Hoog, G.S., Curfs-Breuker, I., *et al.* (2010a) Use of amplified fragment length polymorphism to identify 42 *Cladophialophora* strains related to cerebral phaeohyphomycosis with *in vitro* antifungal susceptibility. *Journal of Clinical Microbiology*, 48, 2350-2356.

Badali, H., Najafzadeh, M.J., van Esbroeck, M., *et al.* (2010b) The clinical spectrum of *Exophiala jeanselmei*, with a case report and *in vitro* antifungal susceptibility of the species. *Medical Mycology*, 48, 318-327.

Badali, H., Chander, J., Bansal, S., *et al.* (2010c) First autochthonous case of *Rhinocladiella mackenziei* cerebral abscess outside the Middle East. *Journal of Clinical Microbiology*, 48, 646-649.

Badali, H., Yazdanparast, S.A., Bonifaz, A., *et al.* (2013) *Veronaea botryosa*: molecular identification with amplified fragment length polymorphism (AFLP) and *in vitro* antifungal susceptibility. *Mycopathologia*, 175, 505-513.

Badali, H., Khodavaisy, S., Fakhim, H., *et al.* (2015) *In vitro* susceptibility profiles of eight antifungal drugs against clinical and environmental strains of *Phaeoacremonium. Antimicrobial Agents and Chemotherapy*, 59, 7818-7822.

Badenoch, R.R., Halliday, C.L., Ellis, D.H., *et al.* (2006) *Ulocladium atrum* keratitis. *Journal of Clinical Microbiology*, 44, 1190-1193.

Bagagli, E., Theodoro, R.C., Bosco, S.M., *et al.* (2008) *Paracoccidioides brasiliensis*: phylogenetic and ecological aspects. *Mycopathologia*, 165, 197-207.

Bagyalakshmi, R., Therese, K.L., Prasanna, S., *et al.* (2008) Newer emerging pathogens of ocular non-sporulating moulds (NSM) identified by polymerase chain reaction (PCR)-based DNA sequencing technique targeting internal transcribed spacer (ITS) region. *Current Eye Research*, 33, 139-147.

Bakuła, Z., Siedlecki, P., Gromadka, R., *et al.* (2021) A first insight into the genome of *Prototheca wickerhamii*, a major causative agent of human protothecosis. *BMC Genomics*, 22, 168-168.

Balajee, S.A., Gribskov, J., Brandt, M., *et al.* (2005a) Mistaken identity: *Neosartorya pseudofischeri* and its anamorph masquerading as *Aspergillus fumigatus*. *Journal of Clinical Microbiology*, 43, 5996-5999.

Balajee, S.A., Gribskov, J.L., Hanley, E., *et al.* (2005b) *Aspergillus lentulus* sp. nov., a new sibling species of *A. fumigatus*. *Eukaryotic Cell*, 4, 625-632.

Balajee, S.A., Nickle, D., Varga, J., *et al.* (2006) Molecular studies reveal frequent misidentification of *Aspergillus fumigatus* by morphotyping. *Eukaryotic Cell*, 5, 1705-1712.

Balajee, S.A., Houbraken, J., Verweij, P.E., *et al.* (2007) *Aspergillus* species identification in the clinical setting. *Studies in Mycolology*, 59, 39-46.

Balajee, S.A., Borman, A.M., Brandt, M.E., *et al.* (2009) Sequence-based identification of *Aspergillus*, *Fusarium*, and *Mucorales* species in the clinical mycology laboratory: where are we and where should we go from here? *Journal of Clinical Microbiology*, 47, 877-884.

Barker, B.M., Litvintseva, A.P., Riquelme, M., *et al.* (2019) *Coccidioides* ecology and genomics. *Medical Mycology*, 57, S21-S29.

Barnett, J.A., Payne, R.W. and Yarrow, D. (1983) *Yeasts: Characteristics and Identification*. Cambridge University Press, Cambridge, UK.

Barron, G.L. (1968) *The Genera of Hyphomycetes from Soil*. Williams & Wilkins Co. Baltimore, Maryland.

Barros, M.B., de Almeida Paes, R. and Schubach, A.O. (2011) *Sporothrix schenckii* and sporotrichosis. *Clinical Microbiology Reviews*, 24, 633-654.

Barrs, V.R., van Doorn, T.M., Houbraken, J., *et al.* (2013) *Aspergillus felis* sp. nov., an emerging agent of invasive aspergillosis in humans, cats, and dogs. *PLoS One*, 8, e64871.

Bastos, R.W., Valero, C., Silva, L.P., *et al.* (2020) Functional characterization of clinical isolates of the opportunistic fungal pathogen *Aspergillus nidulans. mSphere*, 5(2), e00153-20.

Bensch, K., Groenewald, J.Z., Dijksterhuis, J., *et al.* (2010) Species and ecological diversity within the *Cladosporium cladosporioides* complex (Davidiellaceae, Capnodiales). *Studies in Mycology*, 67, 1-94.

Bensch, K., Braun, U., Groenewald, J.Z., *et al.* (2012) The genus *Cladosporium*. *Studies in Mycology*, 72, 1-401.

Beyda, N.D., Chuang, S.H., Alam, M.J., *et al.* (2013) Treatment of *Candida famata* bloodstream infections: case series and review of the literature. *Journal of Antimicrobial Chemotherapy*, 68, 438-443.

Bialek, R., Cirera, A.C., Herrmann, T., *et al.* (2003) Nested PCR assays for detection of *Blastomyces dermatitidis* DNA in paraffin-embedded canine tissue. *Journal of Clinical Microbiology*, 41, 205-208.

Bilman, F.B. and Yetik, M. (2017) *Geotrichum candidum*: A rare infection agent at urinary system and review of the literature. *Journal of Clinical and Experimental Investigations*, 8, 127-129.

Binnicker, M.J., Popa, A.S., Catania, J., *et al.* (2010) Meningeal coccidioidomycosis diagnosed by real-time polymerase chain reaction analysis of cerebrospinal fluid. *Mycopathologia*, 171, 285-289.

Bishop, J.A., Chase, N., Lee, R., *et al.* (2008) Production of white colonies on CHROMagar Candida medium by members of the *Candida glabrata* clade and other species with overlapping phenotypic traits. *Journal of Clinical Microbiology*, 46, 3498-3500.

Boekhout, T., Guého, E., Mayser, P. and Velegraki, A. (eds) (2010) *Malassezia and the Skin. Science and Clinical Practice*. Springer, Heidelberg, Germany.

Booth, C. (1966) The genus *Cylindrocarpon*. *Commonwealth Mycological Institute, Mycological Papers*, 104, 1-56.

Booth, C. (1971) *The genus Fusarium*. Commonwealth Mycological Institute, Kew, Surrey, England.

Booth, C. (1977) *Fusarium: Laboratory Guide to the Identification of the Major Species*. Commonwealth Mycological Institute, Kew, Surrey, England.

Borman, A.M., Petch, R., Linton, C.J., *et al.* (2008) *Candida nivariensis*, an emerging pathogenic fungus with multidrug resistance to antifungal agents. *Journal of Clinical Microbiology*, 46, 933-938.

Borman, A.M., Linton, C.J., Oliver, D., *et al.* (2009) Pyrosequencing analysis of 20 nucleotides of internal transcribed spacer 2 discriminates *Candida parapsilosis*, *Candida metapsilosis*, and *Candida orthopsilosis*. *Journal of Clinical Microbiology*, 47, 2307-2310.

Borman, A.M., Muller, J., Walsh-Quantick, J., *et al.* (2020) MIC distributions for amphotericin B, fluconazole, itraconazole, voriconazole, flucytosine and anidulafungin and 35 uncommon pathogenic yeast species from the UK determined using the CLSI broth microdilution method. *Journal of Antimicrobial Chemotherapy*, 75, 1194-1205.

Bragulat, M.R., Castellá, G., Isidoro-Ayza, M., *et al.* (2017) Characterization and phylogenetic analysis of a *Cunninghamella bertholletiae* isolate from a bottlenose dolphin (*Tursiops truncatus*). *Revista Iberoamericana de Micologia*, 34, 215-219.

Brilhante, R.S.N., Fechinea, M.A.B., Mesquita, J.R.L., *et al.* (2012) Histoplasmosis in HIV-positive patients in Ceará, Brazil: clinical-laboratory aspects and *in vitro* antifungal susceptibility of *Histoplasma capsulatum* isolates. *Transactions of the Royal Society of Tropical Medicine and Hygiene*, 106, 484-488.

Brillowska-Dabrowska, A., Michałek, E., Saunte, D.M., *et al.* (2013) PCR test for *Microsporum canis* identification. *Medical Mycology*, 51, 576-579.

Brown, E.M., McTaggart, L.R., Zhang, S.X., *et al.* (2013) Phylogenetic analysis reveals a cryptic species *Blastomyces gilchristii*, sp. nov. within the human pathogenic fungus *Blastomyces dermatitidis*. *PLoS One*, 8, e59237.

Bshabshe, A.A., Joseph, M.R.P., Hakami, A.M.A., *et al.* (2020) *Basidiobolus haptosporus*-like fungus as a causal agent of gastrointestinal basidiobolomycosis. *Medical Mycology*, 58, 264-267.

Buchta, V. and Otcenasek, M. (1988) *Geotrichum candidum* - an opportunistic agent of mycotic diseases. *Mycoses*, 31, 363-370.

Burges, G.E., Walls, C.T. and Maize, J.C. (1987) Subcutaneous phaeohyphomycosis caused by *Exserohilum rostratum* in an immunocompetent host. *Archives of Dermatology*, 123, 1346-1350.

Burgess, L.W. and Liddell, C.M. (1983) *Laboratory Manual for Fusarium Research*. *Fusarium* Research Laboratory, Department of Plant Pathology and Agricultural Entomology. The University of Sydney.

Burgess, J.W., Schwan, W.R. and Volk, T.J. (2006) PCR-based detection of DNA from the human pathogen *Blastomyces dermatitidis* from natural soil samples. *Medical Mycology*, 44, 741-748.

Buzina, W., Lang-Loidolt, D., Braun, H., *et al.* (2001) Development of molecular methods for identification of *Schizophyllum commune* from clinical samples. *Journal of Clinical Microbiology*, 39, 2391-2396.

Cabanes, F.J., Vega, S. and Castella, G. (2011) *Malassezia cuniculi* sp. nov., a novel yeast species isolated from rabbit skin. *Medical Mycology*, 49, 40-48.

Cabanes, F.J., Coutinho, S.D., Puig, L., *et al.* (2016) New lipid-dependent *Malassezia* species from parrots. *Revista Iberoamericana de Micologia*, 33, 92-99.

Cafarchia, C., Gasser, R.B., Figueredo, L.A., *et al.* (2011) Advances in the identification of *Malassezia*. *Molecular and Cellular Probes*, 25, 1-7.

Cafarchia, C., Iatta, R., Latrofa, M.S., *et al.* (2013) Molecular epidemiology, phylogeny and evolution of dermatophytes. *Infection, Genetics and Evolution*, 20, 336-351.

Calderaro, A., Motta, F., Montecchini, S., *et al.* (2014) Identification of dermatophyte species after implementation of the in-house MALDI-TOF MS database. *International Journal of Molecular Sciences*, 15, 16012-16024.

Campbell, C.K. and Smith, M.D. (1982) Conidiogenesis in *Petriellidium boydii* (*Pseudallescheria boydii*). *Mycopathologia*, 78,145-150.

Cano, J. and Guarro, J. (1990) The genus *Aphanoascus*. *Mycological Research*, 94, 355-377.

Cano, J., Sagues, M., Barrio, E., *et al.* (2002) Molecular taxonomy of *Aphanoascus* and description of two new species from soil. *Studies in Mycology*, 47, 153-164.

Carmichael, J.W. (1962) *Chrysosporium* and some aleuriosporic hyphomycetes. *Canadian Journal of Botany*, 40, 1137-1173.

Casadevall, A. and Perfect, J.R. (1998) *Cryptococcus neoformans*. ASM Press, Washington, DC, USA.

Casadevall, A., Kontoyiannis, D.P. and Robert, V. (2019) On the emergence of *Candida auris*: climate change, azoles, swamps, and birds. *mBio*, 10, 01397-19.

Cassagne, C., Ranque, S., Normand, A.C., *et al.* (2011) Mould routine identification in the clinical laboratory by matrix-assisted laser desorption ionization time-of-flight mass spectrometry. *PLoS One*, 6, e28425.

Castanheira, A., Duncanson, M., Diekema, F.P., *et al.* (2012) Activities of E1210 and comparator agents tested by CLSI and EUCAST broth microdilution methods against *Fusarium* and *Scedosporium* species identified using molecular methods. *Antimicrobial Agents and Chemotherapy*, 56, 352-357.

Cavalier-Smith, T. (1998) A revised six-kingdom system of life. *Biological Reviews of the Cambridge Philosophical Society*, 73, 203-266.

Cendejas-Bueno, E., Kolecka, A., Alastruey-Izquierdo, A., *et al.* (2012) Reclassification of the *Candida haemulonii* complex as *Candida haemulonii* (*C. haemulonii* group I), *C. duobushaemuloni* sp. nov. (*C. haemulonii* group II), and *C. haemulonii* var. *vulnera var. nov*.: three multiresistant human pathogenic yeasts. *Journal of Clinical Microbiology*, 50, 3641-3651.

Chagas-Neto, T.C., Chaves, G.M. and Colombo, A.L. (2008) Update on the genus *Trichosporon*. *Mycopathologia*, 166, 121-132.

Chakrabarti, A., Ghosh, A., Prasad, G.S., *et al.* (2003) *Apophysomyces elegans*: an emerging zygomycete in India. *Journal of Clinical Microbiology*, 41, 783-788.

Chakrabarti, A., Shivaprakash, M.R., Curfs-Breuker, I., *et al.* (2010) *Apophysomyces elegans*: epidemiology, amplified fragment length polymorphism typing, and *in vitro* antifungal susceptibility pattern. *Journal of Clinical Microbiology*, 48, 4580-4585.

Chakrabarti, A., Kaur, H., Rudramurthy, S.M., *et al.* (2016) Brain abscess due to *Cladophialophora bantiana*: a review of 124 cases. *Medical Mycology*, 54, 111-119.

Chandler, F.W., Kaplan, W. and Ajello, L. (1980) *A Colour Atlas and Textbook of the Histopathology of Mycotic Diseases*. Wolfe Medical, London.

Chapman, B., Slavin, M., Marriott, D., *et al.* (2017) Changing epidemiology of candidaemia in Australia. *Journal of Antimicrobial Chemotherapy*, 72, 1103-1108. Erratum in: *Journal of Antimicrobial Chemotherapy*, (2017) 72, 1270.

Chen, S.C., Marriott, D., Playford, E.G., *et al.* (2009) Candidaemia with uncommon *Candida* species: predisposing factors, outcome, antifungal susceptibility, and implications for management. *Clinical Microbiology and Infection*, 15, 662-669.

Chen, C.A., Ellis, D., Sorrell, T.C., *et al.* (2011) Chapter 44, *Trichophyton*. In Liu, D. (ed.) *Molecular Detection of Human Fungal Pathogens.* CRC Press, Florida.

Chen, S.C., Slavin, M.A., Heath, C.H., *et al.* (2012) Clinical manifestations of *Cryptococcus gattii* infection: determinants of neurological sequelae and death. *Clinical Infectious Diseases*, 55, 789-798.

Chen, S.C., Halliday, C.L., Hoenigl, M., *et al.* (2021) *Scedosporium* and *Lomentospora* infections: Contemporary microbiological tools for the diagnosis of invasive disease. *Journal of Fungi*, 7, 23.

Chim, C.S., Ho, P.L. and Yuen, K.Y. (1998) Simultaneous *Aspergillus fischeri* and *Herpes simplex* pneumonia in a patient with multiple myeloma. *Scandinavian Journal of Infectious Diseases*, 30, 190-191.

Chowdhary, A., Randhawa, H.S., Gaur, S.N., *et al.* (2013a) *Schizophyllum commune* as an emerging fungal pathogen: a review and report of two cases. *Mycoses*, 56, 1-10.

Chowdhary, A., Kathuria, S., Singh, P.K., *et al.* (2013b) Molecular characterization and *in vitro* antifungal susceptibility profile of *Schizophyllum commune*, an emerging Basidiomycete in bronchopulmonary mycoses. *Antimicrobial Agents and Chemotherapy*, 57, 2845-2848.
Chowdhary, A., Khaturia, S., Agarwal, K., *et al.* (2014a) Recognizing filamentous basidiomycetes as agents of human disease: A review. *Medical Mycology*, 52, 782-797.
Chowdhary, A., Agarwal, K., Kathuria, S., *et al.* (2014b) Allergic bronchopulmonary mycosis due to fungi other than *Aspergillus*: a global overview. *Critical Reviews in Microbiology*, 40, 30-48.
Chowdhary, A., Prakash, A., Sharma, C., *et al.* (2018) A multicentre study of antifungal susceptibility patterns among 350 *Candida auris* isolates (2009-17) in India: role of the ERG11 and FKS1 genes in azole and echinocandin resistance. *Journal of Antomicrobial Chemotherapy*, 73, 891-899.
Colombo, A., Padovan, A.C.B. and Chaves, G.M. (2011) Current knowledge of *Trichosporon* spp. and Trichosporonosis. *Clinical Microbiology Reviews*, 24, 682-700.
Cooney, D.H. and Emerson, R. (1964) *Thermophilic Fungi; An Account of their Biology, Activities, and Classification.* W.H. Freeman & Co, San Francisco and London.
Cooter, R.D., Lim, I.S., Ellis, D.H., *et al.* (1990) Burn wound zygomycosis caused by *Apophysomyces elegans*. *Journal of Clinical Microbiology*, 28, 2151-2153.
Coriglione, G., Stella, G., Gafa, L., *et al.* (1990) *Neosartorya fischeri* var. *fischeri* (Wehmer) Malloch and Cain 1972 (Anamorph: *Aspergillus fischerianus* Samson and Gams 1985) as a cause of mycotic keratitis. *European Journal of Epidemiology*, 6, 382-385.
Correia, A., Sampaio, P., James, S., *et al.* (2006) *Candida bracarensis sp. nov.*, a novel anamorphic yeast species phenotypically similar to *Candida glabrata*. *International Journal of Systematic and Evolutionary Microbiology*, 56, 313-317.
Cortez, K.J., Roilides, E., Quiroz-Telles, F., *et al.* (2008) Infections caused by *Scedosporium* spp. *Clinical Microbiology Reviews*, 21, 157-197.
Crous, P.W., Slippers, B., Wingfield, M.J., *et al.* (2006) Phylogenetic lineages in the Botryosphaeriaceae. *Studies in Mycology*, 55, 235-253.
Crous, P.W., Braun, U., Schubert, K., *et al.* (2007) Delimiting *Cladosporium* from morphologically similar genera. *Studies in Mycology*, 58, 33-56.
Crous, P., Wingfield, M., Burgess, T., *et al.* (2016) Fungal Planet description sheets: 469-557. *Persoonia*, 37, 218-403.
Crous, P., Wingfield, M., Burgess, T., *et al.* (2017) Fungal Planet description sheets: 558-624. *Persoonia*, 38, 240-384.
Cushion, M.T. (2010) Are members of the fungal genus pneumocystis (a) commensals; (b) opportunists; (c) pathogens; or (d) all of the above? *PLoS Pathogens*, 6, e1001009.
Damasceno, L.S., Teixeira, M.M., Barker, B.M., *et al.* (2019) Novel clinical and dual infection by *Histoplasma capsulatum* genotypes in HIV patients from Northeastern Brazil. *Science Reports*, 9, 11789.
Dannaoui, E. (2009) Molecular tools for identification of Zygomycetes and the diagnosis of zygomycosis. *Clinical Microbiology and Infection*, 15, 66-70.
Davidson, N., Campbell, K., Foroughi, F. *et al.* (2020) Disseminated *Saksenaea* infection in an immunocompromised host associated with a good clinical outcome: a case report and review of the literature. *BMC Infectious Diseases*, 20, 05459-9.
Davis, S.R., Ellis, D.H., Goldwater, P., *et al.* (1994) First human culture-proven Australian case of entomophthoromycosis caused by *Basidiobolus ranarum*. *Journal of Medical and Veterinary Mycology*, 32, 225-230.
da Cunha, K.C., Sutton, D.A., Fothergill, A.W., *et al.* (2012a) Diversity of *Bipolaris* species in clinical samples in the United States and their antifungal susceptibility profiles. *Journal of Clinical Microbiology*, 50, 4061-4066.
da Cunha, K.C., Sutton, D.A., Gene, J., *et al.* (2012b) Molecular identification and *in vitro* response to antifungal drugs of clinical isolates of *Exserohilum*. *Antimicrobial Agents and Chemotherapy*, 56, 4951-4954.
da Cunha, K.C., Sutton, D.A., Fothergill, A.W., *et al.* (2013) *In vitro* antifungal susceptibility and molecular identity of 99 clinical isolates of the opportunistic fungal genus *Curvularia*. *Diagnostic Microbiology and Infectious Disease*, 76, 168-174.
da Cunha, K.C., Sutton, D.A., Gene, J., *et al.* (2014) *Pithomyces* species (Montagnulaceae) from clinical specimens: identification and antifungal susceptibility profiles. *Medical Mycology*, 52, 748-757.
de Almeida, J.N. Jr., Favero Gimenes, V.M., Francisco, E.C., *et al.* (2017) Evaluating and improving Vitek MS for identification of clinically relevant species of *Trichosporon* and the closely related genera *Cutaneotrichosporon* and *Apiotrichum*. *Journal of Clinical Microbiology*, 55, 2439-2444.
de Almeida, S.M., Imano, E.C.M., Vicente, V.A., *et al.* (2020) Primary central nervous system infection by *Histoplasma* in an immunocompetent adult. *Mycopathologia*, 185, 331-338.
de Beer, Z.W., Begerow, D., Bauer, R., *et al.* (2006) Phylogeny of the Quambalariacea fam. nov., including important *Eucalyptus* pathogens in South Africa and Australia. *Studies in Mycology*, 55, 289-298.

de Gruyter, J., Aveskamp, M.M., Woudenberg, J.H.C., *et al.* (2009) Molecular phylogeny of *Phoma* and allied anamorph genera: towards a reclassification of the *Phoma* complex. *Mycological Research*, 113, 508-519.

de Hoog, G.S. (1972) The genera *Beauvaria, Isaria, Tritrachium* and *Acrodontium* gen. nov. *Studies in Mycology*, 1, 1-41.

de Hoog, G.S. and de Vries, G.A. (1973) Two new species of *Sporothrix* and their relation to *Blastobotrys nivea*. *Antonie van Leeuwenhoek*, 39, 515-520.

de Hoog, G.S. (1977) *Rhinocladiella* and allied genera. *Studies in Mycology*, 15, 1-140.

de Hoog, G.S. and Hermanides-Nijhof, E.J. (1977) The black yeasts and allied hyphomycetes. *Studies in Mycology*, 15, Centraalbureau voor Schimmelcultures, The Netherlands.

de Hoog, G.S. (1983) On the potentially pathogenic dematiaceous Hyphomycetes. In Howard, D.H. and Howard, L.F. (eds.) *Fungi Pathogenic for Humans and Animals.* Marcel Dekker, New York, pp. 149-216.

de Hoog, G.S., Rantio-Lehtimaki, A.H. and Smith, M.T.H. (1985) *Blastobotryis; Sporothrix* and *Trichosporiella*; generic delimitation, new species, and a *Stephanoascus* teleomorph. *Antontie van Leeuwenhoek*, 51, 79-109.

de Hoog, G.S., Smith, M.T. and Gueho, E. (1986) A revision of the genus *Geotrichum* and its teleomorphs. *Studies in Mycology*, 29, 1-131.

de Hoog, G.S., Gueho, E., Masclaux, F., *et. al.* (1995) Nutritional physiology and taxonomy of human-pathogenic *Cladosporium-Xylohypha* species. *Journal of Medical and Veterinary Mycology*, 33, 339-347.

de Hoog, G.S., Weenink, X.O. and Gerrits van den Ende, A.H.G. (1999) Taxonomy of the *Phialophora verrucosa* complex with the description of two new species. *Studies in Mycology*, 43, 107-122.

de Hoog, G.S. and Horre, R. (2002) Molecular taxonomy of the *Alternaria* and *Ulocladium* species described from humans and their identification in the routine laboratory. *Mycoses*, 45, 259-276.

de Hoog, G.S., Vicente, V., Caligiorne, R.B., *et al.* (2003) Species diversity and polymorphism in the *Exophiala spinifera* clade containing opportunistic black yeast-like fungi. *Journal of Clinical Microbiology*, 41, 4767-4778.

de Hoog, G.S. and Smith, M.T. (2004) Ribosomal gene phylogeny and species delimitation in *Geotrichum* and its teleomorphs. *Studies in Mycology*, 50, 489-515.

de Hoog, G.S., Adelmann, D., Ahmed, A.O.A., *et al.* (2004a) Phylogeny and typification of *Madurella mycetomatis*, with a comparison of other agents of eumycetoma. *Mycoses*, 47, 121-130.

de Hoog, G.S., Attili, D., Vicente, V.A., *et. al.* (2004b) Molecular ecology and pathogenic potential of *Fonsecaea* species. *Medical Mycology*, 42, 405-416.

de Hoog, G.S., Zeng, J.S., Harrak, M.J., *et al.* (2006) *Exophiala xenobiotica* sp. nov., an opportunistic black yeast inhabiting environments rich in hydrocarbons. *Antonie van Leeuwenhoek*, 90, 257-268.

de Hoog, G.S., Nishikaku, A.S., Fernandez-Zeppenfeldt, G., *et al.* (2007) Molecular analysis and pathogenicity of the *Cladophialophora carrionii* complex, with the description of a novel species. *Studies in Mycology*, 58, 219-234.

de Hoog, G.S. and Smith, M.T. (2011a) Chapter 91, *Geotrichum* Link: Fries (1832). In Kurtzman, C.P., Fell, J.W. and Boekhout, T. (eds) *The Yeasts, a Taxonomic Study*, 5th edition. Elsevier B.V., pp. 1279-1286.

de Hoog, G.S. and Smith, M.T. (2011b) Chapter 97, *Saprochaete* Coker & Shanor ex D.T.S. Wagner & Dawes (1970). In Kurtzman, C.P., Fell, J.W. and Boekhout, T. (eds) *The Yeasts, a Taxonomic Study*, 5th edition. Elsevier B.V., pp. 1317-1327.

de Hoog, G.S. and Smith, M.T. (2011c) Chapter 45, *Magnusiomyces* Zender (1977). In Kurtzman, C.P., Fell, J.W. and Boekhout, T. (eds) *The Yeasts, a Taxonomic Study*, 5th edition. Elsevier B.V., pp. 565-574.

de Hoog, G.S., van Diepeningen, A.D., Mahgoub, el-S., *et al.* (2012) New species of *Madurella*, causative agents of black-grain mycetoma. *Journal of Clinical Microbiology*, 50, 988-994.

de Hoog, G.S., Guarro, J., Gene, J., *et al.* (2015) *Atlas of Clinical Fungi (Version 4.1.2).* Centraalbureau voor Schimmelcultures, Utrecht, The Netherlands.

de Hoog G.S., Dukik, K., Monod, M., *et al.* (2017) Towards a noval multilocus phylogenetic taxonomy for dermatophytes. *Mycopathologia*, 182, 5-31.

de Hoog, G.S., Guarro, J., Gene, J., *et al.* (2020) *Atlas of Clinical Fungi, 4th edition.* Foundation Atlas of Clinical Fungi https://webshop.atlasclinicalfungi.org.

de Respinis, S., Tonolla, M., Pranghofer, S., *et al.* (2013) Identification of dermatophytes by matrix-assisted laser desorption/ionization time-of-flight mass spectrometry. *Medical Mycology*, 51, 514-521.

Desjardins, C.A., Champion, M.D., Holder, J.W., *et al.* (2011) Comparative genomic analysis of human fungal pathogens causing paracoccidioidomycosis. *PLOS Genetics*, 7, e1002345-e1002345.

Desnos-Ollivier, M., Bretagne, S., Dromer, F., *et al.* (2006) Molecular identification of black-grain mycetoma agents. *Journal of Clinical Microbiology*, 44, 3517-3523.

Desnos-Ollivier, M., Ragon, M., Robert, V., *et al.* (2008) *Debaryomyces hansenii* (*Candida famata*), a rare human fungal pathogen often misidentified as *Pichia guilliermondii* (*Candida guilliermondii*). *Journal of Clinical Microbiology*, 46, 3237–3242.

Desnos-Ollivier, M., Blanc, C., Garcia-Hermoso, D., *et al.* (2014) Misidentification of *Saprochaete clavata* as *Magnusiomyces capitatus* in clinical isolates: utility of internal transcribed spacer sequencing and matrix-assisted laser desorption ionization-time of flight mass spectrometry and importance of reliable databases. *Journal of Clinical Microbiology*, 52, 2196-2198.

Diekema, D.J., Petroelje, B., Messer, S.A., *et al.* (2005) Activities of available and investigational antifungal agents against *Rhodotorula* species. *Journal of Clinical Microbiology*, 43, 476–478.

Diekema, D.J., Messer, S.A., Boyken, L.B., *et al.* (2009) *In vitro* activity of seven systemically active antifungal agents against a large global collection of rare *Candida* species as determined by CLSI broth microdilution methods. *Journal of Clinical Microbiology*, 47, 3170-3177.

Diongue, K., Brechard, L., Diallo, M.A., *et al.* (2019) A comparative study on phenotypic versus ITS-based molecular identification of dermatophytes isolated in Dakar, Senegal. *International Journal of Microbiology*, 2019, 6754058-6.

Dixon, D.M. and Polak-Wyss, A. (1991) The medically important dematiaceous fungi and their identification. *Mycoses*, 34, 1-18.

do Espirito Santo, E.P.T., Monteiro, R.C., da Costa, A.R.F., *et al.* (2019) Molecular identification, genotyping, phenotyping, and antifungal susceptibilities of medically important *Trichosporon*, *Apiotrichum*, and *Cutaneotrichosporon* species. *Mycopathologia*, 185, 307-317.

Dolatabadi, S., Walther, G., Gerrits van den Ende, A.H.G., *et al.* (2014) Diversity and delimitation of *Rhizopus microsporus*. *Fungal Diversity*, 64, 145-163.

Domsch, K.H., Gams, W. and Anderson, T.H. (2007) *Compendium of Soil Fungi*. Second Edition, IHW-Verlag, Germany.

Du, H., Bing, J. and Hu, T. (2020) *Candida auris*: Epidemiology, biology, antifungal resistance, and virulence. *PLoS Pathology*, 16, e1008921.

Duboc de Almeida, G.M., Figueiredo Costa, S., Melhem, M., *et al.* (2008) *Rhodotorula* spp. isolated from blood cultures: clinical and microbiological aspects. *Medical Mycology*, 46, 547-556.

Dukik, K., Munoz, J.F., Jiang, Y., *et al.* (2017) Novel taxa of thermally dimorphic systemic pathogens in the Ajellomycetaceae (Onygenales). *Mycoses*, 60, 296-309.

Dukik, K., de Hoog, G.S., Stielow, J.B., *et al.* (2019) Molecular and phenotypic characterization of *Nannizzia* (Arthrodermataceae). *Mycopathologia*, 185, 1, 9-35.

Durie, E.B. and Frey, D. (1957) A new species of *Trichophyton* from New South Wales. *Mycologia*, 49, 401-411.

Dworzack, D.L., Pollock, A.S., Hodges, G.R., *et al.* (1978) Zygomycosis of the maxillary sinus and palate caused by *Basidiobolus haptosporus*. *Archives of Internal Medicine*, 138, 1274-1276.

Edman, J.C., Kovacs, J.A., Masur, H., *et al.* (1988) Ribosomal RNA sequence shows *Pneumocystis carinii* to be a member of the fungi. *Nature*, 334 (6182), 519-522.

El Feghaly, R.E., Sutton, D.A., Thompson, E.H., *et al.* (2012) *Graphium basitruncatum* fungemia in an immunosuppressed child post stem-cell transplantation. *Medical Mycology Case Reports*, 1, 35-38.

Elias, N.A., Cuestas, M.L., Sandoval, M., *et al.* (2012) Rapid Identification of *Histoplasma capsulatum* directly from cultures by multiplex PCR. *Mycopathologia*, 174, 451-456.

Ellis, D.H. (1981) Ascocarp morphology and terminal hair ornamentation in thermophilic *Chaetomium* species. *Mycologia*, 73, 755-773.

Ellis, D.H. and Keane, P.J. (1981) Thermophilic fungi isolated from some Australian soils. *Australian Journal of Botany*, 29, 689-704.

Ellis, D.H. and Kaminski, G.W. (1985) Laboratory identification of *Saksenaea vasiformis*: a rare cause of zygomycosis in Australia. *Sabouraudia*, 23, 137-140.

Ellis, D.H. (2005a) Subcutaneous Zygomycetes - Entomophthoromycosis. In Merz, W.G. and Hay, R.J. (eds) *Topley and Wilson's Microbiology and Microbial Infections: Medical Mycology*, 10th edition. Hodder Arnold, London, pp. 347-355.

Ellis, D.H. (2005b) Systemic Zygomycetes - Mucormycosis. In Merz, W.G. and Hay, R.J. (eds), *Topley and Wilson's Microbiology and Microbial Infections: Medical Mycology* 10 edition. Hodder Arnold, London, pp. 659-686.

Ellis, J.J. (1985) Species and varieties in *Rhizopus arrhizus - Rhizopus oryzae* group as indicated by their DNA complementarity. *Mycologia*, 77, 243-247.

Ellis, J.J. (1986) Species and varieties in the *Rhizopus microsporus* group as indicated by their DNA complementarity. *Mycologia*, 78, 508-510.

Ellis, J.J. and Hesseltine, C.W. (1965) The genus *Absidia*: globose spored species. *Mycologia*, 57, 222-235.

Ellis, J.J. and Hesseltine, C.W. (1966a) Species of *Absidia* with ovoid sporangiospores. *II. Sabouraudia*, 5, 59-77.

Ellis, J.J. and Hesseltine, C.W. (1966b) Two new families of Mucorales. *Mycologia*, 66, 87-95.

Ellis, J.J. and Ajello, L. (1982) An unusual source for *Apophysomyces elegans* and a method for stimulating sporulation of *Saksenaea vasiformis*. *Mycologia*, 74, 144-145.

Ellis, M.B. (1971) *Dematiaceous Hyphomycetes.* Commonwealth Mycological Institute, Kew, Surrey, England.

Ellis, M.B. (1976) *More Dematiaceous Hyphomycetes*. Commonwealth Mycological Institute, Kew, Surrey, England.

Enache-Angoulvant, A. and Hennequin, C. (2005) Invasive *Saccharomyces* infection: a comprehensive review. *Clinical Infectious Diseases*, 41, 1559-1568.

Erhard, M., Hipler, U.C., Burmester, A., *et al.* (2008) Identification of dermatophyte species causing onychomycosis and tinea pedis by MALDI-TOF mass spectrometry. *Experimental Dermatology*, 17, 356-361.

Espinel-Ingroff, A., Boyle, K. and Sheehan, D.J. (2001) *In vitro* antifungal activities of voriconazole and reference agents as determined by NCCLS methods: review of the literature. *Mycopathologia*, 150, 101-115.

Espinel-Ingroff, A. (2003) *In vitro* antifungal activities of anidulafungin and micafungin, licensed agents and the investigational triazole posaconazole as determined by NCCLS methods for 12,052 fungal isolates: review of the literature. *Revista Iberoamericana de Micologia*, 20, 121-136.

Espinel-Ingroff, A., Arendrup, M.C., Pfaller, M.A., *et al.* (2013) Interlaboratory variability of Caspofungin MICs for *Candida* spp. using CLSI and EUCAST methods: should the clinical laboratory be testing this agent? *Antimicrobial Agents and Chemotherapy*, 57, 5836-5842.

Espinel-Ingroff, A., Pfaller, M.A., Bustamante, B., *et al.* (2014) Multilaboratory study of epidemiological cutoff values for detection of resistance in eight *Candida* species to fluconazole, posaconazole, and voriconazole. *Antimicrobial Agents and Chemotherapy*, 58, 2006-2012.

Espinel-Ingroff, A., Chakrabarti, A., Chowdhary, A., *et al.* (2015a) Multicenter evaluation of MIC distributions for epidemiologic cutoff value definition to detect amphotericin B, posaconazole, and itraconazole resistance among the most clinically relevant species of Mucorales. *Antimicrobial Agents and Chemotherapy*, 59, 1745-1750.

Espinel-Ingroff, A., Chowdhary, A., Gonzalez, G.M., *et al.* (2015b) Multicenter study of isavuconazole MIC distributions and epidemiological cutoff values for the *Cryptococcus neoformans-Cryptococcus gattii* species complex using the CLSI M27-A3 broth microdilution method. *Antimicrobial Agents and Chemotherapy*, 59, 666-668.

Espinel-Ingroff, A. and Kidd, S.E. (2015) Current trends in the prevalence of *Cryptococcus gattii* in the United States and Canada. *Infection and Drug Resistance*, 8, 89-97.

Espinel-Ingroff, A., Abreu, D.P., Almeida-Paes, R., Brilhante, R.S., *et al.* (2017) Multicenter, International study of MIC/MEC distributions for definition of epidemiological cutoff values for *Sporothrix* species identified by molecular methods. *Antimicrobial Agents and Chemotherapy*, 61, 10.

Estrada-Barcenas, D.A., Vite-Garín, T., Navarro-Barranco, H., *et al.* (2014) Genetic diversity of *Histoplasma* and *Sporothrix* complexes based on sequences of their ITS1-5.8S-ITS2 regions from the BOLD System. *Revista Iberoamericana Micologia*, 31, 90-94.

Fell, J.W., Scorzetti, G., Connell, L., *et al.* (2006) Biodiversity of micro-eukaryotes in Antarctic dry valley soils with <5% soil moisture. *Soil Biology and Biochemistry*, 38, 3107-3119.

Fischer, J.B., and Kane, J. (1971) The detection of contamination in *Trichophyton rubrum* and *Trichophyton mentagrophytes*. *Mycopathologia*, 43, 169-180.

Fisher, M.C., Koenig, G.L., White, T.J., *et al.* (2002) Molecular and phenotypic description of *Coccidioides posadasii* sp. nov., previously recognized as the non-California population of *Coccidioides immitis*. *Mycologia*, 94, 73-84.

Fonseca, A., Boekhout, T. and Fell, J.W. (2011) Chapter 138, *Cryptococcus* Vuillemin (1901). In Kurtzman, C.P., Fell, J.W. and Boekhout, T. (eds) *The Yeasts, a Taxonomic Study*, 5th edition. pp. 1661-1737.

Formoso, A., Heidrich, D., Felix, C.R., *et al.* (2015) Enzymatic activity and susceptibility to antifungal agents of Brazilian environmental isolates of *Hortaea werneckii*. *Mycopathologia*,180, 345-352.

Frankel, D.H. and Rippon, J.W. (1989) *Hendersonula toruloidea* infection in man. *Mycopathologia*, 105,175-186.

Franzot, S.P., Salkin, I.F. and Casadevall, A. (1999) *Cryptococcus neoformans* var. *grubii*: separate varietal status for *Cryptococcus neoformans* serotype A isolates. *Journal of Clinical Microbiology*, 37, 838-840.

Fraser, M., Brown, Z., Houldsworth, M., *et al.* (2016) Rapid identification of 6328 isolates of pathogenic yeasts using MALDI-ToF MS and a simplified, rapid extraction procedure that is compatible with the Bruker Biotyper platform and database. *Medical Mycology*, 54, 80-88.

Gaitanis, G., Magiatis, P., Hantschke, M., *et al.* (2012) The *Malassezia* genus in skin and systemic diseases. *Clinical Microbiology Reviews*, 25, 106–141.

Gams, W. (1971) *Cephalosporium-artige Schimmelpilze (Hyphomycetes)*. G. Fisher, Stuttgart, p. 262.

Gams, W., Christensen, M. and Onions, A.H. (1985) Infrageneric taxa of *Aspergillus*. In Samson, R.A. and Pitt, J.I. (eds) *Advances in Penicillium and Aspergillus Systematics.* Plenum Press, New York, 55-62.

Garcia-Hermoso, D., Hoinard, D., Gantier, J.C., *et al.* (2009) Molecular and phenotypic evaluation of *Lichtheimia corymbifera* (formerly *Absidia corymbifera*) complex isolates associated with human mucormycosis: rehabilitation of *L. ramosa*. *Journal of Clinical Microbiology*, 47, 3862-3870.

Garcia-Ruiz, J.C., Lopez-Soria, L., Olazabal, I., *et al.* (2013) Invasive infections caused by *Saprochaete capitata* in patients with haematological malignancies: report of five cases and review of the antifungal therapy. *Revista Iberoamericana Micologia*, 30, 248-255.

Geiser, D.M., Klich, M.A., Frisvad, J.C., *et al.* (2007) The current status of species recognition and identification in *Aspergillus*. *Studies in Mycology*, 59, 1-10.

Geiser, D.M., Aoki, T., Bacon, C.W., *et al.* (2013) One fungus, one name: defining the genus *Fusarium* in a scientifically robust way that preserves longstanding use. *Phytopathology*, 103, 400-408.

George, R.B. and Penn, R.L. (1986) Histoplasmosis. In Sarosi, G.A. and Davies, S.F. (eds) *Fungal Diseases of the Lung*. Grune and Stratton Inc.

Gerrits van den Ende, A.H.G. and de Hoog, G.S. (1999) Variability and molecular diagnostics of the neurotropic species *Cladophialophora bantiana*. *Studies in Mycology*, 43, 151-162.

Ghikas, D.V., Kouvelis, V.N. and Typas, M.A. (2010) Phylogenetic and biogeographic implications inferred by mitochondrial intergenic region analyses and ITS1-5.8S-ITS2 of the entomopathogenic fungi *Beauveria bassiana* and *B. brongniartii*. *BMC Microbiology*, 10, 174.

Gilgado, F., Cano, J., Gene, J., *et al.* (2005) Molecular phylogeny of the *Pseudallescheria boydii* species complex: proposal of two new species. *Journal of Clinical Microbiology*, 43, 4930-4942.

Gilgado, F., Gene, J., Cano, J., *et al.* (2010) Heterothallism in *Scedosporium apiospermum* and description of its teleomorph *Pseudallescheria apiosperma* sp. nov. *Medical Mycology*, 48, 122-128.

Giraldo, A., Gene, J., Sutton, D.A., *et al.* (2014a) Phylogenetic circumscription of *Arthrographis* (Eremomycetaceae, Dothideomycetes). *Persoonia*, 32, 102-114.

Giraldo, A., Sutton, D.A., Samerpitak, K., *et al.* (2014b) Occurrence of *Ochroconis* and *Verruconis* species in clinical specimens from the United States. *Journal of Clinical Microbiology*, 52, 4189-4201.

Giraldo, A., Gene, J., Sutton, D.A., *et al.* (2015) Phylogeny of *Sarocladium* (Hypocreales). *Persoonia*, 34, 10-24.

Glenn, A., Bacon, C.W., Price, R., *et al.* (1996) Molecular phylogeny of *Acremonium* and its taxonomic implications. *Mycologia*, 88, 369-383.

Goldschmied-Reouven, A., Shvoron, A., Topaz, M. and Block C. (1989) *Saksenaea vasiformis* infection in a burn wound. *Journal of Medical and Veterinary Mycology*, 27, 427-429.

Gonzalez, G.M., Fothergill, A.W., Sutton, D.A., *et al.* (2005) *In vitro* activities of new and established triazoles against opportunistic filamentous and dimorphic fungi. *Medical Mycology*, 43, 281-284.

Goodman, N.L. and Rinaldi, M.G. (1991) Agents of zygomycosis. In Balows, A., Hausler, W.J., Herrmann, K.L., *et al.* (eds) *Manual of Clinical Microbiology* 5th edition. American Society for Microbiology, Washington, DC.

Gordon, R., Sutton, D., Thompson, E., *et al.* (2012) Cutaneous phaeohyphomycosis caused by *Paraconiothyrium cyclothyrioides*. *Journal of Clinical Microbiology*, 50, 3795-3798.

Gori, S., Pellegrini, G., Filipponi, F., *et al.* (1998) Pulmonary aspergillosis caused by *Neosartorya fischeri* (*Aspergillus fischerianus*) in a liver transplant recipient. *Journal de Mycologie Medicale*, 8, 105-107.

Gramaje, D., Mostert, L., Groenewald, J.Z., Crous, P.W. (2015) *Phaeoacremonium*: from esca disease to phaeohyphomycosis. *Fungal Biology*, 119, 759-783.

Graser, Y., El Fari, M., Presber, W., *et al.* (1998) Identification of common dermatophytes (*Trichophyton*, *Microsporum*, *Epidermophyton*) using polymerase chain reactions. *British Journal of Dermatology*, 138, 576-582.

Graser, Y., El Fari, M., Vilgalys, R., *et al.* (1999a) Phylogeny and taxonomy of the family Arthrodermataceae (dermatophytes) using sequence analysis of the ribosomal ITS region. *Medical Mycology*, 37, 105-114.

Graser, Y., Kühnisch, J. and Presber, W. (1999b) Molecular markers reveal exclusively clonal reproduction in *Trichophyton rubrum*. *Journal of Clinical Microbiology*, 37, 3713-3717.

Graser, Y., Kuijpers, A.F.A., El Fari, M., *et al.* (2000) Molecular and conventional taxonomy of the *Microsporum canis* complex. *Medical Mycology*, 38, 143-153.

Graser, Y., Scott, J. and Summerbell, R. (2008) The new species concept in dermatophytes - a polyphasic approach. *Mycopathologia*, 166, 239-256.

Greer, D.L. and Friedman, L. (1966) Studies on the genus *Basidiobolus* with reclassification of the species pathogenic for man. *Sabouraudia*, 4, 231-241.

Guarro, J., Gams, W., Pujol, I. and Gene, J. (1997) *Acremonium* species: New emerging fungal opportunists - *in vitro* antifungal susceptibilities and review. *Clinical Infectious Diseases*, 25, 1222-1229.

Guarro, J., Kantarcioglus, A.S., Horre, R., *et al.* (2006) *Scedosporium apiospermum*: changing clinical spectrum of a therapy-refractory opportunist. *Medical Mycology*, 44, 295-327.

Guarro, J., Mendiratta, D.K., De Sequeira, H., *et al.* (2007) *Acrophialophora fusispora:* an emerging agent of human mycoses. A report of 3 new clinical cases. *Diagnostic Microbiology and Infectious Disease*, 59, 85-88.

Guarro, J., Chander, J., Alvarez, E., *et al.* (2011) *Apophysomyces variabilis* infections in humans. *Emerging Infectious Diseases*, 17, 134-135.

Guarro, J. (2013) Fusariosis, a complex infection caused by a high diversity of fungal species refractory to treatment. *European Journal of Clinical Microbiology and Infectious Diseases*, 32, 1491-1500.

Gueho, E. (1979) Deoxyribonucleic acid base composition and taxonomy in the genus *Geotrichum* Link. *Antonie van Leeuwenhoek*, 45, 199-210.
Gueho, E., Smith, M.T., de Hoog, G.S., *et al.* (1992) Contributions to a revision of the genus *Trichosporon*. *Antonie van Leeuwenhoek*, 61, 289-316.
Gueho, E., Midgley, G. and Guillot, J. (1996) The genus *Malassezia* with description of four new species. *Antonie van Leeuwenhoek*, 69, 337-355.
Guillot J. and Gueho, E. (1995) The diversity of *Malassezia* yeasts confirmed by rRNA sequence and nuclear DNA comparisons. *Antonie Van Leeuwenhoek*, 67, 297-314.
Guillot, J., Gueho, E., Lesourd, M., *et al.* (1996) Identification of *Malassezia* species. *Journal de Mycologie Medicale*, 6, 103-110.
Guillot, J., Deville, M., Berthelemy, M., *et. al.* (2000) A single PCR-restriction endonuclease analysis for rapid identification of *Malassezia* species. *Letters in Applied Microbiology*, 31, 400-403.
Guo, L.N., Xiao, M., Kong, F., *et al.* (2011) Three-locus identification, genotyping, and antifungal susceptibilities of medically important *Trichosporon* species from China. *Journal of Clinical Microbiology*, 49, 3805-3811.
Gupta, A.K., Horgan-Bell, C.B. and Summerbell, R.C. (1998) Onychomycosis associated with *Onychocola canadensis*: ten case reports and a review of the literature. *Journal of the American Academy of Dermatology*, 39, 410-407.
Hageage, G.J. and Harrington, B.J. (1984) Use of Calcofluor white in clinical mycology. *Laboratory Medicine*, 15, 109-112.
Hagen, F., Khayhan, K., Theelen, B., *et al.* (2015) Recognition of seven species in the *Cryptococcus gattii/Cryptococcus neoformans* species complex. *Fungal Genetics and Biology*, 78, 16-48.
Hainsworth, S., Kucerova, I., Sharma, R., *et al.* (2021) Three-gene phylogeny of the genus *Arthroderma*: Basis for future taxonomic studies. *Medical Mycology*, 59, 355-365.
Hall, M.R., Brumble, L.M., Mayes, M.A., *et al.* (2013) Cutaneous *Microsphaeropsis arundinis* infection initially interpreted as squamous cell carcinoma. *International Journal of Dermatology*, 52, 84-86.
Halliday, C.L., Kidd, S.E., Sorrell, T.C., *et al.* (2015) Molecular diagnostic methods for invasive fungal disease: the horizon draws nearer? *Pathology*, 47, 257-269.
Harrington, B.J. and Hageage, G.J. (2003) Calcofluor white: A review of its uses and applications in clinical mycology and parasitology. *Laboratory Medicine*, 34, 361-367.
Heath, C.H., Slavin, M.A., Sorrell, T.C., *et al.* (2009) Population-based surveillance for scedosporiosis in Australia: epidemiology, disease manifestations and emergence of *Scedosporium aurantiacum* infection. *Clinical Microbiology and Infection*, 15, 689-693.
Heath, C.H., Dyer, J.R., Pang, S., *et al.* (2019) *Candida auris* sternal osteomyelitis in a man from Kenya visiting Australia, 2015. *Emerging Infectious Diseases*, 25, 192-194.
Hedayati, M.T., Pasqualotto, A.C., Warn, P.A., *et al.* (2007) *Aspergillus flavus:* human pathogen, allergen and mycotoxin producer. *Microbiology*, 153, 1677-1692.
Hegedus, D.D. and Khachatourians, G.G. (1996) Identification and differentiation of the entomopathogenic fungus *Beauveria bassiana* using polymerase chain reaction and single-strand conformation polymorphism analysis. *Journal of Invertebrate Pathology*, 67, 289-299.
Hermanides-Nijhof, E.J. (1977) *Aureobasidium* and allied genera. *Studies in Mycology*, 15, 141-177.
Hernandez-Restrepo, M., Maid, H., Tan, Y., *et al.* (2018) Multi-locus phylogeny and taxonomy of *Exserohilum*. *Persoonia*, 41, 71-108.
Hesseltine, C.W. and Ellis, J.J. (1964a) The genus *Absidia*: *Gongronella* and cylindrical-spored species of *Absidia*. *Mycologia*, 56, 568-601.
Hesseltine, C.W. and Ellis, J.J. (1964b) An interesting case of *Mucor*, *M. ramosissimus. Sabouraudia*, 3, 151-154.
Hesseltine, C.W. and Ellis, J.J. (1966) Species of *Absidia* with ovoid sporangiospores. *I. Mycologia*, 58, 173-194.
Hoffman, K., Discher, S. and Voigt, K. (2007) Revision of the genus *Absidia* (Mucorales, Zygomycetes) based on physiological, phylogenetic, and morphological characters, thermotolerant *Absidia* spp. form a coherent group, Mycocladiaceae fam. nov. *Mycological Research*, 111, 1169-1183.
Hoffmann, K., Walther, G. and Voigt, K. (2009) *Mycocladus vs. Lichtheimia*, a correction (Lichtheimiaceae fam. nov., Mucorales, Mucoromycotina). *Mycological Research*, 113, 277-278.
Hohl, P.E., Holley, H.P. Jr., Prevost, E., *et al.* (1983) Infections due to *Wangiella dermatitidis* in humans: report of the first documented case from the United States and a review of the literature. *Reviews of Infectious Diseases*, 5, 854-864.
Holland, J. (1997) Emerging zygomycosis of humans: *Saksenaea vasiformis* and *Apophysomyces elegans*. *Current Topics in Medical Mycology*, 8, 27-34.
Honnavar, P., Prasad, G.S., Ghosh, A., *et al.* (2016) *Malassezia arunalokei* sp. nov., a novel yeast species isolated from seborrheic dermatitis patients and healthy individuals from India. *Journal of Clinical Microbiology* 54, 1826-1834.

Houbraken, J., Verweij, P.E., Rijs, A.J., *et al.* (2010) Identification of *Paecilomyces variotii* in clinical samples and settings. *Journal of Clinical Microbiology*, 48, 2754-2761.

Huguenin, A., Lorot, A. and Zachar, D. (2015) Matrix-assisted laser desorption ionization-time of flight identification of *Schizophyllum commune*: perspectives on the review by Chowdhary *et al. Medical Mycology*, 53, 896-897.

Hung, N., Hsiao, C.H., Yang, C.S., *et al.* (2020a) *Colletotrichum* keratitis: A rare yet important fungal infection of human eyes. *Mycoses*, 63, 407-415.

Hung, T.-Y., Taylor, B., Lim, A., *et al.* (2020b) Skin and soft tissue infection caused by *Basidiobolus* spp. in Australia. *IDCases*, 20, e00731.

Imai, T., Sano, A., Mikami, Y., *et al.* (2000) A new PCR primer for the identification of *Paracoccidioides brasiliensis* on rRNA sequences coding the internal transcribed spacers (ITS) and 5.8S regions. *Medical Mycology*, 38, 323-326.

Inacio, C.A., Pereira-Carvalho, R.C., Morgado, F.G.A., *et al.* (2011) A tomato fruit rot caused by *Trichothecium roseum* in Brazil. *Plant Disease*, 95,1318-1318.

Inderbitzin, P., Davis, R.M., Bostock, R.M., *et al.* (2013) Identification and differentiation of *Verticillium* species and *V. longisporum* lineages by Simplex and Multiplex PCR assays. *PLoS ONE*, 8, e65990.

Irinyi, L., Serena, C., Garcia-Hermoso, D., *et al.* (2015) International Society of Human and Animal Mycology (ISHAM)-ITS reference DNA barcoding database-the quality controlled standard tool for routine identification of human and animal pathogenic fungi. *Medical Mycology*, 53, 313-337.

Irokanulo, E.A., Akueshi, C.O. and Makinde, A.A. (1994) Differentiation of *Cryptococcus neoformans* serotypes A and D using creatinine dextrose bromothymol blue thymine medium. *British Journal of Biomedical Science*, 51, 100-103.

Jackson, L., Klotz, S.A. and Normand, R.E. (1996) A pseudoepidemic of *Sporothrix cyanescens* pneumonia occurring during renovation of a bronchoscopy suite. *Journal of Medical and Veterinary Mycology*, 28, 455-459.

Jagielski, T., Bakuła, Z., Gawor, J., *et al.* (2019) The genus *Prototheca* (Trebouxiophyceae, Chlorophyta) revisited: Implications from molecular taxonomic studies. *Algal Research*, 43, 101639.

Jarv, H., Lehtmaa, J., Summerbell, R.C., *et al.* (2004) Isolation of *Neosartorya pseudofischeri* from blood: first hint of pulmonary Aspergillosis. *Journal of Clinical Microbiology*, 42, 925-928.

Jeffery-Smith, A., Taori, S.K., Schelenz, S., *et al.* (2018) *Candida auris*: a review of the literature. *Clinical Microbiology Reviews*, 31, e00029-17.

Jiang, Y., Dukik, K., Munoz, J.F., *et al.* (2018) Phylogeny, ecology and taxonomy of systemic pathogens and their relatives in Ajellomycetaceae (Onygenales): *Blastomyces, Emergomyces, Emmonsia, Emmonsiellopsis*. *Fungal Diversity*, 90, 245-291.

Kaltseis, J., Rainer, J. and de Hoog, G.S. (2009) Ecology of *Pseudallescheria* and *Scedosporium* species in human-dominated and natural environments and their distribution in clinical samples. *Medical Mycology*, 47, 398-405.

Kandemir, H., Dukik, K., Hagen, F., *et al.* (2020) Polyphasic discrimination of *Trichophyton tonsurans* and *T. equinum* from humans and horses. *Mycopathologia*, 185,113-122.

Kane, J., Summerbell, R., Sigler, L., *et al.* (1997) *Laboratory Handbook of Dermatophytes.* Star Publishing Co, Belmont, California.

Kaneko, T., Makimura, K., Abe, M., *et al.* (2007) Revised culture-based system for identification of *Malassezia* species. *Journal of Clinical Microbiology*, 45, 3737-3742.

Kanj, S.S., Amr, S.S. and Roberts, G.D. (2001) *Ramichloridium mackenziei* brain abscess: report of two cases and review of the literature. *Medical Mycology*, 39, 97-102.

Kano, R., Kimura, U., Kakurai, M., *et al.* (2020) *Trichophyton indotineae* sp. nov.: a new highly terbinafine-resistant anthropophilic dermatophyte species. *Mycopathologia*, 185, 947-958.

Kantarcioglu, A.S., Guarro, J., de Hoog, S., *et al.* (2017) An updated comprehensive systematic review of *Cladophialophora bantiana* and analysis of epidemiology, clinical characteristics, and outcome of cerebral cases. *Medical Mycology*, 55, 579-604.

Kaplan, E., Al-Hatmi, A.M.S., Ilkit, M., *et al.* (2018) Molecular diagnostics of arthroconidial yeasts, frequent pulmonary opportunists. *Journal of Clinical Microbiology*, 56, e01427-17.

Kaplan, E., Aktaş, D., Dogen, A., *et al.* (2021) Development and analysis of qPCR for the identification of arthroconidial yeasts of the genus *Magnusiomyces*. *Mycopathologia*, 186, 41-51.

Kaplan, W. (1977) Protothecosis and infections caused by morphologically similar green algae. *The black and white yeasts. Proceedings of the Fourth International Conference on the Mycoses.* Scientific Publication No. 356. Pan American Health Organization. Washington, DC.

Kathuria, S., Singh, P.K., Meis, J.F., *et al.* (2014) *In vitro* antifungal susceptibility profile and correlation of mycelial and yeast forms of molecularly characterized *Histoplasma capsulatum* strains from India. *Antimicrobial Agents and Chemotherapy*, 58, 5613-5616.

Kathuria, S., Singh, P.K., Sharma, C., *et al.* (2015) Multidrug-resistant *Candida auris* misidentified as *Candida haemulonii*: characterization by matrix-assisted laser desorption ionization- time of flight mass spectrometry and DNA sequencing and its antifungal susceptibility profile variability by Vitek 2, CLSI broth microdilution, and Etest method. *Journal of Clinical Microbiology*, 53, 1823-1830.

Katiraee, F., Kouchak Kosari, Y., Soltani, M., *et al.* (2021) Molecular identification and antifungal susceptibility patterns of dermatophytes isolated from companion animals with clinical symptoms of dermatophytosis. *Journal of Veterinary Research*, 65,175-182.

Katragkou, A., Pana, Z.D., Perlin, D.S., *et al.* (2014) *Exserohilum* infections: review of 48 cases before the 2012 United States outbreak. *Medical Mycology*, 52, 376-386.

Kaufman, L. and Standard, P.G. (1987) Specific and rapid identification of medically important fungi by exoantigen detection. *Annual Review of Microbiology*, 41, 209-225.

Kenyon, C., Bonorchis, K., Corcoran, C., *et al.* (2013) A dimorphic fungus causing disseminated infection in South Africa. *New England Journal of Medicine*, 369, 1416-1424.

Khan, Z.U., Lamdhade, S.J., Johny, M., *et al.* (2002) Additional case of *Ramichloridium mackenziei* cerebral phaeohyphomycosis from the Middle East. *Medical Mycology*, 40, 429-433.

Khan, Z., Gene, J., Ahmad, S., *et al.* (2013) *Coniochaeta polymorpha*, a new species from endotracheal aspirate of a preterm neonate, and transfer of *Lecythophora* species to *Coniochaeta*. *Antonie van Leeuwenhoek*, 104, 243-252.

Khunnamwong, P., Lertwattanasakul, N., Jindamorakot, S., *et al.* (2015) Description of *Diutina* gen. nov., *Diutina siamensis*, f.a. sp. nov., and reassignment of *Candida catenulata, Candida mesorugosa, Candida neorugosa, Candida pseudorugosa, Candida ranongensis, Candida rugosa* and *Candida scorzettiae* to the genus *Diutina*. *International Journal of Systematic Evolutionary Microbiology*, 65, 4701-4709.

Kidd, S.E., Chow, Y., Mak, S., *et al.* (2007a) Characterization of environmental sources of the human and animal pathogen *Cryptococcus gattii* in British Columbia, Canada, and the Pacific Northwest of the United States. *Applied and Environmental Microbiology*, 73, 1433-1443.

Kidd, S.E., Bach, P.J., Hingston, A.O., *et al.* (2007b) *Cryptococcus gattii* dispersal mechanisms, British Columbia, Canada. *Emerging Infectious Diseases*, 13, 51-57.

King, D.S. (1983) Entomophthorales. In Howard, D.H. (ed.) *Fungi Pathogenic for Humans and Animals. Part A Biology*. Marcel Dekker Inc. New York, pp. 61-73.

Kirchhoff, L., Olsowski, M., Rath, P.M., *et al.* (2019) *Exophiala dermatitidis:* Key issues of an opportunistic fungal pathogen. *Virulence*, 10, 984-998.

Kirk, P.M., Cannon, P.F., Minter, D.W. and Stalpers, J.A. (2008) *Ainsworth & Bisby's Dictionary of the Fungi*, 10th ed. CABI Publishing, Wallingford, UK.

Kirkland, T.N. and Fierer, J. (2018) *Coccidioides immitis* and *posadasii;* A review of their biology, genomics, pathogenesis, and host immunity. *Virulence*, 9, 1426-1435.

Klich, M.A. (2002a) *Identification of Common Aspergillus Species*. Centraalbureau voor Schimmelcultures, Utrecht, The Netherlands.

Klich, M.A. (2002b) Biogeography of *Aspergillus* species in soil and litter. *Mycologia*, 94, 21-27.

Kluger, E.K., Della Torre, P.K., Martin, P., *et al.* (2004) Concurrent *Fusarium chlamydosporium* and *Microsphaeropsis arundinis* infections in a cat. *Journal of Feline Medicine and Surgery*, 6, 271-277.

Kolecka, A., Khayhan, K., Groenewald, M., *et al.* (2013) Identification of medically relevant species of arthroconidial yeasts by use of matrix-assisted laser desorption ionization-time of flight mass spectrometry. *Journal of Clinical Microbiology*, 51, 2491-2500.

Kolecka, A., Khayhan, K., Arabatzis, M., *et al.* (2014) Efficient identification of *Malassezia* yeasts by matrix-assisted laser desorption ionization-time of flight mass spectrometry (MALDI-TOF MS). *British Journal of Dermatology*, 170, 332-341.

Kondori, N., Erhard, M., Welinder-Olsson, C., *et al.* (2015) Analyses of black fungi by matrix-assisted laser desorption/ionization time-of-flight mass spectrometry (MALDI-TOF MS): species-level identification of clinical isolates of *Exophiala dermatitidis*. *FEMS Microbiology Letters*, 362, 1-6.

Kordalewska, M., Zhao, Y., Lockhart, S.R., *et al.* (2017) Rapid and accurate molecular identification of the emerging multidrug-resistant pathogen *Candida auris*. *Journal of Clinical Microbiology*, 55, 2445-2452.

Krockenberger, M.B., Martin, P., Halliday, C., *et al.* (2010) Localised *Microsphaeropsis arundinis* infection of the subcutis of a cat. *Journal of Feline Medicine and Surgery*, 12, 231-236.

Kuan, C.S., Yew, S.M., Toh, Y.F., *et al.* (2015) Identification and characterization of a rare fungus, *Quambalaria cyanescens*, isolated from the peritoneal fluid of a patient after nocturnal intermittent peritoneal dialysis. *PLoS One*, 10, e0145932.

Kumar, A., Sachu, A., Mohan, K., *et al.* (2017) Simple low cost differentiation of *Candida auris* from *Candida haemulonii* complex using CHROMagar Candida medium supplemented with Pal's medium. *Revista Iberoamericana de Micologia*, 34, 109-111.

Kurtzman, C., Fell, J.W. and Boekhout, T. (2011) *The Yeasts: A Taxonomic Study*, 5th edition. Elsevier Science, Amsterdam.

Kurtzman, C.P. (2011a) Chapter 42, *Lindnera* Kurtzman, Robnett & Basehoar-Powers (2008). In Kurtzman, C.P., Fell, J.W. and Boekhout, T. (eds) *The Yeasts, a Taxonomic Study*, 5th edition. Elsevier B.V., pp. 521-543.

Kurtzman, C.P. (2011b) Chapter 47, *Meyerozyma* Kurtzman & M. Suzuki (2010). In Kurtzman, C.P., Fell, J.W. and Boekhout, T. (eds) *The Yeasts, a Taxonomic Study*, 5th edition. Elsevier B.V., pp. 621-641.

Kurtzman, C.P. (2011c) Chapter 57, *Pichia* E.C. Hansen (1904). In Kurtzman, C.P., Fell, J.W. and Boekhout, T. (eds) *The Yeasts, a Taxonomic Study*, 5th edition. Elsevier B.V., pp. 685-707.

Kurtzman, C.P. (2011d) Chapter 75, *Torulaspora* Lindner (1904). In Kurtzman, C.P., Fell, J.W. and Boekhout, T. (eds) *The Yeasts, a Taxonomic Study*, 5th edition. Elsevier B.V., pp. 867-874.

Kurtzman, C.P. (2011e) Chapter 80, *Wickerhamomyces* Kurtzman, Robnett & Basehoar-Powers (2008). In Kurtzman, C.P., Fell, J.W. and Boekhout, T. (eds) *The Yeasts, a Taxonomic Study*, 5th edition. pp. 899-917.

Kurtzman, C.P. (2011f) Chapter 82, *Yarrowia* van der Walt & von Arx (1980). In Kurtzman, C.P., Fell, J.W. and Boekhout, T. (eds) *The Yeasts, a Taxonomic Study*, 5th edition. pp. 927-929.

Kwon-Chung, K.J., Polacheck, I. and Bennett, J.E. (1982) Improved diagnostic medium for separation of *Cryptococcus neoformans* var. *neoformans* (serotypes A and D) and *Cryptococcus neoformans* var. *gattii* (serotypes B and C). *Journal of Clinical Microbiology*, 15, 535-537.

Kwon-Chung, K.J. and Bennett, J.W. (1992) *Medical Mycology*. Lea & Febiger, Philadelphia, 861 pp.

Kwon-Chung, K.J., Boekhout, T., Fell, J. and Diaz, M. (2002) Proposal to conserve the name *Cryptococcus gattii* against *C. hondurianus* and *C. bacillisporus* (Basidiomycota, Hymenomycetes, Tremellomycetidae). *Taxon*, 51, 804-806.

Kwon-Chung, K.J. (2011) Chapter 114, *Filobasidiella* Kwon-Chung (1975). In Kurtzman, C.P., Fell, J.W. and Boekhout, T. (eds) *The Yeasts, a Taxonomic Study*, 5th edition, pp. 1443-1455.

Kwon-Chung, K.J., Bennett, J.E., Wickes, B.L., *et al.* (2017) The case for adopting the "species complex" nomenclature for the etiologic agents of *Cryptococcosis*. *mSphere*, 2(1), e00357-16.

Labuda, R., Bernreiter, A., Hochenauer, D., *et al.* (2019) *Saksenaea dorisiae* sp. nov., a new opportunistic pathogenic fungus from Europe. *International Journal of Microbiology*, 2019, 6253829.

Lachance, M.-A. (2011a) Chapter 21, *Clavispora* Ridrigues de Miranda (1979). In Kurtzman, C.P., Fell, J.W. and Boekhout, T. (eds) *The Yeasts, a Taxonomic Study*, 5th edition, pp. 349-353.

Lachance, M.-A. (2011b) Chapter 35, *Kluyveromyces* van der Walt (1971). In Kurtzman, C.P., Fell, J.W. and Boekhout, T. (eds) *The Yeasts, a Taxonomic Study*, 5th edition. pp. 471-481.

Lachance, M.-A., Boekhout, T., Scorzetti, G., *et al.* (2011) Chapter 90, *Candida* Berhout (1923). In Kurtzman, C.P., Fell, J.W. and Boekhout, T. (eds) *The Yeasts, a Taxonomic Study*, 5th edition. Elsevier B.V., pp. 987-1278.

Lackner, M. and de Hoog, G.S. (2011) *Parascedosporium* and its relatives: phylogeny and ecological trends. *IMA Fungus*, 2, 39-48.

Lackner, M., Najafzadeh, M.J., Sun, J., *et al.* (2012) Rapid identification of *Pseudallescheria* and *Scedosporium* strains using rolling circle amplification. *Applied Environmental Microbiology*, 78, 126-133.

Lackner, M., de Hoog, G.S., Yang, L., *et al.* (2014a) Proposed nomenclature for *Pseudallescheria, Scedosporium* and related genera. *Fungal Diversity*, 67, 1-10.

Lackner, M., Hagen, F., Meis, J.F., *et al.* (2014b) Susceptibility and diversity in therapy-refractory genus *Scedosporium*. *Antimicroial Agents and Chemotherapy*, 58, 5877-5885.

Lagrou, K., Chen, S., Masur, H., *et al.* (2021) *Pneumocystis jirovecii* disease: Basis for the revised EORTC/MSGERC Invasive Fungal Disease Definitions in individuals without Human Immunodeficiency Virus. *Clinical Infectious Diseases*, 72, S114-S120.

Lara Oya, A., Medialdea Hurtado, M.E., Rojo Martín, M.D., *et al.* (2016) Fungal keratitis due to *Beauveria bassiana* in a contact lenses wearer and review of published reports. *Mycopathologia*, 181, 745-752.

Lass-Florl, C. and Mayr, A. (2007) Human protothecosis. *Clinical Microbiology Reviews*, 20, 230-242.

Lau, A.F., Drake, S.K., Calhoun, L.B., *et al.* (2013) Development of a clinically comprehensive database and a simple procedure for identification of molds from solid media by matrix-assisted laser desorption ionization-time of flight mass spectrometry. *Journal of Clinical Microbiology*, 51, 828-834.

Lau, A.F., Walchak, R.C., Miller, H.B., *et al.* (2019) Multicenter study demonstrates standardization requirements for mold identification by MALDI-TOF MS. *Frontiers in Microbiology*, 10, 2098-2098.

Lawrence, R.M., Snodgrass, W.T., Reichel, G.W., *et. al.* (1986) Systemic zygomycosis caused by *Apophysomyces elegans*. *Journal of Medical and Veterinary Mycology*, 24, 57-65.

Leach, L., Zhu, Y. and Chaturvedi, S. (2018) Development and validation of a real-time PCR assay for rapid detection of *Candida auris* from surveillance samples. *Journal of Clinical Microbiology*, 56, e01223-17.

Lee, S. and Hanlin, R.T. (1999) Phylogenetic relationships of *Chaetomium* and similar genera based on ribosomal DNA sequences. *Mycologia*, 91, 434-442.

Lee, H.S., Shin, J.H., Choi, M.J., *et al.* (2017) Comparison of the Bruker Biotyper and VITEK MS Matrix-Assisted Laser Desorption/Ionization Time-of-Flight Mass Spectrometry systems using a formic acid extraction method to identify common and uncommon yeast isolates. *Annals of Laboratory Medicine*, 37, 223-230.

Lennon, P.A., Cooper, C.R. Jr., Salkin, I.F., *et al.* (1994) Ribosomal DNA internal transcribed spacer analysis supports synonymy of *Scedosporium inflatum* and *Lomentospora prolificans*. *Journal of Clinical Microbiology*, 32, 2413-2416.

Levesque, S., Dufresne, P.J., Soualhine, H., *et al.* (2015) A side by side comparison of Bruker Biotyper and VITEK MS: Utility of MALDI-TOF MS technology for microorganism identification in a Public Health Reference Laboratory. *PLoS ONE*, 10(12), e0144878.

Li, Y., Xiao, J., de Hoog, G.S., *et al.* (2017) Biodiversity and human-pathogenicity of *Phialophora verrucose* and relatives in *Chaetothyriales*. *Persoonia*, 38, 1-19.

Liu, J.-K., Phookamsak, R., Doilom, M., *et al.* (2012) Towards a natural classification of Botryosphaeriales. *Fungal Diversity*, 57, 149-210.

Liu, J., Ge, L., Mei, H., *et al.* (2021) Comparative genomics and molecular analysis of *Epidermophyton floccosum*. *Mycopathologia*, 186, 487-497.

Liu, X.Z., Wang, Q.M., Goker, M., *et al.* (2015) Towards an integrated phylogenetic classification of the Tremellomycetes. *Studies in Mycology*, 81, 85-147.

Lockhart, S.R., Messer, S.A., Pfaller, M.A., *et al.* (2008) *Lodderomyces elongisporus* masquerading as *Candida parapsilosis* as a cause of bloodstream infections. *Journal of Clinical Microbiology*, 46, 374-376.

Lockhart, S.R., Roe, C.C. and Engelthaler, D.M. (2016) Whole-genome analysis of *Cryptococcus gattii*, Southeastern United States. *Emerging Infectious Diseases*, 22, 1098-1101.

Lockhart, S.R., Etienne, K.A., Vallabhaneni, S., *et al.* (2017a) Simultaneous emergence of multidrug-resistant *Candida auris* on 3 continents confirmed by whole-genome sequencing and epidemiological analyses. *Clinical Infectious Diseases*, 64, 134-140.

Lockhart, S.R., Jackson, B.R., Vallabhaneni, S., *et al.* (2017b) Thinking beyond the common *Candida* species: Need for species-level identification of *Candida* due to the emergence of multidrug-resistant *Candida auris*. *Journal of Clinical Microbiology*, 55, 3324-3327.

L'Ollivier, C., Cassagne, C., Normand, A.C., *et al.* (2013) A MALDI-TOF MS procedure for clinical dermatophyte species identification in the routine laboratory. *Medical Mycology*, 51, 713-270.

Lombard, L., van der Merwe, N.A., Groenewald, J.Z. and Crous, P.W. (2015) Genetic concepts in Nectriaceae. *Studies in Mycology*, 80, 189-245.

Lonial, S., Williams, L., Carrum, G., *et al.* (1997) *Neosartorya fischeri:* an invasive fungal pathogen in an allogeneic bone marrow transplant patient. *Bone Marrow Transplant*, 19, 753-755.

Lorch, J.M., Palmer, J.M., Vanderwolf, K.J., *et al.* (2018) *Malassezia vespertilionis* sp. nov.: a new cold-tolerant species of yeast isolated from bats. *Persoonia*, 41, 56-70.

Lu, X.-L., Liu, Z.-H., Shen, Y.-N., *et al.* (2009) Primary cutaneous zygomycosis caused by *Rhizomucor variabilis*: a new endemic zygomycosis? A case report and review of 6 cases reported from China. *Clinical Infectious Diseases*, 49, e39-e49.

Lu, X.L., Najafzadeh, M.J., Ran, Y.P., *et al.* (2013) Taxonomy and epidemiology *Mucor irregularis*, agent of chronic cutaneous mucormycosis. *Persoonia*, 30, 48-56.

Luangsa-ard, J., Houbraken, J., van Doorn, T., *et al.* (2011) *Purpureocillium*, a new genus for the medically important *Paecilomyces lilacinus*. *FEMS Microbiology Letters*, 321, 141-149.

Lunn, J.A. and Shipton, W.A. (1983) Re-evaluation of taxonomic criteria in *Cunninghamella*. *Transactions of the British Mycological Society*, 81, 303-312.

Luttrell, E.S. (1978) Biosystematics of *Helminthosporium:* impact on agriculture, in Romberger, J.A., Foote, R.H., Knutson, L.C.V., *et al.* (eds) *Biosystematics in Agriculture*. Montclair, Allanheld-Osmon & Co, New Jersey, 193-209.

Lyratzopoulos, G., Ellis, M., Nerringer, R., *et al.* (2002) Invasive infection due to *Penicillium* species other than *P. marneffei*. *Journal of Infection*, 45, 184-207.

Machouart, M., Menir, P., Helenon, R., *et al.* (2012) *Scytalidium* and scytalidiosis: what's new in 2012? *Journal de Mycologie Medicale*, 23, 40-46.

Mackenzie, D.W.R., Loeffler, W., Mantovani, A., *et al.* (1986) *Guidelines for the diagnosis, prevention and control of dermatophytosis in man and animals*. (No. WHO/CDS/VPH/86.67). World Health Organization, Geneva.

Madrid, H., Ruíz-Cendoya, M., Cano, J., *et al.* (2009) Genotyping and *in vitro* antifungal susceptibility of *Neoscytalidium dimidiatum* isolates from different origins. *International Journal of Antimicrobial Agents*, 34, 351-354.

Madrid, H., da Cunha, K.C., Gene, J., *et al.* (2014) Novel *Curvularia* species from clinical specimens. *Persoonia*, 33, 48-60.

Malloch, D. and Cain, R.F. (1972) The Trichocomataceae: Ascomycetes with *Aspergillus, Paecilomyces*, and *Penicillium* imperfect states. *Canadian Journal of Botany*, 50, 2613-2628.

Malloch, D. and Salkin, I.F. (1984) A new species of *Scedosporium* associated with osteomyelitis in humans. *Mycotaxon*, 21, 247-255.

Manamgoda, D.S., Cai, L., McKenzie, E.H.C., *et al.* (2012) A phylogenetic and taxonomic re-evaluation of the *Bipolaris - Cochliobolus - Curvularia* complex. *Fungal Diversity*, 56, 131-144.

Manamgoda, D.S., Rossman, A.Y., Castlebury, L.A., *et al.* (2014) The genus *Bipolaris*. *Studies in Mycology*, 79, 221-288.

Maphanga, T.G., Britz, E., Zulu, T.G., *et al.* (2017) *In vitro* antifungal susceptibility of yeast and mould phases of isolates of dimorphic fungal pathogen *Emergomyces africanus* (formerly *Emmonsia* sp.) from HIV-infected South African patients. *Journal of Clinical Microbiology*, 55, 1812-1820.

Maphanga, T.G., Birkhead, M., Munoz, J.F., *et al.* (2020) Human blastomycosis in South Africa caused by *Blastomyces percursus* and *Blastomyces emzantsi* sp. nov., 1967 to 2014. *Journal of Clinical Microbiology*, 58, e01661-19.

Marimon, R., Cano, J., Gene, J., *et al.* (2007) *Sporothrix brasiliensis*, *S. globosa*, and *S. mexicana*, three new *Sporothrix* species of clinical interest. *Journal of Clinical Microbiology*, 45, 3198-3206.

Marty, F.M., Barouch, D.H., Coakley, E.P., *et al.* (2003) Disseminated trichosporonosis caused by *Trichosporon loubieri*. *Journal of Clinical Microbiology*, 41, 5317-5320.

Matsumoto, T., Padhye, A.A., Ajello, L., *et al.* (1984) Critical review of human isolates of *Wangiella dermatitidis*. *Mycologia*, 76, 232-249.

Maxwell, C.S., Sepulveda, V.E., Turissini, D.A., *et al.* (2018) Recent admixture between species of the fungal pathogen *Histoplasma*. *Evolution Letters*, 2, 210-220.

McCullough, M.J., Clemons, K.V., McCusker, J.H., *et al.* (1998) Intergenic transcribed spacer PCR ribotyping for differentiation of *Saccharomyces* species and interspecific hybrids. *Journal of Clinical Microbiology*, 36, 1035–1038.

McGinnis, M.R. and Padhye, A.A. (1977) *Exophiala jeanselmei*, a new combination for *Phialophora jeanselmei*. *Mycotaxon*, 5, 341-352.

McGinnis, M.R. (1978a) Taxonomy of *Exophiala jeanselmei*. *Mycopathologia*, 65, 79-87.

McGinnis, M.R. (1978b) Human pathogenic species of *Exophiala, Phialophora*, and *Wangiella*. *In the black and white yeasts. Proceedings of the fourth international conference on the mycoses.* Scientific Publication No. 356. Pan American Health Organization, Washington, DC, pp. 37-59.

McGinnis, M.R. (1980) *Laboratory Handbook of Medical Mycology*. Academic Press, New York.

McGinnis, M.R. and Borelli, D. (1981) *Cladosporium bantianum* and its synonym *Cladosporium trichoides. Mycotaxon*, 13, 127-136.

McGinnis, M.R., Schell, W.A. and Carson, J. (1985) *Phaeoannellomyces* and the Phaeococcomycetaceae, new dematiaceous blastomycete taxa. *Journal of Medical and Veterinary Mycology*, 23, 179-188.

McGinnis, M.R., Borelli, D., Padhye, A.A., *et al.* (1986a) Reclassification of *Cladosporium bantianum* in the genus *Xylohypha*. *Journal of Clinical Microbiology*, 23, 1148-1151.

McGinnis, M.R., Rinaldi, M.G. and Winn, R.E. (1986b) Emerging agents of phaeohyphomycosis: pathogenic species of *Bipolaris* and *Exserohilum*. *Journal of Clinical Microbiology*, 24, 250-259.

McGinnis, M.R., Pasarell, L., Sutton, D.A., *et al.* (1997) *In vitro* evaluation of voriconazole against some clinically important fungi. *Antimicrobial Agents and Chemotherapy*, 41, 1832-1834.

McPherson, M.E., Woodgyer, A.J., Simpson, K., Chong, A.H. (2008) High prevalence of tinea capitis in newly arrived migrants at an English-language school, Melbourne. *Medical Journal of Australia*, 189, 13-16.

McTaggart, L., Richardson, S.E., Seah, C., *et al.* (2011) Rapid identification of *Cryptococcus neoformans* var. *grubii*, *C. neoformans* var. *neoformans*, and *C. gattii* by use of rapid biochemical tests, differential media, and DNA sequencing. *Journal of Clinical Microbiology*, 49, 2522-2527.

Michel, J., Maubon, D., Varoquaux, D.A., *et al.* (2015) *Schizophyllum commune*: an emergent or misdiagnosed fungal pathogen in rhinology? *Medical Mycology*, 54, 301-309.

Millner, P.D. (1977) Radial growth responses to temperature by 58 *Chaetomium* species, and some taxonomic relationships. *Mycologia*, 69, 492-502.

Ming, C., Huang, J., Wang, Y., *et al.* (2019) Revision of the medically relevant species of the yeast genus *Diutina*. *Medical Mycology*, 57, 226-233.

Miranda, K.C., de Araujo, C.R., Costa, C.R., *et. al.* (2007) Antifungal activities of azole agents against the *Malassezia* species. *International Journal of Antimicrobial Agents*, 29, 281-284.

Mirhendi, H., Makimura, K., de Hoog, G.S. *et al.* (2015) Translation elongation factor α gene as a potential taxonomic and identification marker in dermatophytes. *Medical Mycology*, 53, 215-224.

Misra, P.C., Srivastava, K.J. and Latas, K. (1979) *Apophysomyces*, a new genus of the Mucorales. *Mycotaxon*, 8, 377-382.

Mochizuki, T., Anzawa, K., Sakata, Y., *et al.* (2013) Simple identification of *Trichophyton tonsurans* by chlamydospore-like structures produced in culture media. *Journal of Dermatology*, 40, 1027-1032.

Mok, W.Y. (1982) Nature and identification of *Exophiala werneckii*. *Journal of Clinical Microbiology*, 16, 976-978.

Molnar, O., Schatzmayr, G., Fuchs, E., *et al.* (2004) *Trichosporon mycotoxinivorans* sp. nov., a new yeast species useful in biological detoxification of various mycotoxins. *Systematic and Applied Microbiology*, 27, 661-671.

Mongkolsamrit, S., Khonsanit, A., Thanakitpipattana, D., *et al.* (2020) Revisiting *Metarhizium* and the description of new species from Thailand. *Studies in Mycology*, 95, 171-251.

Montel, E., Bridge, P.D. and Sutton, B.C. (1991) An integrated approach to *Phoma* systematics. *Mycopathologia*, 115, 89-103.

Moodley, A., Mosam, A., Govender, N.P., *et al.* (2019) *Emergomyces africanus: the mimicking fungus. Dermatopathology*, 6, 157-162.

Moore, M.K. (1986) *Hendersonula toruloidea* and *Scytalidium hyalinum* infections in London, England. *Journal of Medical and Veterinary Mycology*, 24, 219-230.

Morales-Lopez, S.E. and Garcia-Effron, G. (2021) Infections due to rare *Cryptococcus* species. *A literature review. Journal of Fungi*, 7, 279.

Morenz, J. (1964) Taxonomische Untersuchungen zur Gattung *Geotrichum*. *Mykologische Schriftenreihe*, 2, 33-64.

Morjaria, S., Otto, C., Moreira, A., *et al.* (2015) Ribosomal RNA gene sequencing for early diagnosis of *Blastomyces dermatitidis* infection. *International Journal of Infectious Diseases*, 37, 122-124.

Morton, F.J. and Smith, G. (1963) *The genera Scopulariopsis Bainier, Microascus Zukal, and Doratomyces Corda. Mycological Papers, No, 86*. Commonwealth Mycological Institute, Kew, London.

Mostert, L., Groenewald, J.Z., Summerbell, R.C., *et al.* (2005) Species of *Phaeoacremonium* associated with infections in humans and environmental reservoirs in infected woody plants. *Journal of Clinical Microbiology*, 43, 1752-1767.

Mostert, L., Groenewald, J.Z., Summerbell, R.C., *et al.* (2006) Taxonomy and pathology of *Togninia* (*Diaporthales*) and its *Phaeoacremonium* anamorphs. *Studies in Mycology*, 54, 1-115.

Mulet Bayona, J.V., Salvador Garcia, C., Tormo Palop, N., *et al.* (2022) Candida Plus for detection of *Candida auris* and other *Candida* species from surveillance and environmental samples: A multicenter study. *Journal of Fungi*, 8, 281.

Murugaiyan, J., Ahrholdt, J., Kowbel, V., *et al.* (2012) Establishment of a matrix-assisted laser desorption ionization time-of-flight mass spectrometry database for rapid identification of infectious achlorophyllous green micro-algae of the genus *Prototheca*. *Clinical Microbiology and Infection*, 18, 461-467.

Najafzadeh, M.J., Gueidan, C., Badali, H., *et al.* (2009) Genetic diversity and species delimitation in the opportunistic genus *Fonsecaea*. *Medical Mycology*, 47, 17-25.

Najafzadeh, M.J., Badali, H., Illnait-Zaragozi, M.T., *et al.* (2010a) *In vitro* activities of eight antifungal drugs against 55 clinical isolates of *Fonsecaea* spp. *Journal of Clinical Microbiology*, 54, 1636-1638.

Najafzadeh, M.J., Sun, J., Vicente, V., *et al.* (2010b) *Fonsecaea nubica* sp. nov, a new agent of human chromoblastomycosis revealed using molecular data. *Medical Mycology*, 48, 800-806.

Najafzadeh, M.J., Sutton, D.A., Keisari, M.S., *et al.* (2014) *In vitro* activities of eight antifungal drugs against 104 environmental and clinical isolates of *Aureobasidium pullulans*. *Antimicrobial Agents and Chemotherapy*, 58, 5629-5631.

Naranjo-Ortiz, M.A. and Gabaldon, T. (2019) Fungal evolution: diversity, taxonomy and phylogeny of the fungi. *Biological Reviews*, 94, 2101-2137.

Nenoff, P., Erhhar, M., Simon, J.C., *et al.* (2013) MALDI-TOF mass spectrometry - a rapid method for the identification of dermatophyte species. *Medical Mycology*, 51, 17-24.

Ng, K.P., Soo-Hoo, T.S., Na, S.L., *et al.* (2005) The mycological and molecular study *of Hortaea werneckii* isolated from blood and splenic abscess. *Mycopathologia*, 159, 495–500.

Nie, Y., Yu, D.S., Wang, C.F., *et al.* (2020) A taxonomic revision of the genus *Conidiobolus* (Ancylistaceae, Entomophthorales): four clades including three new genera. *MycoKeys*, 66, 55-81.

Nishimura, K. and Miyaji, M. (1983) Studies on the phylogenesis of pathogenic 'black yeasts'. *Mycopathologia*, 81, 135-144.

Nobrega de Almeida, J., de Souza, L.B., Motta, A.L., *et al.* (2014) Evaluation of the MALDI-TOF VITEK MS™ system for the identification of *Candida parapsilosis*, *C. orthopsilosis* and *C. metapsilosis* from bloodstream infections. *Journal of Microbiological Methods*, 105, 105-108.

Noster, J., Koeppel, M.B., Desnos-Olivier, M., *et al.* (2022) Bloodstream infections caused by *Magnusiomyces capitatus* and *Magnusiomyces clavatus*: Epidemiological, clinical, and microbiological features of two emerging yeast species. *Antimicrobial Agents and Chemotherapy*, 66, e0183421.

Nottebrock, H., Scholer, H.J. and Wall, M. (1974) Taxonomy and identification of mucormycosis causing fungi. 1. Synonymity of *Absidia ramosa* with *A. corymbifera*. *Sabouraudia*, 12, 64-74.

Novakova, A., Hubka, V., Dudova, Z., *et al.* (2013) New species in *Aspergillus* section Fumigati from reclamation sites in Wyoming (U.S.A.) and revision of *A. viridinutans* complex. *Fungal Diversity*, 64, 253-274.

Nucci, M. and Anaissie, E. (2007) *Fusarium* infections in immunocompromised patients. *Clinical Microbiology Review*, 20, 695-704.

O'Donnell, K.L. (1979) *Zygomycetes in culture*. Palfrey Contributions in Botany 2. University of Georgia. pp 257.

O'Donnell, K., Sutton, D.A., Fothergill, A., *et al.* (2008) Molecular phylogenetic diversity, multilocus haplotype nomenclature, and *in vitro* antifungal resistance within the *Fusarium solani* species complex. *Journal of Clinical Microbiology*, 46, 2477-2490.

O'Donnell, K., Gueidan, C., Sink, S., *et al.* (2009a) A two-locus DNA sequence database for typing plant and human pathogens within the *Fusarium oxysporum* species complex. *Fungal Genetics and Biology*, 46, 936-948.

O'Donnell, K., Sutton, D.A., Rinaldi, M.G., *et al.* (2009b) Novel multilocus sequence typing scheme reveals high genetic diversity of human pathogenic members of the *Fusarium incarnatum-F. equiseti* and *F. chlamydosporum* species complexes within the United States. *Journal of Clinical Microbiology* 47, 3851-3861.

O'Donnell, K., Whitaker, B.K., Laraba, I., *et al.* (2022) DNA Sequence-Based Identification of *Fusarium*: A Work in Progress. *Plant Disease*, 106, 1597-1609.

Ohst, T., de Hoog, S., Presber, W., *et al.* (2004) Origins of microsatellite diversity in the *Trichophyton rubrum - T. violaceum* clade (Dermatophytes). *Journal of Clinical Microbiology*, 42, 4444-4448.

Okada, G., Kirisits, T., Louis-Seize, G.W., *et al.* (2000) Epitypification of *Graphium penicillioides* Corda, with comments on the phylogeny and taxonomy of *Graphium*-like synnematous fungi. *Studies in Mycology*, 45, 169-188.

Oliveira, M.M.E., Almeida-Paesa, R., Gutierrez-Galhardob, M.C., *et al.* (2013) Molecular identification of the *Sporothrix schenckii* complex. *Revista Iberoamericana de Micologia*, 31, 2-6.

Oliveira, M.M.E., Santos, C., Sampaio, P., *et al.* (2015) Development and optimization of a new MALDI-TOF protocol for identification of the *Sporothrix* species complex. *Research in Microbiology*, 166, 102-110.

Onions, A.H.S., Allsopp, D. and. Eggins, H.O.W. (1981) *Smith's Introduction to Industrial Mycology*, 7th edition. Edward Arnold, London.

Ozgun, H.D., Jacobs, D.L. and Toms, S.A. (2019) *Cladophialophora bantiana*. In *Fungal Infections of the Central Nervous System.* Springer, Cham, Switzerland, pp. 195-204.

Ozhak-Baysan, B., Ogunc, D., Dogen, A., *et al.* (2015) MALDI-TOF MS-based identification of black yeasts of the genus *Exophiala*. *Medical Mycology*, 53, 347-52.

Packeu, A., Hendrickx, M., Beguin, H., *et al.* (2013) Identification of the *Trichophyton mentagrophytes* complex species using MALDI-TOF mass spectrometry. *Medical Mycology*, 51, 580-585.

Packeu, A., De Bel, A., L'Ollivier, C., *et al.* (2014) Fast and accurate identification of dermatophytes by matrix-assisted laser desorption ionization-time of flight mass spectrometry: validation in the clinical laboratory. *Journal of Clinical Microbiology*, 52, 3440-3443.

Padhye, A.A. and Carmichael, J.W. (1972) *Arthroderma insingulare* sp. nov., another gymnoascaceous state of the *Trichophyton terrestre* complex. *Sabouraudia*, 10, 47-51.

Padhye, A.A. and Ajello, L. (1988) Simple method of inducing sporulation by *Apophysomyces elegans* and *Saksenaea vasiformis*. *Journal of Clinical Microbiology*, 26, 1861-1863.

Padhye, A.A., Koshi, G., Anandi, V., *et al.* (1988) First case of subcutaneous zygomycosis caused by *Saksenaea vasiformis* in India. *Diagnostic Microbiology and Infectious Disease*, 9, 69-77.

Padhye, A.A., Godfrey, J.H., Chandler, F.W., *et al.* (1994) Osteomyelitis caused by *Neosartorya pseudofischeri*. *Journal of Clinical Microbiology*, 32, 2832-2836.

Padhye, A.A., Verghese, S., Ravichandran, P., *et al.* (2003) *Trichosporon loubieri* infection in a patient with adult polycystic kidney disease. *Journal of Clinical Microbiology*, 41, 479-482.

Park, J.H., Oh, J., Sang, H., *et al.* (2019) Identification and antifungal susceptibility profiles of *Cyberlindnera fabianii* in Korea. *Mycobiology*, 47, 449-456.

Pastor, F.J. and Guarro, J. (2008) *Alternaria* infections: laboratory diagnosis and relevant clinical features. *Clinical Microbiology and Infection*, 14, 734-746.

Patel, R. (2019) A moldy application of MALDI: MALDI-ToF Mass Spectrometry for fungal identification. *Journal of Fungi*, 5, 4.

Pendle, S., Weeks, K., Priest, M., *et al.* (2004) Phaehyphomycotic soft tissue infections caused by the coelomycetous fungus *Microsphaeropsis arundis*. *Journal of Clinical Microbiology*, 42, 5315-5319.

Perdomo, H., Sutton, D.A., Garcia, D., *et al.* (2011a) Spectrum of clinically relevant *Acremonium* species in the United States. *Journal of Clinical Microbiology* 49, 243-256.

Perdomo, H., Sutton, D.A., Garcia, D., *et al.* (2011b) Molecular and phenotypic characterization of *Phialemonium* and *Lecythophora* isolates from clinical samples. *Journal of Clinical Microbiology*, 49, 1209-1216.

Perdomo, H., Cano, J., Gene, J., *et al.* (2013) Polyphasic analysis of *Purpureocillium lilacinum* isolates from different origins and proposal of the new species *Purpureocillium lavendulum*. *Mycologia*, 105, 151-161.

Perez-Cantero, A. and Guarro, J. (2020) *Sarocladium* and *Acremonium* infections: New faces of an old opportunistic fungus. *Mycoses*, 63, 1203-1214.

Peterson, S.W. and Sigler, L. (1998) Molecular genetic variation in *Emmonsia crescens* and *Emmonsia parva*, etiologic agents of adiaspiromycosis, and their phylogenetic relationship to *Blastomyces dermatitidis* (*Ajellomyces dermatitidis*) and other systemic fungal pathogens. *Journal of Clinical Microbiology*, 36, 2918-2925.

Peterson, S.W. (2000) Phylogenetic relationships in *Aspergillus* based on rDNA sequence analysis. In Samson, R.A. and Pitt, J.I. (eds) *Integration of Modern Taxonomic Methods for Penicillium and Aspergillus Classification.* CRC Press, London, pp. 323-355.

Peterson, S.W. (2008) Phylogenetic analysis of *Aspergillus* species using DNA sequences from four loci. *Mycologia*, 100, 205-226.

Pfaller, M.A. and Diekema, D.J. (2010) Epidemiology of invasive mycoses in North America. *Critical Reviews in Microbiology*, 36, 1-53.

Pfaller, M.A., Messer, S.A., Rhomberg, P.R., *et al.* (2013a) *In vitro* activities of isavuconazole and comparator antifungal agents tested against a global collection of opportunistic yeasts and molds. *Journal of Clinical Microbiology*, 51, 2608-2616.

Pfaller, M.A., Messer, S.A., Woosley, L.N., *et al.* (2013b) Echinocandin and triazole antifungal susceptibility profiles for clinical opportunistic yeast and mould isolates collected from 2010 to 2011: Application of new CLSI clinical breakpoints and epidemiological cutoff values for characterization of geographic and temporal trends of antifungal resistance. *Journal of Clinical Microbiology*, 51, 2571-2581.

Pfaller, M.A., Rhomberg, P.R., Messer, S.A., *et al.* (2015) Isavuconazole, micafungin, and 8 comparator antifungal agents' susceptibility profiles for common and uncommon opportunistic fungi collected in 2013: temporal analysis of antifungal drug resistance using CLSI species-specific clinical breakpoints and proposed epidemiological cutoff values. *Diagnostic Microbiology and Infectious Diseases*, 82, 303-313.

Pfaller, M.A., Rhomberg, P.R., Wiederhold, N.P., *et al.* (2018) *In vitro* activity of Isavuconazole against opportunistic fungal pathogens from two mycology reference laboratories. *Antimicrobial Agents and Chemotherapy*, 62, e01230-18.

Phillips, A.J., Alves, A., Abdollahzadeh, J., *et al.* (2013) The Botryosphaeriaceae: genera and species known from culture. *Studies in Mycology*, 76, 51-167.

Pinjon, E., Sullivan, D., Salkin, I., *et al.* (1998) Simple, inexpensive, reliable method for differentiation of *Candida dubliniensis* from *Candida albicans*. *Journal of Clinical Microbiology*, 36, 2093-2095.

Pinto, A., Halliday, C., Zahra, M., *et al.* (2011) Matrix-assisted laser desorption ionization-time of flight mass spectrometry identification of yeasts is contingent on robust reference spectra. *PloS One*, 6, e25712.

Pitt, J.I. (1979) *The genus Penicillium and its teleomorphic states Eupenicillium and Talaromyces*. Academic Press, London.

Pore, R.S. (1985) *Prototheca* taxonomy. *Mycopathologia*, 129, 129-139.

Pottier, I., Gente, S., Vernoux, J.P., *et al.* (2008) Safety assessment of dairy microorganisms: *Geotrichum candidum*. *International Journal of Food Microbiology*, 126, 327-332.

Prakash, H., Ghosh, A.K., Rudramurthy, S.M., *et al.* (2016) The environmental source of emerging *Apophysomyces variab*ilis infection in India. *Medical Mycology*, 54, 567-575.

Pritchard, R.C., Muir, D.B., Archer, K.H. *et al.* (1986) Subcutaneous zygomycosis due to *Saksenaea vasiformis* in an infant. *Medical Journal of Australia*, 145, 630-631.

Pryor, B.M. and Gilberton, R.L. (2000) Molecular phylogenetic relationships amongst *Alternaria* species and related fungi based upon analysis of nuclear ITS and mt SSU rDNA sequences. *Mycological Research*, 104, 1312-1321.

Punithalingam, E. (1979) Sphaeropsidales in culture from humans. *Nova Hedwigia*, 31, 119-158.

Rainer, J. and de Hoog, G.S. (2006) Molecular taxonomy and ecology of *Pseudallescheria, Petriella* and *Scedosporium prolificans* (Microascaceae) containing opportunistic agents on humans. *Mycological Research*, 110, 151-160.

Ramani R. and Chaturvedi, V. (2007) Antifungal susceptibility profiles of *Coccidioides immitis* and *Coccidioides posadasii* from endemic and non-endemic areas. *Mycopathologia*, 163, 315-319.

Raper, K.B. and Fennell, D.I. (1965) *The genus Aspergillus*. William & Wilkins Co., Baltimore, Maryland.

Raper, K.B. and Thom, C.H. (1949) *A manual of the penicillia.* William & Wilkins Co., Baltimore, Maryland.

Ramirez, C. (1982) *Manual and Atlas of the Penicillia.* Elsevier Biomedical Press, Amsterdam.

Ramos, L.S., Figueiredo-Carvalho, M.H., Barbedo, L.S., *et al.* (2015) *Candida haemulonii* complex: species identification and antifungal susceptibility profiles of clinical isolates from Brazil. *Journal of Antimicrobial Chemotherapy*, 70, 111-115.

Rebell, G. and Taplin, D. (1970) *Dermatophytes; their Recognition and Identification*. University of Miami Press, Florida.

Reblova, M., Jaklitsch, W.M., Reblova, K., *et al.* (2015) Phylogenetic reconstruction of the Calosphaeriales and Togniniales using five genes and predicted RNA secondary structures of ITS, and *Flabellascus tenuirostris gen. et sp. nov*, *PloS One*, 10, e0144616.

Rehner, S.A. and Buckley, E. (2005) A *Beauveria* phylogeny inferred from nuclear ITS and EF1-alpha sequences: evidence for cryptic diversification and links to *Cordyceps* teleomorphs. *Mycologia*, 97, 84-98.

Reppas, G., Gottlieb, T., Krockenberger, M., *et al.* (2015) *Microsphaeropsis arundinis* an emerging cause of phaeohyphomycosis in cats and people. *Microbiology Australia*, 36, 74-78.

Revankar, S.G. and Sutton, D.A. (2010) Melanized fungi in human disease. *Clinical Microbiology Reviews*, 23, 884-928.

Riddell, R.W. (1950) Permanent stained mycological preparations obtained by slide culture. *Mycologia*, 42, 265-270.

Rippon, J.W., Arnow, P.M., Larson, R.A., *et. al.* (1985) "Golden tongue" syndrome caused by *Ramichloridium schulzeri*. *Archives of Dermatology*, 121, 892-894.

Rippon, J.W. (1988) *Medical Mycology: the Pathogenic Fungi and the Pathogenic Actinomycetes*, 3rd edition. WB. Saunders & Co., Philadelphia.

Rodrigues, A.M., Beale, M.A., Hagen, F., *et al.* (2020a) The global epidemiology of emerging *Histoplasma* species in recent years. *Studies in Mycology*, 97, 100095.

Rodrigues, A.M., Della Terra, P.P., Gremiao, I.D., *et al.* (2020b) The threat of emerging and re-emerging pathogenic *Sporothrix* species. *Mycopathologia*, 18, 813-842.

Rodriguez-Tudela, J.L., Diaz-Guerra, T.M., Mellado, E., *et al.* (2005) Susceptibility patterns and molecular identification of *Trichosporon* species. *Antimicrobial Agents and Chemotherapy*, 49, 4026-4034.

Rodriguez-Tudela, J.L., Berenguer, J., Guarro, J., *et al.* (2009) Epidemiology and outcome of *Scedosporium prolificans* infection, a review of 162 cases. *Medical Mycology*, 47, 359-370.

Rojas, F.D., Sosa, M. de los A. and Fernandez, M.S. (2014) Antifungal susceptibility of *Malassezia furfur*, *Malassezia sympodialis*, and *Malassezia globosa* to azole drugs and amphotericin B evaluated using a broth microdilution method. *Medical Mycology*, 52, 641-646.

Romeo, O., Scordino, F. and Criseo, G. (2011) New insight into molecular phylogeny and epidemiology of *Sporothrix schenckii* species complex based on calmodulin encoding gene analysis of Italian isolates. *Mycopathologia*, 172, 179-186.

Ruggiero, M.A., Gordon, D.P., Orrell, T.M., *et al.* (2015) A higher level classification of all living organisms. *PLoS One*, 10(4), e0119248.

Sabatelli, F., Patel, R., Mann, P.A., *et al.* (2006) *In vitro* activities of posaconazole, fluconazole, itraconazole, voriconazole, and amphotericin B against a large collection of clinically important molds and yeasts. *Antimicrobial Agents and Chemotherapy*, 50, 2009-2015.

Sakamoto, Y., Kano, R., Nakamura, Y., *et al.* (2001) Case report. First isolation of *Trichosporon domesticum* from a cat. *Mycoses*, 44, 518-520.

Saksena, S.B. (1953) A new genus of Mucorales. *Mycologia*, 45, 426-436.

Salah, H., Al-Hatmi, A.M.S., Theelen, B., *et al.* (2015) Phylogenetic diversity of human pathogenic *Fusarium* and emergence of uncommon virulent species. *Journal of Infection*, 71, 658-666.

Salkin, I.F., McGinnis, M.R., Dykstra, M.J. and Rinaldi, M.G. (1988) *Scedosporium inflatum*, an emerging pathogen. *Journal of Clinical Microbiology*, 26, 498-503.

Samaddar, A. and Sharma, A. (2021) Emergomycosis, an emerging systemic mycosis in immunocompromised patients: current trends and future prospects. *Frontiers in Medicine*, 8, 670731-670731.

Samerpitak, K., van der Linde, E., Choi, H.-J., *et al.* (2014) Taxonomy of *Ochroconis*, genus including opportunistic pathogens on humans and animals. *Fungal Diversity*, 65, 89-126.

Sampaio, J.P. (2011) Chapter 155, *Rhodotorula* Harrison (1928). In Kurtzman, C.P., Fell, J.W. and Boekhout, T. (eds) *The Yeasts, a Taxonomic Study*, 5th edition. Elsevier B.V., pp. 1873-1927.

Samson, R.A. (1974) *Paecilomyces* and some allied hyphomycetes. *Studies in Mycology*, 6, 1-120.

Samson, R.A. and Pitt, J.I. (1990) *Modern Concepts in Penicillum and Aspergillus classification*. Plenum Press, New York.

Samson, R.A., Hoekstra, E.S., Frisvad, J.C., *et al.* (1995) *Introduction to Food-borne Fungi. Centraalbureau voor Schimmelcultures*, P.O.Box 273, 3740 AG BAARN, The Netherlands.

Samson, R.A. and Pitt, J.I. (2000) *Integration of Modern Taxonomic Methods for Penicillium and Aspergillus Classification*. CRC Press, London.

Samson, R.A., Varga, J., Witiak, S.M., *et al.* (2007) The species concept in *Aspergillus:* recommendations of an international panel. *Studies in Mycology*, 59, 71-73.

Samson, R.A., Varga, J. and Frisvad, J.C. (2011a) *Taxonomic Studies on the Genus Aspergillus*. CBS-KNAW Fungal Biodiversity Centre, Utrecht, The Netherlands.

Samson, R.A., Yilmaz, N., Houbraken, J., *et al.* (2011b) Phylogeny and nomenclature of the genus *Talaromyces* and taxa accommodated in *Penicillium* subgenus *Biverticillium. Studies in Mycology*, 70, 159-183.

Samson, R.A., Visagie, C.M., Houbraken, J., *et al.* (2014) Phylogeny, identification and nomenclature of the genus *Aspergillus*. *Studies in Mycology*, 78, 141-173.

Sandoval-Denis, M., Sutton, D.A., Fothergill, A.W., *et al.* (2013) *Scopulariopsis*, a poorly known opportunistic fungus: spectrum of species in clinical samples and *in vitro* responses to antifungal drugs. *Journal of Clinical Microbiology*, 51, 3937-3943.

Sandoval-Denis, M., Giraldo, A., Sutton, D.A., *et al.* (2014a) *In vitro* antifungal susceptibility of clinical isolates of *Arthrographis kalrae*, a poorly known opportunistic fungus. *Mycoses*, 57, 247-248.

Sandoval-Denis, M., Sutton, D.A., Cano-Lira, J.F., *et al.* (2014b) Phylogeny of the clinically relevant species of the emerging fungus *Trichoderma* and their antifungal susceptibilities. *Journal of Clinical Microbiology*, 52, 2112-2125.

Sandoval-Denis, M., Gene, J., Sutton, D.A., *et al.* (2015a) *Acrophialophora*, a poorly known fungus with clinical significance. *Journal of Clinical Microbiology*, 53, 1549-1555.

Sandoval-Denis, M., Sutton, D.A., Martin-Vicente, A., *et al.* (2015b) *Cladosporium* species recovered from clinical samples in the United States. *Journal of Clinical Microbiology*, 53, 2990-3000.

Sandoval-Denis, M, Gene, J., Sutton, D.A., *et al.* (2016) Redefining *Microascus*, *Scopulariopsis* and allied genera. *Persoonia*, 36, 1-36.

Scheel, C.M., Zhou, Y., Theodoro, R.C., *et al.* (2014) Development of a loop-mediated isothermal amplification method for detection of *Histoplasma capsulatum* DNA in clinical samples. *Journal of Clinical Microbiology*, 52, 483-488.

Schell, W.A., McGinnis, M.R. and Borelli, D. (1983) *Rhinocladiella aquaspora* a new combination for *Acrotheca aquaspersa*. *Mycotaxon*, 17, 341-348.

Schipper, M.A.A. (1976) On *Mucor circinelloides*, *Mucor racemosus* and related species. *Studies in Mycology*, 12, 1-40.

Schipper, M.A.A. (1978) *1. On certain species of Mucor with a key to all accepted species. 2. On the genera Rhizomucor and Parasitella. Studies in Mycology No.17*. Centraalbureau voor Schimmelcultures, Baarn, The Netherlands.

Schipper, M.A.A. (1984) A revision of the genus *Rhizopus* 1. The *Rhizopus stolonifer*group and *Rhizopus oryzae*. *Studies in Mycology*, 25, 1-19.

Schipper, M.A.A. and Staplers, J.A. (1984) A revision of the genus *Rhizopus* II. The *Rhizopus microsporus* group. *Studies in Mycology*, 25, 30-34.

Schipper, M.A.A. and Staplers, J.A. (2003) Zygomycetes: The Order Mucorales: In Howard, D.H. (ed.) *Pathogenic Fungi in Humans and Animals, 2nd edition*. Marcel Dekker Inc., New York, pp. 67-125.

Schipper, M.A.A., Maslen, M.M., Hogg, G.G., *et. al.* (1996) Human infection by *Rhizopus azygosporus* and the occurrence of azygospores in Zygomycetes. *Journal of Medical and Veterinary Mycology*, 34, 199-203.

Scholer, H.J., Muller, E. and Schipper, M.A.A. (1983) Mucorales. In Howard, D.H. (ed.) *Fungi Pathogenic for Humans and Animals.* Marcel Dekker Inc., New York, pp. 9-59.

Schroedl, W., Heydel, T., Schwartze, V.U., *et al.* (2012) Direct analysis and identification of pathogenic *Lichtheimia* species by matrix-assisted laser desorption ionization-time of flight analyzer-mediated mass spectrometry. *Journal of Clinical Microbiology*, 50, 419-427.

Schroers, H-J., O'Donnell, K., Lamprecht, S.C., *et al.* (2009) Taxonomy and phylogeny of the *Fusarium dimerum* species group. *Mycologia*, 101, 44-70.

Schubert, K., Groenewald, J.Z., Braun, U., *et al.* (2007) Biodiversity in the *Cladosporium herbarum* complex (Davidiellaceae, Capnodiales), with standardisation of methods for *Cladosporium* taxonomy and diagnostics. *Studies in Mycology*, 58, 105-156.

Schwartz, I.S., Sanche, S., Wiederhold, N.P., *et al.* (2018a) *Emergomyces canadensis*, a dimorphic fungus causing fatal systemic human disease in North America. *Emerging Infectious Disease*, 24, 758-761.

Schwartz, I.S., Maphanga, T.G. and Govender, N.P. (2018b) *Emergomyces*: a new genus of dimorphic fungal pathogens causing disseminated disease among immunocompromised persons globally. *Current Fungal Infections Report*, 12, 44-50.

Schwartz, I.S., Govender, N.P., Sigler, L., *et al.* (2019) *Emergomyces*: The global rise of new dimorphic fungal pathogens. *PLoS Pathogens*, 15, e1007977.

Sepulveda, V.E., Marquez, R., Turissini, D.A., *et al.* (2017) Genome sequences reveal cryptic speciation in the human pathogen *Histoplasma capsulatum. mBio*, 8(6), e01339-17.

Seth, H.K. (1970) *A Monograph of the genus Chaetomium*. J. Cramer, Lehre, Germany.

Seyedmousavi, S., Samerpitak, K., Rijs, A.J.M.M., *et al.* (2014) Antifungal susceptibility patterns of opportunistic fungi in the genera *Verruconis* and *Ochroconis*. *Antimicrobial Agents and Chemotherapy*, 58, 3285-3292.

Shaikh, N., Hussain, K.A., Petraitiene, R., *et al.* (2016) Entomophthoramycosis: a neglected tropical mycosis. *Clinical Microbiology and Infection*, 22, 688-694.

Shipton, W.A. and Zahari, P. (1987) Sporulation media for *Basidiobolus* species. *Journal of Medical and Veterinary Mycology*, 25, 323-327.

Sidamonidze, K., Peck, M.K., Perez, M., *et al.* (2012) Real-time PCR assay for identification of *Blastomyces dermatitidis* in culture and in tissue. *Journal of Clinical Microbiology*, 50, 1783-1786.

Sigler, L. and Carmichael, J.W. (1976) Taxonomy of *Malbranchea* and some other hyphomycetes with arthroconidia. *Mycotaxon*, 4, 349-488.

Sigler, L. and Congly, H. (1990) Toenail infection caused by *Onychocola canadensis gen. et. sp. nov. Journal of Medical and Veterinary Mycology*, 28, 405-417.

Sigler, L., Abbott, S.P. and Woodgyer, A.J. (1994) New records of nail and skin infection due to *Onychocola canadensis* and description of its teleomorph *Arachnomyces nodosetosus sp. nov. Journal of Medical and Veterinary Mycology*, 32, 275-285.

Sigler, L., de la Maza, L.M., Tan, G., *et. al.* (1995) Diagnostic difficulties caused by a nonclamped *Schizophyllum commune* isolate in a case of fungus ball of the lung. *Journal of Clinical Microbiology*, 33, 1979-1983.

Sigler, L., Hambleton, S. and Pare, J.A. (2013) Molecular characterization of reptile pathogens currently known as members of the *Chrysosporium* anamorph of *Nannizziopsis vriesii* complex and relationship with some human-associated isolates. *Journal of Clinical Microbiology*, 51, 3338-3357.

Simmons, E.G. (1967) Typification of *Alternaria, Stemphylium*, and *Ulocladium*. *Mycologia*, 59, 67-92.

Simmons, E.G. (2007) *Alternaria, an Identification Manual*. Centraalbureau voor Schimmelcultures, Biodiversity Series, 6, 1-775.

Simpson, J.A. (2000) *Quambalaria*, a new genus of eucalypt pathogens. *Australasian Mycologist*, 19, 57-62.

Sinclair, K., Wakefield, A.E., Banerji, S., and Hopkin, J.M. (1991) *Pneumocystis carinii* organisms derived from rat and human hosts are genetically distinct. *Molecular and Biochemical Parasitology*, 45, 183-184.

Sitterle, E., Giraud, S., Leto, J., *et al.* (2014) Matrix-assisted laser desorption ionization time of flight mass spectrometry for fast and accurate identification of *Pseudallescheria / Scedosporium* species. *Clinical Microbiology and Infection*, 20, 929-935.

Sivanesan, A. (1987) Graminicolous species of *Bipolaris, Curvularia, Drechslera, Exserohilum* and their teleomorphs, 261 pp. Mycological Paper No. 158. CAB International, Wallingford, U.K.

Sleiman, S., Halliday, C.L., Chapman, B., *et al.* (2016) Performance of matrix-assisted laser desorption ionization-time of flight mass spectrometry for identification of *Aspergillus*, *Scedosporium*, and *Fusarium* spp. in the Australian clinical setting. *Journal of Clinical Microbiology*, 54, 2182–2186.

Smith, R.M., Schaefer, M.K., Kainer, M.A., *et al.* (2013) Multistate fungal infection outbreak response team. Fungal infections associated with contaminated methylprednisolone injections. *New England Journal of Medicine*, 369,1598-1609.

Sorrell, T.C. (2001) *Cryptococcus neoformans* variety *gattii. Medical Mycology*, 39, 155-168.

Spiliopoulou, A., Anastassiou, E.D. and Christofidou, M. (2012) *Rhodotorula* fungemia of an intensive care unit patient and review of published cases. *Mycopathologia*, 174, 301-309.

Spivak, E.S. and Hanson, K.E. (2018) *Candida auris*: an emerging fungal pathogen. *Journal of Clinical Microbiology*, 56, e01588-17.

Staib, F., Seibold, M., Antweiler, E., *et al.* (1989) Staib agar supplemented with a triple antibiotic combination for the detection of *Cryptococcus neoformans* in clinical specimens. *Mycoses*, 32, 448-454.

Stchigel, A.M., Sutton, D.A., Cano-Lira, J.F., *et al.* (2013) Phylogeny of chrysosporia infecting reptiles: proposal of the new family Nannizziopsiaceae and five new species. *Persoonia*, 31, 86-100.

Steinbach, W.P., Latge, J.-P. and Stevens, D.A. (2005) Advances against aspergillosis. Proceedings of a meeting. September 9-11, 2004. San Francisco, California. *Medical Mycology*, 43 Suppl 1, S1-319.

Stevens, D.A. (1980) *Coccidioidomycosis: A text*. Springer, New York.

Strinivasan, M.C. and Thirumalachar, M.J. (1965) *Basidiobolus* species pathogenic for man. *Sabouraudia*, 4, 32-34.

Suetrong, S., Hyde, K.D., Zhang, Y., *et al.* (2011) Trematosphaeriaceae fam. nov. (Dothideomycetes, Ascomycota). *Cryptogamie Mycologie*, 32, 343-358.

Sugar, A.M. and Stevens, D.A. (1985) *Candida rugosa* in immunocompromised infection. Case reports, drug susceptibility, and review of the literature. *Cancer*, 56, 318-320.

Sugar, A.M. and Liu, X.P. (1996) *In vitro* and *in vivo* activities of SCH 56592 against *Blastomyces dermatitidis*. *Antimicrobial Agents and Chemotherapy*, 40, 1314-1316.

Sugita, T., Nishikawa, A., Ikeda, R., *et al.* (1999) Identification of medically relevant *Trichosporon* species based on sequences of internal transcribed spacer regions and construction of a database for *Trichosporon* identification. *Journal of Clinical Microbiology*, 37, 1985-1993.

Sugita, T., Ikeda, R. and Nishikawa, A. (2004) Analysis of *Trichosporon* isolates obtained from the houses of patients with summer-type hypersensitivity pneumonitis. *Journal of Clinical Microbiology*, 42, 5467-5471.

Sugita, T. (2011) Chapter 161, *Trichosporo*n Behrend (1890), in *The Yeasts, a Taxonomic Study*, *5th edition* eds Kurtzman, C.P., J.W. Fell and T. Boekhout, Elsevier B.V. pp 2015-2061.

Sugita, T., Cho, O. and Takashima, M. (2017) Current status of taxonomy of pathogenic yeasts: *Cryptococcus neoformans/Cryptococcus gattii* and the genus *Trichosporon*. *Medical Mycology Journal*, 58, J77-J81.

Sugiura, Y. and Hironaga, M. (2010) *Arthrographis kalrae,* a rare causal agent of onychomycosis, and its occurrence in natural and commercially available soils. *Medical Mycology*, 48, 384-389.

Summerbell, R.C., Rosenthal, S.A. and Kane, J. (1988) Rapid method for differentiation of *Trichophyton rubrum*, *Trichophyton mentagrophytes*, and related dermatophyte species. *Journal of Clinical Microbiology*, 26, 2279-2282.

Summerbell, R.C., de Repentigny, L., Chartrand, C., *et al.* (1992) Graft-related endocarditis caused by *Neosartorya fischeri* var. *spinosa*. *Journal of Clinical Microbiology*, 30, 1580-1582.

Summerbell, R.C. and Schroers, H.J. (2002) Analysis of phylogenetic relationship of *Cylindrocarpon lichenicola* and *Acremonium falciforme* to the *Fusarium solani* species complex and a review of similarities in the spectrum of opportunistic infections caused by these fungi. *Journal of Clinical Microbiology*, 40, 2866-2875.

Summerbell, R.C., Gueidan, C., Schroers, H.J., *et al.* (2011) *Acremonium* phylogenetic overview and revision of *Gliomastix, Sarocladium,* and *Trichothecium*. *Studies in Mycology*, 68, 139-162.

Summerbell, R.C., Gueidan, C., Guarro, J., *et al.* (2018) The protean *Acremonium. A. sclerotigenum/egyptiacum*: Revision, food contaminant, and human disease. *Microorganisms*, 6, 1-21.

Sun, Q.N., Fothergill, A.W., McCarthy, D.I., *et. al.* (2002) *In vivo* activities of posaconazole, itraconazole, voriconazole, amphotericin B, and fluconazole against 37 clinical isolates of zygomycetes. *Antimicrobial Agents Chemotherapy*, 46, 1581-1582.

Sutton, B.C. (1980) *The Coelomycetes, Fungi imperfecti with Pycnidia, Acervuli and Stromata.* Commonwealth Mycology Institute, Kew, London.

Sutton, B.C. and Dyko, B.J. (1989) Revision of *Hendersonula*. *Mycological Research*, 93, 466-488.

Sutton, D.A., Slifkin, M., Yakulis, R. and Rinaldi, M.G. (1998) U.S. case report of cerebral phaeohyphomycosis caused by *Ramichloridium obovoideum* (*R. mackenziei*): criteria for identification, therapy, and review of other known dematiaceous neurotropic taxa. *Journal of Clinical Microbiology*, 36, 708-715.

Suzuki, M., Prasad, G.S. and Kurtzman, C.P. (2011) Chapter 24, *Debaryomyces* Lodder & Kreger-van Rij (1952). In Kurtzman, C.P., Fell, J.W. and Boekhout, T. (eds) *The Yeasts, a Taxonomic Study*, 5th edition, pp. 361-372.

Taj-Aldeen, S.J., Almaslamani, M., Alkhalf, A., *et al.* (2010) Cerebral phaeohyphomycosis due to *Rhinocladiella mackenziei* (formerly *Ramichloridium mackenziei*): a taxonomic update and review of the literature. *Medical Mycology*, 48, 546-556.

Tambini, R., Farina, C., Fiocchi, R., *et al.* (1996) Possible pathogenic role for *Sporothrix cyanescens* isolated from a lung lesion in a heart transplant patient. *Journal of Medical and Veterinary Mycology*, 34, 195-198.

Tanaka, K., Hirayama, K., Yonezawa, H., *et al.* (2015) Revision of the Massarineae (Pleosporales, Dothideomycetes). *Studies in Mycology*, 82, 75-136.

Tang, C., Kong, X., Ahmed, S.A., *et al.* (2021) Taxonomy of the *Trichophyton mentagrophytes/T. interdigitale* species complex harboring the highly virulent, multiresistant genotype *T. indotineae*. *Mycopathologia*, 186, 315-326.

Tavanti, A., Davidson, A.D., Gow, N.A., *et al.* (2005) *Candida orthopsilosis* and *Candida metapsilosis spp. nov*. to replace *Candida parapsilosis* groups II and III. *Journal of Clinical Microbiology*, 43, 284-292.

Teixeira, M.M., Theodoro, R.C., Oliveira, F.F., *et al.* (2014) *Paracoccidioides lutzii* sp. nov.: biological and clinical implications. *Medical Mycology*, 52, 19-28.

Theodoro, R.C., Teixeira, M., Soares Felipe, M.S., *et al.* (2012) Genus *Paracoccidioides*: species recognition and biogeographic aspects. *Plos One*, 7, e37694.

Tintelnot, K., de Hoog, G.S., Antweiler, E., *et al.* (2007) Taxonomic and diagnostic markers for identification of *Coccidioides immitis* and *Coccidioides posadasii*. *Medical Mycology*, 45, 385-393.

Torres-Cruz, T.J., Whitaker, B.K., Proctor, R.H., *et al.* (2021) FUSARIUM-ID v.3.0: An updated, downloadable resource for *Fusarium* species identification. *Plant Disease*, PDIS-09-21-2105-SR.

Tortorano, A.M., Richardson, M., Roilides, E., *et al.* (2014) ESCMID & ECMM joint guidelines on diagnosis and management of hyalohyphomycosis: *Fusarium* spp, *Scedosporium* spp, and others. *Clinical Microbiology and Infection*, 20, Suppl 3, 27-46.

Tragiannidis, A., Bisping, G., Koehler, G., *et al.* (2010) Minireview: *Malassezia* infections in immunocompromised patients. *Mycoses*, 53, 187-195.

Tuon, F.F. and Costa, S.F. (2008) *Rhodotorula* infection. A systematic review of 128 cases from literature. *Revista Iberoamericana de Micologia*, 25, 135-140.

Ueno, R., Hanagata, N., Urano, N., *et al.* (2005) Molecular phylogeny and phenotypic variation in the heterotrophic green algal genus *Prototheca*. *Journal of Phycology*, 41, 1268-1280.

Valero, C., Buitrago, M.J., Gago, S., *et al.* (2018) A matrix-assisted laser desorption/ionization time of flight mass spectrometry reference database for the identification of *Histoplasma capsulatum*. *Medical Mycology*, 56, 307-314.

Vanbreusegham, R., de Vroey, C.H. and Takashio, M. (1978) *Practical Guide to Medical and Veterinary Mycology.* Masson Publishing, USA.

van den Bossche, H., Cauwenbergh, G. and MacKenzie, D.W.R. (1988) *Aspergillus and Aspergillosis*. Springer, New York.

Vandepol, N., Liber, J., Desiro, A., *et al.* (2020) Resolving the Mortierellaceae phylogeny through synthesis of multi-gene phylogenetics and phylogenomics. *Fungal Diversity*, 104, 267-289.

van Diepeningen, A.D., Brankovics, B., Iltes, J., *et al.* (2015) Diagnosis of *Fusarium* infections: approaches to identification by the clinical mycology laboratory. *Current Fungal Infection Reports*, 9, 135-143.
van Oorschot, C.A.N. (1980) A revision of *Chrysosporium* and allied genera. *Studies in Mycology*, 20, 1-89. Centraalbureau voor Schimmelcultures, Baarn, The Netherlands.
Varga, J., Frisvad, J.C. and Samson, R.A. (2011) Two new aflatoxin producing species, and an overview of *Aspergillus* section Flavi. *Studies in Mycology*, 69, 57-80.
Vaughan-Martini, A. and Martini, A. (2011) Chapter 61, *Saccharomyces* Meyen ex Reess (1870). In Kurtzman, C.P., Fell, J.W. and Boekhout, T. (eds) *The Yeasts, a Taxonomic Study*, 5th edition. Elsevier B.V., pp. 733-746.
Vaux, S., Criscuolo, A., Desnos-Ollivier, M., *et al.* (2014) Multicenter outbreak of infections by *Saprochaete clavata*, an unrecognized opportunistic fungal pathogen. *mBio*, 5, e02309-14.
Velegraki, A., Alexopoulos, E.C., Kritikou, S., *et. al.* (2004) Use of fatty acid RPMI 1640 media for testing susceptibilities of eight *Malassezia* species to the new triazole posaconazole and six established antifungal agents by a modified NCCLS M27-A2 microdilution method and Etest. *Journal of Clinical Microbiology*, 42, 3589-3593.
Vidal, P., Vinuesa, M.A., Sanchez-Puelles, J.A., *et al.* (2000) Phylogeny of chrysosporia infecting reptiles: proposal of the new family *Nannizziopsiaceae* and five new species. *Revista Iberoamericana de Micologia*, 17, 24-31.
Vieira, W.A. dos S., Bezerra, P.A., Silva, A.C., *et al.* (2020) Optimal markers for the identification of *Colletotrichum* species. *Molecular Phylogenetics and Evolution*, 143, 106694-106694.
Vijaykrishna, D., Mostert, L., Jeewon, R., *et al.* (2004) *Pleurostomophora*, an anamorph of *Pleurostoma* (*Calosphaeriales*), a new anamorph genus morphologically similar to *Phialophora. Studies in Mycology*, 50, 387-395.
Vikram, H.R., Smilack, J.D., Leighton, J.A., *et al.* (2012) Emergence of gastrointestinal basidiobolomycosis in the United States, with a review of worldwide cases. *Clinical Infectious Diseases*, 54, 1685-1691.
Vilela, R., Silva, S.M., Riet-Correa, F., *et al.* (2010) Morphologic and phylogenetic characterization of *Conidiobolus lamprauges* recovered from infected sheep. *Journal of Clinical Microbiology*, 48, 427-432.
Vilela, R. and Mendoza, L. (2018) Human pathogenic entomophthorales. *Clinical Microbiology Reviews*, 31, e00014-18.
Vilela, R., Huebner, M., Vilela, C., *et al.* (2021) The taxonomy of two uncultivated fungal mammalian pathogens is revealed through phylogeny and population genetic analyses. *Scientific Reports*, 11, 18119-18119.
Visagie, C.M., Houbraken, J., Frisvad, J.C., *et al.* (2014) Identification and nomenclature of the genus *Penicillium. Studies in Mycology*, 78, 343-371.
Vitale, R.G. and de Hoog, G.S. (2002) Molecular diversity, new species and antifungal susceptibilities in the *Exophiala spinifera* clade. *Medical Mycology*, 40, 545-556.
Voigt, K., Cigelnik, E. and O'Donnell, K. (1999) Phylogeny and PCR identification of clinically important zygomycetes based on nuclear ribosomal-DNA sequence data. *Journal of Clinical Microbiology*, 37, 3957-3964.
von Arx, J.A., Guarro, J. and Figueras, M.J. (1986) *The Ascomycete genus Chaetomium*. J. Cramer, Berlin.
Wahyuningsih, R., SahBandar, I.N., Theelen, B., *et al.* (2008) *Candida nivariensis* isolated from an Indonesian human immunodeficiency virus-infected patient suffering from oropharyngeal candidiasis. *Journal of Clinical Microbiology*, 46, 388-391.
Walther, G., Pawlowska, J., Alastruey-Izquierd, A., *et al.* (2012) DNA barcoding in *Mucorales*: an inventory of biodiversity. *Persoonia*, 30, 11-47.
Walther, G., Wagner, L. and Kurzai, O. (2019) Updates on the taxonomy of Mucorales with an emphasis on clinically important taxa. *Journal of Fungi*, 5, 106.
Wang, L., Yokoyama, K., Miyaji, M., *et al.* (1998) The identification and phylogenetic relationship of pathogenic species of *Aspergillus* based on the mitochondrial cytochrome b gene. *Medical Mycology*, 36, 153-164.
Wang, P., Kenyon, C., de Hoog, S., *et al.* (2017) A novel dimorphic pathogen, *Emergomyces orientalis* (Onygenales), agent of disseminated infection. *Mycoses*, 60, 310-319.
Wang, X., Fu, Y-F., Wang, R-Y., *et al.* (2014) Identification of clinically relevant fungi and *Prototheca* species by rRNA gene sequencing and multilocus PCR coupled with electrospray ionization mass spectrometry. *PLoS One*, 9, e98110-e98110.
Wang, X.W., Lombard, L., Groenewald, J.Z., *et al.* (2016) Phylogenetic reassessment of the *Chaetomium globosum* species complex. *Persoonia*, 36, 83-133.
Wang, X.W., Han, P.J., Bai, F.Y., *et al.* (2022) Taxonomy, phylogeny and identification of Chaetomiaceae with emphasis on thermophilic species. *Studies in Mycology*, 101, 121-243.
Weitzman, I. and Crist, M.Y. (1980) Studies with clinical isolates of *Cunninghamella* II. *Physiological and morphological studies. Mycologia*, 72, 661-669.
Weitzman, I. (1984) The case for *Cunninghamella elegans*, *C. bertholletiae and C. echinulata as separate species. Transactions of the British Mycological Society*, 83, 527-529.
Weitzman, I., McGinnis, M.R., Padhye, A.A., and Ajello, L. (1986) The genus *Arthroderma* and its later synonym *Nannizzia*. *Mycotaxon*, 25, 505-505.

Wieden, M.A., Steinbronn, K.K., Padhye, A.A., *et. al.* (1985) Zygomycosis caused by *Apophysomyces elegans*. *Journal of Clinical Microbiology*, 22, 522-526.
Wilson, C.M., O'Rourke, E.J., McGinnis, M.R., *et al.* (1990) *Scedosporium inflatum*: Clinical spectrum of a newly recognised pathogen. *Journal of Infectious Diseases*, 161, 102-107.
Woldeamanuel, Y., Leekkassa, R., Petrini, B., *et al.* (2005) White variants of *Trichophyton violaceum* isolated in Ethiopia. *Acta Pathologica, Microbiologica et Immunologica Scandinavica*, 113, 708-712.
Won, E.J., Shin, J.H., Lim, S.C., *et al.* (2012) Molecular identification of *Schizophyllum commune* as a cause of allergic fungal sinusitis. *Annals of Laboratory Medicine*, 32, 375-379.
Woudenberg, J.H., Groenewald, J.Z., Binder, M., *et al.* (2013) *Alternaria* redefined. *Studies in Mycology*, 75, 171-212.
Xi, L., Sun, J., Lu, C., *et al.* (2009) Molecular diversity of *Fonsecaea* (Chaetothyriales) causing chromoblastomycosis in southern China. *Medical Mycology*, 47, 27-33.
Xiao, M., Guo, L.N., Kong, F., *et al.* (2013) Practical identification of eight medically important *Trichosporon* species by reverse line blot hybridization (RLB) assay and rolling circle amplification (RCA). *Medical Mycology*, 51, 300-308.
Yan, J., Deng, J., Zhou, C.J., *et al.* (2010) Phenotypic and molecular characterization of *Madurella pseudomycetomatis* sp. nov., a novel opportunistic fungus possibly causing black-grain mycetoma. *Journal of Clinical Microbiology*, 48, 251-257.
Yanagihara, M., Kawasaki, M., Ishizaki, H., *et al.* (2010) Tiny keratotic brown lesions on the interdigital web between the toes of a healthy man caused by *Curvularia* species infection and a review of cutaneous *Curvularia* infections. *Mycoscience*, 51, 224-233.
Yilmaz, N., Visagie, C.M., Houbraken, J., *et al.* (2014) Polyphasic taxonomy of the genus *Talaromyces*. *Studies in Mycology*, 78, 175-341.
Yogo, N., Shapiro, L. and Erlandson, K.M. (2014) *Sepedonium* intra-abdominal infection: a case report and review of an emerging fungal infection. *Journal of Antimicrobial Chemotherapy*, 69, 2583-1585.
Yoshida, E., Makimura, K., Mirhendi, H., *et al.* (2006) Rapid identification of *Trichophyton tonsurans* by specific PCR based on DNA sequences of nuclear ribosomal internal transcribed spacer (ITS) 1 region. *Journal of Dermatological Science*, 42, 225-230.
Yu, J., Walther, G., Van Diepeningen, A.D., *et al.* (2015) DNA barcoding of clinically relevant *Cunninghamella* species. *Medical Mycology*, 53, 99-106.
Yu, S.Y., Guo, L.N., Xiao, M., *et al.* (2020) Clinical and microbiological characterization of invasive pulmonary aspergillosis caused by *Aspergillus lentulus* in China. *Frontiers in Microbiology*, 11, 1672.
Yuan, G.F. and Jong, S.C. (1984) A new obligate azygosporic species of *Rhizopus*. *Mycotaxon*, 20, 397-400.
Zalar, P., de Hoog, G.S., Schroers, H.J., *et al.* (2007) Phylogeny and ecology of the ubiquitous saprobe *Cladosporium sphaerospermum*, with descriptions of seven new species from hypersaline environments. *Studies in Mycology*, 58, 157-183.
Zalar, P., Gostincar, C., de Hoog, G.S., *et al.* (2008) Redefinition of *Aureobasidium pullulans* and its varieties. *Studies in Mycology*, 61, 21-38.
Zeng, J.S., Sutton, D.A., Fothergill, A.W., *et al.* (2007) Spectrum of clinically relevant *Exophiala* species in the United States. *Journal of Clinical Microbiology*, 45, 3713-3720.
Zeng, J. and de Hoog, G.S. (2008) *Exophiala spinifera* and its allies: diagnostics from morphology to DNA barcoding. *Medical Mycology*, 46, 193-208.
Zeng, J., Feng, P., Gerrits van den Ende, A.H.G., *et al.* (2014) Multilocus analysis of the *Exophiala jeanselmei* clade containing black yeasts involved in opportunistic disease in humans. *Fungal Diversity*, 65, 3-16.
Zhang, Y., Liu, F., Wu, W., *et al.* (2015a) A phylogenetic assessment and taxonomic revision of the thermotolerant hyphomycete genera *Acrophialophora* and *Taifanglania*. *Mycologia*, 107, 768-779.
Zhang, Y., Hagen, F., Stielow, B., *et al.* (2015b) Phylogeography and evolutionary patterns in *Sporothrix* spanning more than 14,000 human and animal case reports. *Persoonia*, 35, 1-20.
Zhao, Y., Petraitiene, R., Walsh, T.J., *et al.* (2013) A real-time PCR assay for rapid detection and quantification of *Exserohilum rostratum*, a causative pathogen of fungal meningitis associated with injection of contaminated methylprednisolone. *Journal of Clinical Microbiology*, 51, 1034-1036.
Zheng, R. and Chen, G. (2001) A monograph of *Cunninghamella. Mycotaxon*, 80, 1-75.
Zheng, H., Blechert, O., Mei, H., *et al.* (2020) Assembly and analysis of the whole genome of *Arthroderma uncinatum* strain T10, compared with *Microsporum canis* and *Trichophyton rubrum*. *Mycoses*, 63, 683-693.
Zhu, Y., O'Brien, B., Leach, L., *et al.* (2020) Laboratory analysis of an outbreak of *Candida auris* in New York from 2016 to 2018: Impact and lessons learned. *Journal of Clinical Microbiology*, 58, e01503-19.
Zycha, H., Siepmann, R. and Linnemann, G. (1969) *Mucorales: Eine Beschreibung Aller Gattungen und Arten Dieser Pilzgruppe*. Cramer Lehre, Germany, 355 pp.

Index

www.ingramcontent.com/pod-product-compliance
Lightning Source LLC
LaVergne TN
LVHW070106021026
849160LV00026B/1696